Logistics 4.0

Digital Transformation of Supply Chain Management

Editors

Turan Paksoy

Department of Industrial Engineering
Konya Technical University
Konya-Turkey

Çiğdem Koçhan

Operations Research and Supply Chain Management
College of Business Administration
Ohio Northern University, USA

Sadia Samar Ali

Department of Industrial Engineering
King Abdul Aziz University
Jeddah, Kingdom of Saudi Arabia

CRC Press is an imprint of the
Taylor & Francis Group, an **informa** business
A SCIENCE PUBLISHERS BOOK

CRC Press
Taylor & Francis Group
6000 Broken Sound Parkway NW, Suite 300
Boca Raton, FL 33487-2742

Version Date: 20200720

International Standard Book Number-13: 978-0-3673-4003-2 (Hardback)

Library of Congress Cataloging-in-Publication Data

Names: Paksoy, Turan, editor. | Koçhan, Çiğdem, editor. | Ali, Sadia Samar, editor.
Title: Logistics 4.0 : digital transformation of supply chain management / editors, Turan Paksoy, Department of Industrial Engineering, Konya Technical University, Konya-Turkey, Çiğdem Koçhan, Operations Research and Supply Chain Management, College of Business Administration, Ohio Northern University, USA, Sadia Samar Ali, Department of Industrial Engineering, King Abdul Aziz University, Jeddah, Kingdom of Saudi Arabia.
Other titles: Logistics four point oh
Description: First edition. | Boca Raton, FL : CRC Press, 2021. | Summary: "Manufacturing and service industry has been broadly affected by the past industrial revolutions. From the invention of the steam engine to digital automated production, the first Industrial Revolution and the following revolutions conduced to significant changes in operations and supply chain management (SCM) processes. Swift changes in manufacturing and service systems caused by industrial revolutions led to phenomenal improvements in productivity for the companies. This fast-paced environment brings new challenges and opportunities for the companies that are associated with the adaptation to the new concepts such as Internet of Things and Cyber Physical Systems, artificial intelligence, robotics, cyber security, data analytics, block chain and cloud technology. These emerging technologies facilitated and expedited the birth of Logistics 4.0. The Industry 4.0 initiatives in SCM has attracted stakeholder's attentions due to it is ability to empower using a set of technologies together that helps to execute more efficient production and distribution systems. This initiative has been called Logistics 4.0 as the fourth Industrial Revolution in SCM due to its high potential. Connecting entities, machines, physical items and enterprise resources to each other by using sensors, devices and the Internet along the supply chains are the main attributes for Logistics 4.0. The context of the Internet of Things (IoT) enables customers to make more suitable and valuable decisions due to the data-driven structure of the Industry 4.0 paradigm"-- Provided by publisher.
Identifiers: LCCN 2020027404 | ISBN 9780367340032 (hardcover)
Subjects: LCSH: Business logistics.
Classification: LCC HD38.5 .L6125 2021 | DDC 658.7--dc23
LC record available at https://lccn.loc.gov/2020027404

Visit the Taylor & Francis Web site at
http://www.taylorandfrancis.com
and the CRC Press Web site at
http://www.routledge.com

Preface

The past three industrial revolutions have not only brought the terms of "the steam engine, the age of science and mass production, and the digitalization" to our lives but also imposed fundamental changes in our society. Manufacturing and supply chain operations have been radically altered and transformed into a new shape as industrial revolutions progressed. Rapid changes in manufacturing and service systems caused by industrial revolutions have led to improvements in business productivity and efficiency for companies over the years.

Now, we are on the edge of the Fourth Industrial Revolution that is powered by the rapid technological improvements and emerging technologies that are transforming the way companies do their business for decades. These fast-paced technological changes impose unprecedented challenges and create opportunities for companies who adopt emerging technologies such as the Internet of Things, Cyber-Physical Systems, Artificial Intelligence, Robotics, Cyber Security, Data Analytics, The Block Chain, and Cloud Computing Systems.

In recent years, globalization, increasing global competition and technological growth rate, diversity in customer demands, and increasing complexity in supply chain processes urged companies to adopt and intensely use emerging technologies in their business operations. The Fourth Industrial Revolution, also known as Industry 4.0, was coined for the first time in 2011 in Germany and it is an innovative paradigm that has the aim of intensely integrating technologies into the production and distribution processes.

The birth of Logistics 4.0 is accelerated by the emergence of these innovative technologies. Logistics 4.0 is an emerging logistics paradigm that can connect entities, machines, physical items, products, and enterprise resources by using sensors, devices, and the Internet within supply chains. This paradigm enables more efficient production and distribution systems which have attracted stakeholder's attention due to its potential leading to high-performance supply chains.

The Internet of Things (IoT) is at the core of this digital transformation in SCM. The IoT's ability to collect and analyze real-time data and help supply chains to adapt rapidly changing markets add an unusual value to the SCM processes. The IoT's role on the collaboration between the supply chain partners and the coordination of supply chain activities enable data-driven, flexible and agile, and operationally efficient supply chains. The merits of IoT can be applied from real-time product tracking and warehouse condition monitoring activities to precise forecasting, and product delivery date and delay estimation.

In this context, our book "*Logistics 4.0: Digital Transformation of Supply Chain Management*" presents the state-of-art research in the digital transformation of supply chains. The book targets audiences who are interested in the history of the past industrial revolutions and their impacts on our lives, while covering the most recent developments in disruptive technologies used in the transformation process of today's supply chains.

The contribution of our books includes but not limited to:

- A detailed literature review on the Fourth Industrial Revolution and the Digital Transformation in SCM
- The Role of the Internet of Things and Cyber-Physical Systems on the Digital Transformation of Supply Chains
- Decision Making with the Machine Learning Algorithms
- Smart Factories and the Transformation of the Conventional Production Systems
- The Use of Artificial Intelligence and Augmented Reality in SCM
- Advances in the Robotics and Autonomous Systems in SCM
- Smart Operations and Block Chain in SCM

This peer-reviewed book consists of 12 sections and 22 chapters, while bringing researchers together from all over the world on Logistics 4.0 and Industry 4.0 tools in SCM. I am very pleased and honored to announce the release of our book entitled "Logistics 4.0: Digital Transformation of Supply Chain Management". I want to present my gratitude to all expert authors in their fields from all over the world contributed to our book and also give my special thanks to the wonderful team of CRC Press.

Turan Paksoy

Contents

SECTION 1

Introduction and Conceptual Framework

Chapter 1

A Conceptual Framework for Industry 4.0 (How is it Started, How is it Evolving Over Time?)

Sercan Demir,[1,*] *Turan Paksoy*[2] and *Cigdem Gonul Kochan*[3]

1. Introduction

Manufacturing and service industry has been broadly affected by the past industrial revolutions. Swift changes in manufacturing and service systems caused by industrial revolutions led to improvements in productivity for the companies. This fast-paced environment brings new challenges for the companies that are associated with adaptation to the new concepts such as industrial internet, cyber-physical systems, adaptive robotics, cybersecurity, data analytics, artificial intelligence, and additive manufacturing. These emerging technologies facilitated and expedited the birth of Industry 4.0, the latest industrial revolution era (Salkin et al. 2018).

From the invention of the steam engine to digital automated production, the First Industrial Revolution and the following revolutions have led to significant changes in the manufacturing process. As a result, ever more complex, automated and sustainable manufacturing systems have emerged. In the European Union, the industry is accountable for approximately 17% of the total GDP that creates 32 million jobs (Qin et al. 2016). The Industry 4.0 initiative has attracted stakeholder's attention due to its ability to apply a bundle of technologies to execute more efficient production systems. This initiative has been accepted as the Fourth Industrial Revolution by many due to its high potential. Connecting physical items such as sensors, devices, and enterprise resources to the internet are major attributes for industrial manufacturing in Industry 4.0. The context of the Internet of Things (IoT) enables customers to make more suitable and valuable decisions due to

[1] Department of Industrial Engineering, Faculty of Engineering, Harran University, Sanliurfa, Turkey.
[2] Department of Industrial Engineering, Faculty of Engineering, Konya Technical University, Konya, Turkey.
Email: tpaksoy@yahoo.com
[3] Department of Management and Marketing, College of Business and Management, Northeastern Illinois University, Chicago, Illinois, USA.
* Corresponding author: sercanxdemir@gmail.com

the data-driven structure of the Industry 4.0 paradigm. Besides that, the system's ability to gather and analyze information about the environment at any given time and adapt itself to the rapid changes adds significant value to the manufacturing process (Alexopoulos et al. 2016).

The organization of the rest of this chapter is as follows. In the second section, the history of the first three Industrial Revolutions and their impacts are presented. The framework of the Fourth Industrial Revolution and the newly emerging technologies that are reshaping the manufacturing are discussed in the third section. Section four provides a review of the relevant literature. The final section concludes the chapter with a discussion and suggests future research directions.

2. First Three Industrial Revolutions: Industry 1.0–3.0

In the literature, the term "industrial revolution" and "industrialization" are used interchangeably. The appearance of many industrial revolutions throughout history raises questions related to their type, nature, and concept (Coleman 1956).

The Industrial Revolution refers to the rise of modern economic growth, such as a sustained and substantial increase of GDP per capita in real terms, during the transition from a pre-industrial to an industrial society. The process of revolution by its own nature is not abrupt and rapid, but it is deep and extensive. Great Britain was the first industrial nation, and its transition took almost a century from the 1750s to the 1850s. However, the real per capita income has started growing after the 1840s over one percent per year. Many new industrial sectors had reached significant increases in productivity at an early stage. However bad harvests, frequent wars, a high population increase, and changes in the economic structure had a negative effect on the growth rate, especially in the pioneer country, Great Britain. Countries that industrialized later, overall, had a faster pace of development and a higher rate of growth (Vries 2008).

Although the industrial revolution is not considered a historical episode by itself, it was the most important development in human history over the past three centuries. The phenomenon began about two and a half centuries ago. With new methods for producing goods, the industrial revolution has reshaped where people live, how they work, how they define political issues, and more. It continues to shape the contemporary world. While the oldest industrial nations are still adapting themselves to its impact, the newer industrial societies, such as China, repeat elements of the original process but extend its range in new directions (Stearns 2012).

Industrialization was the major force that brought changes in world history that began in the 19th and 20th centuries and continues to shape the 21st century and our lives. Industrial revolutions took place in three waves. The first occurred in Western Europe and the United States beginning with developments in Great Britain in the 1770s, while the second wave hit Russia and Japan, some parts of eastern and southern Europe, plus Canada and Australia from the 1880s onward. The most recent wave began in the 1960s in the Pacific Rim, and two decades later it reached Turkey, India, Brazil, and other parts of Latin America. Each major wave of industrialization quickly engulfed other countries that were not industrialized outright and converted their basic social and economic relationships (Stearns 2012).

The first three industrial revolutions stretched over nearly a 200-year time period. Starting with the steam engine driven mechanical looms in the late 1700s, the fabric production moved to central factories from private homes causing an extreme increase in productivity. Nearly 100 years later, Ohio marked the beginning of the Second Industrial Revolution by using the conveyor belts in the slaughterhouses in Cincinnati. Following years saw the peak point of this era with the production of the Ford T model in the United States. The introduction of the continuous production lines and the conveyor belts led to the extreme increase in productivity due to the advantage of mass production. The breakthrough that enabled the digital programming of automation systems came with presentation of the first programmable logic controller by Modicon in 1969, marking the beginning of the third Industrial Revolution. The programming paradigm still governs today's modern automation system engineering that leads to highly flexible and efficient automation systems (Drath and Horch 2014). Figure 1 presents an overview of the industrial revolutions.

The Fourth Industrial Revolution has emerged by means of CPS. These systems are industrial automation systems that connect the physical operations with computing and communication infrastructures via their networking and accessibility to the cyber world (Jazdi 2014).

The integration of physical operations in industrial production, information, and communication technologies is called Industry 4.0. Industry 4.0 has recently gained more attention from academics. The term "Industry 4.0" is used for the next industrial revolution, which has been preceded by three other industrial revolutions in history. The First Industrial Revolution started with the introduction of mechanical production facilities in the second half of the 18th century and accelerated over the 19th century. Electrification and the division of labor (i.e., Taylorism) induced the Second Industrial Revolution starting from the 1870s. The progress in the automation of the production process with the help of advanced electronics and information technology started the Third Industrial Revolution (the digital revolution) around the 1970s (Hermann et al. 2016).

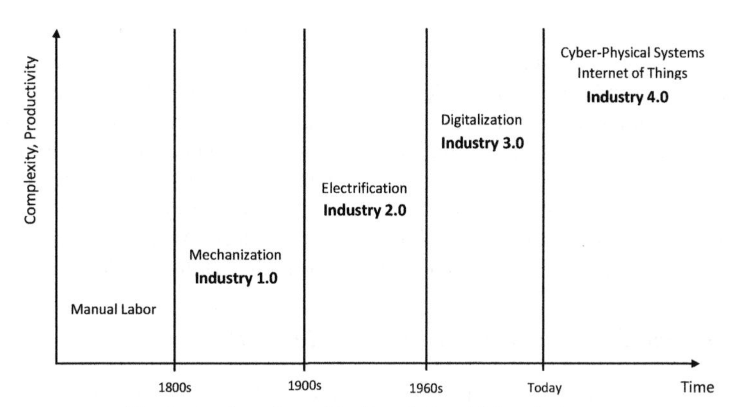

Fig. 1: An Overview of the Four Industrial Revolutions.

2.1 How it began: The First Industrial Revolution

The Wealth of Nations was written by Adam Smith in 1776, at the very beginning of the First Industrial Revolution. Smith's ideas and the views were phenomenal; however, he did not conceive of the following events. As workers in the industrializing countries shifted from farms to factories, societies were reformed beyond expectations in this fast-paced environment. This transformation impacted the distribution of the labor force across economic sectors dramatically. For instance, 84% of the U.S. workforce participated in agriculture, compared to an inconsiderable 3% in manufacturing in 1810. However, the manufacturing market share climbed to almost 25 percent while agriculture market share gradually diminished to just 8 percent over the years until the year 1960. As of today, the agriculture market share is under 2 percent. The revolution significantly impacted people's lives, education, the organization of businesses, the forms and practices of government (Blinder 2006).

There have been many important industrial innovations even before the First Industrial Revolution; however, the innovations of the late eighteenth century (at the time of the First Industrial Revolution) can be differentiated from those that affected the processes of production. The impact of these innovations was so profound because of the extensive application of new sources of power and heat on the production processes. As a result of these innovations, fossil fuel (coal) replaced the traditional power resources such as the power of man, wind, water, animals, and the heat of a wood fire, etc. Coal became a major energy source that led to a tremendous increase in throughput and dropped the cost of basic industrial processes (Chandler 1980).

Three basic technological innovations set the stage for the First Industrial Revolution. First, James Watt's steam engine, patented in 1769, which permitted the transformation of heat energy into steam and mechanical energy. Second, the spinning machines of Arkwright and Crompton, which were patented in 1770 and 1779—were too large and cumbersome to be moved by a man or an animal—made possible the mass production of thread and yarn. Third, Henry Cort's reverberatory furnace, invented in 1784, fabricated a high volume of iron, the most widely used industrial metal of all time. The impact of these three fundamental innovations hit Great Britain at the same time during the last fifteen years of the eighteenth century (Chandler 1980). Subsequently, a series of inventions began to shift cotton manufacturing toward a factory system in the 1730s. The improved accuracy of the flying shuttle was one of the key developments in the industrialization of weaving during the early industrial revolution. Flying shuttle was retouched over the next thirty years to make it possible to work with new power sources other than human power. The Spinning Jenny device, the early multiple-spindle machine for spinning wool and cotton invented by James Hargreaves in 1764, mechanically drew out and twisted the fibers into threads. Similar to the flying shuttle, the Spinning Jenny device also utilized human power and not a new power source when it was used for the first time (Stearns 2012). Richard Arkwright patented the Water Frame (aka. Arkwright Frame) in 1769. This new machine used water as a power source and produced a better thread than the Spinning Jenny. The Water Frame was a machine with a series of cogs linked to a large wheel that turned by running water. This invention led to the building of a majority of mills in Britain (Newlanark.org 2019).

At first, the users of the Arkwright Frame and Crompton Mule relied on waterpower to run their machines. Therefore, in order to operate those machines, mills were built at the spots where a powerful steady flow of water was located, and these spots were not common in Britain. However, after James Watt and his associates had optimized the steam engine, new spinning factories, with a central source of power, batteries of expensive machines, and large permanent working force moved out of hills to lowland towns located close to markets, sources of supply, and labor. Manchester had its first steam mill in 1787. By 1800 dozens of great mills were in operation. Manchester had already become the prototype of the modern industrial city with dozens of mills in operation by the year 1800. Coal became one of the most important sources for industrial power and heat that led to the swift spread of factories and industrial towns, causing the birth of an enlarged urban middle class, an industrial bourgeoisie, and a much larger working class—an industrial proletariat in Great Britain. Moreover, the route of international trade was remodeled, as Britain turned into the workshop of the world with the help of new coal-powered factories. India was the larger exporter of cotton for Great Britain before the First Industrial Revolution; however, it became the major market for the British textile industry after the advancements in production in Britain. Another big impact of the First Industrial Revolution was on the economy of the United States. The country rapidly turned into the most important buyer for British textile and hardware products. At the same time, the United States became the largest supplier of the raw materials for the spinning and weaving mills in Britain's industrialized cities (Chandler 1980).

2.2 How it Advances: The Second Industrial Revolution

The core of the industrial revolution was the application of new sources of power to the production by means of the transmission equipment necessary to apply this power to manufacturing. This core also includes an increased scale in a human organization that assisted the progress of specialization and coordination of work done at levels which the preindustrial groupings had rarely achieved. As a result of the Industrial Revolution, the early power sources of production, humans, and animals, were replaced with motors powered by fossil fuels. Watt's steam engine enabled to harness the energy potential of coal, which was considered as the essential invention for Europe's industrial revolution. Electric motors, internal combustion motors, developed by the 1870s, and petroleum products were used by the next industrial revolutions later on (Stearns 2012).

The period of the second Industrial Revolution is usually assumed to be between 1870 and 1914. While many characteristic events of this period dated back to the 1850s, the fast-paced rate of pioneering inventions of the First Industrial Revolution era slowed down after 1825 until it picked up the speed again in the last quarter of the century. The First Industrial Revolution and most technological developments preceding it had little or no scientific base. The natural process involved in the production was not fully understood causing the difficulty in removing defects, improving quality, and having user-friendly products and processes. On the other hand, the Second Industrial Revolution set the stage for mutual feedbacks between science and technology (Mokyr 1998).

Many new revolutionary technologies, including electricity and the internal combustion engine, were invented during the period from 1860 to 1900—the Second Industrial Revolution. These ground breaking inventions opened a door for a transition that continued for decades and led to a swift technical change in production that brought a quick transformation into the new economy. Many believe that the invention of electricity during the Second Industrial Revolution has helped to advance technological developments even after the end of this revolution. The adoption of the electricity was very slow among the manufacturers. Since it took time for manufacturers to fully conceive the best utilization of the electricity, the use of electricity did not yield instant results in improving productiveness in the US manufacturing companies (Atkeson and Kehoe 2001).

The First Industrial Revolution resulted in the integration of new energy sources into the process of production. The Second Industrial Revolution brought a massive revision in production techniques with the presentation of modern transportation and communication facilities, including the railroad, telegraph, steamship, and cable systems. These inventions promoted mass production and distribution systems in the late 1800s and early 1900s (Jensen 1999).

It is argued by the researchers that the transition to a new economy brought by the Second Industrial Revolution had three main characteristics. First, the time interval between the rise in the momentum of technological developments and the increase in the growth rate of measured productivity during this period was long which was called productivity paradox. Next, the adoption rate of new technologies by the manufacturers was slow. Finally, some manufacturers continued to invest in old technologies instead of switching to new technologies during the transition period of the Second Industrial Revolution. Interestingly, these characteristics of the transition period after the Second Industrial Revolution showed similarities with the transition period that occurred after the Third Industrial Revolution (Atkeson and Kehoe 2007).

In the literature, many studies questioned the slow transition. Technological constraints were considered as main challenges for the slow transition. First of all, plants were the entities that internalize new technologies, and they had to go through a massive change in order to adopt new technologies and tools. However, improvements in these technologies were continuous, and plants had required a reasonable time frame to learn and absorb these new technologies and use them.

Atkeson and Kehoe (2007) devised a quantitative model to measure the criticality of technological constraints when transitioning to new technology and discovered that the learning curve is one of the major critical technological constraints. If the learning process in the old technological revolution persisted, the productivity paradox was triggered when transitioning to the next technology. Such a long learning process prompts firms to accumulate a large stock of knowledge of the new technology from the beginning of the transition. Once a firm passes through this troublesome process, it would be less willing to adopt the new technology and will not quickly discard existing technology practice. Rather, the firm continues to spend a long time learning about the existing technology before transitioning into new technology. This practice will cause a long interval between the increase in the speed of the technological transformation and increase in the measured productivity rate produced by this new technology.

2.3 Shifting from Mechanical Technology to Digital Electronics: The Third Industrial Revolution

The sudden explosion of US companies beyond national limits led to the beginning of the Third Industrial Revolution in the last half of the 1950s (Leighton 1986). The First Industrial Revolution's impact lasted over two centuries, while the Second Industrial Revolution has offered rapid diffusion of new technology and innovative techniques over a couple of decades. The impact of the Third Industrial Revolution in terms of the time for adaptation was overwhelming. The time available for the adaptation to the innovations was so short, and the pace of the change threatened both individuals and institutions. According to Finkelstein (1984), six major changes in the production process and markets in the Third Industrial Revolution era are inventions of microprocessors, computer-aided design and manufacturing (CAD/CAM), fiber optics, biogenetics, lasers, and holography.

The invention of the integrated circuit, the processor, or the chip in 1958 is one of the technologies that shaped the Third Industrial Revolution and is recognized as one of the essential inventions of the 20th century. The invention of the microprocessor has reduced the cost of computers while gradually improving computing power. The affordability of computing power accelerated the spread of computers. As the microprocessor has continually developed, Gordon Moore made his world-famous observation, known as Moore's Law: "the computing power of the microprocessor doubled every 18 to 24 months while the costs are halved" (Smith 2001). Figure 2 shows some of the Intel microprocessors, their year of introduction, and the number of transistors on them. The number of transistors shows a positive trend year by year with a monotonous increasing count of transistors.

The graphical explanation of the increase in the number of transistors demonstrated in Figure 3. The rate of change in the number of transistors among each time interval has a positive increasing slope. Especially after the year 2000, the increase in each year is tremendous.

In order to automate the production, water, and steam power were used in the First Industrial Revolution. Mass production became widespread by the use of electric energy during the Second Industrial Revolution. The Third Industrial Revolution took advantage of the rise of electronics and utilized electronics and information technology to automate the production process. During this era, telecommunications and computer technology had stepped up to the next level. Production of miniaturized device components followed, which later contributed to the advancements in space research and biotechnology. In the second half of the 20th century, nuclear energy also took its place at the core of the Third Industrial Revolution (Sentryo 2019). Subsequently, programmable logic controllers (PLCs) and robot technology boosted the high-level automation in production during the Third Industrial Revolution era.

One of the most crucial technological changes in American manufacturing during the Third Industrial Revolution was the debut of programmable automation. Programmable automation standardized machines and processes to perform different operations. This technology introduced robots such as programmable machine tools that can manipulate and move materials and parts through versatile motions; numerically controlled (NC) machine tools that shape or cut metal according to programmed instructions; and automated materials handling, storage, and retrieval systems. Flexible manufacturing systems controlled by a central computer system connect multiple workstations (e.g., NC machines with transfer robots). Computer-integrated manufacturing was born as the integration of programmable automation with the design, manufacturing, and management. The adaptation to computer-based manufacturing technology has brought benefits such as improvements in product quality and reliability. A human being during work is not flawless since the accuracy of the work being done varies throughout the day. However, a programmable machine iterates the same standardized job impeccably. Increased productivity, reduced waste and cost, time-saving, safer and healthier workplaces are results of the introduction of computer-based manufacturing technology. Flexible production systems that can respond to the market demand shifts promptly was the greatest long-term benefit of computer-based manufacturing technology (Helfgott 1986).

3. The Industry 4.0 Framework

The Fourth Industrial Revolution is built upon the Third Industrial Revolution and the Digital Revolution, both of which were initiated in the middle of the 20th century. The Fourth Industrial Revolution is a melting pot in which the physical, digital, and biological areas are merged and promotes exponential growth tendency for industry worldwide. The Fourth Industrial Revolution brought changes in production, management, and governance systems around the globe.

Microprocessor	Year of Introduction	Transistors
4004	1971	2,300
8008	1972	2,500
8080	1974	4,500
8086	1978	29,000
Intel286	1982	134,000
Intel386™ processor	1985	275,000
Intel486™ processor	1989	1,200,000
Intel® Pentium® processor	1993	3,100,000
Intel® Pentium® II processor	1997	7,500,000
Intel® Pentium® III processor	1999	9,500,000
Intel® Pentium® 4 processor	2000	42,000,000
Intel® Itanium® processor	2001	25,000,000
Intel® Itanium® 2 processor	2003	220,000,000
Intel® Itanium® 2 processor (9MB cache)	2004	592,000,000

Fig. 2: Intel® Microprocessor Transistor Count Chart (Intel.com 2019).

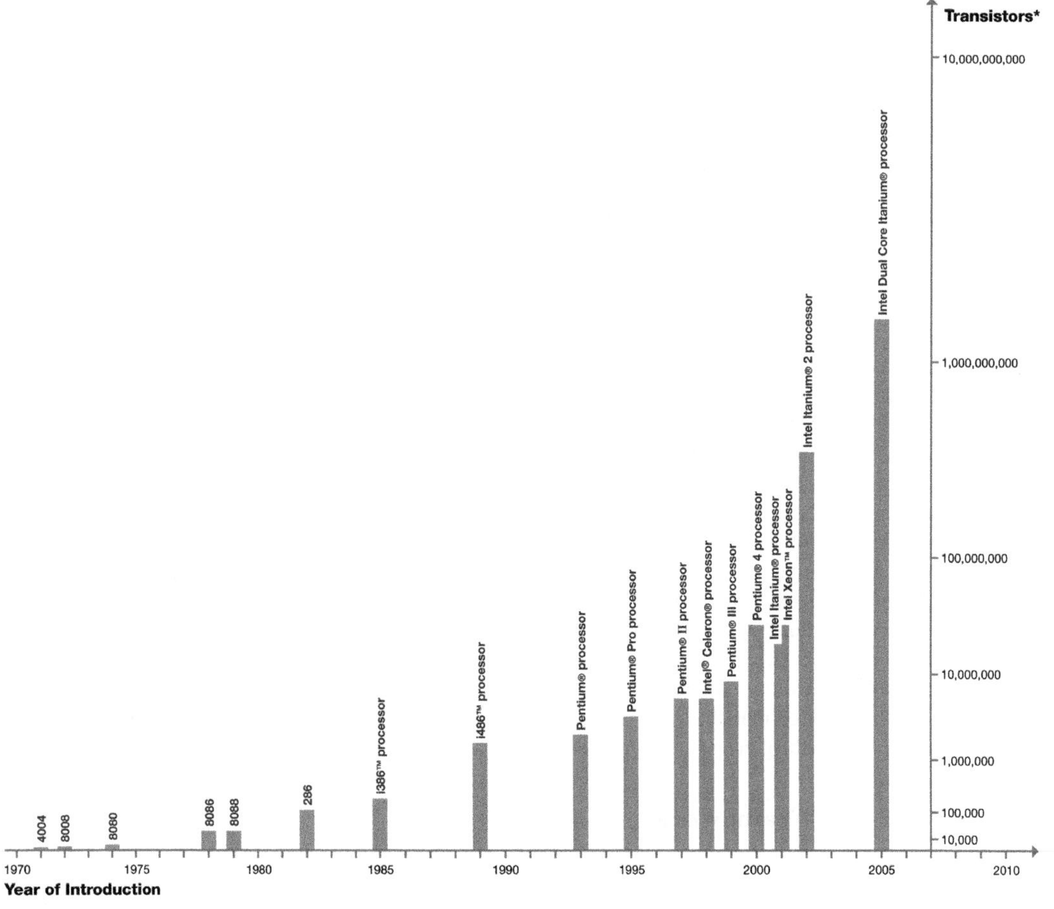

Fig. 3: Moore's Law Microprocessor Chart (Intel.com 2019).

The provenance of the Fourth Industrial Revolution dates back to the emergence of the Internet at the dawn of the new Millennium. The inception period of the first three industrial revolutions has started with the emergence of a new type of energy; however, the Fourth Industrial Revolution is the first revolution that initiated a new technological phenomenon, namely digitalization, rather than giving birth to a new type of energy (Sentryo 2019).

The concept of "Industrial 4.0" came into sight for the first time in an article published by the German government in 2011 to highlight Germany's high-tech strategy for 2020. The fourth stage of industrialization was named "Industry 4.0" after identifying the first three stages as mechanization, electrification, and information, respectively. The term "Industry 4.0" reappeared in 2013 at an industry fair in Hannover and subsequently, Industry 4.0 rapidly became a national strategy for Germany. Currently, Industry 4.0 draws the attention of many global industries, and it is a hot topic worldwide. It is predicted that Industry 4.0 will construct the foundation of the new industrial revolution and as such affect the international industry on a large scale (Zhou et al. 2016).

Industry 4.0 (Industry 4.0 or I40) is a national strategic initiative from the German government through the Federal Ministry of Education and Research (BMBF) and the Ministry for Economic Affairs and Energy (BMWI). Its goal is to move (drive) the digital manufacturing forward by increasing digitization and the interconnection of products, value chains, and business models. It also aims to support research, the networking of industry partners and standardization (Digital transformation monitor 2019).

Industry 4.0 also refers to a network system that encloses smart components and machines that are part of a standardized network based on the well-established internet standards. Industry 4.0 describes the thriving integration of Information and Communication Technologies (ICT) into production. VDMA, Bitkom, and ZVEI, three leading German companies of mechanical engineering, ICT, and electrical industry announced a definition of Industry 4.0 in spring 2014. According to VDMA, Bitkom, and ZVEI, Industry 4.0 aims for the optimization of value chains by implementing an autonomously controlled dynamic production (Kolberg and Zühlke 2015). The Industry 4.0 initiative is believed to change the design, manufacturing, operation, and service of products and production systems entirely. The connectivity and interaction among items, machines, and humans will expedite the pace of the processes in production systems up to 30 percent, increase the efficiency of processes up to 25 percent, and improve mass customization (Rüßmann et al. 2015).

Many digital technologies such as IoT, autonomous robots, and big data analytics are at the heart of Industry 4.0, and these technologies continue to revamp production and assist the progress of the digitalization of the basic production processes. These technologies have been implemented by leading companies to facilitate operational development plans. In order to build a quick momentum and achieve a strategic vision, implementation of these technologies by the companies should generate quick returns and yield long-term gains by implemented. Many companies have already taken advantage of implementing Industry 4.0. However,new ways to create values from the Industry 4.0 are still being explored. As new methods and techniques are uncovered, the value generated from this new approach will rise. Figure 4 demonstrates nine technologies that remodel the production process (Brunelli et al. 2017).

Industrial productivity has undergone an impressive development since the beginning of the Industrial Revolution. Starting with the invention of the steam engine in the 19th century, other power sources and production methods such as electricity-powered assembly lines in the first part of the 20th century and automated production in the 1970s led to a dramatic increase in productivity. Over the years, technology innovations proliferated and transformed information technology (IT), mobile communication, and e-commerce.

Industry 4.0, the newest digital industrial technology, is a ground breaking advancement powered by nine technology pillars. The connected systems are at the core of this transformation. Sensors, machines, IT systems, and work pieces are connected through the whole value chain, and these connected systems can analyze the data and communicate with each other via internet-based protocols. Data analysis helps systems to predict failure, configure itself, and adapt to the sudden changes. Industry 4.0 renders data collection and analysis across machines and enables a fast, flexible, and efficient system that generates higher-quality products at a lower production cost. Consequently, Industry 4.0 improves manufacturing productivity as the industrial growth rate increases. The improvement in productivity leads a company to gain a competitive advantage compared to others (Rüßmann et al. 2015). Table 1 shows the foundational nine technologies of Industry 4.0 that revamp the production.

Some of the nine core Industry 4.0 technologies are already in use in today's manufacturing systems. However, they are designed to reconstruct the production process. For instance, isolated and optimized cells would come together to form a fully integrated, automated, and optimized production flow that leads to greater productivity by altering the conventional production relationships between suppliers, manufacturers, and end customers. Figure 5 demonstrates how Industry 4.0 changes the manufacturing process.

Fig. 4: Nine Technologies that are Transforming the Industrial Production (Rüßmann et al. 2015).

Table 1: The Nine Technologies that are Reshaping the Production (Brunelli et al. 2017).

Technology	Impact and Contribution to Manufacturing
Advanced robots	— Autonomous, cooperating industrial robots, with integrated sensor and standardized interfaces
Additive Manufacturing	— 3D printers, used predominantly to make spare parts and prototypes — Decentralized 3D printing facilities, which reduce transport distances and inventory
Augmented Reality	— Digital enhancement, which facilities maintenance, logistics, and SOPs Display devices, such as glasses
Simulation	— Network simulation and optimization, which use real-time data from intelligent systems
Horizontal and vertical system integration	— Data integration within and across companies using a standard data transfer protocol — A fully integrated value chain (from supplier to customer) and organization structure (from management to shop floor)
The Industrial Internet of Things	— A network of machines and products — Multidirectional communications among networked objects
Cloud Computing	— The management of huge volumes of data in open systems — Real-time communication for production systems
Cyber Security	— The management of heightened security risks due to a high level of networking among intelligent machines, products, and systems
Big data and analytics	— The comprehensive evaluation of available data (from CRM, ERP, and SCM systems, for example, as well as from an MES and machines — Support for optimized real-time decision making

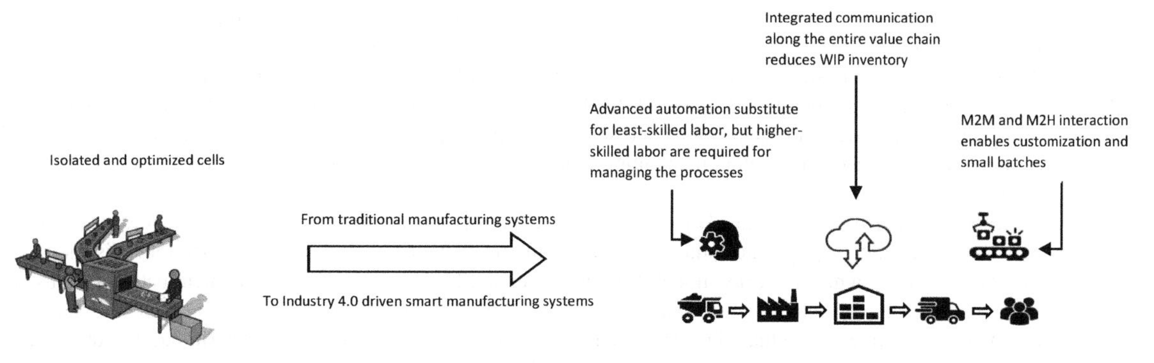

Fig. 5: Industry 4.0 Changing Traditional Manufacturing Relationships (Rüßmann et al. 2015).

4. A review of the Industry 4.0 Literature

Finkelstein (1984) discusses the dramatic changes experienced in new products, processes, and markets. The author classifies six significant technological changes during the Second Industrial Revolution era. According to Leighton (1986), changes in businesses after the second part of the 20th century could be depicted as the "Third Industrial Revolution", and the main feature of this revolution was the rise of corporations with extraordinary size, complexity, extent, and globalized structure. Kanji (1990) investigates the relationship between the quality revolution after the 1950s and the Second Industrial Revolution. The author concludes that the quality revolution through the process of "Total Quality Management" led to the "Second Industrial Revolution" for the survival of the fittest. Atkeson and Kehoe (2007) study technological advancements of the Second Industrial Revolution and adaptation problems to these technologies and devise a quantitative model to capture the technological constraints that slow down the adaptation process. Their model discovers the critical technology constraints that cause the delay between the new technological diffusion and measured productivity that is the result of adapting to new technology.

Several studies compare industrial revolutions in terms of many aspects, including economic impact, adaptation process, similarities, and differences. Mokyr (1998) exhibits the differences and the similarities of the First and the Second Industrial Revolutions. The author states that the Second Industrial Revolution was the direct continuation of the first one in many industries, and he discusses the important aspects that both revolutions differ from each other. Jensen (1999) investigates similarities between the Second Industrial Revolution and the Third Industrial Revolution. The author acquaints the readers with the dynamics of the Third Industrial Revolution in light of the outcomes of the Second Industrial Revolution.

Other studies investigate multiple industrial revolutions and their impact on economic, social, and technological development. Von Tunzelmann (1997) explores the contribution of engineers and the field of engineering to the industrialization process indifferent countries such as the UK, the United States, and Japan and investigates how engineers helped to advance the technology during industrial revolutions. Blinder (2006) studies the first three industrial revolutions and their impact on offshoring on today's economy. Kasa (1973) explores the relationships between macro-level technological improvements due to industrial revolutions and their negative impacts on the environment. Stearns (2013) investigates the extent and the history of industrial revolutions. The findings of the Stearns' (2012) study explain the scope, social and economic impact of industrial revolutions on many different societies worldwide.

Recent studies focus on Industry 4.0 and technological advancements that mark the start of the Fourth Industrial Revolution. Cooper and James (2009) present different types of data that increase the potential of the IoT and discuss the challenges for database management in the IoT platform. The authors provide scenarios to demonstrate some cases that will be possible through the use of IoT. Drath and Horch (2014) focus on the background and technical drivers of the new industrial revolution and describe the levels that form CPS in Industry 4.0. Brettel et al. (2014) analyze the developments of Industry 4.0 in the context of individualized production, end-to-end engineering in a virtual process chain and production networks. They present managerial insights for adopting or refusing decisions for Industry 4.0.

Kolberg and Zühlke (2015) discuss lean automation technology while linking them to the Industry 4.0 foundations. The authors claim that the collaboration of Industry 4.0 and lean production systems add value to the companies. Lee et al. (2015) propose a unified 5-level architecture for Cyber-Physical Systems in Industry 4.0 manufacturing systems. Roblek et al. (2016) present a theoretical framework for Industry 4.0 and discuss the influence of Industry 4.0 and the Internet-connected technologies on organizations and society. Hermann et al. (2016) investigate four design principles companies should take into account when implementing Industry 4.0 solutions, and authors consolidate these principles with a case study review.

Schumacher et al. (2016) propose a novel model to assess the Industry 4.0 maturity phase of the companies that operate in the field of discrete manufacturing. The authors extend existing readiness and maturity models, and tools discussed in the literature by developing a new maturity model. Vuksanovic et al. (2016) explore the development paths of Industry 4.0 and the future perception of smart factories. The authors further discuss the fundamental technologies behind Industry 4.0 and the impact of the Internet on manufacturing technologies. Erol et al. (2016) offer a scenario-based Industry 4.0 Learning Factory concept to overcome challenges in industrial practice that slow down the transformation process of Industry 4.0. They help in the understanding of abstract perception of Industry 4.0. Zezulka et al. (2016) explain two models developed by three German companies (BITCOM, VDMA, and ZWEI) for Industry 4.0 platform, namely, the Reference Architecture Model Industry 4.0 (RAMI 4.0) and Industry 4.0 component model.

Rojko (2017) discusses the concepts of Industry 4.0, its drivers and the Reference Architecture Model (RAMI 4.0) in detail. Santos et al. (2017) review major European industrial guidelines, roadmaps, and scientific literature to evaluate the Industry 4.0 vision. Ivanov et al. (2018) depict important issues that characterize the dynamics of supply chains, operations, and Industry 4.0 networks. The authors assert that a comprehensive collaboration between control engineers and supply chain experts may improve the performance of supply chains and Industry 4.0 networks. Kamble et al. (2018) review the current status of the research in domains of Industry 4.0 and classify Industry 4.0 research categories. The authors propose a sustainable Industry 4.0 framework based on the findings of their literature review.

Gunasekaran et al. (2019) examine improvements in quality management in the era of Industry 4.0 in terms of economic, human and technological aspects. Verba et al. (2019) propose a novel approach to load and delay optimization through application for migration between the edge, e.g., piece of hardware that controls data flow at the boundary between two networks, and the cloud. The authors validate the effectiveness of their proposed model using an Industry 4.0 based case study.

Table 2 summarizes research papers discussed in this section by their publication year, scope, type of research (quantitative/qualitative), keywords and the industrial revolutions studied in the paper.

5. Conclusion and Future Research

This chapter aimed to present a comprehensive literature review of recent journal and conference papers that study the first three industrial revolutions in the history of mankind and the last wave among all, Industry 4.0, and their impact on production, living and working conditions, and economic growth.

Germany is the world's third, Europe's biggest commodity exporter, specifically in automotive, chemical, electronic, and mechanical products. During Europe's debt crisis, Germany had managed to manufacture outstanding products. Germany became one of the leading global manufacturers, and the country gained a competitive advantage among the other giant manufacturers by implementing Industry 4.0 (Wang 2016).

Today, we are at the Fourth Industrial Revolution era, and it was initiated by the improvement of ICT. Smart automation of the CPS with decentralized control and IoT constitutes the core of Industry 4.0 paradigm. ICT allow reorganization of classical hierarchical automation systems into the self-organizing cyber-physical production systems. CPS provide flexible mass custom production and production quantity flexibility (Rojko 2017).

The basic concept of Industry 4.0 was first presented at the Hannover Fair in Germany in the year 2011. Industry 4.0 has gained popularity in many areas of academic research, and industry communities since its debut. The main idea behind this phenomenon is to utilize the potential of emerging technologies and concepts. Some of which are;

- Availability and use of the Internet and the Internet of Things (IoT),
- Integration of technical processes and business processes in companies,
- Digital mapping and virtualization of the real world,
- 'Smart' factory, including 'smart' means of industrial production and 'smart' products.

Table 2: Summary of the Literature Review.

Authors/Year	Qualitative	Quantitative	Conceptual Framework	Survey/Questionnaire /Interview	Case Study	Mathematical Model	Proposed New Model	Keywords	Industrial Revolution Studied
Finkelstein (1984)	x								Second
Leighton (1986)	x								Third
Kanji (1990)									Second
Von Tunzelmann (1997)	x		x		x				First, Second, Third
Mokyr (1998)	x								First, Second
Blinder (2006)	x		x						First, Second, Third
Atkeson and Kehoe (2007)	x	x				x	x		Second
Kasa (1973)	x							Internet, Internet of things, costumer behavior, Industry 4.0	First, Second, Third
Cooper and James (2009)	x				x			Database management challenges, IoT, Road map, Technical priorities	Industry 4.0
Drath and Horch (2014)	x								Industry 4.0
Brettel et al. (2014)	x		x						Industry 4.0
Kolberg and Zühlke (2015)	x		x					I40, CPS, Lean Automation, Lean Production	Industry 4.0
Lee et al. (2015)			x					CPS, I40, Health management and prognostics; Time machine	Industry 4.0
Roblek et al. (2016)			x						Industry 4.0
Hermann et al. (2016)		x	x		x				Industry 4.0
Schumacher et al. (2016)	x		x		x		x		Industry 4.0
Erol et al. (2016)	x				x			I40, smart manufacturing, learning factory, scenario-based learning	Industry 4.0
Zezulka et al. (2016)	x		x					RAMI 4.0, I40 component, administrative shell, communication, virtual	Industry 4.0

Table 2 contd. ...

...Table 2 contd.

Authors/Year	Qualitative	Quantitative	Conceptual Framework	Survey/Questionnaire /Interview	Case Study	Mathematical Model	Proposed New Model	Keywords	Industrial Revolution Studied
Vuksanovic et al. (2016)	x							I40, factory of the future, smart factory, IoT, augmented (virtual) reality	Industry 4.0
Rojko (2017)	x		x					I40, CPS, ERP, Manufacturing Execution System	First, Second, Third, Industry 4.0
Santos et al. (2017)	x							I40, technology roadmap, convergence	Industry 4.0
Kamble et al. (2018)	x							I40, Smart manufacturing, IoT, Process safety, Augmented reality, Sustainability Big data	Industry 4.0
Ivanov et al. (2018)	x		x	x				Control, SC, Operations, I40, Dynamics Planning, Scheduling Optimal program control Model-predictive control Adaptation, Resilience Digital Supply Chain	Industry 4.0
Gunasekaran et al. (2019)	x							Quality management, I40, technology, quality culture, behavioral aspects	Industry 4.0
Verba et al. (2019)		x	x		x	x		Cloud computing, Fog computing, Application model, Migration	Industry 4.0

According to the McKinsey & Company's (2016) report, many companies come across obstructions while implementing Industry 4.0, such as coordination problems among the different organizational units, cybersecurity and data ownership concerns with the third-party providers, resisting to a dramatic transformation, and lack of necessary capabilities (Bauer et al. 2016). However, there are many advantages of Industry 4.0 including but not limited to, cost reduction for production, logistics, and quality management, short launch time of products, improved customer responsiveness, custom mass production capability, and flexibility in the working environment (Rojko 2017).

The Industry 4.0 concept emphasizes global networks of connected machines (aka. Cyber-Physical Systems) in a smart factory environment that can communicate, autonomously exchange information, and send commands to each other. CPS and IoT enable autonomously operated smart factories. Some of the digital technology advancements that are integrated in smart factories are: (1) advanced robotics and artificial intelligence, (2) hi-tech sensors, (3) cloud computing, the Internet of Things, (4) data capture and analytics, (5) digital fabrication including 3D printing, (6) software-as-a-service and other new marketing models, (7) mobile devices, (8) platforms that use algorithms to direct motor vehicles including navigation tools, (9) ride-sharing apps, delivery and ride services, and autonomous vehicles, and (10) the integration of all these elements in an interoperable global value chain shared by many companies from many countries (Tjahjono et al. 2017).

Interconnected machines and smart devices are reshaping the way value is created in manufacturing and many other areas and in advancing manufacturing and computer technologies. Companies are in search of the ways of adopting Industry 4.0 to enjoy a more productive, flexible and sustainable production systems. Companies realized that production, control, and monitoring processes of smart and connected products will replace conventional labor centered production by fully automated and computerized production (Salkin et al. 2018).

References

Alexopoulos, K., S. Makris, V. Xanthakis, K. Sipsas and G. Chryssolouris. 2016. A concept for context-aware computing in manufacturing: the white goods case. International Journal of Computer Integrated Manufacturing, 29(8): 839–849.

Atkeson, A. and P.J. Kehoe. 2001. The transition to a new economy after the second industrial revolution (No. w8676). National Bureau of Economic Research.

Atkeson, B.A. and P.J. Kehoe. 2007. Transition to a new economy : lessons from two technological. The American Economic Review, 97(1): 64–88.

Bauer, H., C. Baur, D. Mohr, A. Tschiesner, T. Weskamp, K. Alicke and D. Wee. 2016. Industry 4.0 After the Initial Hype–Where Manufacturers are Finding Value and How they can Best Capture it. McKinsey Digital.

Blinder, A.S. 2006. Offshoring: The next industrial revolution? Foreign Affairs, 85(2): 113–128.

Brettel, M., N. Friederichsen, M. Keller and M. Rosenberg. 2014. How virtualization, decentralization and network building change the manufacturing landscape: An Industry 4.0 Perspective. International Journal of Mechanical, Industrial Science and Engineering, 8(1): 37–44.

Brunelli, J., V. Lukic, T. Milon and M. Tantardini. 2017. Five Lessons from the Frontlines of Industry 4.0 [Electronic resource]. BCG–Mode of access: http://image-src. bcg. com/Images/BCG-Five-Lessons-from-the-Frontlines-of-Industry-4.0-Nov-2017_tcm9-175989. pdf.

Chandler, A.D. 1980. Industrial revolutions and institutional arrangements. Bulletin of the American Academy of Arts and Sciences, 33(8): 33–50.

Coleman, D.C. 1956. Industrial growth and industrial revolutions. Economica, 23(89): 1–22.

Cooper, J. and A. James. 2009. Challenges for database management in the internet of things. IETE Technical Review, 26(5): 320. Available at: http://tr.ietejournals.org/text.asp?2009/26/5/320/55275.

Digital transformation monitor. 2019. Digital transformation monitor. [online] Available at: https://ec.europa.eu/growth/tools-databases/dem/monitor/ [Accessed 4 Dec. 2019].

Drath, R. and A. Horch. 2014. Industrie 4.0: Hit or hype?[industry forum]. IEEE Industrial Electronics Magazine, 8(2): 56–58.

Erol, S., A. Jäger, P. Hold, K. Ott and W. Sihn. 2016. Tangible Industry 4.0: a scenario-based approach to learning for the future of production. Procedia CiRp, 54(1): 13–18.

Finkelstein, J. 1984. Revolution : a special challenge to managers. Organizational Dynamics, 13(1): 53–65.

Gunasekaran, A., N. Subramanian and W. Ting Eric Ngai. 2019. Quality management in the 21st century enterprises: Research pathway towards Industry 4.0. International Journal of Production Economics, 207: 125–129.

Helfgott, R.B. 1986. America's third industrial revolution. Challenge, 29(5): 41–46.

Hermann, M., T. Pentek and B. Otto. 2016. Design principles for industrie 4.0 scenarios. Proceedings of the Annual Hawaii International Conference on System Sciences, 2016–March, pp. 3928–3937.

Intel.com. 2019. PRESS KIT—Moore's Law 40th Anniversary. [online] Available at: https://www.intel.com/pressroom/kits/events/moores_law_40th/[Accessed 4 Dec. 2019].

Ivanov, D., S. Sethi, A. Dolgui and B. Sokolov. 2018. A survey on control theory applications to operational systems, supply chain management, and Industry 4.0. Annual Reviews in Control, 46: 134–147.

Jazdi, N. 2014, May. Cyber physical systems in the context of Industry 4.0. In 2014 IEEE International Conference on Automation, Quality and Testing, Robotics (pp. 1–4). IEEE.
Jensen, M.C. 1999. The modern industrial revolution, exit, and the failure of internal control systems. Journal of Finance, XLVIII (3), July, 831–80. International Library of Critical Writings in Economics, 106(3): 188–237.
Kamble, S.S., A. Gunasekaran and S.A. Gawankar. 2018. Sustainable Industry 4.0 framework: A systematic literature review identifying the current trends and future perspectives. Process Safety and Environmental Protection, 117: 408–425.
Kanji, G.K. 1990. Total quality management: The second industrial revolution. Total Quality Management, 1(1): 3–12.
Kasa, S. 1973. Industrial revolutions and environmental problems. Confluence, 1941, p. 70.
Kolberg, D. and D. Zühlke. 2015. Lean Automation enabled by Industry 4.0 Technologies. IFAC-PapersOnLine, 48(3): 1870–1875.
Lee, J., B. Bagheri and H.A. Kao. 2015. A Cyber-Physical Systems architecture for Industry 4.0-based manufacturing systems. Manufacturing Letters, 3: 18–23. Available at: http://dx.doi.org/10.1016/j.mfglet.2014.12.001.
Leighton, D.S.R. 1986. The Internationalization of American Business—The Third Industrial Revolution, 34(3): 3–6.
Mokyr, J. 1998. The second industrial revolution. The Economic Journal, 41(161): 1. Available at: http://www.jstor.org/stable/10.2307/2224131?origin=crossref.
Newlanark.org. 2019. Children & Cotton - Learning Zone for Social Studies & Citizenship. [online] Available at: https://www.newlanark.org/learningzone/clitp-ageofinvention.php [Accessed 4 Dec. 2019].
Qin, J., Y. Liu and R. Grosvenor. 2016. A Categorical Framework of Manufacturing for Industry 4.0 and beyond. Procedia CIRP, 52: 173–178. Available at: http://dx.doi.org/10.1016/j.procir.2016.08.005.
Roblek, V., M. Meško and A. Krapež. 2016. A complex view of industry 4.0. Sage Open, 6(2): 2158244016653987.
Rojko, A. 2017. Industry 4.0 concept: background and overview. International Journal of Interactive Mobile Technologies (iJIM), 11(5): 77–90.
Rüßmann, M., M. Lorenz, P. Gerbert, M. Waldner, J. Justus, P. Engel and M. Harnisch. 2015. Industry 4.0: The future of productivity and growth in manufacturing industries. Boston Consulting Group, 9(1): 54–89.
Salkin, C., M. Oner, A. Ustundag and E. Cevikcan. 2018. A conceptual framework for Industry 4.0. In Industry 4.0: Managing The Digital Transformation (pp. 3–23). Springer, Cham.
Santos, C., A. Mehrsai, A.C. Barros, M. Araújo and E. Ares. 2017. Towards Industry 4.0: an overview of European strategic roadmaps. Procedia Manufacturing, 13: 972–979.
Schumacher, A., S. Erol and W. Sihn. 2016. A maturity model for assessing Industry 4.0 readiness and maturity of manufacturing enterprises. Procedia CIRP, 52: 161–166. Available at: http://dx.doi.org/10.1016/j.procir.2016.07.040.
Sentryo. 2019. Industrial revolutions: the 4 main revolutions in the industrial world. [online] Available at: https://www.sentryo.net/the-4-industrial-revolutions/[Accessed 4 Dec. 2019].
Smith, B.L. 2001. The third industrial revolution: Policymaking for the Internet. Colum. Sci. & Tech. L. Rev. 3: 1.
Stearns, P.N. 2012. The Industrial Revolution in World History. Westview press.
Tjahjono, B., C. Esplugues, E. Ares and G. Pelaez. 2017. What does industry 4.0 mean to supply chain? Procedia Manufacturing, 13: 1175–1182.
Von Tunzelmann, G.N. 1997. Engineering and innovation in the industrial revolutions. Interdisciplinary Science Reviews, 22(1): 67–77.
Verba, N., K.M. Chao, J. Lewandowski, N. Shah, A. James and F. Tian. 2019. Modeling industry 4.0 based fog computing environments for application analysis and deployment. Future Generation Computer Systems, 91: 48–60.
Vries, P. 2008. The industrial revolution. Encyclopaedia of the Modern World, 4: 158–161.
Vuksanovic, D., J. Ugarak and D. Korčok. 2016. Industry 4.0: The future concepts and new visions of factory of the future development. Conference Sinteza 2016.
Wang, L. 2016, July. Comparative Research on Germany Industry 4.0 and Made in China 2025. In 2016 2nd International Conference on Humanities and Social Science Research (ICHSSR 2016). Atlantis Press.
Zezulka, F., P. Marcon, I. Vesely and O. Sajdl. 2016. Industry 4.0–An Introduction in the phenomenon. IFAC-PapersOnLine, 49(25): 8–12.
Zhou, K., T. Liu and L. Zhou. 2016. Industry 4.0: Towards future industrial opportunities and challenges. 2015 12th International Conference on Fuzzy Systems and Knowledge Discovery, FSKD 2015, pp. 2147–2152.

CHAPTER 2

Logistics 4.0: SCM in Industry 4.0 Era (Changing Patterns of Logistics in Industry 4.0 and Role of Digital Transformation in SCM)

Sercan Demir,[1,*] *Turan Paksoy*[2] and *Cigdem Gonul Kochan*[3]

1. Introduction

Supply chains and logistics operations experienced important and rapid changes during the 1990s and early 2000s. These changes imposed significant challenges on the freight shipping industry. Just-in-time practices and the necessity of customer responsiveness were two of the main challenges faced by the industry. As the economies and markets have become globalized, the procurement and distribution of goods have been affected by this swift trend. Electronic Data Interchange (EDI) and the Internet, Global Positioning Systems via satellites (GPS), and Decision Support Systems (DSS) were the new information technologies that emerged as a response to these challenges in the logistics industry. Operations capacity and real-time decision-making capability of freight forwarders were substantially increased as they adopted these new technologies (Roy 2001).

The integration of physical and digital technologies, such as sensors, embedded systems, cloud computing, and the Internet of Things (IoT) has launched the Fourth Industrial Revolution. The main idea behind the industrial revolution is to increase resource utilization and productivity that leads to gaining a competitive edge for companies. The current industrial transformation is not only reshaping core business processes but also uncovering novel concepts of smart and connected technologies (Onar and Ustundag 2017).

Companies experience complex processes and incur high costs during their transformation to Industry 4.0 practices due to the newly emerged technologies that affect process input and output. The support from the top management becomes increasingly more important since Industry 4.0 transformation changes a company's core methods of production and requires a broad perspective on a company's vision, strategy, organization, and products (Akdil et al. 2018).

The shift from computers to smart devices that use infrastructure services based on cloud computing was one of the significant advances in the last decade. Computer-based automation systems become connected to the wireless network in today's internet era. The interconnection of humans, machines, and platforms that allow machines to communicate with each other are the advancements that are emerging now as a merit of the Internet. The implementation of this technology on production and business operations is described as Industry 4.0 (Tjahjono et al. 2017).

Industry 4.0 is imposing foundational changes in the current manufacturing process of companies. The integration of digitalization and the Internet to the manufacturing process is leading to a global transformation of the manufacturing industry. The factory of the future is envisioned as Cyber-Physical Systems (CPS) that connects machines and human beings. These smart factories will come to existence as technological advancements are adopted and used in harmony

[1] Department of Industrial Engineering, Faculty of Engineering, Harran University, Sanliurfa, Turkey.
[2] Department of Industrial Engineering, Faculty of Engineering, Konya Technical University, Konya, Turkey.
Email: tpaksoy@yahoo.com
[3] Department of Management and Marketing, College of Business and Management, Northeastern Illinois University, Chicago, Illinois, USA.
* Corresponding author: sercanxdemir@gmail.com

to produce intelligent products in industrial processes. Some of the digital technologies include, but are not limited to, advanced robotics and artificial intelligence, hi-tech sensors, cloud computing, the Internet of Things, digital fabrication (including 3D printing), data capture and analytics (Tjahjono et al. 2017).

2. Fundamentals of Logistics: Definitions and Terminology

Logistics is fundamentally a planning orientation, coordination or scheme that seeks to create an effective plan for the flow of products and information through a business. Supply chain management builds upon this framework and attempts to connect and coordinate the processes between entities in the pipeline, i.e., suppliers and customers, and the organization itself (Christopher 2016).

The most common belief about the term "logistics" is that this term was used by the Swiss General Baron de Jomini (1779–1869) for the first time. The word "logistics" has two roots, both of which are French in origin. "Logistique" comes from military rank, and it addresses the organization of the military support troops. "Loger" refers to a spatial military organization, i.e., camping. The US Army started to use the term "logistics" at the end of the nineteenth century, referring to the practices of military support service, i.e., transport and supply for the Armed Forces. During the Second World War, "logistics" was used to describe the planning and management process of providing, repopulating, and supplying the Allied military. Logistics was first used in the civilian sector in the trade industry in the 1960s. Logistics means planning and performing the physical distribution of goods in the US. Logistics was evolved into science by Hans Christian Pfohl in 1974 when characteristic areas of logistics tasks were defined and its axioms conceptualized (Tepić et al. 2011).

Logistics comprises of a complex set of activities that require a collection of metrics to adequately measure performance (Caplice and Sheffi 1995). The Seven R's of Logistics is one of the commonly accepted definitions of logistics. Logistics involves ensuring the availability of the right product, in the right quantity, and in the right condition, at the right place, at the right time, for the right consumer, at the right cost. Logistics is defined as "part of the supply chain process that plans, implements, and controls the efficient, effective flow and storage of goods, services and related information from point of origin to point of consumption in order to meet customers' requirements" by the Council of Logistics Management (Rutner and Langley 2000).

As its functions and interest areas are diversified, the definition of logistics has evolved over time. Introduced into the military for the first time, logistics eventually influenced many sectors in the economy. Transportation of agricultural goods led to the introduction of a non-military logistics concept known as "physical distribution". Advancements in industry and IT technologies as wells as technological, economic, political, social or environmental factors also impact the development of logistics. The need for fast action and rapid decision making, performing time-sensitive service, and being flexible enough to meet customers' needs are some of the main challenges that logistics operations experience. Proper and efficient implementation of modern technologies can overcome the challenges mentioned above. Companies are becoming more interested in these technological developments as they need to improve their business performance and gain a competitive edge in the market through the implementation of these advancements (Szymańska et al. 2017).

2.1 Inbound and Outbound Logistics

Inbound logistics refers to the flow of raw materials from suppliers to manufacturers. Receiving, storing, and distributing raw materials or goods that are coming into a business internally are inbound logistics activities. Freight consolidation, selection of carrier and mode of transportation, materials handling, warehousing, and backhaul management are management decisions associated with inbound logistics. Outbound logistics covers physical distribution activities of finished goods such as collecting, storing and distributing products from manufacturers to buyers. Warehousing of finished goods, materials handling, network planning and management, order processing, and vehicle scheduling and routing are all considered as outbound logistics activities. The main difference between inbound and outbound logistics are product characteristics. While materials handled in inbound logistics are raw materials or unfinished goods, the materials handled in outbound logistics are finished goods. Outbound logistics includes more complex processes than inbound logistics due to the higher production values and strict customer requirements such as on-time delivery (Wu and Dunn 1995).

Physical distribution is the area of business management responsible for the movement of raw materials and finished products and the development of movement systems. Even though physical distribution is usually associated with outbound product movements from a firm, it covers a broader concept that includes both inbound and outbound movements (Ballou 2007). Inbound and outbound logistics activities are shown in Figure 1 below.

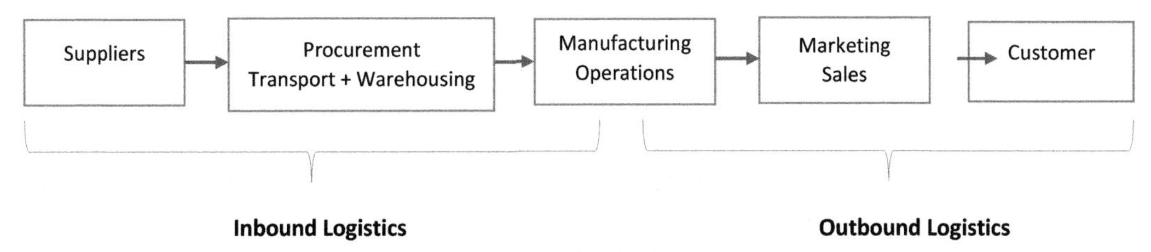

Fig. 1: Inbound and Outbound Logistics Activities.

2.2 Globalization and Liberalization and their Impact on Supply Chains

Globalization, trade liberalization, and opening borders to trade have generally led to an increased inflow of foreign investment, the establishment of multinational companies in developing countries, and the integration of these countries into global supply chains (Minten et al. 2007). Economic integration and progress of a nation are highly dependent on the successful establishment of logistics service; hence, trade liberalization is increasingly supported by efforts to liberalize logistics services. Trade liberalization of logistics services is an essential stage of a broader strategy to expand the potential of exports and achieve economic development (Tongzon 2012).

Supply chain management cannot be thought of as a domestic phenomenon since today's supply chains exceed national boundaries and spread across different countries. The expanse of supply chains brings about new challenges of globalization to companies who enjoy geographically distributed supply chains for their existing or new product lines (Meixell and Gargeya 2005).

Globalization offers enormous opportunities as well as increased risks in the development of supply chains. While some supply chains take advantage of these opportunities, some others are inflicted damage by the risks emerge from globalization. Hence, both opportunities and uncertainties should be taken into account when designing a global supply chain network. While globalization offers companies the opportunity of reaching new markets where they can advertise to potential customers, it also presents significant cost reduction opportunities by letting them expand operations to low-cost countries. However, these opportunities are usually accompanied by potential risks that might disrupt the flow in a supply chain. Some of these risk factors are natural disasters, shortage of skilled resources, geopolitical uncertainty, terrorist infiltration of cargo, volatility of fuel prices, currency fluctuation, etc. (Chopra and Meindl 2013). Spatial fragmentation is considered as one of the main engines of globalization. Many companies break down their business operations into various stages and move these stages across different regions. Several business activities that form a company's supply chain are organized and performed in distinct locations or different countries. Companies target to take advantage of technology, wage, and other cost differences by adopting the spatial fragmentation that is accepted as one of the main factors of economic globalization (Fujita and Thisse 2006).

3. Digitalization of Logistics and Challenges in Logistics 4.0

3.1 Inventory Control Systems (ICS)

The primary competitive edge was "cost" for manufacturers during the 1960s. Thus, companies predominantly focused on high volume production and cost minimization during this period. Inventory control systems (ICS) such as computerized reorder point (ROP) systems were sufficient for the basic manufacturing and planning needs of many companies. These systems used to include economic order quantity (EOQ) and economic reorder quantity functions (Jacobs and Weston 2007). In addition, ICSs were designed to manage basic conventional inventory management process. ICSs were one of the earliest business applications, which did not belong to the areas of finance and accounting. (Shehab et al. 2004).

3.2 Materials Requirement Planning (MRP)

The late 1960s witnessed the birth of Materials Requirement Planning (MRP) in response to the need for a state-of-art system capable of planning and scheduling materials for the manufacturing of sophisticated products. Manufacturing Resource Planning (MRP 2) and Enterprise Resource Planning (ERP) were derived from the MRP and were the successors of it. The very first MRP solutions required large technical support officers to support the mainframe computers; thus, they were costly, slow, and hard to handle. The development of more integrated business information systems was enabled by the emergence of faster and higher capacity disk storage (Jacobs and Weston 2007).

In the late 1970s, the primary competitive edge of manufacturers shifted from cost towards "marketing". At that time, the manufacturers adopted unique target-market strategies by putting emphasis on production planning. In other words, they focused on production integration while identifying their target market by focusing on a particular group of consumers at which their product or service was aimed. MRP systems met the requirements of companies during the late 1970s since they enabled the integration of core business functions, such as forecasting, master production planning, procurement, production, and inventory control. Many software corporations, such as SAP, Oracle, J.D. Edward, who would become the major ERP companies in the following decades, were founded during the mid-1970s as a response to the need for enterprise technology solutions (Jacobs and Weston 2007).

MRP systems were production-oriented information systems based on a time-phased order release system. These systems distribute activities, tasks, and resources over a planned time scale based on scheduled completion of a plan, task, or project. Manufacturing work orders and purchase orders are scheduled and released based on a master production schedule (MPS) in order to ensure that components and parts are received when they are needed in a production line. Inventory reduction, customer service improvement, and increment in productivity and efficiency are some of the major benefits of MRP systems (Shehab et al. 2004).

3.3 Manufacturing Resource Planning (MRP II)

The primary competitive edge of companies during the 1980s shifted to "quality" after the appearance of world-famous Total Quality Management (TQM) founders, such as W. Edwards Deming, Joseph M. Juran, Philip B. Crosby, and Kaoru Ishikawa. In this decade, manufacturing strategies of companies mainly focused on strict control of their processes, high-quality manufacturing, and attempts to reduce overhead costs. The implementation of world-class manufacturing techniques was the most important advancement in this decade. Companies wanted their goods, services, and processes to be ranked among the best by their customers and industry experts. These changes in companies' primary competitive edge brought about the need for a revision in the scope of the existing enterprise technology solutions (Jacobs and Weston 2007).

As a result of increasing competition among the companies on the market and product sophistication, MRP was developed and revised to capture more business functions such as product costing and marketing. The former material planning and control system had become a company-wide system capable of planning all the resources of a company. This new system was called Manufacturing Resource Planning (MRP II) at that time (Shehab et al. 2004).

A major purpose of MRP II was to integrate primary functions of a business such as production, marketing, and finance, and other functions such as personnel, engineering, and purchasing into the planning process. MRP II was a company-wide system, and it often had a built-in simulation that was capable of running "what-if" scenarios (Chen 2001). Manufacturing Resources Planning (MRP II) systems integrated the financial accounting system and the financial management system along with the manufacturing and materials management systems. This integrated business system enabled companies to make robust decisions about the material and capacity requirements pertaining to planned operations, elaborate on the activities and operations, and translate all activities into financial statements (Umble et al. 2003).

3.4 Enterprise Resource Planning (ERP)

Continuing improvements in technology allowed MRP II to be expanded to incorporate all resource planning activities for the entire business by the early 1990s. Besides the existing main functionalities, some business areas such as product design, information warehousing, capacity planning, human resources, finance, project management, and marketing are integrated into the new system (Umble et al. 2003). These critical business areas impact the companies that seek to obtain a competitive advantage by utilizing their assets, including information, effectively. Unlike previous versions, the ERP software companies made it possible to implement these critical business systems to not only manufacturing companies but also non-manufacturing companies (Ptak and Schragenheim 2003).

Enterprise Resource Planning (ERP) software systems are composed of a wide range of software products supporting daily business operations and decision-making process of a corporation. ERP integrates and automates operations of supply chain management, inventory control, manufacturing scheduling, and production, sales support, customer relationship management, financial and cost accounting, human resources, and many other business processes (Hitt et al. 2002).

Historically, ERP systems derived from MRP II systems are designed to manage a company's inventory orders, schedule production plans, and organize inventories. In addition to these functions, ERP systems integrate inventory data with financial, sales, and human resources data to enable an organization to price their products, generate financial statements, manage the workforce, materials, and money efficiently (Markus et al. 2000). The expansion of MRP II into ERP in the 1990s aspired to further improve resource planning by including the components of the supply chain in the scope of the planning phase. Hence, the main difference between MRP II and ERP is that MRP II focuses on the planning and scheduling of a company's internal resources, while ERP plans and schedules a company's supplier resources in addition to its internal resources, by taking the dynamic customer demands and schedules into account (Chen 2001).

ERP has added a new range of capabilities to the MRP II, including finance, distribution, and human resources development, and all of these existing and newly added functionalities were integrated to handle global business operations. These functionalities have been extended to cover many other "back-office" functions through the mid-1990s. Some of these extensions include order management, financial management, warehousing, distribution production, quality control, asset management, and human resources management. Later, the functionalities of ERP systems further broadened to engulf "front office" jobs, e.g., sales force and marketing automation, electronic e-commerce, and supply chain systems (Rao 2000). Figure 2 below presents the timeline of the development stages of ERP starting from the ICS.

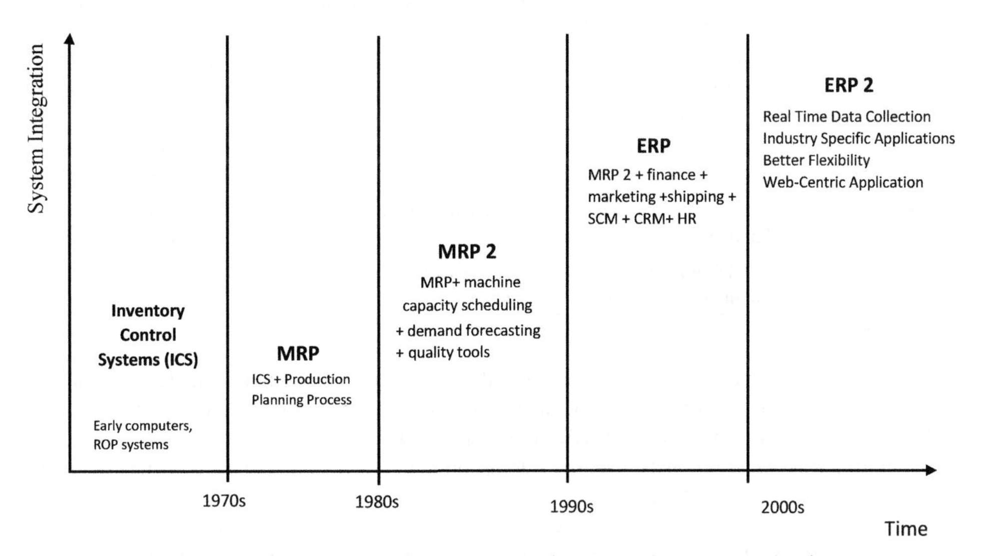

Fig. 2: Evolution of Inventory Control Systems into Enterprise Resource Planning.

4. Industry 4.0, Logistics 4.0 & Supply Chain 4.0

The importance of logistics in business has increased over time. Supply chain integration, being time-sensitive to customer orders (quick response), and just-in-time applications on inventory management transformed companies' business strategies, including their logistics strategies. Core business activities, such as customer and vendor selection, product design, and strategic alliance have been affected by logistics strategies (Caplice and Sheffi 1995).

A company's competency involves learning how to differentiate from competitors since a new trend can create progress that improves business performance dramatically. An innovation that initiates a new business model by discarding the current one is called a disruptive technology. The new technologies might not meet the immediate needs of a company's customers; however, the ability of these emerging technologies usually brings about performance improvement to the companies over a period of time. Disruptive technologies have the ability to improve the business performance above the performance levels of the incumbent technologies that are discarded. Hence, ignoring emerging technologies during their earliest stages can result in serious consequences for companies (Angeleanu 2015).

Disruptive technologies are transforming the core competencies of many companies and the business models of many industries. Digitalization in manufacturing processes and rapid growth in information technologies have been impacting the supply chains dramatically during the era of the Fourth Industrial Revolution. These emerging technologies should be comprehended thoroughly by the management in order to build a strong strategic path for the future. Industry 4.0 is the sum of all disruptive technologies that are implemented in a value chain. The seven characteristics of Industry 4.0 are discussed in Table 1 (Pfohl et al. 2015).

CPS are at the core of the Fourth Industrial Revolution. CPS are physical mechanisms that can be monitored, controlled and coordinated by a communication system through the internet. The development of CPS and their integration into the manufacturing systems has contributed to the birth of Industry 4.0. CPS enable the interaction with the physical world via network agents such as sensors, actuators, control processing units, and communication devices. The high growth rate of cyber technologies and the integration of digital devices into the supply chains contributed to the development of many

Table 1: The Seven Characterizing Features of Industry 4.0 (Pfohl et al. 2015).

Characteristics	**Description**
1. Digitalization	Key aspects of the supply chain, such as internal processes, product components, communication channels, are undergoing an accelerated digitalization process.
2. Autonomization	Industry 4.0 technologies enables machines and algorithms to make decisions and perform learning-activities independently. This autonomous decision-making and learning ability enables manufacturing facilities to work with minimum human-machine interaction.
3. Transparency	Industry 4.0 technologies are increasing the transparency of the value creation process of a firm that results in a more collaborative and efficient decision-making mechanism, and transparency in corporate partners' and customers' behavior.
4. Mobility	The spread of mobile devices streamline communication and data sharing globally. The mobility of the devices is modifying the interaction of customers and companies, and communication between machines in the production process.
5. Modularization	Industry 4.0 technologies enables the modularization of products and the whole value creation process. Adjustable modular production facilities increases the flexibility of the production processes.
6. Network-Collaboration	Companies' processes will be defined and activities will be decided through the interaction of machines and human beings within specific networks in and out of the companies' organizational borders.
7. Socializing	The collaboration in networks enables machines to start communicating and interacting with each other and human beings in a socialized manner.

areas such as manufacturing processes, logistics sector, health services, and autonomous vehicles. The Internet of Things (IoT) accelerated the integration of CPS into the manufacturing and service operations, leading to the revolutionary steps into the production, service, and logistics sectors (Barreto et al. 2017).

The communication between products, machines, transportation systems, and humans is transforming the current production systems. Smart factories will be the framework for future manufacturing systems. Smart factories are flexible, cost-efficient, and individualized mass production systems where the products flow independently through the manufacturing process with minimal human intervention (Hofmann and Rüsch 2017). Information and Communication Technologies (ICTs) have become an important and inevitable component of industrial manufacturing in recent years. New challenges have appeared in the logistics sector as a result of the rise of the Industrial Internet of Things (IIoT). These challenges have required technological modifications such as transparency in supply chains, and integrity control that guarantees the delivery of the right products to the right location at the right time. Logistics 4.0 concept is introduced and shaped by the developments in ICT and IIOT (Barreto et al. 2017).

4.1 Development of Logistics

Logistics went through three revolutionary steps before reaching its latest breakthrough, Logistics 4.0. The first step **(Logistics 1.0)** was launched by the "**mechanization of transportation**" starting in the late 19th and early 20th century. In this period, ships and trains equipped with steam engines were the main mode of transport for moving goods and containers, and they replaced human and animal power. The capacity of transportation grew significantly in the 20th century marking the beginning of the **mass transportation** age. The second step **(Logistics 2.0)** was the result of "**the automation of handling systems**" during the 1960s. This second innovation in Logistics was initiated by the invention of electric power and the spread of the mass-production techniques in manufacturing and completed by the automation of cargo handling. In this period, the automated warehousing and sorting systems, and automated loading and unloading systems were substituted with the conventional warehousing, and the heaviest work was starting to be done by electrically driven machines. At this time, container ships became dominant in ports, and they transformed the port cargo handling system. The third step **(Logistics 3.0)** appeared as "**the system of logistics management**" in the 1980s. Computers and Information Technology (IT) led the systemization of logistics activities and initiated the third innovation in logistics. Automation and logistics management capabilities were significantly developed as a result of integrating IT systems, such as Warehouse Management Systems (WMS) and Transport Management System (TMS), into logistics. The fourth step **(Logistics 4.0)** is now in its early stages. Internet of Things and Internet of Services (IoT & IoS) are the main drivers of Logistics 4.0 (Wang 2016). Figure 3 below presents the development process of logistics over time.

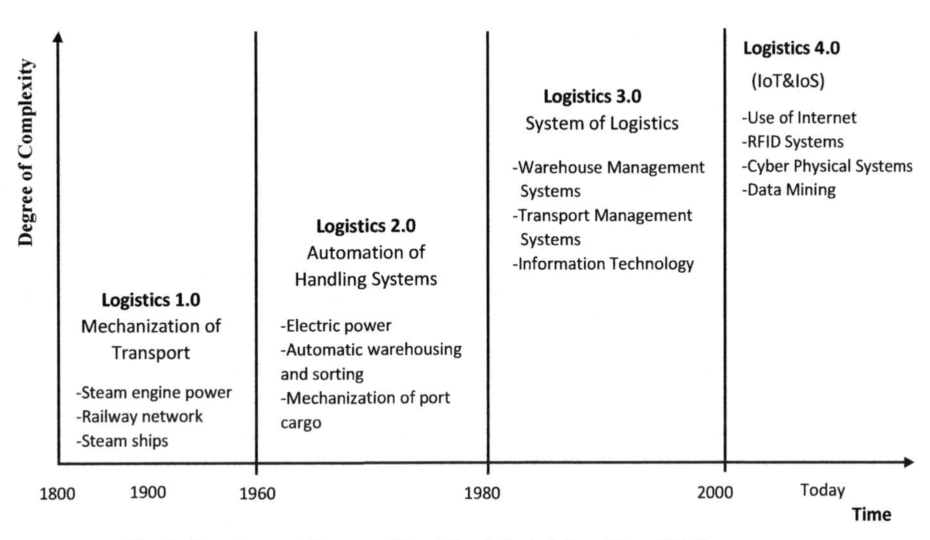

Fig. 3: Development Process of Logistics (adopted from Wang 2016).

4.2 Logistics 4.0

Logistics 4.0 is a collective term for technologies and concepts of value chain organization. Many of the emerging technologies take a crucial part in Logistics 4.0. For instance, CPS monitor physical processes, generate virtual copies of the physical world and make decentralized decisions within the logistical processes while Data Mining (DM) discovers the required knowledge to assist in decision-making processes. CPS communicate and cooperate with each other and humans in real-time via the IoT. The collective use of barcodes, radio frequency identification technology (RFID), sensors, global positioning systems (GPS), and other advanced network technologies for information processing and network communication purposes compose the Logistics 4.0 concept. These advanced technologies are used extensively in logistical operations, such as freight transportation, warehousing, distribution, packaging, handling, and related activities. Automated logistical operations and carrying out efficient transportation processes increase the service level and customer satisfaction and reduce the overall costs and the consumption of natural resources (Wang 2016).

The term Logistics 4.0 indicates the specific application of Industry 4.0 in the area of logistics. Logistics 4.0 is induced by new technologies and their application in logistics. Each information and communication technology causes a novel solution in a specific area of logistics (Glistau and Machado 2018). Logistics 4.0 refers to logistics systems consisting of independent subsystems where the behavior of the subsystems depends on other surrounding subsystems. The Logistics 4.0 definition combines two aspects: processual (supply chain processes are a subject of the Logistics 4.0 actions) and technical tools and technologies that support internal processes in supply chains (Szymańska et al. 2017). Table 2 depicts important technologies and their application area within Logistics 4.0 in the form of a morphological box.

The needs and complexity in the manufacturing industry show a monotonically increasing trend in recent years. The fast-paced evolution of technology, harsh competition, fast-growing volatility on the international markets, rising demand for highly individualized products, and products with short life cycles impose crucial challenges to firms. It is unlikely that the current approaches will buildcost-efficient, flexible, adaptable, stable, and sustainable supply chains. Strong industrial nations will be required to adopt Industry 4.0 to maintain their positions. Adjustment to this new initiative will introduce highly flexible mass production, real-time system coordination and optimization, cost reduction, and new business models. Major new trends in logistics are anticipated as a result of this new initiative, as well. Real-time monitoring of the material flows, enhanced material handling, and risk management are some of the prospective applications of Industry 4.0 on logistics (Hofmann and Rüsch 2017).

Some studies (e.g., Hofmann and Rüsch 2017; Strandhagen et al. 2017a) investigate the implication of Industry 4.0 on logistics management. Hofmann and Rüsch (2017) examine the two dimensions of logistical operations: (1) Physical supply chain dimension and (2) Digital data value chain dimension. The physical dimension involves autonomous and self-controlled logistics systems (e.g., autonomous trucks), automated material handling systems (e.g., piece picking robots), and autonomous order processing systems (e.g., smart contracts on the blockchain technology) that are connected and interacting with each other. The digital dimension, on the other side, encompasses sensor and machine data that are

Table 2: Important Technologies of Logistics 4.0 (Glistau and Machado 2018).

Technology/Criterion	Characteristics				
Identification	Smart card	Bar code	RFID	Sensor technologies	Biometrics
Mobile Communication	5G network	UMTS/LTE	GSM/GPRS	WLAN	Satellite communication
Localization	Geo reference point based	5G	UMTS/LTE GSM/GPRS	WLAN	Satellite-based
Electronic Data Interchange	Electronic data processing medium	EDI	XML	Internet	Telematics
Terminals	Smartphones	Tablets	Special hand-held units	On-board computer	
Architecture Paradigm	Centralized	Decentralized, Agent-based	Decentralized, Blockchain		
Architecture	Network	Hardware (Server, Cloud, Storage)	Software (Operating system, open-source)	Database	Virtualization
Data Analysis Methods	Descriptive	Inferential (Point and interval estimate)	Explorative (Big Data)	Regression, casual analysis	
Data Analytics Processing	Data Access	OLAP	Data Mining		

collected from the physical dimension of a supply chain and it is a crucial input for strategic business decisions. The model proposed is shown in Figure 4.

Both inbound and outbound logistics have to comply with the rapidly changing market dynamics as the demand for highly customized products and services is constantly increasing. Because of its complex nature, traditional planning and control methods are not useful. The term "Logistics 4.0" signifies the integration of logistics and the emerging innovations and applications of CPS. Logistics 4.0 is similar to the "Smart Products" and "Smart Services" in terms of technology-driven approach. Smart products and services carry out tasks that are repetitive and do not require intelligence; therefore, the employees can focus on tasks that require intelligence.

"Smart Logistics" is a system that has merits, such as improving the supply chain flexibility, quick adaptation to the volatile markets, and the accuracy in meeting the customer needs. All these will lead to higher customer service levels, production optimization, and reduced storage and manufacturing cost. The increasing use of the Internet that enables the

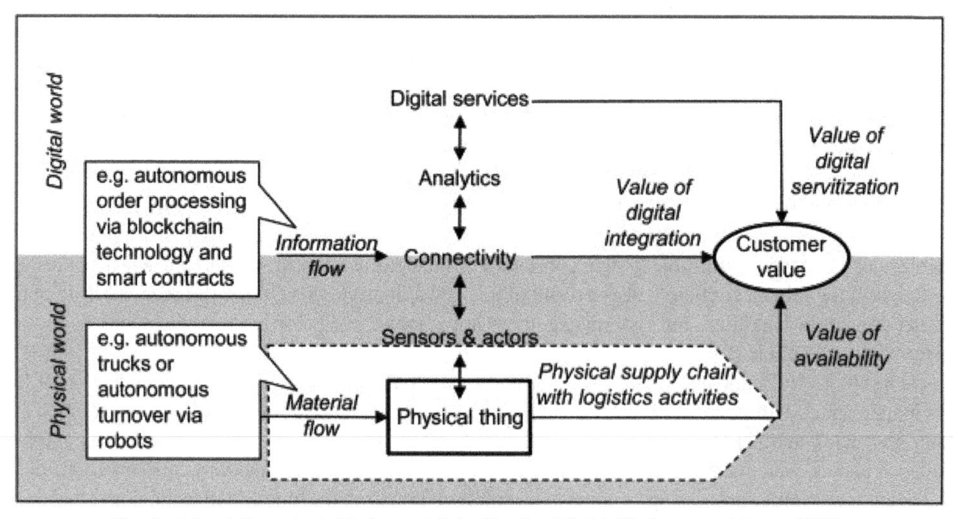

Fig. 4: A Logistics-oriented Industry 4.0 Application Model (Hofmann and Rüsch 2017).

real-time interaction between man and machines and the rapid digitalization in recent years led to the birth of this new concept (Barreto et al. 2017). An effective and resilient Logistics 4.0 requires to possess the technological applications shown in Table 3.

Table 3: Essential Technological Applications for Logistics 4.0 (adopted from [1]Barreto et al. 2017; [2]Min 2018; [3]Schuldt et al. 2010; [4]Coetzee and Eksteen 2011; [5]Sun 2012; [6]Strandhagen et al. 2017a).

Technological Application	Rationale
1) Resource Planning	Along with the adoption of Industry 4.0 and the implementation of CPS, a proper resource planning management system improves the overall productivity, flexibility, and agility of a supply chain. A capable resource planning system will build a robust forecasting model for the resources of an organization (e.g., people, materials, and equipment). This will lead to optimize resources and processes, reduce time to market, increase customer satisfaction.[1]
2) Warehouse Management Systems	The introduction of smart systems and the implementation and integration of these systems into the Warehouse Management Systems (WMSs) cause a radical transformation in warehouse activities. The location and estimated arrival time of transporters can be monitored by the intelligent WMSs through the use of CPS. Hence, intelligent WMSs will be able to optimize just-in-time and just-in-sequence delivery by deciding and preparing the proper docking area. At the same time, the delivery data (e.g., quantity, size, and price) will be sent to the entire supply chain by the RFID sensors. In order to move the incoming goods, the appropriate material handling equipment will be requested and available storage space will be assigned immediately by the WMSs based on the specifics of the delivery.[1]
3) Transportation Management Systems	Transportation Management Systems (TMSs) facilitate the interaction between an order management system (OMS) and a distribution center (DC) or a warehouse. Advanced TMSs can be integrated into other supply chain technologies such as Warehouse Management Systems and Global Trade Management Systems. TMSs can communicate electronically with customers, trade partners, and carriers. Logistics 4.0 utilizes real-time and inline data to achieve efficient and effective logistics processes. A TMS allows a company to accurately pinpoint the location of its transporters by the use of GPS technology while they're on the road, monitor and track the movements of goods, negotiate with carriers, consolidate shipments, and interact with Intelligent Transportation Systems (ITSs). Therefore, a TMS is one of the essential elements for Logistics 4.0 concept.[1]
4) Intelligent Transportation Systems	Intelligent Transportation Systems (ITSs) provide solutions for a reliable platform for transportation operations. ITSs interoperate in different fields of transportation systems, such as transportation management, control, infrastructure, and operations. Computing hardware, positioning system, sensor technologies, telecommunications, data processing, virtual operation, and planning techniques are some of the technologies adopted by ITSs. A fully operational ITS supports intelligent truck parking and delivery areas management, multimodal shipping, i.e., planning and coordinating different transport modes during the various logistics operations, CO_2 emission calculation and monitoring, operations priority and vehicle speed guidance (e.g., reduce fuel consumption, lower emissions and reducing heavy vehicles operations in urban areas), and eco-drive support (e.g., less fuel consumption and CO_2 emission by embracing energy-efficient methods of driving).[1]
5) Information Security	The way businesses are done by organizations has been affected by the emerging internet-based technologies such as cloud systems, CPS, IoT, and Industry 4.0. Innovative technologies with potential impact for businesses become more important day by day for the companies that are interested in lowering their operating cost and gaining competitive advantage in the market. This tendency brings information security issues. Unexpected security risks are inherent in the new technologies. Therefore protecting IT infrastructure and information assets is one of the main concerns for organizations. Promoting the continuous attempt to build a security culture and acknowledging that all technological devices, applications, and systems have their inherent vulnerabilities will help organizations to reach the desired level of security and foster their business goals. Companies should identify, implement, monitor, and control their desired security requirements in order to achieve a required level of information security.[1]

Table 3 contd. ...

...Table 3 contd.

Technological Application	Rationale
6) Blockchain Technology	Blockchain Technology (BCT) can be applied to some specific areas of supply chains such as smart contracts, asset tracking, secure and error-free order fulfillment, and cybersecurity. Smart contracts help companies to exchange money property, shares, or anything valuable in a reliable, transparent, and conflict-free way. Hence, the transaction time and costs will be minimized. One of the main functions of the BCT is to track and record all the supply chain activities of a particular asset from its origin to its final destination. This is called the asset tracking feature of blockchain. This feature of the BCT hedges companies against fake transactions and make it easier to track goods throughout the supply chain. The asset tracking capability of BCT reduces the risk of loss and damage during transit. BCT can expedite the order fulfillment process with its merits such as rapid confirmation of customer credit history, quick inventory status analysis, order/shipping status notification, and offering transparency throughout the order fulfillment process. Growing cybercrime threats in recent years imposes a high risk on supply chain networks. BCT, with its visibility, privacy and, non-stop information verification features, is an outstanding technology that mitigates the cybercrime risks in supply chain networks.[2]
7) Cloud Technology	Cloud computing is a prominent technology in order to implement autonomous control in logistics. Autonomous logistics solutions with intelligent software agents can be set up on a scalable IT infrastructure that is offered by cloud service providers. Some services offered by these providers may range from scalable hardware platforms to complete process control of the logistics operations. Autonomous logistics control systems enable users to focus on their core business operations and they are no longer require to invest in IT infrastructures.[3]
8) Internet of Things (IoT)	The Internet can connect physical objects to each other and create smarter services for the environment. These benefits of the Internet are considered as the main drivers of the IOT.[4] The IoT enables companies to track their goods at each stage of the logistics process in real-time, and manage their logistics architecture. While the flow of goods is being monitored, a company can analyze the data generated at each stage of the logistics process and distribute data to all parties in the process. The use of real-time data in forecasting allows companies to realize future trends and the probability of unexpected events. Therefore, preventive measures or policies can be adopted in advance. Thus, companies gain a competitive edge since they become more responsive to the market.[5]
9) Augmented Reality	Augmented reality (AR) systems can take part in logistics, manufacturing, training, and maintenance operations. AR combines computer-generated data with the physical world to help workers. For instance, pick-by-vision is one of an innovative order picking method based on AR technology. This logistics solution offers a fast and effective way of picking of products while reducing operating time.[6]

5. Conclusion and Future Direction

This chapter introduces an extensive review of logistics, its changing patterns within the Industry 4.0 era, and its role in the digital transformation in SCM. The historical development of ICSs, the evolution of logistics, the interaction between Industry 4.0 and smart logistics, and potential challenges of Logistics 4.0 are broadly discussed.

Logistics is one of the core pillars in the value chain for suppliers, manufacturers, and retailers since they cannot be competitive in the market without getting the right good at the right time in the right place. It is getting more difficult to fulfill these requirements as logistics networks become uncertain and volatile. New methods, technologies, and services are required by companies as the complexity increases in logistics networks. Many challenges and opportunities for logistics emerge as a result of today's consumer behavior. Therefore, the shift from traditional logistics to new logistics solutions is inevitable. Industry 4.0 will transform and improve the traditional logistics and its self-perception (Wang 2016). Logistics 4.0 is a complex system composed of many emerging technologies that are connected and able to communicate with each other, and it is responsible for fulfilling customers' needs and meeting the requirements of increasing complexity in logistical operations. Logistics 4.0 is an element of Industry 4.0 and these two cannot be thought of as independent from each other (Wang 2016).

As an element of Industry 4.0, Logistics 4.0 creates possibilities for new business models. Instantaneous information exchange, computerized business solutions, and real-time big data analysis capability are some of the features that companies enjoy as a result of adopting Logistics 4.0. The combined use of all these features of Logistics 4.0 is changing the way companies do their business. Today, service-oriented businesses, where the customer involvement starts at the product or service design stage, are at the focus of the companies. This transformation is expedited by the Industry 4.0 technologies such as CPS, IoT, IoS, smart products and smart processes (Strandhagen et al. 2017b).

Traditional logistics will undergo a transformation as Industry 4.0 takes over today's conventional production process in use. This change in logistics will not be easy or effortless. However, it will substantially change the way companies perform their logistics operations and improve process efficiency, productivity and customer satisfaction. Companies who embrace Logistics 4.0 will gain a competitive advantage since Logistics 4.0 will equip them with a flexible, sustainable, and highly responsive supply chain. Transforming conventional logistics systems into smart logistics systems and optimizing the logistics process will lead to agile supply chains, improved cost-saving, higher customer service levels, and satisfaction.

References

Akdil, K.Y., A. Ustundag and E. Cevikcan. 2018. Maturity and readiness model for industry 4.0 strategy. In Industry 4.0: Managing The Digital Transformation (pp. 61–94). Springer, Cham.

Angeleanu, A. 2015. New technology trends and their transformative impact on logistics and supply chain processes. International Journal of Economic Practices and Theories, 5(5): 413–419.

Ballou, R.H. 2007. The evolution and future of logistics and supply chain management. European Business Review, 19(4): 332–348.

Barreto, L., A. Amaral and T. Pereira. 2017. Industry 4.0 implications in logistics: an overview. Procedia Manufacturing, 13: 1245–1252.

Caplice, C. and Y. Sheffi. 1995. A review and evaluation of logistics performance measurement systems. The International Journal of Logistics Management, 6(January 1995), pp. 61–74.

Chen, I.J. 2001. Planning for ERP systems: analysis and future trend. Business Process Management Journal, 7(5): 374–386.

Chopra, S. and P. Meindl. 2013. Supply Chain Management: Strategy, Planning, and Operation, 5/e. Pearson Education India.

Christopher, M. 2016. Logistics & Supply Chain Management. Pearson UK.

Coetzee, L. and J. Eksteen. 2011. May. The Internet of Things-promise for the future? An introduction. In 2011 IST-Africa Conference Proceedings (pp. 1–9). IEEE.

Fujita, M. and J.F. Thisse. 2006. Globalization and the evolution of the supply chain: who gains and who loses? International Economic Review, 47(3): 811–836.

Glistau, E. and N.I. Coello Machado. 2018. Industry 4.0, logistics 4.0 and materials-Chances and solutions. In Materials Science Forum. Trans. Tech. Publications, 919: 307–314.

Hitt, L.M., D.J. Wu and X. Zhou. 2002. Investment in enterprise resource planning: Business impact and productivity measures. Journal of Management Information Systems, 19(1): 71–98.

Hofmann, E. and M. Rüsch. 2017. Industry 4.0 and the current status as well as future prospects on logistics. Computers in Industry, 89: 23–34.

Jacobs, R. and F.C.T. Weston. 2007. ERP—a brief history. Journal of Operations Management 25(2): 357–363.

Markus, M.L., C. Tanis and P.C. Van Fenema. 2000. Enterprise resource planning: multisite ERP implementations. Communications of the ACM, 43(4): 42–46.

Meixell, M.J. and V.B. Gargeya. 2005. Global supply chain design: A literature review and critique. Transportation Research Part E: Logistics and Transportation Review, 41(6): 531–550.

Min, H. 2019. Blockchain technology for enhancing supply chain resilience. Business Horizons, 62(1): 35–45.

Minten, B., L. Randrianarison and J.F.M. Swinnen. 2007. 12 Global Supply Chains, Poverty and the Environment: Evidence from Madagascar. Global supply chains, standards and the poor: How the globalization of food systems and standards affects rural development and poverty, p. 147.

Onar, S.C. and A. Ustundag. 2018. Smart and connected product business models. In Industry 4.0: Managing The Digital Transformation (pp. 25–41). Springer, Cham.

Pfohl, H.-C., B. Yahsi and T. Kurnaz. 2015. The impact of Industry 4.0 on the Supply Chain. Innovations and Strategies for Logistics and Supply Chains (August), pp. 32–58.

Ptak, C.A. and E. Schragenheim. 2003. ERP: tools, techniques, and applications for integrating the supply chain. Crc Press.

Rao Siriginidi, S. 2000. Enterprise resource planning in reengineering business. Business Process Management Journal, 6(5): 376–391.

Roy, J. 2001. Recent trends in logistics and the need for real-time decision tools in the trucking industry. Proceedings of the Hawaii International Conference on System Sciences, 00(c): 76.

Rutner, S.M. and C. John Langley. 2000. Logistics Value: Definition, Process and Measurement. The International Journal of Logistics Management, 11(2): 73–82.

Schuldt, A., K. Hribernik, J.D. Gehrke, K.D. Thoben and O. Herzog. 2010. Cloud computing for autonomous control in logistics. INFORMATIK 2010. Service Science–Neue Perspektiven für die Informatik. Band 1.

Shehab, E.M., M.W. Sharp, L. Supramaniam and T.A. Spedding. 2004. Enterprise resource planning: An integrative review. Business Process Management Journal, 10(4): 359–386.

Strandhagen, J.W., E. Alfnes, J.O. Strandhagen and L.R. Vallandingham. 2017a. The fit of Industry 4.0 applications in manufacturing logistics: a multiple case study. Advances in Manufacturing, 5(4): 344–358.

Strandhagen, J.O., L.R. Vallandingham, G. Fragapane, J.W. Strandhagen, A.B.H. Stangeland and N. Sharma. 2017b. Logistics 4.0 and emerging sustainable business models. Advances in Manufacturing, 5(4): 359–369.

Sun, C. 2012. Application of RFID technology for logistics on internet of things. AASRI Procedia, 1: 106–111.

Szymańska, O., M. Adamczak and P. Cyplik. 2017. Logistics 4.0-a new paradigm or set of known solutions? Research in Logistics & Production, 7.

Tepić, J., L. Tanackov and G. Stojić. 2011. Ancient logistics–historical timeline and etymology. Tehnički vjesnik, 18(3): 379–384.

Tjahjono, B., C. Esplugues, E. Ares and G. Pelaez. 2017. What does industry 4.0 mean to supply chain? Procedia Manufacturing, 13: 1175–1182.

Tongzon, J. 2012. The challenge of globalization for the logistics industry: evidence from Indonesia. Transportation Journal, 51(1): 5–32.

Umble, E.J., R.R. Haft and M.M. Umble. 2003. Enterprise resource planning: Implementation procedures and critical success factors. European Journal of Operational Research, 146(2): 241–257.

Wang, K. 2016. November. Logistics 4.0 Solution-New Challenges and Opportunities. In 6th International Workshop of Advanced Manufacturing and Automation. Atlantis Press.

Wu, H.J. and S.C. Dunn. 1995. Environmentally responsible logistics systems. International Journal of Physical Distribution & Logistics Management, 25(2): 20–38.

SECTION 2

Internet of Things and Cyber-Physical Systems in SCM

CHAPTER 3

The Internet of Things in Supply Chain Management

Volkan Ünal,[1] *Mine Ömürgönülşen,*[2] *Sedat Belbağ*[3,*] and *Mehmet Soysal*[2]

1. Introduction

Supply chain management (SCM) is the management of the flow of materials, services and information among the partners (e.g., suppliers, manufacturers, retailers, etc.) along the supply chain. In a traditional supply chain management, companies are confronted with various challenges such as reducing cost, improving efficiency, ensuring coordination and managing uncertainty. Rapid advances in technology lead companies to integrate new technological developments into supply chains to maintain their competitive advantage. The concept of Industry 4.0 has the potential to change the structure of today's traditional production and transportation processes to a great extent. Real-time collaboration can be built among the partners in the supply chain utilizing the recent technologies introduced by Industry 4.0 such as radio frequency identification (RFID), cyber-physical systems and internet of things (IoT).

Among these new technologies, IoT is a dynamic network system where each technological device has an identity, physical attribute and virtual personality with self-configuring capabilities based on standard and communication protocols (van Kranenburg 2008). That is to say, IoT is a connectivity network of smart devices at anytime and anywhere, which will affect the structure of any industry, as well as daily life. IoT provides an opportunity to attach technology to regular devices (e.g., home appliances, microwave oven, home theatre, etc.) and make them online (Whitmore et al. 2015). IoT offers considerable potential to public and private sectors by enabling innovative applications to overcome common challenges faced in many industries and is capable of gathering and transporting information from all devices that can connect internet via Wi-Fi, sensors, Bluetooth, cellular networks, Global Positioning System (GPS) and RFID technologies.

[1] Senior Software Engineer, STM Savunma Teknolojileri Mühendislik ve Ticaret A.Ş., Email: unal.vlkn@gmail.com
[2] Assoc. Prof. Dr., Hacettepe University, Emails: mergun@hacettepe.edu.tr; mehmetsoysal@hacettepe.edu.tr
[3] Ankara Hacı Bayram Veli University.
* Corresponding author: sedat.belbag@hbv.edu.tr

Morgan Stanley's report predicts that there will be nearly 75 billion smart devices connected by 2020 (Danova 2013). These devices will generate trillions of bytes per day that should be collected, stored, analyzed and transmitted. IoT offers companies to make smart devices more visible, traceable, adaptable and flexible in a data-driven environment. Furthermore, IoT enables (i) to monitor and control the manufacturing system, (ii) to analyze the big data accurately, and (iii) to share necessary information between people and things (Lee and Lee 2015). Several IoT applications can be observed in agriculture, food processing, retailing, healthcare, home appliance, security, recycling, and manufacturing industries (see, e.g., Li et al. 2012; Xu et al. 2014).

IoT related applications will inevitably lead companies to redesign their supply chains. Along with providing smart and dynamic manufacturing processes, IoT technologies can contribute to forecast and rapidly react to unexpected changes that occur throughout the supply chain (Fatorachian and Kazemi 2018). IoT enables more visibility and real-time information among partners in a supply chain with horizontal integration of business processes (e.g., inbound logistics, production, and outbound logistics). Particularly, enhanced visibility and up-to-date information provide significant improvements in logistics and supply chain management (Flügel and Gehrmann 2008). The mitigation of bullwhip effect (Yan and Huang 2009) and advances in product traceability can be given as examples for such improvements (Zhengxia and Laisheng 2010).

In line with the increasing awareness of Industry 4.0 related technology practices, there has been a growing interest in studies related to IoT applications in SCM. The current study aims to reveal the tendencies and interests in the integration of IoT into the SCM. To the best of our knowledge, this study will be the first attempt to address the potential effects of IoT applications in SCM.

2. The Internet of Things

The first industrial revolution began with the implementation of steam power into production processes at the end of the 18th century. This revolution was immediately followed by the second one with the introduction of electricity which, in turn, leads to a division of labor at the beginning of the 20th century. During the early 1970s a new industrial era, namely the third industrial revolution, has been started with the integration of information technologies into operational processes. Robots, Computer Numerical Control (CNC) machines, computer-aided manufacturing, and electronic devices facilitate the automation of manufacturing processes, as machines have taken over not only a substantial proportion of the labor work but also some of the brainpower (Yin et al. 2018).

In recent years, a new industrialization stage has begun, as the industry evolves even faster with remarkable advances in the internet, computer and software technologies. Industry 4.0 was firstly introduced by German practitioners during the Hannover Fair event in 2011. The developments introduced by Industry 4.0 had been published in a report by a working group in 2013 (Kagermann et al. 2013). The philosophy behind the Industry 4.0 is twofold that can be summarized as follows: (i) to change the current manufacturing devices with the fully automated ones and (ii) to minimize human interaction and control for gaining advantages in a highly competitive business environment. Several countries have started to initiate similar attempts in their industries. A smart system, "Society 5.0" was suggested by the Japanese Government that aims to integrate smart technologies into the industry and the community to meet the needs of each individual (Government of Japan, Public Relations Office, 2019).[1] China established a strategic initiative, "*Made in China 2025*", which aims to attain self-reliance for the key components in the selected high-tech industries including new energy vehicles, industrial robotics, and semiconductors (European Chamber Report 2017).[2]

The concept of Industry 4.0 is based on the emergence of new digital technologies. As a consequence of the latest technological improvements, each device becomes "*smart*" over time with the integration of the internet. The internet provides instant communication among these smart devices without human control and interaction. When smart devices communicate with each other, transactions generate a large amount of data, which needs to be stored and processed. Industry 4.0 includes technologies of many disciplines and makes extensive use of the big data, artificial intelligence, cyber-physical systems, smart factories, system integration, RFID, sensors, simulation, robots, 3-D printers, cloud computing, cyber-security, simulation, and IoT.

IoT is a technological term that provides a connection among many devices, at anytime and anywhere. According to the definition of the European Commission Information Society[3] (2008), IoT is defined as "*things having identities and virtual personalities operating in smart spaces using intelligent interfaces to connect and communicate within social, environmental, and user contexts*". According to a definition, IoT is a network of digitally connected devices to communicate with each other and facilitate planning, control, and collaboration of supply chain processes among supply chain partners (Ben-Daya et al. 2019).

[1] https://www8.cao.go.jp/cstp/english/society5_0/index.html, Online accessed: September 2019.

[2] https://www.europeanchamber.com.cn/en/china-manufacturing-2025, Online accessed: September 2019.

[3] https://ec.europa.eu/digital-single-market/en/internet-of-things, Online accessed: August 2019.

IoT is one of the fastest developing technology related to the Industry 4.0 concept (Szozda 2017). It is estimated that the value of IoT will be between $4 trillion and $11 trillion globally by 2025 (Bauer et al. 2015). IoT is capable of collecting information by detection technologies and allows all physical objects to be connected by using the internet, RFID, and sensors. IoT has great potential to affect both daily life and the industrial environment. Different types of communication protocols and the internet allow end-users to connect to the corresponding systems. IoT related technologies have already been used in various areas, such as sensors, GPS, Wi-Fi and Bluetooth in mobile devices.

IoT related technologies lead companies to incorporate smart devices into production systems. In smart production systems, these devices enable real-time data flow and generate a large amount of data. As a consequence, IoT related technologies help companies to collect, store and analyze big data emitted from smart devices. Big data is defined as a large volume of complex and variable data set that requires advanced techniques and technologies to get, store and analyze raw data (TechAmerica Foundation's Federal Big Data Commission, 2012). IoT related technologies enable decision-makers to discover and control several issues in business such as changes in customer behavior and providing valuable services (Lee and Lee 2015).

Although IoT is an important component of the Industry 4.0 concept, it is also related to various new technological improvements, such as cyber-physical systems, machine learning, and cloud computing. These technologies can be briefly explained as follows. Cyber-physical systems refer to a system where computer-based algorithms monitor and control the physical and software components (Lee 2008). Lu (2017) states that IoT is integrated into a complex cyber-physical system by using various devices equipped with the detection, identification, processing, communication and networking capabilities. Machine learning is an important factor in the integration process between cyber-physical systems and IoT in recent years. Machine learning enables smart devices to learn without human intervention and perform autonomously (Mahdavinejad et al. 2018). Cloud computing is an on-demand network system that can be reached by one or many users (Mell and Grance 2011). Cloud computing enables users to reach IoT applications by providing a network with unlimited storage and computation capacity (Atlam et al. 2017).

As a result of recent advances in IoT, the structure of industrial systems rapidly adapts new technological improvements. IoT does not only focus on the structure of the factory itself but also is adapted into various processes such as distribution and customer service. For instance, machine learning-based approaches support and improve the decision-making process in distribution utilizing proper big data mining. These IoT embedded systems directly affect the structure of production planning, maintenance scheduling, inventory planning and control (Upasani et al. 2017). IoT based data has been transmitted by wired or wireless networks into the industrial cloud systems. Manufacturers store and analyze IoT based data to optimize the production process and produce high-quality products.

3. Challenges of The Internet of Things

Companies are confronted with new challenges in their organizational structures with the integration of the IoT process. Installing new technologies into a fully working IoT-based system requires a considerable amount of investment, effort and time. Any delay or malpractice during the integration process of new technologies into production lines may result in a heterogeneous network system (including both wired and wireless connections). Furthermore, the new network system connects various smart devices by using a wide range of applications. Such event-oriented sensors and augmented reality displays increase system complexity and reveal additional challenges in the management of the network (Pereira et al. 2018).

The number of interconnected physical devices will significantly increase the system complexity and these devices will constantly interact with each other. Smart production systems will obtain, understand and convert machine-generated data into a piece of meaningful information in the decision-making process. Storing and analyzing big data is a major challenge, although big data create valuable business opportunities in terms of providing a competitive advantage. Besides, analyzing big data in the IoT environment requires different types of structures, processes, and technologies. Companies should allow fundamental changes in the production systems and make all the necessary adjustments. Specific technologies, such as big data analytics (i.e., a method for collecting and analyzing a large amount of data to solve real-world problems) and cloud computing may help companies to extract relevant information from big data (Fatorachian and Kazemi 2018).

Monostori et al. (2016) state that security is an important problem related to IoT related technologies. The security system should be considered as an independent process in the smart production system. Wireless sensors and RFID devices may cause problems for security systems. For instance, a vulnerable production system can be a subject of a cyber-attack by hackers. Many companies are not aware of the security threats against their smart production systems. Economic and production losses are the possible outcomes of these cyber-attacks (Tuptuk and Hailes 2018). Other potential outcomes of security breaches would be injuries, loss of life, damage to physical infrastructure, equipment, and the environment, unauthorized access, data modification and forgery (Cardoso et al. 2017). Cybersecurity deserves more attention to decrease vulnerability against industrial espionage and sabotage.

Although wireless network devices (e.g., sensors, RFID, etc.) may cause vulnerable production systems, various security devices provide additional protection to these wireless technology embedded systems. Growing technological advances in security systems decrease the number of potential cyber-attacks to IoT related technologies with highly secure wireless network technologies. Cybersecurity aims to alter the intended behavior of these networks or connected devices to protect them from malicious interventions (Abomhara and Køien 2015). Thus, all smart devices are connected by the integration of secure cloud systems.

Connected smart devices require international communication protocols and standards. Several organizations (e.g., International Organization for Standardization and International Electro-Technical Commission) develop technical standards to improve the effectiveness of the IoT-based systems (Li et al. 2015). Standardization supports the integration process of IoT-related technologies in smart production systems (Trappey et al. 2017). However, different standards or lack of common international standards may cause serious problems such as incompatibility among IoT-related technologies. Another important problem is to transform the traditional factory layout into a technological one that enables the deployment of a huge amount of data from numerous smart devices (Lin and Yang 2018). Cloud computing helps smart production systems to handle processing loads and big data for avoiding long delays in production processes.

The major challenges of transformation in production systems related to IoT related technologies can be summarized as (i) difficulties in technical integration, (ii) difficulties in storing and analyzing big data, (iii) challenge of enabling consistent cybersecurity, (iv) different international standards, and (v) challenge of redesigning the traditional layout.

4. Changes in Business Models and Production Processes

The total integration of IoT related technologies into production systems will deeply change the way of the business environment soon (Lee et al. 2018). The IoT has a major impact on the business models of manufacturing companies in various industries. It will be the key factor in the transformation process of traditional production systems into smart production systems by facilitating the transfer of knowledge. Therefore, IoT-related technologies may reveal business opportunities to work with a significant amount of real-time data (Kumar et al. 2018). IoT affects the structure of the industrial environment concerning five key factors; (i) design and innovation, (ii) asset utilization and revenue planning, (iii) supply chain and logistics design, (iv) resource efficiency and (v) stakeholder experience (Kamble et al. 2019).

Traditional production systems become less efficient with new technological developments. Lee (2008) states that customer expectations are no longer met by traditional production systems within a short period. However, IoT enables better control over smart production systems by transmitting and analyzing big data. IoT provides remote access and control over the production system from all over the world. Companies can easily obtain the necessary data related to production processes with the wide usage of sensors and wireless devices. Big data are transferred to the super-computers to be analyzed for potential product improvements (Gierej 2017). Digital simulation models (e.g., augmented reality) analyze the collected data to improve the quality of processes and products. In conclusion, the productivity of manufacturing systems increases, when the cycle times and the number of defects decrease. IoT provides efficiency in resource usage and reducing cycle time. Especially, resource inefficiency is a major problem for food supply chains due to the mismanagement of resources such as food losses and waste (Jagtap and Rahimifard 2017).

Core competencies of a company evolve to satisfy customer expectations in a better way. IoT related applications provide control over resources and core competencies to develop capabilities in the product development process. The IoT enables manufacturers to create valuable, flexible and customized products. Customers might have a chance to be involved in the decision-making process regarding the product design process with IoT related technologies (Lu 2017). Manufacturing companies can greatly benefit from the IoT by developing value-added applications. For instance, a cement manufacturing company has implemented an IoT related technology to estimate the energy consumption trend. Along with optimizing energy consumption amounts, IoT application reduces the energy consumption of the company by 10% (Xu and Li 2018).

IoT mainly improves the effectiveness of production processes. Moreover, it is also a useful tool for companies to deal with changes in customer behavior, product design, packaging, and distribution. IoT directly affects the distribution-related decisions (e.g., delivery plans and delivery times) in a highly dynamic, uncertain and complex environment (Wang 2016). To maintain a competitive advantage, companies focus on responsiveness and delivery times. Accordingly, IoT has the potential to bring significant changes and improvements in traditional logistics systems. Smart machines can detect real-time data, be sensitive to the content of data and provide value-added information to help managers to make better decisions in logistics.

Investment in new technologies and the employment of a highly skilled and flexible workforce will also provide a competitive advantage to the companies (Strange and Zucchella 2017). IoT related technologies create specialized departments and jobs in human resources management. Additional technical assistance by smart devices will also decrease the routine workload of the employees. Qualified workers are required to operate smart devices, especially in planning, monitoring and controlling production processes. The integration of IoT related technologies allows employees to work in safe conditions and to boost employee productivity. For instance, drones are commonly used in the inspection of oil zones and natural gas pipelines that prevent workers from potential exposure of hazardous gases or chemicals (Sissini et al. 2018).

Even though IoT offers numerous benefits for companies, it should be carefully integrated into the existing system. It will be easier for a warehouse management system to cope with the changes in production orders and there can be an increase in the efficiency of operations with the integration of new technologies. In a classical warehouse management system, the main problem of a worker is the high workload. The entire process is considerably time-consuming and the workload is relatively higher compared to the workload in an automatic warehouse (Lee et al. 2018). A worker may randomly place a product and the collection process typically depends on the employee's memory and experience. There is always a possibility to miscalculate the correct amount of the inventory. However, IoT related technologies, such as RFID, make inventory more visible and significantly decrease the workload in the entire system.

Due to technological advances, global competition evolves into an IoT-based competition. Many companies face serious challenges during the transformation process of traditional production systems into the IoT embedded ones. Proper integration requires close cooperation among companies to increase the effectiveness of the supply chain. Inter-company integration and collaboration among supply chain partners will result in more visible and controllable storage and distribution systems. IoT helps decision-makers to analyze the current state and the structure of the company by enabling more visibility through connection among smart devices. Especially, the structure of the production systems is completely differentiated from the traditional ones with the integration of IoT related technologies. The smart production systems provide autonomy for the entire system rather than the autonomy in the manufacturing floor alone (Kusiak 2018). Therefore, all partners in the supply chain should adapt their processes to maximize the benefits of new technologies for the end-users.

5. The Effects of The Internet of Things on Supply Chain Management

Today's highly competitive environment leads companies to integrate Industry 4.0 related technologies into their supply chain management. The structure of supply chains needs to be redefined with these radical changes. According to Pereira et al. (2018), Industry 4.0 enables self-organized supply chains by implementing IoT. Through Industry 4.0 related technologies, supply chain management evolves into a "smart supply chain management" or "SCM 4.0".

IoT redesigns both the structure of the supply chain and the relationship among suppliers, manufacturers and customers. The structure of the new supply chain becomes more transparent, flexible and customer-oriented. Dunke et al. (2018) state that IoT related technologies positively affect supply chain performance and help supply chain partners to cope with real-time challenges such as uncertain demand and lead time. IoT provides numerous advantages to supply chain partners, such as increased visibility, improved collaboration among supply chain partners, additional agility and adaptability and reduced supply chain risk.

Real-time visibility allows companies to observe and control both internal and external processes through the supply chain. IoT focuses on each phase of the supply chain from production to distribution and contributes to the increased operational efficiency. The flow of goods, services, information, and funds can be monitored by IoT-based technologies which could enable higher supply chain performance (Sun et al. 2018). Furthermore, increased order visibility enables companies to track items through the entire supply chain. Due to the real-time visibility and instant communication among smart devices, supply chain partners will rarely need to keep additional inventory for unexpected demands. IoT related technologies will also help to decrease inventory costs in each stage of a supply chain.

More visibility increases communication among supply chain partners. IoT related technologies are appropriate for effective communication tools between transmitter and receiver partners in a supply chain. Smart communication tools provide a more visible network among the partners in a supply chain. The IoT forces companies to work collaboratively to increase supply chain surplus. Better information sharing and improved foresight lead a company to develop a new collaboration with its suppliers. As a result, more valuable goods and services are offered to customers (Zheng and Wu 2017). Although IoT related technologies promote strong collaboration among supply chain partners, these technologies increase infrastructure costs (e.g., reader, tag, and server costs) and operational costs (Bardaki et al. 2012).

Along with providing more visible structure and strong collaboration, IoT related technologies may also transform the structure of companies into more agile ones and make them adaptable against unexpected changes in the environment. Instant communication among smart devices provided by IoT leads companies to quickly respond to customer requirements at an acceptable cost (White et al. 2005). Thus, the adaptation performance of companies becomes more efficient to market-driven changes (Shen and Liu 2010). An agile supply chain provides superior value to all partners as well as manages and mitigates supply chain risk (Braunscheidel and Suresh 2009).

Supply chain risk may arise from environmental, organizational and network-related factors and these factors may deeply affect the structure of the supply chain (Jüttner et al. 2003). However, the emergence of IoT related technologies makes it possible to mitigate the negative effects of supply chain risk by providing well-developed communication channels and visible processes. Instant information exchange and strong collaboration among supply chain partners strengthen companies against unforeseen events.

In addition to connecting all physical entities involved in a supply chain (Kiel et al. 2017), IoT also provides the opportunity of having smart production lines and considerable cost savings for manufacturers (Sun et al. 2018). The smart machines are inter-communicated to each other and able to transfer real-time data. IoT is a necessary element of the new production systems to collect and manage a large amount of information obtained from smart devices (Canizares and Alarcon 2018). Zheng and Wu (2017) propose a model which horizontally integrates IoT into the production system. The model uses real-time and available orders to predict the amount of spare parts usage.

6. Conclusion

Rapid advances in technology lead companies to integrate new technological developments into supply chains to maintain a competitive advantage. IoT has the potential to significantly change the structure of today's traditional production and transportation processes. Many industries (e.g., retailing, automobile, electronics, etc.) will integrate IoT related technologies into production systems. The structure of supply chains has the potential to be altered with the new technologies and become more visible, agile and risk-free.

Traditional supply chains become more "*smarter*" with the integration of Industry 4.0 based technologies. IoT, which is one of the technologies introduced by Industry 4.0, has the potential to be used in companies and may easily cope with complexities confronted by traditional supply chains. Although companies may encounter various challenges throughout the integration process of IoT related technologies into their supply chain systems, these technologies provide considerable opportunities to outperform the competitors. Recently, IoT has been applied in numerous industries including agriculture, healthcare, retailing, manufacturing, and logistics.

IoT applications in SCM is a new research area which requires interest from both academics and practitioners. IoT provides an effective and real-time communication system among supply chain partners. A better communication network enables to present innovative products to customers within a short period that contributes to the supply chain responsiveness. For instance, a proper big data analysis may have a huge impact on strategic decisions such as mass customization (Saniuk and Saniuk 2018). Furthermore, smart devices enable to plan and control of the entire supply chain system, which prevents the supply chain partners from serious problems (e.g., reduction in the amount of inventory due to bullwhip effect). IoT helps supply chains to cope with uncertain changes in demand, to improve product quality, to design new products, to prevent production failures and to deliver products on time. Some other benefits include increased visibility, traceability, transparency, adaptability and flexibility in a supply chain.

Although companies can obtain numerous benefits from IoT technologies, IoT may complicate supply chain management. Obtaining and analyzing processes of the machine-generated big data requires a considerable amount of time and effort. Additionally, a supply chain system may become more vulnerable to the cyber-attacks with potential security holes by the integration of Wi-Fi and RFID technologies, if necessary protection has not been provided.

References

Abomhara, M. and G.M. Køien. 2015. Cyber security and the internet of things: vulnerabilities, threats, intruders and attacks. Journal of Cyber Security and Mobility, 4(1): 65–88.

Atlam, H.F., A. Alenezi, A. Alharthi, R.J. Walters and G.B. Wills. 2017. Integration of cloud computing with internet of things: challenges and open issues. In 2017 IEEE International Conference on Internet of Things (iThings) and IEEE Green Computing and Communications (GreenCom) and IEEE Cyber, Physical and Social Computing (CPSCom) and IEEE Smart Data (SmartData) (pp. 670–675). IEEE.

Bardaki, C., P. Kourouthanassis and K. Pramatari. 2012. Deploying RFID-enabled services in the retail supply chain: Lessons learned toward the Internet of Things. Information Systems Management, 29(3): 233–245.

Bauer, H., M. Patel and J. Veira. 2015. Internet of Things: Opportunities and challenges for semiconductor companies, McKinsey Insights, https://www.mckinsey.com/industries/semiconductors/our-insights/internet-of-things-opportunities-and-challenges-for-semiconductor-companies.

Ben-Daya, M., E. Hassini and Z. Bahroun. 2019. Internet of things and supply chain management: a literature review. International Journal of Production Research, 57(15-16): 4719–4742.

Braunscheidel, M.J. and N.C. Suresh. 2009. The organizational antecedents of a firm's supply chain agility for risk mitigation and response. Journal of Operations Management, 27(2): 119–140.

Canizares, E. and V. Alarcon. 2018. Analyzing the effects of applying IoT to a metal-mechanical company. Journal of Industrial Engineering and Management-JIEM, 11(2): 308–317.

Cardoso, W., W.A. Azzolini, J.F. Bertosse, E. Bassi and E.S. Ponciano. 2017. Digital manufacturing, Industry 4.0, cloud computing and thing internet: Brazilian contextualization and reality. Independent Journal of Management & Production, 8(2): 459–473.

Danova, T. 2013. Morgan Stanley: 75 billion devices will be connected to the Internet of Things by 2020. Business Insider.

Dunke, F., I. Heckmann, S. Nickel and F. Salhanda-Ha-Gama. 2018. Time traps in supply chains: Is optimal still good enough? European Journal of Operational Research, 264(3): 813–829.

European Chamber Report. 2017. https://www.europeanchamber.com.cn/en/china-manufacturing-2025, online accessed: September 2019.

Fatorachian, H. and H. Kazemi. 2018. A critical investigation of Industry 4.0 in manufacturing: theoretical operationalization framework. Production Planning & Control, 29(8): 633–644.

Flügel, C. and V. Gehrmann. 2008. Scientific Workshop 4: Intelligent objects for the Internet of Things: Internet of Things–application of sensor networks in logistics. In European Conference on Ambient Intelligence, 16–26.

Gierej, S. 2017. Big data in the industry—Overview of selected issues. Management Systems in Production Engineering, 25(4): 251–254.

Government of Japan, Public Relations Office. 2019. https://www8.cao.go.jp/cstp/english/society5_0/index.html, Online accessed: September 2019.

Jagtap, S. and S. Rahimifard. 2017. Utilisation of Internet of Things to improve resource efficiency of food supply chains. Proceeding of the 8th International Conference on Information and Communication Technologies in Agriculture, Food and Environment (HAICTA 2017), Chania, Greece, 21–24 September, 2017.

Jüttner, U., H. Peck and M. Christopher. 2003. Supply chain risk management: Outlining an agenda for future research. International Journal of Logistics: Research and Applications, 6(4): 197–210.

Kagermann, H., J. Helbig, A. Hellinger and W. Wahlster. 2013. Recommendations for implementing the strategic initiative Industrie 4.0: Securing the future of German manufacturing industry; final report of the Industrie 4.0 Working Group, Forschungsunion.

Kamble, S.S., A. Gunesakaran, H. Parekh and S. Joshi. 2019. Modeling the internet of things adoption barriers in food retail supply chains. Journal of Retailing and Consumer Services, 48: 154–168.

Kiel, D., J.M. Muller, C. Arnold and K.I. Voigt. 2017. Sustainable industrial value creation: benefits and challenges of industry 4.0. International Journal of Innovation Management, 21(8): 1–34.

Kumar, S., V. Mookerjee and A. Shubham. 2018. Research in operations management and information systems interface. Production and Operations Management, 27(11): 1893–1905.

Kusiak, A. 2018. Smart manufacturing. International Journal of Production Research, 56(1-2): 508–517.

Lee, E.A. 2008. Cyber-Physical Systems: Design Challenges. 11th IEEE International Symposium on Object and Component-Oriented Real-Time Distributed Computing (ISORC).

Lee, I. and K. Lee. 2015. The Internet of Things (IoT): Applications, investments, and challenges for enterprises. Business Horizons, 58(4): 431–440.

Lee, C.K.M., L. Yaqiong, K.K.H. Ng, W. Ho and K.L. Choy. 2018. Design and application of Internet of Things-based warehouse management system for smart logistics. International Journal of Production Research, 56(8): 2753–2768.

Li, S., L. Da Xu and S. Zhao. 2015. The Internet of Things: A survey. Information Systems Frontiers, 17(2): 243–259.

Li, Y., M. Hou, H. Liu and Y. Liu. 2012. Towards a theoretical framework of strategic decision, supporting capability and information sharing under the context of Internet of Things. Information Technology and Management, 13(4): 205–216.

Lin, C.C. and J.W. Ya. 2018. Cost-Efficient deployment of fog computing systems at logistics centers in Industry 4.0. IEEE Transactions on Industrial Informatics, 14(10): 4603–4611.

Lu, Y. 2017. Industry 4.0: A survey on technologies, applications and open research issues. Journal of Industrial Information Integration, 6: 1–10.

Mahdavinejad, M.S., M. Rezvan, M. Barekatain, P. Adibi, P. Barnaghi and A.P. Sheth. 2018. Machine learning for Internet of Things data analysis: A survey. Digital Communications and Networks, 4(3): 161–175.

Mell, P. and T. Grance. 2011. The NIST definition of cloud computing. National Institute of Standards and Technology. US Department of Commerce.

Monostori, L., B. Kadar, T. Bauernhansl, S. Kondoh, S.S. Kumara, G. Reinhart, O. Sauer, G. Schuh, W. Sihn and K. Ueda. 2016. Cyber-physical systems in manufacturing. Cirp Annals-Manufacturing Technology, 65(2): 621–641.

Pereira, A., E.O. Simoneto, G. Putnik and H.C.G.A. Castro. 2018. How connectivity and search for producers impact production in Industry 4.0 networks. Brazilian Journal of Operations & Production Management, 15(4): 528–534.

Saniuk, S. and A. Saniuk. 2018. Challenges of Industry 4.0 for production enterprises functioning within cyber industry networks. Management Systems in Production Engineering, 26(4): 212–216.

Shen, G. and B. Liu. 2010. Research on application of Internet of Things in electronic commerce. In 2010 Third International Symposium on Electronic Commerce and Security, 13–16.

Sissini, E., A. Saifullah, S. Han, U. Jennehag and M. Gidlund. 2018. Industrial Internet of Things: Challenges, opportunities, and directions. IEEE Transactions on Industrial Informatics, 14(11): 4724–4734.

Strange, R. and A. Zucchella. 2017. Industry 4.0, global value chains and international business. Multinational Business Review, 25(3): 174–184.

Sun, J.N., M.L. Gao, Q.F. Wang, M.J. Jiang, X. Zhang and R. Schmitt. 2018. Smart services for enhancing personal competence in Industrie 4.0 digital factory. Logforum, 14(1): 51–57.

Szozda, N. 2017. Industry 4.0 and its impact on the functioning of supply chains. Logforum, 13(4): 401–414.

TechAmerica Foundation's Federal Big Data Commission. 2012. Demystifying big data: A practical guide to transforming the business of Government. Retrieved from http://www.techamerica.org/Docs/fileManager.cfm?f=techamericabigdatareport-final.pdf.

Trappey, A.J., C.V. Trappey, U.H. Govindarajan, A.C. Chua and J.J. Sun. 2017. A review of essential standards and patent landscapes for the Internet of Things: A key enabler for Industry 4.0. Advanced Engineering Informatics, 33: 208–229.

Tuptuk, N. and S. Hailes. 2018. Security of smart manufacturing systems. Journal of Manufacturing Systems, 47: 93–106.

Upasani, K., M. Bakshi, V. Pandhare and B.K. Lad. 2017. Distributed maintenance planning in manufacturing industries. Computers & Industrial Engineering, 108: 1–14.

Van Kranenburg, R. 2008. The Internet of Things: A critique of ambient technology and the all-seeing network of RFID. Institute of Network Cultures.

Wang, K.S. 2016. Logistics 4.0 solution: New challenges and opportunities. 6th International Workshop of Advanced Manufacturing and Automation, Manchester, United Kingdom, 10–11.

White, A.E.D.M., E.M. Daniel and M. Mohdzain. 2005. The role of emergent information technologies and systems in enabling supply chain agility. International Journal of Information Management, 25(5): 396–410.

Whitmore, A., A. Agarwal and X.L. Da. 2015. The Internet of Things-A survey of topics and trends. Information Systems Frontiers, 17(2): 261–274.

Xu, L., W. He and S. Li. 2014. Internet of Things in industries: A survey, IEEE Transactions on Industrial Informatics, 10(4): 2233–2243.

Xu, L.D. and L.L. Li. 2018. Industry 4.0: State of the art and future trends. International Journal of Production Research, 56(8): 2941–2962.

Yan, B. and G. Huan. 2009. Supply chain information transmission based on RFID and Internet of Things in 2009. ISECS International Colloquium on Computing, Communication, Control, and Management, 4: 166–169. IEEE.

Yin, Y., K.E. Stecke and D. Li. 2018. The evolution of production systems from Industry 2.0 through Industry 4.0. International Journal of Production Research, 56(1-2): 848–861.

Zheng, M.M. and K. Wu. 2017. Smart spare parts management systems in semiconductor manufacturing. Industrial Management & Data Systems, 117(4): 754–763.

Zhengxia, W. and X. Laisheng. 2010. Modern logistics monitoring platform based on the Internet of Things in 2010. International Conference on Intelligent Computation Technology and Automation 2: 26–731. IEE E.

Chapter 4

The Impact of the Internet of Things on Supply Chain 4.0

A Review and Bibliometric Analysis

Sema Kayapinar Kaya,[1,*] *Turan Paksoy*[2] and *Jose Arturo Garza-Reyes*[3]

1. Introduction

Supply Chain (SC) has a very extensive and dynamic structure that incorporates new business models, new customer expectations, market searches, and technological developments. With the emergence of Industry 4.0, SC had to bring about some changes to keep up with the innovations that Industry 4.0 has brought. Industry 4.0 has relocated the SC and logistics into a digital environment and restructured it. All the processes in SC have been restructured within the framework of Industry 4.0, from raw material procurement to production line and till the last step that the product reaches the final customer. With the Industry 4.0, SC is digitized and renewed with more advanced technological equipment. Today, 28% of SC companies seem to have advanced digital technology. Digital Supply Chain (DSC) and logistics sectors have a share of 41%, particularly in the automotive industry and 45% in the electronics sector. Within the framework of Industry 4.0, many SC companies have planned to invest 5% of their annual revenues in technological investments until the year 2020 (Zuberer 2016).

The most significant change in SC has occurred with the tracking of objects throughout SC. This new concept, called IoT, shortly Radio Frequency Identification System (RFID), is defined as objects that communicate and share information with each other through sensors and various communication protocols. It is assumed that the camera system was the first step in IoT. This camera system was set up for the first time in 1991 by about 15 researchers at Cambridge University to monitor the coffee machine from their rooms. Then, in 1999, with Kevin Ashton using RFID technology in the Auto-ID Laboratory of Massachusetts Institute of Technology (MIT), the IoT was used for the first time (Ashton 2009). Procter & Gamble implemented IoT technology in the SC industry for the first time in 1999. Thanks to RFID placed on the products, product tracking was made instantaneously throughout the SC. On this topic, DHL (logistics service provider) and Cisco (Information server provider) prepared a new trend report on the IoT. According to this report, by the year 2020, 50 billion devices will be connected via the internet, which is expected to lead to a significant development in business technology. According to Cisco's economic analysis, IoT will generate $ 8 trillion in worldwide revenues over the next decade, with revenue of $ 1.9 trillion for supply chains and SC activities. According to Cisco's report, the number of devices connected to the IoT is estimated to be 3.47 million in 2015, while the number of devices connected per capita is expected to be 6.58 million in 2020 (Cisco 2015). The number of devices connected per capita is shown in Figure 1.

With the IoT, SC operators, corporate customers, and end consumers can be provided with remote access. Thus, problems arising in operational services, transportation safety, customer satisfaction, and new business models can be easily detected. The IoT in the SC Sector has been examined in four different structural processes as production design, customers, suppliers, and equipment procurement, which are shown in Figure 2.

[1] Department of Industrial Engineering, Munzur University, Tunceli, Turkey.
[2] Department of Industrial Engineering, Konya Technical University, Konya, Turkey.
[3] Centre for Supply Chain Improvement, University of Derby, Derby, United Kingdom.
* Corresponding author: kysem24@gmail.com

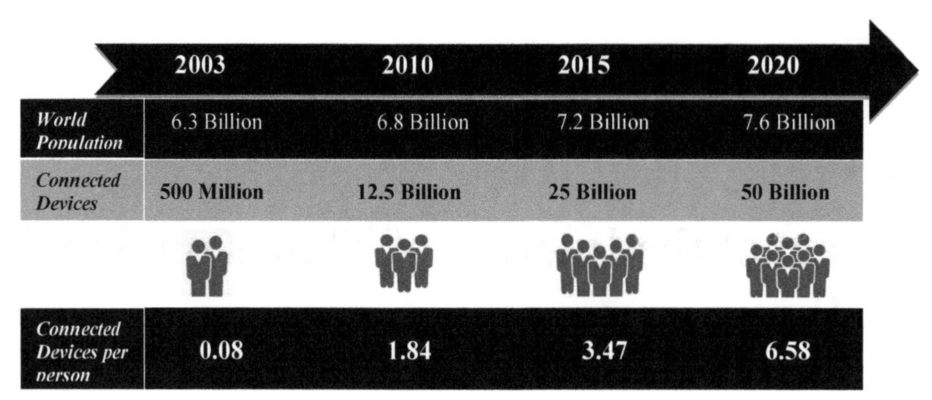

Fig. 1: Number of devices connected to the IoT.

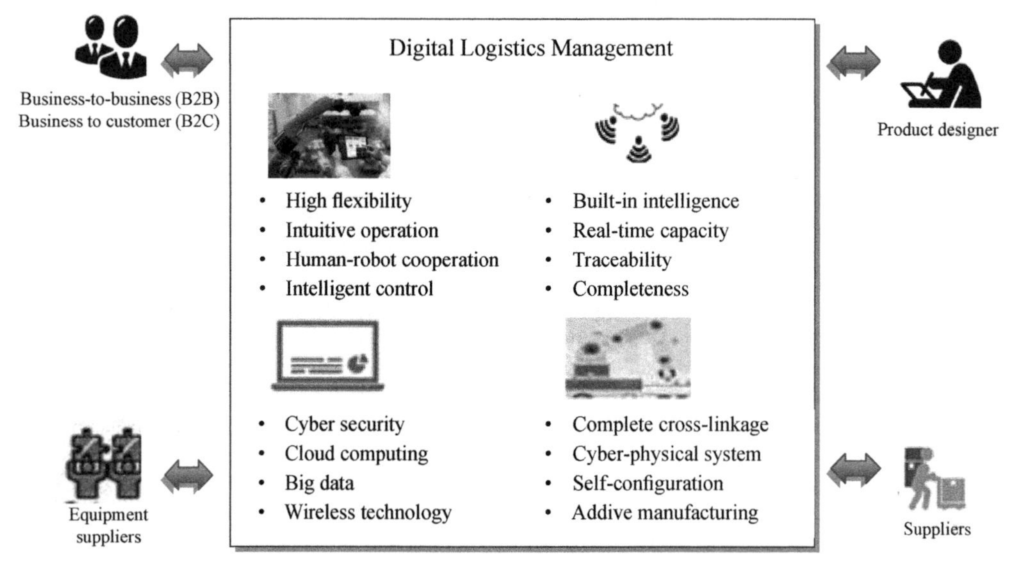

Fig. 2: IoT Logistics Management (Sadıkzade 2016).

Together with the IoT, companies will have a low-cost opportunity in storage, transportation, and all other SC activities. In following the storage, pallets, and vehicles in communication with each other, there can be a smaller, more efficient storage policy. International transport, tracking, and monitoring of products can be faster, more precise, more reliable, and errors can automatically be detected with the product tracking system. The material flows within the SC are monitored instantaneously, making transportation and handling processes easier, minimizing the risks in transportation. With the IoT, SC will be digitized, which will significantly contribute to the delivery of the products to the customer at the right time, the right place, and the right quality, and the SC process will be facilitated in all its aspects.

This study begins by asking how Industry 4.0 affects SC and what kind of roles IoT and big data play in SC industry. Section 2 presents the Industrial Revolution and its historical development and the emergence of Industry 4.0. Section 3 offers the work-study and application areas of Supply Chain 4.0. Then, the literature review of IoT and Supply Chain within the scope of Industry 4.0 are included in Section 4. The finding of Bibliometric mapping and clustering analysis are presented in the Section 5. Finally, Section 6 concludes the results and provide some directions of the future studies.

2. Historical Development of the Industrial Revolution and The Emergence of Industry 4.0

Four different industrial revolutions have emerged up until today. The first Industrial Revolution began with the mechanization of weaving looms in England in the early 1760s. The most important source of energy in the transition from simple workshop production to factory production was coal and steam power. Textile, steam machine, and iron are the three most important elements of this period. During this period, steam machines started to be used in the textile industry, and raw materials were supplied with steam trains and ships. With the increase in transportation means, the spread of the Industrial Revolution to Europe gained speed. The First Industrial Revolution, based in England, contributed to the increase in national income in these countries by creating new wealth holders in the USA, especially in Europe. With the increase in labor and capital needed, migration from rural to urban areas increased, and urbanization was accelerated (Jensen 1993).

The Second Industrial Revolution covers a period starting from 1870 up until 1914. This period began with the widespread use of cheap steel production methods invented by British Inventor H. Bessemer. In this period, steel, electricity, petroleum, and chemical substances were started to be used instead of steam and coal in manufacturing. Henry Ford, who is known as the father of mass production, left his mark on this period and enabled the widespread use of the manufacturing assembly line system in the automotive sector. During the same period, electronic computers were used for the first time, and Graham Bell expanded the communication network with the invention of the telephone. Railway transportation and trade accelerated by using steel instead of iron in production. The use of electricity in factories and cities began with Edison in 1882. Then, electric machines took part in production (Engelman 2018).

The Third Industrial Revolution covers the period starting from 1970 until a decade ago. During this period, automation in production began with the development of a Programmable Logic Controller (PLC). The production process has been enriched with computer-aided machines and automation-based systems. The automobile industry, mobile phone, internet, aviation and space technologies, computer-aided design, computer-aided machinery (CNC), and robots have started to be used in production. With the development of scientific fields such as telecommunication, nuclear energy, laser, fiber optics, and biogenetics, many innovations have emerged in the field of production (Jacinto 2015). The increase in fossil energy resources and the rapid depletion of world resources have brought about the issue of environmental awareness and the use of renewable energy sources. New technologies related to the use of sustainable energy resources (solar, wind, geothermal, hydraulic energy, etc.) in production have been introduced.

The Fourth Industrial Revolution, known as the "Industry 4.0", was introduced for the first time at the Hannover Fair in Germany in 2011. Supported by the German Government, this technology has received the support of many countries, especially the USA and Japan (Pfeiffer 2017). Industry 4.0 has become increasingly concrete and has been implemented in many areas, such as production, in particular, supply chain, food, health, etc. With the transition to the Fourth Industrial Revolution, rapid automation in production, robotic systems, and digitization has positively affected the global economy. The biggest goal of the Industry 4.0 is to develop a robotic-based manufacturing system in which various machines within a factory can communicate with each other, detect ambient conditions (heat, humidity, energy, weather, etc.), and by analyzing the data they gather, detects the needs of the system. In this way, it aims to make high quality, more flexible, and low-cost production in a swift manner. Industry 4.0 has created an impact on cost, human resources, management efficiency, and benefit in terms of technology for the ever-evolving and growing SC sector. Industry 4.0 is the fourth industrial revolution that consists of many innovations such as the transfer of the production process entirely to the robots, the development of artificial intelligence and Internet technology, the use of three and four-dimensional printers in production, the sorting out and evaluation of massive data by data analysis, and smart objects communicating with each other. Industry 4.0 brings many benefits, such as higher degrees of integration, facilitation of transmission, higher throughput in a given time delay, and greater process transparency in the entire system for production, inventory management, SC, and a quality economic system. Although Industry 4.0 provides many advantages to production, business models, and technology, it has a negative effect on employment, economic conditions, and data security. Kovacs (2018) has analyzed the dark corners of the development of Industry 4.0 and its effects on the digital economy.

Cyber-Physical Production System (CPPS) provides the integration of the physical environment with the virtual environment (Hermann et al. 2016). CPPS is the complex dimensional structure that works together with the IoT.

Industry 4.0 consists of nine main components, which are given in Figure 3. Industry 4.0 is a comprehensive Industrial Revolution comprising all these components. The main components forming the Fourth Industrial Revolution is illustrated in Figure 4. Industry 4.0 is a collection of systems consisting of many different technological components (Hermann et al. 2016).

2.1 Industry 4.0 Components

Cyber-Physical Production Systems (CPPS): CPPS aims to connect the physical world to the virtual information system with the help of sensors and actuators. Data is communicated between computer terminals, wireless devices, and cloud systems. Thanks to the complex and dynamic CPPS, production process activities (planning, analysis, modeling, design, implementation, and maintenance) can work together. With CPPS, the physical work environment and the virtual information system are synchronized with each other. In this way, the monitoring and control of the production process can be more

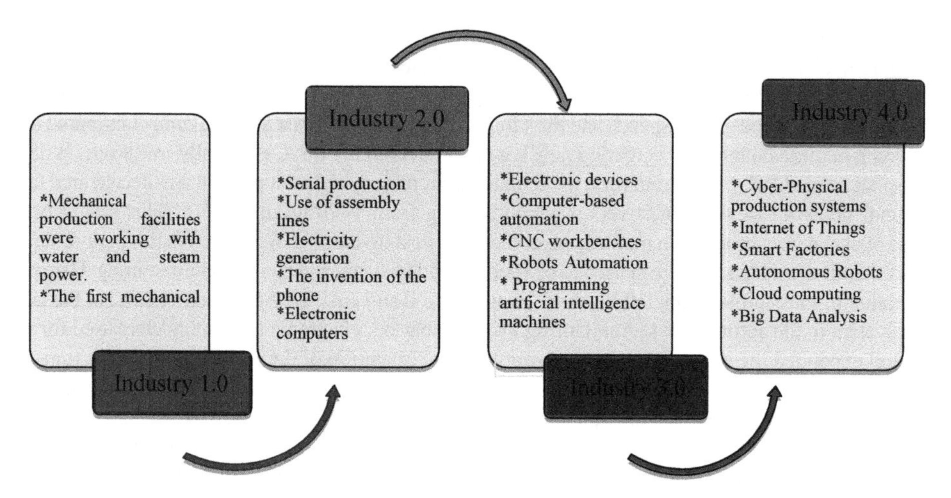

Fig. 3: Historical Development of Industrial Revolution (Industry 4.0 2015).

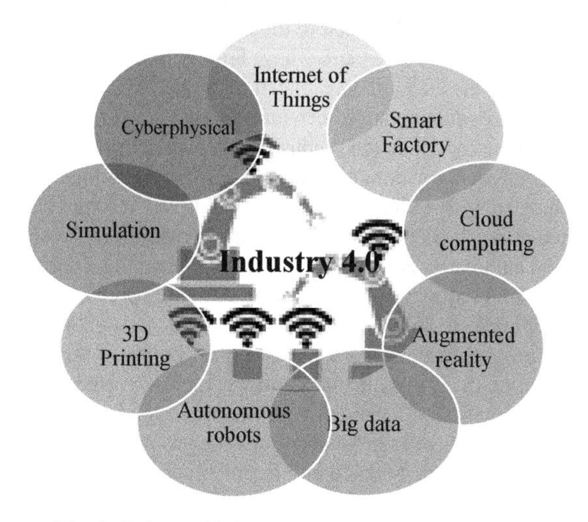

Fig. 4: Industry 4.0 Components (Hermann et al. 2016).

transparent and effective. The development of CPPS is evaluated in three major stages. In the first stage, identifying technologies, such as RFID tags, are developed. Thus, a centralized service provider does the storage and analysis of the data. In the second stage, dynamic data can be collected in a limited range with sensors and actuators. In the third stage of development, the data are stored and analyzed with multiple sensors and actuators, and a smarter network system can be received. CPPS fulfills active and dynamic requirements in manufacturing and plays a major role in the overall efficiency of the industry (Lu 2017).

Internet of Things: The term IoT, which emerged in the early 21st century, is the most important technological component of the basic philosophy of Industry 4.0. The IoT is also referred to as the communication network in which physical objects are interconnected with each other or with larger systems. The IoT and "smart products" are two terms that are used interchangeably (McFarlane et al. 2003). Smart products can communicate with each other with the help of embedded RFID or sensors and store and analyze the data they receive from the environment. Different researchers have defined smart products over time. Accordingly, McFarlane and others (2003) defined smart products as both physical and information-based products. Tags and RFID readers do the data flow between physical products and information. Venta (2007) refers to smart products as products with the ability to make decisions. Smart products can interpret and analyze the data they have. They interacts with the environment and can present the information in their environment to the user as instant visual information, when necessary. Today, smart products supported by new technologies can inform the user about all the processes from production to final consumption.

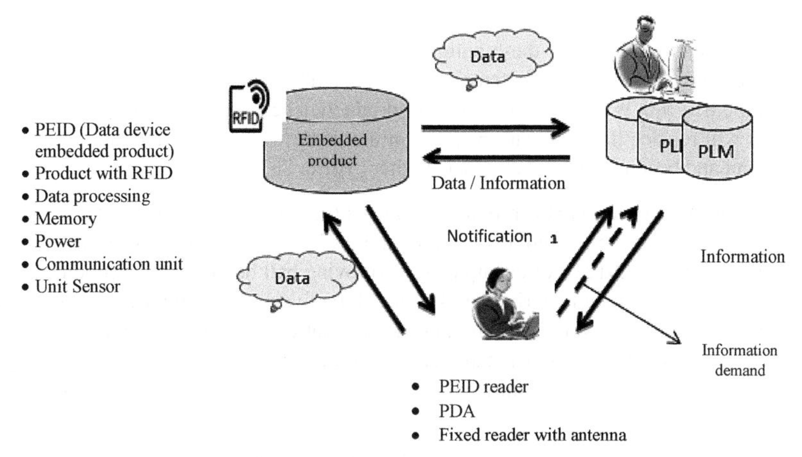

Fig. 5: Product lifecycle process (Kiritsis 2011).

Hribernik et al. (2011) illustrated the working principle of smart products as the product life-cycle process (Figure 5). RFID and sensor-embedded devices collect information from their environment wirelessly, with the help of readers and wireless technology. These devices monitor the entire life-cycle process from product assembly to the final use stage. The product has a built-in driver, display screen, main unit, ISDN modem, processor, and motherboard. Each item can be monitored independently, and the instantaneous information is collected and stored. Product tracking and storage are done using Electronic Code Information Services (EPCIS) or PROMISE Message Interface (PMI). EPCIS and PMI devices record the production time, location, assembly, and disassembly processes of the product. Product lifecycle information is collected with personalized mobile devices and product information readers installed in computers. The collected information is sent to the product life-cycle system from each point (retailer, distributor, recycler, etc.) via the internet. With this system, product life-cycle information can be supplied individually or collectively at any time (Kiritsis 2011).

Smart Factories: Smart Factories are digital-based factories that emerge at the point where data processing meets with the production process. The manufacturing process envisaged by Industry 4.0 is a fully automated production system that can run fully automatically without human intervention. In these factories, smart robots carry out production. These robots recognize the materials that are moving on the traditional production line with RFID and sensor tags. They also know which processes they need to go through (Thoben et al. 2017). Machines can communicate with each other and can obtain any data via a central computer. In this way, a product can be processed on the same production line and tracked without any error. IoT technology plays an important role in establishing smart factories. The virtualized factory with the Internet of objects is integrated into the system digitized with CPPS. The IoT platform acts as a cloud computing system that collects real-time data and can track the data in the factory at any moment (Lee and Lee 2015).

Big Data: Big data is the general name for voluminous, gigantic, defined or undefined data. Vast amounts of data that are beyond imagination are produced daily in a variety of sectors, such as health, management, social networks (Facebook, MSN, etc.), marketing, finance, and so on. Since this collected data is nothing but piles of data, unless interpreted, it is extremely important to analyze this data quickly and in a comprehensible way. Previously, businesses did not prefer retaining their data in their archives for long periods, and they did not analyze their data sets. However, with new technological developments, data can be analyzed, stored and made available in a safe environment. In this way, companies can see the important competitive data, develop new insights, and customize the services they provide to their customers (Mazzei and Noble 2017). As an example of the work done on big data in the SC sector, the data from the vehicle that is tracked using sensors, the wireless adapter, and the GPS is collected in an internet environment. Thanks to this data, the Supply chain department can monitor drivers and guide them by determining the shortest route. In addition, bus companies can analyze the data they receive from the passengers, design a more efficient transportation plan, and determine the travel frequency and optimum travel time. With big data mining method, they can categorize the estimated number of passengers and make more accurate predictions about the estimated demands (Oussous et al. 2017).

Cloud Computing: Cloud computing is the general name for Internet-based computing services that provide computing resources that can be used at any time and shared among users, for computers and other devices. It is the general name of the system that users can access from anywhere with an internet (Schouten 2014). The most well-known cloud-computing example is the Office 365 service that organizes and stores MS office documents.

Autonomous Robots: Autonomous robots are robotic systems with a certain intelligence, supported by artificial intelligence technology. Based on artificial intelligence, these autonomous robots, which can detect the environment, that can be implemented very comfortably in production systems and which can make their own decisions with this technology, are the key technology for Industry 4.0. With autonomous robots, production benches that can be positioned at any point within the factory on demand will replace the stationary benches in the factories of future and humanoid robots will emerge, perceive the environment and become able to talk to each other (Yazıcı 2016).

3D Printers: Three-dimensional printers are the production tools of the new era that work with laser or inkjet printer logic, transforming products from digital media into a solid three-dimensional object in a "layer-based" structure defined as "additive manufacturing" (Berman 2012). For example, a free-moving ball bearing can be produced in a 3D-printer as a single piece with its balls. Additive manufacturing, unlike the subtractive production process, allows production without resorting to any cutting, drilling or grinding process. This means that even producing complex objects becomes much easier (Berman 2012). The technology that most manufacturers use in prototype production, especially since it provides flexibility, low cost and time saving, has now initiated a revolution that will enable final consumers to manufacture in their homes (Çallı and Taşkın 2015).

Simulation and Virtual Reality: Simulation is the imitation of the operation of a real-world system or process in a computer platform. Simulation enables the generation of an artificial history of the system and the observation of that artificial history to draw inferences concerning the operating characteristics of the real system. Simulation or Virtual Reality began in 1962 with a device called Sensora, developed by Morton Heilig, and it has been extended to the daily Google Glass project. Virtual reality is a term used for computer-aided 3D environments where users experience the feeling of being in the designed environment. With Industry 4.0, a virtual copy of the smart production facilities is made in 3D, and the data coming from the sensors is transferred into the simulation environment. In this context, the dark factories of Siemens, HTC's virtual reality glasses, and Caterpillar augmented reality demo can be given as examples. With virtual reality and simulation, the physical systems of the factories will be monitored through web-based systems, and smart technology applications will proliferate (History of Virtual Reality 2017).

3. Towards Supply Chain 4.0

Although Industry 4.0 first emerged as a manufacturing-based approach, it has also affected many industries associated with manufacturing. Considering that manufacturing and SC form an inseparable whole, it is envisaged that Industry 4.0 will also reshape the SC Sector and bring a different perspective to it. The fact that machines and objects are in communication with one another with instant access to data within Industry 4.0 makes SC services more efficient. In the SC and manufacturing sector, computer-aided team systems and autonomous robots will decrease the time spent on production, and the resources will be used more efficiently. Remote-controlled vehicles and products will be able to reach the customers in a shorter time. Especially the IoT technology, one of the components of Industry 4.0, leads to great innovations in transportation and the SC Sector. The IoT contributes positively to all stages of SC, starting from production to the delivery of the final product to the customer. With the aid of RFID, sensors, Global Positioning Systems (GPS), data collected during the SC process can be tracked through an internet-based system. In this way, SC activities can be faster, more flexible, and transparent (Tadejko 2015). There are many kinds of research on the use of IoT in the supply chain industry in various fields. Kong et al. (2018), Leng et al. (2018), Accorsi et al. (2018), Tian (2018), Pal and Kant (2018), Yan et al. (2017), Zhang et al. (2017), and Yan (2017) focused on the perishable and fresh agricultural food supply chain by combining IoT technologies. Tsai et al. (2018), Lue et al. (2016), and Chan et al. (2014) developed an intelligent tracking system to enhance the cold supply chain.

Supply Chain 4.0 (also known as Digital Supply Chain) came out with the emergence of Industry 4.0 for factories. Mainly IoT and Big Data drive it. With the mix, complementary technologies such as RFID, sensors, GPS, Electronic Data Interchange (EDI), and information-sensing equipment, can be easily tracked throughout SC activities. It is important for them to have seven major requirements that are the right quality, at the right time, at the right place and the right good with the right quantity and in the right condition and at the right price for SC management. By estimating the information from the products and materials, the accidents that can occur in the SC process can be predicted, and the warning can be given ahead (Gnimpieba et al. 2015).

IoT is used in all processes from supplier supply to material handling, transportation of materials, production to reaching latest customers (Figure 6). IoT can optimize whole process of Transportation Management System (TMS). By integrating GPS technology placed into transport vehicles, it can monitor and learn all the relevant information (e.g., route, shipping conditions, and status of shipment) related to the smart goods being transported. The Internet of Objects detects

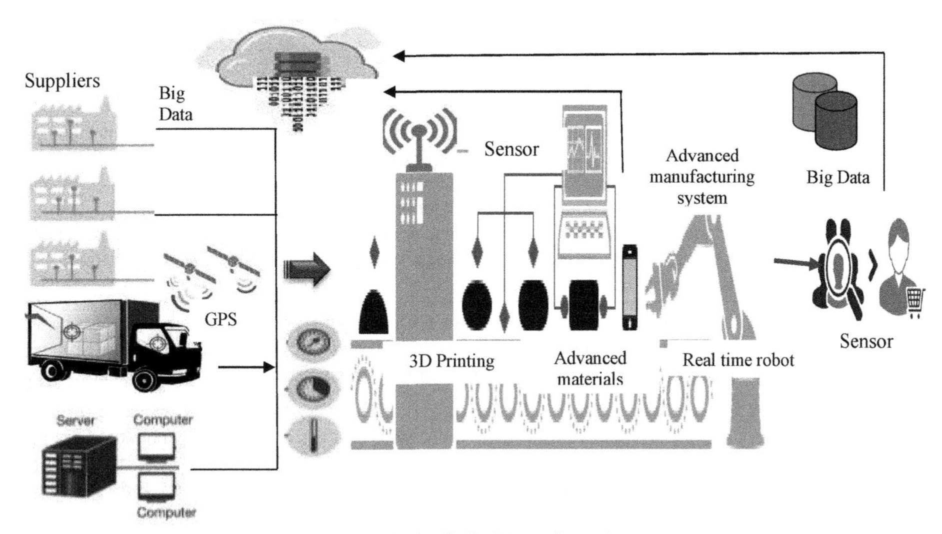

Fig. 6: Supply Chain 4.0 configuration.

return errors and reduces return and damaged product costs (Macaulay et al. 2015). The material coming to the factory starts the production process with real-time production plan. IoT technologies make decisions that are more optimistic and faster than the other producers and can make these decisions faster for both their staff and their employees. When the machines are connected to each other (M2M), the generated data is rapidly transferred to each other through high-speed internet support via software that generates the data of each device, and it is possible to make faster and more effective decisions by looking at the results obtained from the data. These decisions are transferred both to the staff at work and the manager who follows the job and to all the devices, and thus a synchronized working environment is provided. IoT plays a big role in the customer relations after the product reaches the customer. It provides data and information that can be used to enhance the consumer experience, provide insight into consumer behavior, which would result in a better understanding of the consumers and help enhance interaction and engagement with consumers. The application areas of Supply Chain 4.0 are presented in Table 1.

4. Literature Reviews of Supply Chain 4.0 based on IoT

The potential use of IoT technology across the supply chain is huge. The IoT enabled supply chain can be visualized as a smart interconnected network that binds together many levels of suppliers, manufacturers, service providers, distributors, and customers physically located across different areas of the world.

In this section, we categorized literature reviews of IoT and SC, and present several summary statistics into five categories based on sources, publications type, published place, journals and conferences type, application fields, and authors. As part of the expansive concept of Industry 4.0, "Supply Chain 4.0" is called a new paradigm by numerous authors and experts. In this review, these keywords are "Internet of Thing", "IoT", and "Supply Chain", respectively. The search was performed in the Scopus Database on 10 November 2018. It was possible to download the "title", "key-words", and "abstract" from all the 806 documents, including article, conference paper, book chapter, etc., between 2007–2018 years. The reviewed literature included journal articles, conference papers, and edited volumes.

4.1 Literature Review over time

The distribution of papers based on publication year is presented with respect to the years in Figure 7 from 2007 to 2018, the number of relevant publications gradually increased, except for the year 2015. This analysis indicates that while the research area is still in the stage of development, the concern of this research has progressively grown in recent years. In the following sections, all articles are summarized and reviewed based on various criteria including; literature sources, publication type, geographic location, journals and conference type, application fields, and most cited authors.

Table 1: Contributions and applications of the IoT to Supply Chain 4.0.

Field	Applications
Manufacturing	❖ With the IoT, real-time Material Requirements Planning is done by obtaining real-time information from the operator, pallet, material, machine, etc., in the manufacturing process. Since the material order time is known, the procurement period can also be estimated from the supplier. ❖ Production lines are synchronized by sending information to each other, failures, delays, and errors are minimized in the production process. Reduction of waste, loss, and scrap is aimed. ❖ Multiple and different kinds of products can be produced at the same time with manufacturing line systems that are able to exchange information with the other components; thus customer satisfaction will be increased (Shariatzadeh et al. 2016).
Vehicle tracking	❖ With the information collected by sensors, GPS, and RFID, vehicles can be monitored instantaneously. In this way, transportation time and transportation routes are optimized. With Google Maps and API smartphone, the image of the car and its location will be visually recorded by logistics providers (Lee et al. 2009).
Warehouse Management	❖ The quantity and amount of raw materials remaining in the warehouse are monitored instantaneously, and the movement of the objects can be monitored easily. In this way, the quantity of the remaining product, the number of products ordered, the number of products and materials needed can be obtained in an up-to-date and swift manner. ❖ Smart heat and lighting systems are added to the warehouse to save energy and expenses. ❖ The speed of the forklift used in product transport in the warehouse can be controlled with sensors, and the risk of accidents can be minimized. ❖ With intelligent conveyor and separating systems, products and materials can be easily separated, and components are placed in the places where they belong more easily and in a shorter time (Lee 2015).
Risk Management	❖ Tracking of products carried throughout the Supply Chain is done with embedded RFID and sensors. With the signals coming from these devices, it is easy to know at which stage and where the product is. This will reduce the loss of value of the products that are perishable and have a short lifetime. The risk share from deterioration will be minimized, and the risk cost will be minimized (Lee 2015).
Reverse Logistics Activities	❖ By managing all the data about the product lifecycle, many uncertainties in logistics activities are eliminated. Thanks to RFID, sensors, and similar devices, it will be possible to determine which recycling stage (repair, disassembly, waste, etc.) the expired product should be exposed to. In this way, most of the uncertainties in reverse logistics activities will be eliminated, and the logistics cost will be minimized (Gu and Liu 2013).
Informatics	❖ It prevents the loss of information by providing all kinds of information about the products that move through the Supply Chain processes and by making it possible for the products to be stored and shared by the Supply Chain elements. As the information obtained is up to date, the bullwhip effect in the Supply Chain is decreased.
Fleet Management	❖ Trucks and containers can be monitored with sensors. Thanks to effective fleet management and sensors, it is an important factor in increasing profitability by providing substantial fuel savings (Sadıkzade 2016).
Environmental Awareness	❖ With the IoT, the carbon footprint of the products is easily recorded. This allows access to the commercial history record of the carbon loan that provides compliance with environmental regulations. This is an important step in terms of Green Supply Chain (GSC) (Gu and Liu 2013).

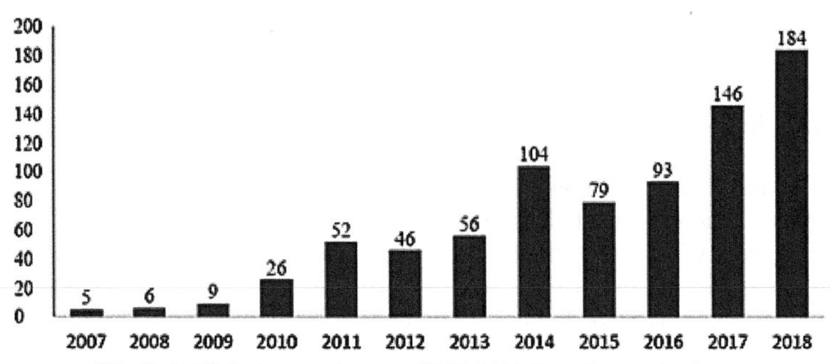

Fig. 7: Published papers between 2007–2018 from Scopus database.

4.2 Contribution to the Literature from Various Sources

This review will help us determine the significant concern of research areas where gaps are obvious. As shown in Figure 8, much of the work being done on the IoT and SC is widespread among conference papers (50.5%). Most of these papers were presented at technical and engineering conferences that were abstracted by the IEEE Xplore Library. Such papers are much more commonplace and encouraged in engineering than they are in other fields. Many of the articles are published in information and computer fields. These papers were reviewed and presented the most common sources that are represented in Figure 9. The most contributed sources are Applied Mechanical and Materials, Advanced Materials Research, Communications in Computer and Information Science. After an in-depth analysis of the case articles, the studies carried out in the field area of SC is present in Figure 9, in which the most studied subjects are "RFID" and "SC and supply chain operations", and the least studied subjects are production, cold SC, and inbound SC activities.

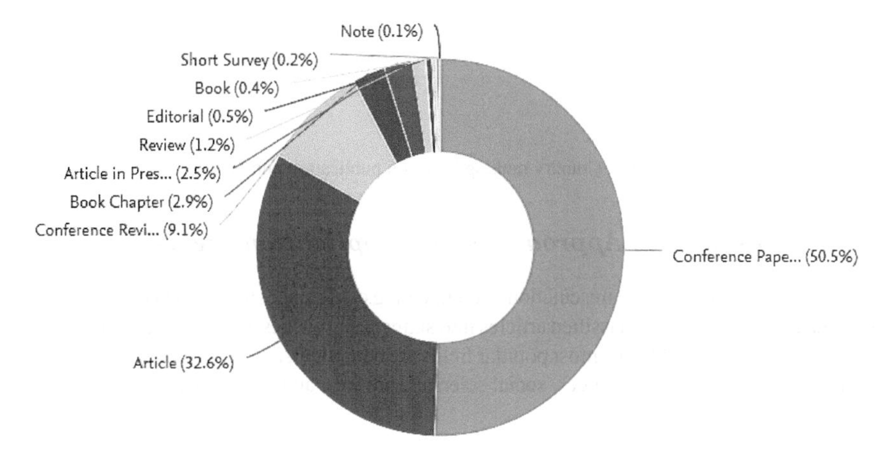

Fig. 8: Review literature by publication type.

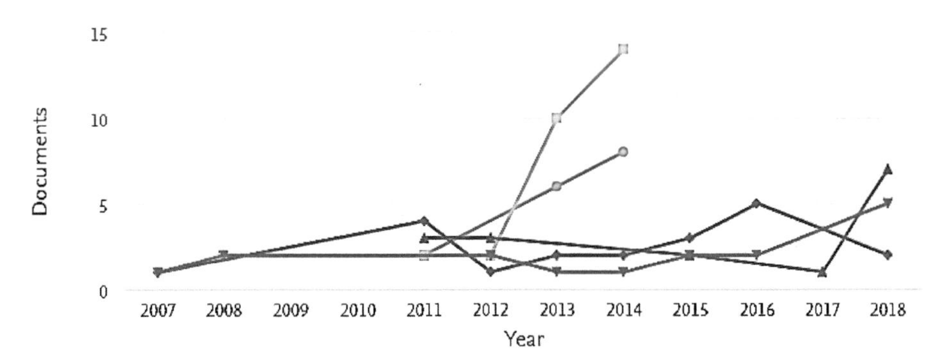

Fig. 9: Papers per year by sources.

4.3 Distribution of Papers by Geography

Investigation of the geographic location for academic research containing IoT and SC is presented in Figure 10. According to this study, the vast majority of the IoT and SC focused articles and conference papers are held in China, the United States, and the United Kingdom. Although the Industry 4.0 concepts emerged for the first time in Germany, many studies on this field have been published mostly in China. Most papers are published in Far East Asia and Europe, with very limited representation in South America, the Middle East region and Africa.

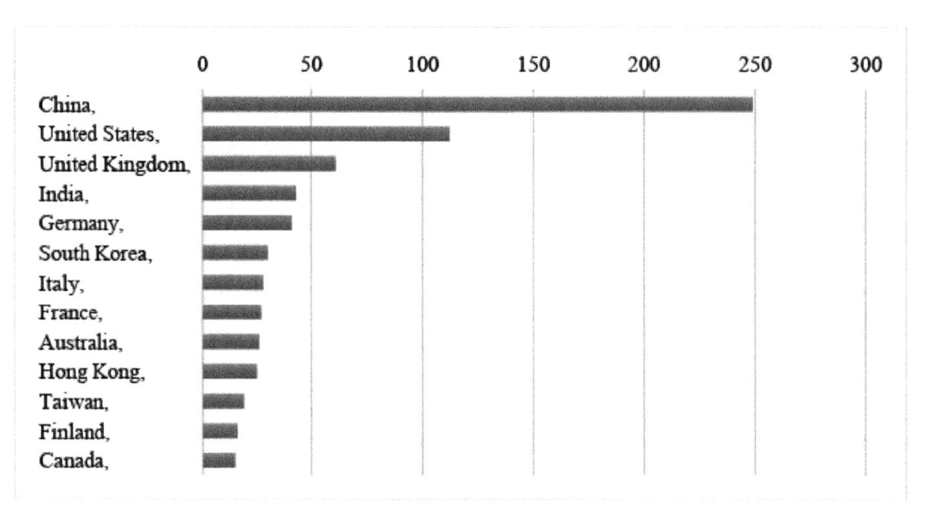

Fig. 10: Country ranking based on publication numbers.

4.4 Distribution of Paper by Approaches and Application Fields

This corresponds to the disproportionate representation of engineering conferences and journals that are currently developing the IoT literature. Based on datasets, we classified articles into sixteen approaches and application fields, which are the most published fields presented in Figure 11. The most popular fields are computer sciences, engineering, business management and accounting, mathematics, decision sciences, social sciences, physics, and astronomy.

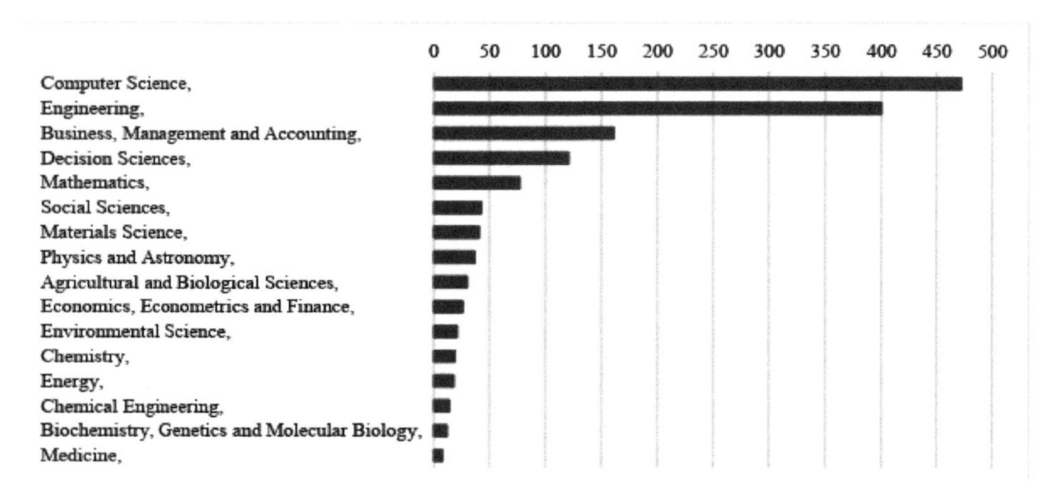

Fig. 11: Distribution of papers with the highest contributions in publishing IoT and logistics topics.

4.5 Citation Report of Authors

One of the most important criteria of the research is the impact of the paper on other scientific environments. The author's analysis can be a good measure to evaluate the impacts of the publications. Figure 12 represents the most contributing authors in the fields. According to results, Li Rang Zheng, George Huang is the most productive authors in this field. Zheng reviewed the highest number of papers according to other literature reviews. Zheng and Han cooperate together, and they work on the food supply chain, agriculture, IoT. George Q. Huang focuses on IoT, Big data, decision making, and production Supply Chain system; Yang's studies are related to IoT, RFID, and agriculture supply chain.

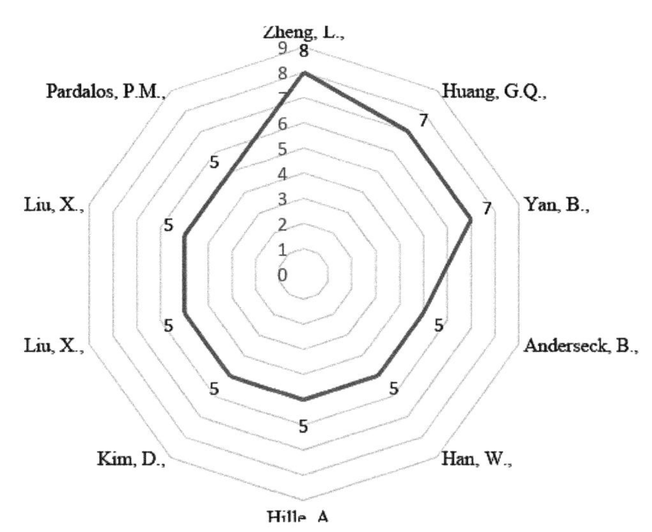

Fig. 12: Document counts for top authors.

5. Bibliometric Mapping and Clustering Analysis

Analysis has a similar property, that it provides an insight into the structure of a network, which is used in bibliometric research. These techniques are based on similar principles and play a significant role in identifying and classify groups of publications, terms, authors, and journals. When we deal with a great number of data, the clustering and mapping solution can easily analyze and interpret many millions of publications and their related terms. Clustering analysis is the classification of a set of elements into subsets so that elements in the same cluster are in common. It is a statistical method and can be efficiently used in many fields, such as big data, machine learning, text mining, pattern recognition, image analysis, and bioinformatics (Van Eck and Waltman 2018). In this study, VOS viewer 1.6.9 software was used to construct and illustrated for bibliometric networks based on keywords, and all information was exported in the CSV file in Excel data analysis.

5.1 Keywords Co-occurrence Analysis

Keywords co-occurrence can effectively indicate the hot research topics in the discipline fields. All the 806 Supply chain and IoT–related publications have a total of 176 keywords; among them, only 59 selected keywords appeared in Figure 13. The node with the same color belongs to the same cluster; therefore, the VOS viewer divided the keywords of publications into three clusters. The node size expresses the magnitude of the occurrence of keywords, and the line between two keywords represents that there is a relationship between them. The length of the link between the two keywords indicates the intensity of the relationship between them. A shorter distance means a stronger relationship (Perianes-Rodriguez et al. 2016). A qualitative index indicates the occurrence of the use of keywords. The most used keywords include "Internet of Things" (243), "RFID" (115), "Supply Chain" (89), "IoT" (80), "Industry 4.0" (28), "big data" (27), "traceability" (26), and "cloud computing" (25). According to the keywords network, the relationships between "supply chain", "big data", and "blockchain" reflect that development trends in security and transparency in supply chain management. The relationships between "blockchain", "cryptography" and "privacy" show that researchers are greatly interested in how blockchains might convert the supply chain management. "IoT", "big data", "industry 4.0", "industry IoT", and "cyber-physical systems" have an increasing importance on supply chain management for the last two years. Researchers focus on how the "industrial" application of the Internet of Things (IIoT) is transforming supply chains.

5.2 Co-authorship Analysis

Country co-authorship analysis can help to understand the degree of communication between countries and as well as the influential countries in "Supply chain" and "IoT" fields. The collaboration network of publishing during the period from 2007 to 2018 is presented in Figure 14. Node size indicates the publication rate by each country, and lines represent the level of cooperation among countries (Reyes et al. 2016). The top research center in the field is in China and the United States. The link strength between China and The United States is 20, between the USA and Germany it is 6, between the

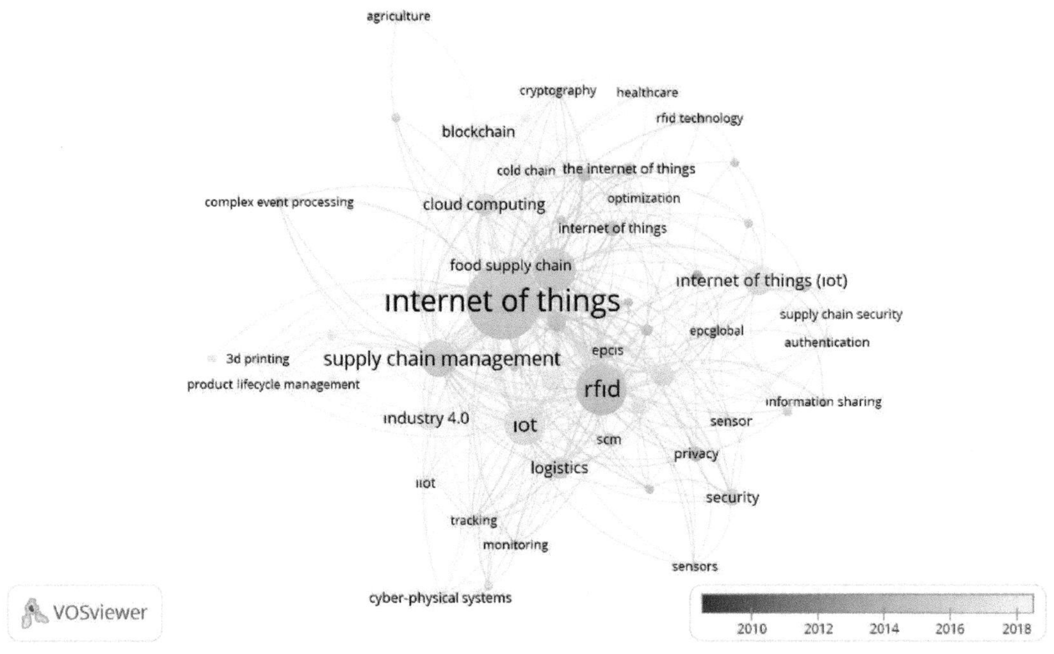

Fig. 13: Combined mapping and clustering publications of keywords.

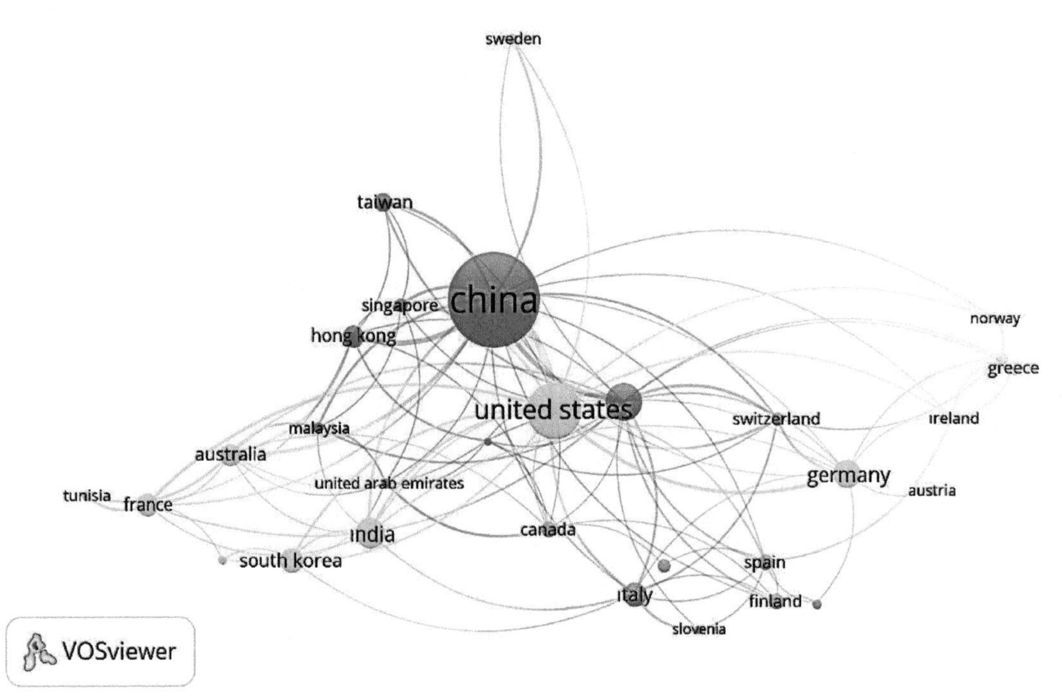

Fig. 14: Co-authorship analysis based on the country.

USA and the UK it is 7, between Australia and China, it is 6. This result shows that countries that have the same geographic do not affect the authors' collaboration. The researchers located in South-America, the Middle East, and Africa have an inadequate publication based on these fields.

5.3 The Cited Publication Analysis Based on Supply Chain and IoT

Table 2 indicates the most cited or influential papers in terms of the author's name, publication year, publication title, and citation numbers. Two highly cited papers are related to the "Internet of Things" and its challenges. All these papers were made individually. Overlay visualization of cited authors is shown in Figure 15, node size represents the number of citations, and the proximity of the nodes is related to the partnership of the authors.

Table 2: The top-cited articles.

Authors	Year	Publications name	Citations number
Weber	2010	Internet of Things-new security and privacy challenges	492
Lee	2015	The internet of things (IoT): applications, investments, and challenges for enterprises	212
Xu	2011	Information architecture for supply chain quality management	166
Wolfert	2017	Big data in smart farming—a review	95
Theisse	2009	Technology, standards, and real-world deployments of the epc network	81

5.4 Co-citations Analysis Based on Sources

The co-citations analysis reveals the total strength of the co-citation links with other sources. Figure 16 presents the journal co-citations network with 69 nodes. These node sizes represent the greatest link strength and the number of published papers. The distance between the nodes indicates the citation frequency. According to co-citation analysis, all these journals are divided into six clusters having different colors. While the green cluster represents information and computer journals, the red cluster consists of operational and production journals. International journal of production economics has the most links (62) to other journals and citations (277).

6. Conclusion and Future Studies

This summation will point out some of the most important findings of the research and show some directions for further studies. Studies on the IoT in the SC industry have gradually increased over years. Since the concept of IoT emerged in 2009, literature studies have increased by almost 20 times. The accelerated growth in IoT in the SC sector means SC 4.0 will continue to spread to every part of business operations, especially in the field of the SC.

With the introduction of IoT in the SC industry, major changes have emerged in almost all areas, especially in computer science, engineering, and business management. There are few studies focused on the field of reverse SC, social and business sciences, especially on application studies. There is a need for case studies focusing on sustainable and eco-friendly concepts for reverse SC management.

While studies on RFID, sensors, and SC activities are emphasized, this chapter identifies four research gaps in the literature of the digital supply chain, which are a cold chain, food and beverage supply chain, and inbound supply chain.

Through the IoT, all processes and operations in SC will interact with each other by connecting to a network. In this way, by creating a smart SC, it is possible to increase efficiency and productivity in the supply processes to provide the products/services demanded by the customer, to gain customer satisfaction, to reduce costs and to keep the quality high. IoT affects all SC processes. It provides more efficient use and optimization of the Supply Chain 4.0. With the Internet of objects, SC data management is made more transparent so that processes can be monitored instantaneously. With the digitalization of the SC industry, unnecessary SC activities will be eliminated, the efficiency of processes will increase, and the costs will be reduced. Customers' purchasing behavior will be examined, and the individual expectations and customer changes will be responded to more quickly. The feedback from customers will be received quickly, and after-sales services will develop. With real-time data, SC performance can be better monitored, and problems that can be experienced in processes can be detected quickly. With the devices used within the IoT, transportation and distribution costs will be reduced. IoT will be ensured in sustainable SC, and the negative effects on the environment will be reduced.

In the upcoming years, the transportation and SC industry is predicted to have vehicles without drivers, ships without captains, and planes without pilots. With SC 4.0, smart-talking systems, and new technologies will closely affect our lives and the existing SC system will leave its place to integrated new systems digitized with state-of-the-art technology.

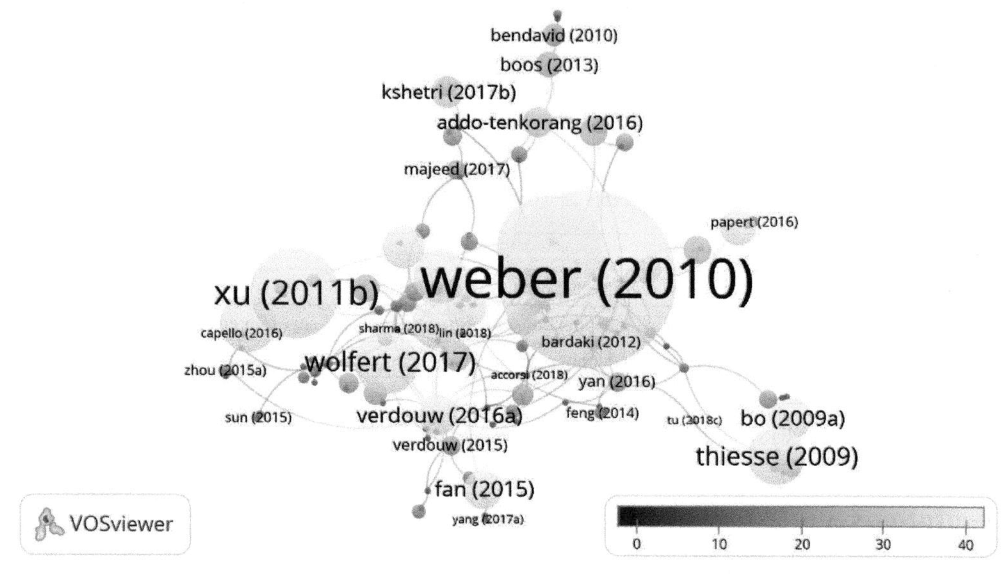

Fig. 15: Most cited authors.

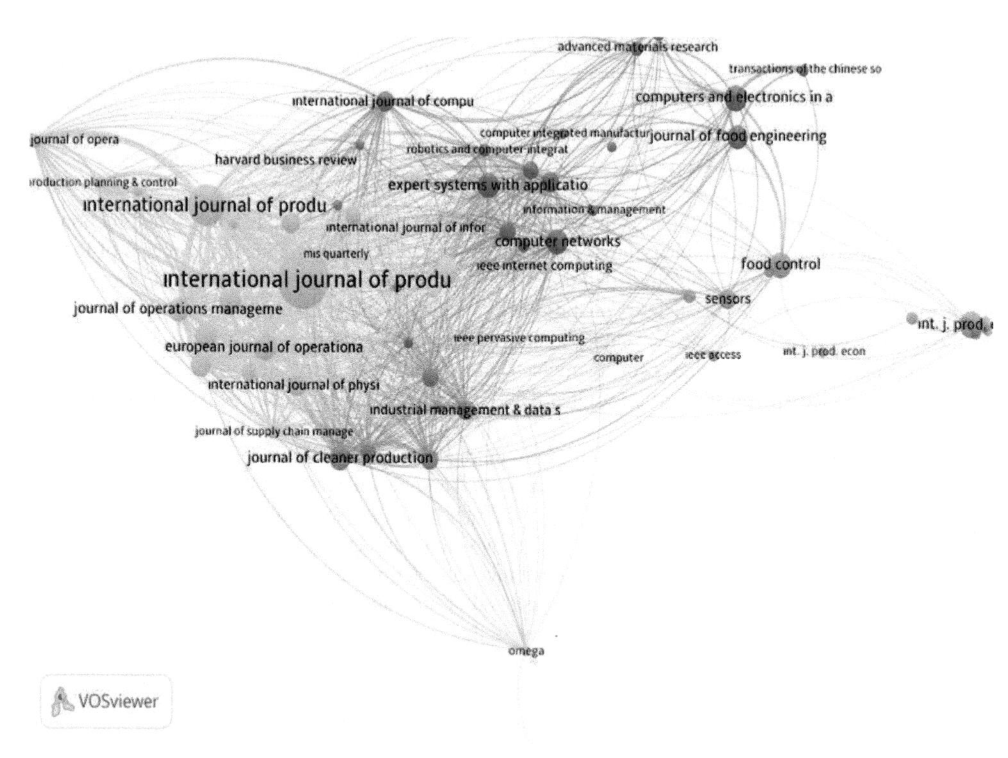

Fig. 16: The journal co-citations network.

References

Accorsi, R., S. Cholette, R. Manzini and A. Tufano. 2018. A hierarchical data architecture for sustainable food supply chain management and planning. Journal of Cleaner Production, 203: 1039–1054.

Ashton, K. 2009. That 'IoT' Thing. Available at: http://www.rfidjournal.com/articles/view?4986.

Berman, B. 2012. 3D printing: The new industrial revolution. Business Horizons, 552: 155–162.

Çallı, L. and K. Taşkın. 2015. New Markets and Marketing Practices by the 3D Printer Industry, ICEB, Macadonia.
Chen, Y.Y., Y.J. Wang and J.K. Jan. 2014. A novel deployment of smart cold chain system using 2G-RFID-Sys. Journal of Food Engineering, 141: 113–121.
Cisco, 2015. Internet of Things Will Deliver $1.9 Trillion Boost to Supply Chain and Supply Chain Operations. Available at: https://newsroom.cisco.com/press-release content?type=webcontent&articleId=1621819.
Engelman, R. 2018. The second Industrial Revolution, Available at: http://ushistoryscene.com/article/second-industrial-revolution/.
Gnimpieba, Z.D.R., A. Nait-Sidi-Moh, D. Durand and J. Fortin. 2015. Using Internet of Things technologies for a collaborative supply chain: Application to tracking of pallets and containers. Procedia Computer Science, 56: 550–557.
Gu, Y. and Q. Liu. 2013. Research on the application of the IoT in reverse SC information management. Journal of Industrial Engineering and Management, 64: 963–973.
Hermann, M., T. Pentek and B. Otto. 2016. Design principles for industrie 4.0 scenarios. In 2016 49th Hawaii International Conference on pp. 3928–3937. IEEE.
History of Virtual Reality. 2017. Available at: https://www.vrs.org.uk/virtual-reality/history.html.
Hribernik, K.A., Z. Ghrairi, C. Hans and K.D. Thoben. 2011. Co-creating the Internet of Things-First experiences in the participatory design of Intelligent Products with Arduino, 17th International Conference on Concurrent Enterprising ICE, Aachen, Germany.
Industry 4.0. 2015. Available at: https://www.allaboutlean.com/industry-4-0/industry-4-0-2/.
Jacinto, J. 2015. Industry 4.0: Industrial Computers and the Industrial Internet of Things. Available at: https://www.totallyintegratedautomation.com/2015/07/industry-4-0-industrial-computers-and-the-industrial-internet-of-things/.
Jensen, M.C. 1993. The modern industrial revolution, exit, and the failure of internal control systems. The Journal of Finance, 483: 831–880.
Kiritsis, D. 2011. Closed-loop PLM for intelligent products in the era of the IoT Computer Aided Design: 479–501.
Kong, X., G.Q. Huang, H. Luo and B.P. Yen. 2018. Physical-internet-enabled auction logistics in perishable supply chain trading: State-of-the-art and research opportunities. Industrial Management & Data Systems, 1188: 1671–1694.
Kovacs, O. 2018. The dark corners of Industry 4.0-Grounding economic governance 2.0. Technology in Society, 55: 140–145.
Lee, A.H.I., H.-Y. Kang, C.-F. Hsu and H.C. Hung. 2009. A green supplier selection model for high-tech industry. Expert Systems with Applications, 364: 7917–7927.
Lee, I. and K. Lee. 2015. The Internet of Things IoT: Applications, investments, and challenges for enterprises. Business Horizons, 584: 431–440.
Lee, K. 2015. How the IoT will Change Your World, United State: Pearson.
Leng, K., L. Jin, W. Shi and I. Van Nieuwenhuyse. 2018. Research on agricultural products supply chain inspection system based on internet of things. Cluster Computing, 1–9.
Lu, Y. 2017. Industry 4.0: A survey on technologies, applications and open research issues. Journal of Industrial Information Integration, 6: 1–10.
Luo, H., M. Zhu, S. Ye, H. Hou, Y. Chen and L. Bulysheva. 2016. An intelligent tracking system based on internet of things for the cold chain. Internet Research, 262: 435–445.
Mazzei, M.J. and D. Noble. 2017. Big data dreams: A framework for corporate strategy. Business Horizons, 60: 405–414.
McFarlane, D., S. Sarma, J.L. Chirn, C. Wong and K. Ashton. 2003. Auto ID systems and intelligent manufacturing control. Engineering Applications of Artificial Intelligence, 164: 365–376.
Macaulay, J., L. Buckalew and G. Chung. 2015. Internet of Things in Supply Chain. Available at: www.dpdhl.com.
Oussous, A., F.Z. Benjelloun, A.A. Lahcen and S. Belfkih. 2017. Big Data technologies: A Survey, Journal of King Saud University- Computer and Information Sciences, 30: 431–448.
Pal, A. and K. Kant. 2018. IoT-Based sensing and communications infrastructure for the fresh food supply chain. Computer, 512: 76–80.
Perianes-Rodriguez, A., L. Waltman and N.J. Van Eck. 2016. Constructing bibliometric networks: A comparison between full and fractional counting. Journal of Informetrics, 104: 1178–1195.
Pfeiffer, S. 2017. The vision of "Industrie 4.0" in the making a case of future told, tamed, and traded. Nanoethics, 111: 107–121.
Reyes, G.L., C.N.B. Gonzalez and F. Veloso. 2016. Using co-authorship and citation analysis to identify research groups: A new way to assess performance. Scientometrics, 108: 1171–1191.
Sadıkzade, M. 2016. Industry 4.0 Transformation and Digital Supply Chain. Available at: https://tr.linkedin.com/pulse/end.
Schouten, E. 2014. Cloud computing defined: Characteristics & service levels. Available at: https://www.ibm.com/blogs/cloud-computing/2014/01/31/cloud-computing-defined-characteristics-service-levels/.
Shariatzadeh, N., T. Lundholm, L. Lindberg and G. Sivard. 2016. Integration of digital factory with smart factory based on IoT. Procedia CIRP, 50: 512–517.
Tadejko, P. 2015. Application of IoT in SC–current challenges. Economics and Management, 74: 54–64.
Thiesse, F., C. Floerkemeier, M. Harrison, F. Michahelles and C. Roduner. 2009. Technology, standards, and real-world deployments of the EPC network. IEEE Internet Computing, 13: 36–43.

Thoben, K.D., S. Wiesner and T. Wuest. 2017. Industrie 4.0 and smart manufacturing—a review of research issues and application examples. Int. J. Autom. Technol., 111: 4–19.

Tian, F. 2018. An information System for Food Safety Monitoring in Supply Chains based on HACCP, Dissertation, Vienna University of Economics and Business, Wien, Austria.

Tsai, K.M. and K.S. Pawar. 2018. Special issue on next-generation cold supply chain management: research, applications and challenges. The International Journal of Logistics Management, 293: 786–791.

Van Eck, N.J. and L. Waltman. 2018. VOS viewer Manual. Available at: https://www.vosviewer.com/documentation/Manual_VOSviewer_1.5.4.pdf.

Venta, O. 2007. Intelligent products and systems. Technology Theme-Final Report. Espoo: VTT Publications 304, Finland.

Weber, R.H. 2010. Internet of Things–New security and privacy challenges. Computer Law & Security Review, 26: 23–30.

Wolfert, S., L. Ge, C. Verdouw and M.J. Bogaardt. 2017. Big data in smart farming–a review. Agricultural Systems, 153: 69–80.

Xu, Li Da. 2011. Information architecture for supply chain quality management. International Journal of Production Research, 49: 183–198.

Yan, B., X.H. Wu, B. Ye and Y.W. Zhang. 2017. Three-level supply chain coordination of fresh agricultural products in the Internet of Things. Industrial Management & Data Systems, 1179: 1842–1865.

Yan, R. 2017. Optimization approach for increasing revenue of perishable product supply chain with the Internet of Things. Industrial Management & Data Systems, 1174: 729–741.

Yazıcı, A. 2016. Industry 4.0 and otonom robots. Journal of Electrical Engineering, 39: 459.

Zhang, Y., L. Zhao and C. Qian. 2017. Modeling of an IoT-enabled supply chain for perishable food with two-echelon supply hubs. Industrial Management & Data Systems, 1179: 1890–1905.

Zuberer, S. 2016. Industry 4.0: Companies Worldwide Are Investing over $US 900 Billion per Year until 2020. Available at: https://press.pwc.com/News-releases/Industry-4.0--companies-worldwide-are-investing-over-US-900-billion-per-year-until-2020/s/09b6f9d5-bf5f-4933-a14f-a0a48d7aa37b.

CHAPTER 5

The New Challenge of Industry 4.0

Sustainable Supply Chain Network Design with Internet of Things

Sema Kayapinar Kaya,[1,*] *Turan Paksoy*[2] and *Jose Arturo Garza-Reyes*[3]

1. Introduction

Today, environmental pollution is considered one of the major reasons that may lead to the extinction of humanity. Consequentially, "environmental awareness" has developed environmental control consciousness within the industrial cycles of enterprises. One force encouraging or forcing enterprises to implement green policies is the state power and laws; another force is the negative financial and legal results that they might experience because of wrong administrative approaches in terms of the environment. As economic and technological developments increased environmental values degenerated or were destroyed, which led to problems such as famine, hunger, greenhouse effect, global warming. Although the urban communities developed in the particular second half of the 20th century, attentions were drawn increasingly on environmental issues, and green management concept has emerged.

"Supply-chain Sustainable" becomes an important issue by force of not only economic effects but also environmental and social effects, as one of the most important factors causing global warming disaster is that carbon emission, CO_2, has reached higher rates. It can be said that supply chain activities are the main source of carbon emissions. Logistics and transportation industries have a great part in the Carbon emission cake. According to the IPCC-2007 study, logistic, including passenger transportation, has a big part. Similarly, logistic constitutes 24% of global greenhouse gas emissions. It is accepted that even reducing the carbon footprints of commercial customers of big logistics companies will play a key role in reducing the general CO_2 emission.

An important part of the ecological problem is ineffective transportation methods in modern Supply Chain Management (SCM). The report by "Eyes for transport" showed that around 75% of a company's carbon footprint results from transportation and logistics alone. To tackle the environmental problems in the supply chain, enterprises have implemented Sustainable Supply Chain Management (SSCM), which involves environmentally and financially viable practices into the complete sustainable chain lifecycle, from product design and development, manufacturing, transportation, consumption, return and disposal.

By increasing our digital sophistication, sustainable supply chain management can lead to innovation during the digital transformation. Emerging sensor-embedded products can transform SSCM to future levels. In reverse flow, the EOLP can be recovered with various processes such as reuse, recycle, repair, or dispose. Reverse flow leads to many uncertainties. The products are returned from customers because they do not meet definite standard requirements. Their amount, their date of expiration, the number of recyclable components of the product, and the model of the product are uncertain. This condition always causes changes and uncertainties in developing options for the returned products. Ambiguities are largely resolved with the sophisticated digital applications such as the Internet of Things (IoT); products are followed up

[1] Department of Industrial Engineering, Munzur University, Tunceli, Turkey.
[2] Department of Industrial Engineering, Konya Technical University, Konya, Turkey.
[3] Centre for Supply Chain Improvement, University of Derby, Derby, United Kingdom.
* Corresponding author: kysem24@gmail.com

along different stages of the supply chain by means of the planted devices (Vermesan and Friess 2013). Radio-Frequency Identification (RFID) and sensor labels integrated with the products follow the life cycle of critical parts in products when the lifetime of the products expires. They include not only static information such as the price of products, their serial numbers, place, repair instructions, but also dynamic information such as the working conditions of products, their error rates, environmental effects, etc. (Ondemir and Gupta 2014a). Parlikad and McFarlane (2007) stated that RFID-based descriptive technologies have positive effects for the retrieval options of the returned products and that they provide sufficient information. Therefore, the decision about which improvement option a product which has expired should be subject to is taken more precisely and within a shorter time, and it makes it possible to decrease the expensive processes such as preliminary examination or full mounting, which are required for the quality level of the returned products.

This chapter presents a novel mathematical model that developed an environmental impact on SSC design via the Internet of Things. IoT provides information about a product when they return and plays a significant role in the recovery process of SCM. This information by reducing and eliminating uncertainty regarding the condition and remaining lives of components in EOLPs IoT technologies such as asset tracking solutions, has become one of the biggest trends in SSC network configurations. Using sensors, RFID, tags, and other IoT devices to track goods through the global supply chain is one of the first use cases for the IoT. Due to the uncertainty of reverse logistics, we creatively provide a new forecast application by using IoT.

The rest of the paper is organized as follows. In the next section, we outline the literature review of Sustainable Supply Chain Management and Reverse supply chain based on IoT. Problem definition and model assumptions are presented in Section 3. In Section 4, a case study illustrates a computation experiment and then, model results are discussed in Section 5. Finally, the conclusion and future studies suggested in the Section 6.

2. Literature Reviews

In this section, we probe the literature and categorize studies into two. The first one is the SSCs, and the second one is the Internet of Things, using the reverse supply chain.

2.1 The Sustainable Supply Chain Management

SSCM is considered a subtopic of SCM and has been gaining in importance day by day. For that reason, many scientific types of research have been carried out in this field.

Bouzembrak et al. (2011) developed a green supply chain network configuration by considering environmental concerns and proposed a multi-objective decision optimization model that trade-off between the total cost and environmental cost. The environmental aspect of the issue is related to the total CO_2 emission in all the supply chain. Coskun et al. (2016) presented an integrated model that combines the Analytic Network Process (ANP) and multi-objective programming methodology by considering the green partner selection. ANP methodology is capable of balancing between green and business criteria to select favorable green partners.

Elhedhli and Merrick (2012) proposed a concave minimization model that takes CO_2 emissions into account. This model was solved by a Lagrangian relaxation that generates a feasible solution for each iteration. The objective of this model is to minimize logistics costs and the environmental cost of carbon emissions simultaneously. Paksoy et al. (2010) designed a green supply chain network that includes suppliers and recycle centers. That model deals with CO_2 gas emission, the different recyclable ratio of raw materials and the opportunity price to sell recyclable products. They developed a multi-objective model whose first objective is to minimize the total cost of transportation via different trucks. In addition, the second objective is to minimize transportation costs. Paksoy and Özceylan (2013) focused on the environmental issue for the optimization of supply chain configurations and constructed an integer non-linear programming model. The main purpose of their model was to consider gas emissions, the noise of vehicles, fuel consumption between facilities, transportation times of vehicles and road roughness. They concluded that consumed fuel, produced noise, transportation time, and emitted carbon emission are affected by vehicle speed when designing a supply chain network. Pinto et al. (2010) addressed the trade-off between profit and environmental impacts on the design of the supply chain network to maximize the annual profit. Profit and environmental impacts are balanced with an optimization approach adapted from Symmetric Fuzzy Linear Programming (SFLP) and formulated a mixed-integer linear programming model using the Resource-Task-Network application. Memari et al. (2015) developed a novel mathematical model in green supply chain management. The objectives of the model was to minimize the total costs and also to minimize the environmental impact on the logistics network. The model determines the green economic production quantity by using Just-in-Time logistics. Cao and Zhao (2014) proposed a green supply chain network considering healthy, low-carbon transportation systems. The penalty function coefficient helped convert a multi-objective optimization problem consisting of objectives such as profitability, service level, and environmental protection into a single objective. Chibeles-Martins et al. (2016) proposed a mixed linear multi-objective programming model in which the first objective minimizes the sum of all environmental impacts from diesel and electricity consumption, while the second objective function maximizes the total profit. This problem was tested for large problems which take a long time to find a solution. Thus, Simulated Annealing (SA) algorithm was adapted to improve the algorithm's efficiency and effectiveness. Memari et al. (2015) designed a supply chain network for multi-manufacturers, distribution centers, multi-products and multi-periods. They developed a multi-objective mathematical model in which the first objective aimed to minimize the total cost including production cost, holding cost of the distribution center, and transportation costs between echelons, while the second objective function is to minimize the total carbon emission in the whole network. Multi Objective Genetic Algorithm (MOGA) and Goal attainment technique was used to solve for their proposed model. Govindan et al. (2015) formulated a mixed-integer programming model which consists of five echelons: suppliers, plants, distribution centers, cross-docks, and retailers. Their study takes the pollution rate of CO_2, CFC, NO_x gases into account as the most harmful emission. There are three types of environmental impacts; the impact of shipping the products among the network, the impact caused by opening facilities and manufacturing impacts of plants related to technology. A hybrid multi-objective metaheuristics algorithm was proposed to solve this model and some experimental designs were analyzed and tested. According to the result show that the Multi-Objective Hybrid Approach (MOHEV) is a better solution approach compared with the others.

2.2 *Usage of the Internet of Things in Reverse Supply Chain*

Since the emerge of IoT by the Massachusetts Institute of Technology in 1999, the field in "Internet of Thing" has got significant attention from practices and researchers. IoT is the configuration of physical devices, mechanical and electronic machines, and the other items embedded with RFID, sensor, chip, network connectivity that provide these objects to identify, collect and transfer data over this network. IoT technology is the new communication platform including the number of mobile devices and network connected equipment that can communicate with each other simultaneously (Torğul et al. 2016). It offers communications of each object by covering a variety of protocols, domains, and applications. IoT has positively influenced supply chain management and this revolutionary technology allows to control the external and internal environment of the supply chain. With IoT, a smart product produced by a factory can be easily monitored during each echelon of the supply chain process, such as production, distribution, inventory, and consumption. IoT also has many advantages for SSCM. In reverse logistics information resources, it plays a vital role in the enterprise's operation, because there are some uncertain and inaccurate data about the property of products in reverse logistics management. All general processes such as collection, inspection, re-processing, disposal associated with reverse activity include uncertain information. Thanks to IoT technology, the enterprises can get more accurate information, which incorporates data ranging from production in the factory to consumption of final users. The devices of IoT embedded into the product's components in the production process can record information during its whole life recycling period. Lifecycle information is collected and stored into RFID, sensor, or PML server. It challenges uncertainty in the return of reverse logistics and gets a more accurate prediction about reverse logistics operations related to the reuse of products and materials.

Reverse supply chain (RSC) management, which is oriented towards the entire lifecycle of products, has been concerned with the emerging of green production. With the rapid development of IoT and supply chain digitization, a smart reverse supply chain (SRSC) will be formed based on RFID, sensors, high-performance computing, intelligent technologies. These technologies have been widely used by many researchers in the reverse supply chain.

Kiritsis et al. (2008) developed models and RDIF application closed the gap of information in closed cycle supply chain. Kiritsis (2011) categorized product life management into three-stage: Beginning of the stage (BOL), Middle of the life cycle (MOL) and last cycle life (LCL). By using IoT, all information can be collected and analyzed at every life-stage. Hribernil et al. (2011) put PEID in the returned plastic automotive parts and conducted a sample study in order to follow up the life cycle of the products. Ilgın and Gupta (2011) used sensors in order to measure the performance values in dismantling lines in the control of kanban, and they concluded that the sensor led to a decrease in cost and an increase in profit. Gu and Liu (2013) adapted IoT application to information management in reverse logistics. In conclusion, they suggested that precise and accurate information is an important means for the success of reverse logistics management. Ondemir et al. (2012) investigated how RFID and sensors estimate the demands for products, components and returning behaviors. In this way, they collected static and dynamic information from IoT and developed a mixed-integer linear model, which was to minimize the total cost. Ondemir and Gupta (2014a) developed the linear physical programming model. They determined that one objective function is to minimize the total cost and amount of disposed of wastes while the other objective is to maximize product sale profit and total quality level, and solved it using lexicographic goal programming. Ondemir and Gupta (2014b) developed a multi-purpose model, in the products having a sensor embedded, for minimization of total cost and amount of wastes and maximization of sale profit and customer satisfaction level, considering the mounting and repair recycling options. The relationship between environmental consciousness, cost and satisfaction as well as sale profit are taken into consideration.

3. Problem Definition and Assumptions

The problem being analyzed concerns a producer who meets sales and collection center demand with new, repaired and remanufactured products. It is assumed that the factory, distribution center, repair center, sales, and collection center in the sustainable supply chain network belong to one enterprise.

The sales and collection center sell the products it receives from the factory and purchases EOLPs from customers. IoT provides information about a product when they are returned and plays a significant role in the recovery process and backup. This information reduces and eliminates the uncertainty of the condition and remaining lives of components in an EOLP. The price paid for EOLPs is based on their level of value. The returned products are processed via one of three recycling procedures: repair, disassembly, or disposal. The products that have little damage and can be reused are repaired. The broken components of these products are replaced with new components received from the factory, and the products are sent to the distribution center as needed to meet demand. If the returned products cannot be repaired, they are disassembled. The components salvaged from disassembly are sorted by value and sent to the sales and collection center to be sold to the factory for use in new products or to the waste center for disposal. The factory must outsource for new components to meet the demand for products and components and optimize total production costs by making the correct recycling decision for each returned product.

New parts and spare parts are also sold in the sales and collection center. When the demand is high for spare parts, it is met with new parts. Sales and collection centers bear the cost of stocking new parts.

This present model is a single product, the multi-stage model that includes the cost and recycling evaluation process of CO_2 emission for different types of vehicles. It is a linear mathematical model. This network supply chain network design presents in Figure 1.

Model Assumptions

1. This product is modular and contains lifecycle information.
2. There is no difference between new products and recovered products.

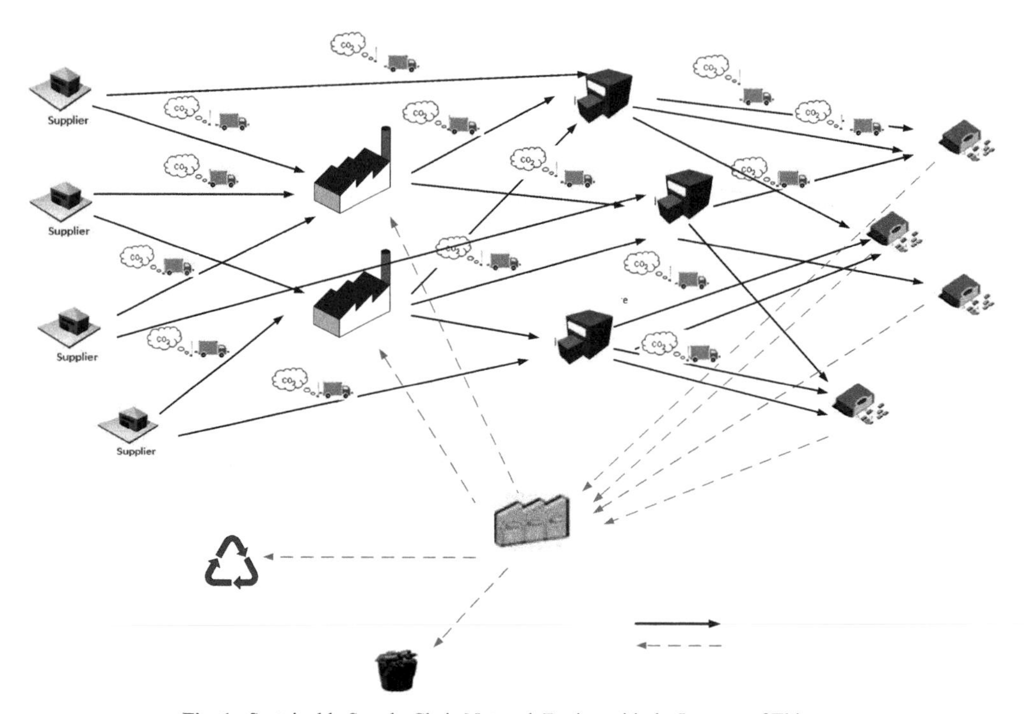

Fig. 1: Sustainable Supply Chain Network Design with the Internet of Things.

3. All cost and sale prices are known.
4. End-of-life products (EOLPs) contain both functional and non-functional components.
5. The components of the product have importance weights that range from new to one, and the sum of them is 1.
6. It is assumed that the product value level is calculated according to the current component value levels as follows:

$$\text{Product value rate } (diot_j) = \sum \text{Component value level } (diot)_j * \text{Importance weight of component}$$

7. New products and components are sold in the sales and collection center.
8. Inventory costs exist and no shortage cost is allowed.
9. Sales and collection centers have the capacity for the product/components.
10. The factory has a production and modular capacity.
11. Sales from the inventory of the sales collection center are not allowed.
12. For a single product type, a dynamic, multi-stage, and mixed linear model has been developed.
13. Carbon emission amounts of the vehicles were considered, and it was punished by a penalty coefficient.
14. It is assumed that vehicle capacities are limitless.

3.1 Model Indices and Parameters

Based on the above assumptions, a uniform product (automobile, computer, phone, etc.) is selected that is modular. A multi-stage, single-product and mixed whole number model was developed by considering environmental factors.

Notations

i Returned products (i = 1, 2, …, I)
j Parts (j = 1, 2, …, J)
l Returned product value range (l = 1, 2, …, L)
s Number of supplier (s = 1,……, S)
p Number of factory (p = 1,……, P)
d Number of distribution centers (d = 1,……..D)
st Number of sales and collection center (st = 1,……..,ST)
t Vehicle types (t = 1, 2, …, T)

Model Parameters

a_j : Assembly cost of component j in factory and sales collection center
$d1_j$: Disassembly cost of component j
e_j : Waste cost of component j
$g2_j$: Unit transportation cost of component j
$p1_{sj}$: Ordering cost of component j from s. supplier
pp_j : Purchasing value of j component obtaining by disassembly
rr_l : Purchasing cost of an l-level product from sales and collection center
$f_{st,j}$: Cost of replacing the j used component in st. sales center
$h2_{st,j}$: Unit sales price of j. original component in st. sales collection center
$h3_{st,j}$: Unit sales price of j. used component in st. sales collection center
rc_j : Number of component j in a product
$R_{st,l}$: Amount of used product in the l-level are purchased from sales and collection center
bro_{ij} : 1 if component j of the returned product I is the quality deficit in period t, zero otherwise
C_i : Value level of the returned product
$aa_{st,i}$: 1 if the returned product i is purchased, 0 otherwise in period t
dm_{st} : Product demand of st. sales and collection center
$dms_{st,j}$: j. new component demand of st. sales and collection center

$dmc_{st,j}$	:	j. used component demand of st. sales and collection center
$Capmax_{st,j}$	:	j. the component capacity of st. sales and collection center
$CapmaxU_{st}$	:	Capacity of st. sales and collection center for product
$Capp1_s$	:	Capacity of s. supplier
$Capp2_p$	:	Capacity of p. factory
$Capp3_d$	:	Capacity of d. distribution center
$Capp4_{dj}$	:	j component capacity in d distribution center
$K1_{spt}$	:	Unit transportation cost of t vehicle from s supplier to p factor
$K2_{sdt}$	:	Unit transport cost of t vehicle from s supplier to p factory
$K3_{pdt}$	:	Unit transport cost of t vehicle from s supplier to p factory
$K4_{d,st,t}$	:	Unit transport cost of t vehicle from d distribution center to p factory
$CC1_{spt}$	:	Amount of CO_2 emission of truck t during the transportation of s supplier to p factory
$CC2_{sdt}$	:	Amount of CO_2 emission of truck t during the transportation from s supplier to d distribution center
$CC3_{d,st,t}$	:	Amount of CO_2 emission truck t during the transportation from p factory to d distribution center
$CC4_{d,st,t}$	:	Amount of CO_2 emission truck t vehicle the transportation from d distribution center to st sales collection center
$h1_{st}$	:	Unit product sales price in the sales collection centers
Q_{st}	:	Amount of products repaired in the sales collection center
M_{jt}	:	Amount of components can be disposed
g1	:	Transportation cost of unit product
rw_j	:	Weight of j component per product
$teta_j$	:	Rate of the material to can be recycled for j component (%)
tww	:	recycling rate for product one another (%)
Weight	:	Unit weight of a product (kg)
Recycling	:	Amount of product can be recycled
$Recy_j$	:	Return gained in recycling of sale of j component
Recy1	:	Unit recycling return of a product
Wcost	:	Unit disposal cost of a product
pn	:	Unit penalty cost of CO_2 emission

Expression of some parameters

Binary parameters bro_{ijt} determine how to process components according to their value levels. Accordingly, components whose value level is sufficient are refurbished and reused after the disassembly process. If the value level of a component is insufficient, the component is sent for disposal. The threshold value to determine if components should be refurbished is represented by n.

$$bro_{ijt} = \begin{cases} 1 \text{ if component } j \text{ is non-functional}, & 0 \le diot_j \le n \\ 0 \text{ component } j \text{ is functional (Again used)}, & diot_j \ge n \end{cases} \quad \forall i \text{ and } 0 < n \tag{1}$$

Assuming the value level of components and their importance weights of return are known, there are three value ranges that can be calculated for product value levels ($diot_i$). C_i (the product value range) is determined where n_1 and n_2 represent value range limits. According to the information on which recovery process (repair, disassembly or disposal) is applied for each EOLP, the parameter C_i can be calculated as follows:

$$C_i = \begin{cases} 1 \text{ (Disposal)} & 0 \le diot_i < n_1 \\ 2 \text{ (Disassembly)} & n_1 \le diot_i < n_2 \\ 3 \text{ (Repair)} & diot_i \ge n_2 \end{cases} \quad \forall i \text{ and } 0 < n_1 < n_2 \tag{2}$$

(Ondemir and Gupta 2014b)

Returned products are repaired in sales and collection points if they are purchased in t period and at 3rd level of importance. Therefore;

$$C_i.\bar{a}_{st,i} = 3 \quad \text{then} \quad yy_{st,i} = 1 \tag{3}$$

If returned products are purchased in t period and at 2nd importance value, they are disassembled. Disassembled products are determined in sales and collection centers and they are brought to decomposition centers. Then, they are subjected to the disassembly process. Therefore;

$$C_i.\bar{a}_{st,i} = 2 \quad \text{then} \quad xx_{st,i} = 1 \tag{4}$$

If returned products are in 1st importance value, they are directly eliminated. Product parts are eliminated during disassembly and repair. Therefore;

$$C_i = 1 \quad \text{then} \quad \Sigma_t\, xx_{st,i} + yy_{st,i} = 0 \tag{5}$$

Amount of total renewed parts after disassembly (rfc_{jt}) is calculated as the following:

$$rfc_{st,j} = \sum_{\{i \in I | \bar{a}_{st,i} = 1\}} xx_{st,i}.(1 - bro_{ij}) \tag{6}$$

Amount of product repaired in sales and collection centers;

$$Q_{st} = \sum_i yy_{st,i} \tag{7}$$

Components with inadequate levels came out of repair and disassembly periods are directly sent to waste in decomposition centers.

$$M_j - \sum_i \sum_{st} bro_{ij} \,.\, (xx_{st,i} + yy_{st,i}) = 0 \; \forall j \tag{8}$$

$Geridon_j$: Amount of material of j component to be sent to recycling in kg;

$$Geridon_j = M_j \,.\, rw_j \,.\, teta_j \tag{9}$$

Amount of product to be recycled;

$$Recycling = \sum_{st} \sum_i TW_{st,i} \,.\, weight.tww \tag{10}$$

Model Decision Variables

b_j : Auxiliary decision variable used to decide the value of $SS_{st,j}$ precisely (takes the value 0 or 1)
X_{spjt} : Amount of j. new component transported by t truck from s supplier to p factory
Y_{sdjt} : Amount of j. unused component transported by t truck from s supplier to d distribution center
Z_{pdt} : Amount of component transported by t truck from p factory to d distribution center
$W_{d,st,j,t}$: Amount of j. component transported by t truck from d distribution center to st. sales and collection center
$V_{d,st,t}$: Amount of product transported by t truck from s supplier to p factory
$SS_{st,j}$: Amount of j. used component transported from st. sales and collection center to dismantling center

N_{pj} : Amount of j. renewed components transported from p factory to dismantling center
$Icc_{st,j}$: Amount of j component stored in the st. sales and collection center

3.2 Objective Functions

The objective function is the Maximization of the total revenue. Total revenue (TR) and Total Cost (TC) and Objective Function (Z) is shown below:

$$Z = TR - TC$$

Total Revenue (TR): Companies regard the sale of a brand-new product, original components and used components as income for the company. The sales of used components, which are in good condition after the disassembly process, returns to the company as revenue.

$$TR = \sum_{st}\sum_{t} dm_{t,st}.h1_{st} + \sum_{t}\sum_{st}\sum_{j}(dms_{t,st,j} + dmc_{t,st,j} - SS_{st,j,t}).h2_{st,j} + \sum_{st}\sum_{j}\sum_{t} SS_{st,j,t}.h3_{st,j} + \sum_{p}\sum_{j} N_{pj}.pp_j + \sum_{j}\sum_{t} Geridon_{jt}.recy_j + \sum_{t} recy1.recyling_t \quad (11)$$

Total Cost (TC): Total cost consists of total purchasing cost (TPC), total manufacturing cost (TMC), total transport cost (TTC), Total Waste Cost (TWC), Total Renting Car cost (TRC), total emission Cost (TCC), and Total Fixed Cost (TFC).

$$TC = TPC + TMC + TTC + TWC + TRC + TCC + TFC \quad (12)$$

Total Purchasing Cost (TPC): Purchasing cost is defined as the sum of purchasing costs made in order to meet the demand of factory and sales collection centers in original components from the suppliers and purchasing costs of the products returned from customers to sales collection centers.

$$TPC = \sum_{s}\sum_{p}\sum_{j}\sum_{t} pl_{sj} X_{spjt} + \sum_{s}\sum_{d}\sum_{j}\sum_{t} pl_{sj} Y_{sdjt} + \sum_{l} rr_l \sum_{st}\sum_{t} R_{st,l,t} \quad (13)$$

Total Manufacturing Cost (TMC): In factories, manufacturing costs are composed of repair costs in sales and collection centers and assembly costs in disassembly centers and remanufacturing costs in the plant.

$$TMC = \sum_{j} a_j rc_j \sum_{p}\sum_{d}\sum_{t} Z_{pdt} + \sum_{st}\sum_{i}\sum_{j}\sum_{t}(dl_j + a_j + \sum_{s} pl_{sj})bro_{ijt} yy_{st,i,t} + \sum_{st}\sum_{i}\sum_{t} xx_{st,i,t} \sum_{j} dl_j + f_j(1 - bro_{ijt})) \quad (14)$$

Total Transport Cost (TTC): Transport costs consists of both product and component cost. Products are transported from factory to distribution centers and from distribution center to sales collection center.

$$TTC = \sum_{p}\sum_{d}\sum_{t} Z_{pdt}.g1 + \sum_{d}\sum_{st}\sum_{t} V_{d,st,t}.g1 + \sum_{d}\sum_{st}\sum_{j}\sum_{t} w_{d,st,j,t}.g2_j + \sum_{st}\sum_{j} SS_{st,j}.g \quad (15)$$

Total Waste Cost (TWC): Original components are disposed if they are broken in the repair and disassembly processes. These original components are sent to the disposal center.

$$TWC = \sum_{i}\sum_{j} e_j bro_{ij} \quad (16)$$

Total Renting Truck Cost (TRC): Renting costs of trucks associated with the amount of transported products and components.

$$TRC = \sum_{s}\sum_{p}\sum_{j}\sum_{t} X_{spjt}.K1_{spt} + \sum_{s}\sum_{d}\sum_{j}\sum_{t} Y_{sdjt}.K2_{sdt} + \sum_{p}\sum_{d}\sum_{t} Z_{pdt}.K3_{pdt} + \sum_{d}\sum_{st}\sum_{t} V_{d,st,t}.K4_{d,st,t} + \sum_{d}\sum_{st}\sum_{j}\sum_{t} W_{d,st,j,t}.K4_{d,st,t} \quad (17)$$

Total Emission Cost (TCC): Carbon emission cost varies by the type of trucks and it is multiplied by a penalty cost coefficient as well.

$$\begin{aligned} TCC = \sum_s \sum_p \sum_j \sum_t X_{spjt}.CC1_{spt} + \sum_s \sum_d \sum_j \sum_t Y_{sdjt}.CC2_{sdt} + \sum_p \sum_d \sum_t Z_{pdt}.CC3_{pdt} \\ + \sum_d \sum_{st} \sum_t V_{d,st,t}.CC4_{d,st,t} + \sum_d \sum_{st} \sum_j \sum_t W_{d,st,j,t}.CC4_{d,st,t} \end{aligned} \tag{18}$$

Total Fixed Cost of Facilities (TFC): The fixed cost of facilities includes the selection cost of suppliers, the opening cost of plant and distribution centers.

$$TFC = \sum_s f1_s k11_s + \sum_p f2_p k12_p + \sum_d f3_d k13_d \tag{19}$$

3.3 Constraints

The number of original components coming from factories to distribution centers and the number of original components going from distribution centers to sales and collection centers should be equal to each other.

$$\sum_s \sum_t Y_{sdjt} = \sum_{st} \sum_t W_{s,st,jt} \forall d, j \tag{20}$$

The number of products coming from factories to distribution centers and the number of products going from distribution centers to sales and collection centers should be equal.

$$\sum_p \sum_t Z_{pdt} = \sum_{st} \sum_t V_{d,st,t} \forall d \tag{21}$$

Products are disassembled in decomposition center and then useable parts are sent to sales and collection centers in order to meet the demand of usable components.

$$\sum_{st} rfc_{st,jt} = \sum_{st} SS_{st,j,t} + \sum_p N_{pjt} \forall j \tag{22}$$

A certain part of components used in the factory to remanufacture is supplied by the dismantling center and the other parts are supplied by suppliers. Therefore, the sum of the parts that arrived should meet the need for the component of the factory.

$$\sum_s \sum_t X_{spjt} + N_{pj} - rc_j \sum_d \sum_t Z_{pdt} = 0, \forall p, j \tag{23}$$

Product demand of sales and collection centers supplied by final products in distribution centers.

$$\sum_d \sum_t V_{d,st,t} = dm_{st}, \forall st \tag{24}$$

The customer has two different demands as an original component and used components. The amount of original components coming from distribution centers to sales and collection centers is equal to original component stock remaining from the previous period in sales and collection centers, demand in original component and stock amount remain at the end of the period.

$$S_{st,j} + \sum_d \sum_t W_{d,st,jt} = dms_{st,j} + dmc_{st,j} \forall st, j \tag{25}$$

Certain components are renewed after the dismantling process is sent to sales and collection centers in order to meet the demand in usable and new components are send to remanufacture in the factory. As priority is to meet the demand in used component, it must be as either total amount of refurbished components after the disassembly or used component demand.

$$\sum_{st} SS_{st,j,t} = \sum_{st} rfc_{st,j,t}.b_{jt} + \sum_{st} dmc_{st,j,t}(1 - b_{jt}), \forall j, t \tag{26}$$

Capacity constraints for sales and collection centers are considered for each product and component.

$$\sum_{d}\sum_{t} V_{d,st,t} \leq \text{Cap}\max U_{st}, \forall st \tag{27}$$

$$\sum_{d}\sum_{t} W_{d,st,j,t} + SS_{jt} \leq \text{Cap}\max_{st,j}, \forall st, j \tag{28}$$

Capacity constraint for each supplier.

$$\sum_{p}\sum_{d}\sum_{t} Z_{pdt} \leq \text{Capp2}_{p}, \quad \forall p, t \tag{29}$$

Capacity constraint both product and components in the distribution center.

$$\sum_{st}\sum_{t} V_{d,st,t} \leq \text{Capp3}_{d}, \quad \forall d \tag{30}$$

$$\sum_{st}\sum_{t} W_{d,st,j,t} \leq \text{Capp4}_{djt}, \quad \forall d, j \tag{31}$$

The variables should be greater than zero.

$$X_{spjt}, Y_{sdjt}, Z_{pdt}, W_{d,st,jt}, V_{d,st,t}, SS_{st,j}, N_{pj}, Icc_{st,j} \geq 0, \qquad \forall s,p,j,t,d,st \tag{32}$$

4. A Computation Experiment

4.1 General Information

We proposed a sustainable supply chain network design based on hypothetical data. A small example illustrates the properties of this problem and this model. This network design is divided into two parts; the first part is forward logistics and consist of four suppliers, two factories, three distribution centers, four sales and collect centers, while the reverse direction of this network includes in four sales and collections centers (collection proses), dismantling and recycling and disposal center. And, network configurations consider on environmental focused on CO_2 gas emission released by trucks and recycling process of end-of-life products. It is assumed that outsourcing is used only for transportations. The third-party logistics (3PL) firms provide service with two types of trucks for transportation, which is between 0–5 years, 5–10 years old, respectively. As aging trucks, rental free will be cheaper. Thus, choosing the oldest trucks is a good option for the firms, but CO_2 emission also increase because the engines are old. Rental costs of all trucks according to their ages, are given in Table 2, Table 3 and Table 4, respectively. The added deterrent penalty cost (pn CO_2 = 10.5 $/ gr than more 2000 kg CO_2) in model puts the decision maker into another trade-off situation which is penalty cost versus CO_2 emissions (Table 5, Table 6 and Table 7). Suppliers, factories and distribution centers have their own components and product capacities, which give in Table 1. In addition, the demand for new parts and components of the sales and collection center are given in Table 8.

The CO_2 gas emissions from all trucks between each echelon have demonstrated in Tables 5, 6, and 7, respectively.

Table 1: The component and product capacities of Suppliers-Factory- Distribution Centers (DC)-Sales and collection center (SCC)

SCC	Capacity	Suppliers	Capacity	F	Capacity	DC	Capacity
1	497	1	1520	1	127	1	1075
2	480	2	1500	2	142	2	1045
3	370	3	1530			3	1032
4	315	4	1450				

Table 2: The rental costs ($/unit) of each truck during the transportation between Suppliers (S) – Factories (F) – Distribution centers (DC).

	Truck-1		Truck-2		Truck-1			Truck-2		
S	F1	F2	F1	F2	DC1	DC2	DC3	DC1	DC2	DC3
1	16	11	24	15	8	11	10	9	15	10
2	23	14	16	16	12	15	14	12	15	13
3	23	11	16	16	8	8	10	11	9	11
4	20	15	18	18	12	10	10	9	15	14

Table 3: The rental costs ($/unit) of each truck during the transportation between factory and distribution centers and sales collection centers.

	Truck-1			Truck-2		
F	DC1	DC2	DC3	DC1	DC2	DC3
1	18	25	28	17	30	21
2	29	27	16	25	22	25

Table 4: The rental costs ($/unit) of each truck during the transportation between distribution centers and sales and collection centers.

	Truck-1				Truck-2			
DC	SCC1	SCC2	SCC3	SCC4	SCC1	SCC2	SCC3	SCC4
1	8	12	12	15	14	12	10	8
2	8	18	7	14	7	16	11	10
3	8	8	11	7	14	9	8	14

Table 5: The CO_2 gas emissions (gr/unit) during the transportation between factories and distribution centers.

	Truck-1		Truck-2		Truck-1			Truck-2		
S	F1	F2	F1	F2	DC1	DC2	DC3	DC1	DC2	DC3
1	1.87	1.32	1.64	0.09	1.84	1.74	0.7	0.6	1.37	0.66
2	0.74	0.78	0.24	1.59	1.18	0.59	0.23	1.9	0.87	1.52
3	2	0.85	1.34	0.1	0.17	0.4	1.53	1.29	1.29	0.08
4	1.49	0.26	1.94	0.38	0.37	1.55	0.64	0.56	1.13	1.39

Table 6: The CO_2 gas emissions (gr/unit) during the transportation between factories and distribution centers.

	Truck-1			Truck-2		
F	DC1	DC2	DC3	DC1	DC2	DC3
1	1.52	1.51	0.21	1.72	1.01	0.22
2	0.9	0.94	1.29	0.15	1.66	1.66

Table 7: The CO_2 gas emissions (gr/unit) during the transportation between distribution centers and sales and collection centers.

	Truck-1				Truck-2			
DC	SSC1	SSC2	SSC3	SSC4	SSC1	SSC2	SSC3	SSC4
1	0.27	1.44	0.27	0.06	0.55	1.06	0.92	1.38
2	1.78	0.07	0.79	1.56	0.82	1.84	1.47	0.98
3	0.98	1.07	0.17	1.51	0.09	0.15	0.57	1.2

Table 8: The demand for new components (nc) and used components (uc)

SSC	nc1	nc2	nc3	nc4	nc5	uc1	uc2	uc3	uc4	uc5
1	5	12	11	6	12	18	15	17	15	13
2	12	11	9	5	7	13	18	13	14	16
3	8	6	5	4	8	17	17	16	14	17
4	4	7	5	6	7	16	14	14	16	16

4. Model Result

The mathematical model is solved by using GAMS—Cplex 24.3 solver subroutine in 0.047 second. All the experiments are conducted on a notebook with Intel Pentium i7-55000 CPU 2.40 GHz and 8 GB Ram. The following result indicates below tables.

According to result, as shown in Table 10; the total revenue is 1332558.10 $, the total cost is 128324.54 $ and the optimal objective is 1204233.56 $. Renting cost and fixed cost of facilities have one of the highest costs in the total cost and objective values are shown in Figure 2. CO_2 emissions of all trucks are totally 11931.04 $ during the all transportation in forward logistics. According to data obtained GAMS package program, objective function values are given Table 9 and result of decision variables presents below Table 9.

Table 9: The Optimal solution of numerical example.

$Y_{4,3,1,1}$	80	$X_{2,1,2,2}$	95	$W_{3,3,5,1}$	25	$Z_{2,3,1}$	142
$Y_{2,3,2,1}$	89	$X_{2,1,3,2}$	380	$W_{3,4,5,1}$	6	$Z_{1,3,2}$	95
$Y_{4,3,3,1}$	73	$X_{2,1,4,2}$	190	$W_{3,1,1,2}$	23	$V_{3,3,1}$	61
$Y_{2,3,4,1}$	49	$X_{2,1,5,2}$	95	$W_{3,2,1,2}$	25	$V_{3,4,1}$	74
$Y_{4,3,4,1}$	19	$W_{3,3,1,1}$	25	$W_{3,1,2,2}$	27	$V_{3,1,2}$	57
$Y_{2,3,5,1}$	79	$W_{3,4,1,1}$	7	$W_{3,2,2,2}$	29	$V_{3,2,2}$	45
$X_{4,2,1,1}$	142	$W_{3,3,2,1}$	23	$W_{3,1,3,2}$	28	$SS_{4,1}$	13
$X_{4,2,2,1}$	142	$W_{3,4,2,1}$	10	$W_{3,2,3,2}$	22	$SS_{4,2}$	11
$X_{4,2,3,1}$	568	$W_{3,3,3,1}$	21	$W_{3,1,4,2}$	21	$SS_{4,3}$	17
$X_{4,2,4,1}$	284	$W_{3,4,3,1}$	2	$W_{3,2,4,2}$	19	$SS_{4,4}$	12
$X_{4,2,5,1}$	142	$W_{3,3,4,1}$	18	$W_{3,1,5,2}$	25	$SS_{4,5}$	17
$X_{2,1,1,2}$	95	$W_{3,4,4,1}$	10	$W_{3,2,5,2}$	23		

Table 10: Objective Function values (OBJ).

Objective Function	**1204233.56**
Total Revenue (OBJ1)	1332558.1
Total Cost (OBJ2)	128324.54
Purchasing Cost (OBJ21)	12102
Manufacturing Cost (OBJ22)	11756.5
Assemble Cost (OBJ221)	5451
Repair Cost (OBJ222)	5918
Disassemble Cost (OBJ223)	387.5
Transportation Cost (OBJ23)	1164.3
Waste Cost (OBJ24)	58.7
Renting Cost (OBJ25)	48542
Carbon emission Cost (OBJ26)	11931.04
Fixed Cost (OBJ27)	42770

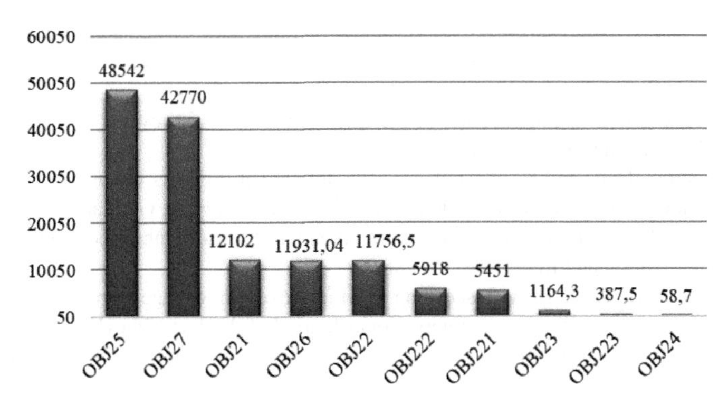

Fig. 2: Total costs of each objective functions.

5. Conclusion and Future Studies

IoT influences in the sustainable supply chain is still at the initial stage where the study suggests that supply chain industry should step up to extract the benefits of next-generation technologies.

This study introduces an SSC network design with respect to the Internet of Thing application. IoT devices monitored and collected static and dynamic information about the product life cycle. Sensor and RFID embedded products are emerging communication and data storage technology products that have provided an advantage in SSC operations, including disassembly, remanufacturing and repair process. These devices eliminate ambiguous information due to the uncertainty involved in the reverse flow of the returned products.

We proposed a multi-components, multi-echelon, and capacitated linear mathematical model to solve the SSC problems which consider the sustainable impact on SCN design. The effectiveness of the generated optimization model is tested by solving an example. It was solved using the GAMS/CPLEX 23.3 optimization tool. Both environmental and economic objectives can be achieved with the same system by utilizing a sustainable supply chain model.

In this chapter, we contribute a new sustainable supply chain model considering carbon emission during transportation and IoT applications, which provide exact recovery information. The contributions of this study are:

- ✓ Minimizing the transportation costs, purchasing costs of components from suppliers, dismantling costs/remanufacturing cost, opening costs of facilities, disposal cost of products, renting costs of trucks,
- ✓ Minimizing the total CO_2 emission costs,
- ✓ This study discusses the importance of digitalization and the influence of IoT in the overall SSCM.
- ✓ In the sustainable supply chain network, it was the first mathematical model established using the IoT.

Further researches indicate that this proposed model will be a more complex network by adding a new supplier, distribution center and plants. Therefore, Metaheuristics such as genetic algorithms can be developed to solve this mixed-integer-programming model in a reasonable time with increasing problem sizes. Also, this model can be enhanced with a fuzzy modeling approach to overcome uncertainty in customer demands. The multi-objective mathematical model should be applied to solve this model by considering maximizing the customer satisfaction level.

References

Bouzembrak, Y., H. Allaoui, G. Goncalves and H. Bouchriha. 2011. A multi-objective green supply chain network design. 4th International Conference on Logistics (LOGISTIQUA), Hammanet, Tunisia.

Cao, C.Z. and G.H. Zhao. 2014. Decision-making model of network optimization in green supply chain. Advanced Materials Research, 962: 2277–2282.

Chibeles-Martins, N., T. Pinto-Varela, A.P. Barbosa-Póvoa and A.Q. Novais. 2016. A multi-objective meta-heuristic approach for the design and planning of green supply chains. Expert Systems with Applications, 47: 71–84.

Coskun, S., L. Ozgur, O. Polat and A. Gungor. 2016. A model proposal for green supply chain network design based on consumer segmentation. Journal of Cleaner Production, 110: 149–157.

Elhedhli, S. and R. Merrick. 2012. Green supply chain network design to reduce carbon emissions. Transportation Research Part D: Transport and Environment, 17: 370–379.

Govindan, K., A. Jafarian and V. Nourbakhsh. 2015. Bi-objective integrating sustainable order allocation and sustainable supply chain network strategic design with stochastic demand using a novel robust hybrid multi-objective metaheuristic. Computers & Operations Research, 62: 112–130.

Gu, Y. and Q. Liu. 2013. Research on the application of the Internet of Things in reverse logistics information management. Journal of Industrial Engineering Management, 6: 963–973.

Hribernik, K., M. Stietencron C. Hans and K.-D. Thoben. 2011. Intelligent products to support closed-loop reverse logistics. 18th CIRP International Conference on Life Cycle Engineering. Braunschweig, Germany.

Ilgin, M. A. and S.M. Gupta. 2011. Performance improvement potential of sensor embedded products in environmental supply chains. Resources Conservation and Recycling, 55: 580–592.

Kiritsis, D., V. Nguyen and J. Stark. 2008. How closed-loop PLM improves knowledge management over the complete product lifecycle and enables the factory of the future. International Journal of Product Lifecyle Management, 3(1): 54–77.

Kiritsis, D. 2011. Closed-loop PLM for intelligent products in the era of the Internet of Things. Computer Aided Design, 43.5 : 479–501.

Memari, A., A.R.A. Rahim and R.B. Ahmad. 2015. An integrated production-distribution planning in green supply chain: a multi-objective evolutionary approach. Procedia CIRP, 26: 700–705.

Ondemir, O., M.A. Ilgin and S.M. Gupta. 2012. Optimal end of life management in closed loop supply chain using RFID and sensors. IEEE Transactions on Industrial Informatics, 8: 719–728.

Ondemir, O. and S.M. Gupta. 2014a. A multi-criteria decision-making model for advanced repair to order and disassembly to order system. European Journal of Operational Research, 223: 408–419.

Ondemir, O. and S.M. Gupta. 2014b. Quality management in product recovery using the Internet of Things: An optimization approach. Computers in Industry, 65(3): 491–504.

Paksoy, T., E. Ozceylan, G.-W. Weber, N. Barsoum, G. Weber and P. Vasant. 2011. A multi objective model for optimization of a green supply chain network. Glob. J. Technol. Optim., 2: 84–96.

Paksoy, T. and E. Özceylan. 2013. Environmentally conscious optimization of supply chain networks. Journal of the Operational Research Society, 65(6): 855–872.

Parlikad, A.K. and D. McFarlane. 2007. RFID-based product information in end-of-life decision making. Control Engineering Practice, 15(11): 1348–1363.

Pinto, T., A.P. Barbosa-Póvoa and A.Q. Novais. 2010. Supply chain network optimization with environmental impacts. In 2nd International Conference on Engineering Optimization. Lisbon, Portugal.

Torğul, B., L. Şağbanşua and F.B. Balo. 2016. Internet of things: a survey. International Journal of Applied Mathematics Electronics and Computers, Special Issue-1: 104–110.

Vermesan, O. and P. Friess. 2013. Internet of Things: Converging Technologies for smart enviroments and Integrated Ecosystems. Aalborg, Denmark.

SECTION 3

Fuzzy Decision Making in SCM

Chapter 6

Fuzzy Decision Making in SCM

Fuzzy Multi Criteria Decision Making for Supplier Selection

Belkız Torğul,[1,*] *Turan Paksoy*[1] and *Sandra Huber*[2]

1. Introduction

In recent years, mindfulness of the essential role of purchasing has increased for protection of competition by a company. The most important task of the purchasing process is the selection of an innovative supplier with high quality, low cost, on-time delivery capabilities (Tidwell and Sutterfield 2012). In the new global market, managers have recognized the importance of selecting the most suitable suppliers among alternatives and emphasized it is a critical factor for the success of companies (Galankashi et al. 2016a). This decision significantly affects an organization's overall supply chain performance. Selecting the right suppliers will reduce purchasing costs, enhance sustainable relationships, decrease production lead-time, increase customer satisfaction, profitability and quality of products and competitiveness in the market. Consequently, selecting the right suppliers requires much more than scanning a set of price lists and depends on a wide range of factors, both quantitative and qualitative (Wang and Yang 2009; Ho et al. 2010; Mavi et al. 2016).

Actually, two types of supplier selection are prevalent; single supplier and multiple supplier. In the single supplier type, a supplier meets the needs of all buyers, so the buyer selects only the best supplier.

In the multiple suppliers' type, which is more common, more than one supplier is selected. Therefore, businesses should select the best suppliers and determine how much to order from each of the suppliers to create a stable competitive environment (Kannan et al. 2013). Supplier selection requires multiple goals and criteria to be considered, so it is a complex multi-criteria decision-making (MCDM) process that takes into account both quantitative and qualitative factors (Junior et al. 2014; Govindan et al. 2015).

However, many decision makers choose suppliers based on their experience, knowledge and intuition. In a real situation, many inputs are not exactly known. In the evaluation of criteria for supplier selection, fuzzy logic approach is used in cases where it is not possible to determine the uncertainty of information and judgments by deterministic methods.

[1] Konya Technical University, Department of Industrial Engineering, Konya, Turkey.
Emails: tpaksoy@yahoo.com

[2] Operations Research Institute, Planungsamt, München, Germany, Email: hubersandra100@googlemail.com

* Corresponding author: belkistorgul@gmail.com

Uncertainty in critical information creates problems in reflecting the real situation to the model. Supplier selection in a fuzzy environment is the case where there is no fuzziness in the set of decision-makers and alternatives, but the objectives and decision criteria may contain fuzziness. The decision maker may determine the access level of the objective function as fuzzy. In addition, parameter values (profit, cost, etc.) can be defined in fuzzy numbers. Fuzzy set theory in combination with MCDM methods have been widely used to deal with uncertainty in supplier selection decision-making, because it provides a suitable language to handle fuzzy criteria that can integrate the analysis of qualitative and quantitative factors (Amid et al. 2006; Kumar et al. 2006). The most popular Fuzzy MCDM approaches adopted in the supplier evaluation and selection literature are, Fuzzy AHP, Fuzzy TOPSIS, Fuzzy ANP, Fuzzy VIKOR, Fuzzy DEA, Fuzzy mathematical programming and their hybrids.

The current literature on supplier selection is extensive. Previous studies focused on defining the criteria used in supplier selection, such as Dickson (1966), which is one of the earliest studies in supplier selection, identified 23 supplier criteria. Weber et al. (1991) reviewed 74 articles from 1966 to 1991 on supplier selection criteria and methods. Stamm and Golhar (1993) identified common supplier evaluation criteria. Taherdoost and Brard (2019) provides a comprehensive picture of research relating to supplier selection criteria and supplier evaluation methods.

Further studies generally focused on the MCDM approaches and mathematical models supporting decision-making used in the supplier selection process. Such as De Boer et al. (2001), Ho et al. (2010), Chai et al. (2013), Wetzstein et al. (2016), etc. Among these, Lee et al. (2011), Genovese et al. (2013), Igarashi et al. (2013), Govindan et al. (2015) and Jenssen and de Boer (2019) focus only on green supplier selection. Zimmer et al. (2016) focus only on sustainable supplier selection.

Karsak and Dursun (2016), Keshavarz Ghorabaee et al. (2017a), Simić et al. (2017), Ozkok and Kececi (2019) reviewed fuzzy set theory and models, MCDM approaches and mathematical models in fuzzy environments for supplier selection. Karsak and Dursun (2016) reviewed stochastic methods too.

All of the papers where we have found from review studies made so far on supplier evaluation and selection are listed in Table 1.

The main aim of this chapter is to provide a detailed overview of Fuzzy Logic and Fuzzy Decision Making models for the supplier selection process. Moreover, we aim to analyses the articles with respect to date of publication, the journal title, studied industry, supplier type and fuzziness type.

The remainder of this chapter is organized as follows: Section 2 presents classification of literature review with research methodology. This section provides a detailed literature review on the single and hybrid fuzzy decision-making approaches for selecting suppliers. In Section 3, some analyses of reviewed papers are made to show the most frequent approach, the dates of publication and distribution of papers by journals in this field. Section 4 provides observations and discussions about the results of this study and conclusions are presented in Section 5.

2. Classification of Literature Review

This paper extends existent literature reviews and provides an up-to-date version by surveying the supplier evaluation and selection literature from 2000 to 2019 while focusing on fuzzy logic and fuzzy decision making in SCM. In this paper, only English written 310 scientific papers (225 Journal paper, 18 Book Chapter and 67 Conference paper) published in refereed journals, books and conference proceedings between 2000 and 2019 are reviewed. There is a remarkable growth in the number of papers published between 2008 and 2019, and 96% of the papers considered in this survey were published during about the last 10–year period of the review. Data were sought through various sources including Web of Science, Science Direct, Springer, Scopus, Taylor & Francis, IEEE Explore Digital Library and Google Scholar database. The following keywords were used in some form: supplier evaluation, supplier selection, vendor selection and fuzzy. After collecting the literature, the published papers were categorized into two main categories based on the type and frequency of fuzzy MCDM approaches and other methods; single and hybrid approaches. Then some subcategories were defined for these two categories.

As found in literature, the single approaches (242 papers—78%) were slightly more popular than the hybrid approaches (68 papers—22%). The next two sub-sections present single fuzzy approaches and hybrid fuzzy approaches in detail.

2.1 Single Approaches

Researchers have developed many multi-criteria decision-making (MCDM) and mathematical programming (MP) approaches in fuzzy environment. In this section, single fuzzy approaches applied to supplier selection are reviewed. The Fuzzy AHP, Fuzzy ANP, Fuzzy TOPSIS, Fuzzy VIKOR, Fuzzy MOORA, Fuzzy ELECTRE, Fuzzy DEMATEL, Fuzzy DEA, Fuzzy PROMETHEE, Fuzzy BWM, Fuzzy QFD and Fuzzy MP, which appear more frequently in the literature, are

Table 1: Existent literature reviews on supplier selection.

S. no	Author(s)	Year	Number of reviewed articles	Time range	Type of publication	Research subject
1	(Dickson)	1966	-	-	Journal paper	Vendor selection systems and decisions
2	(Weber et al.)	1991	74	1996–1990	Journal paper	Vendor selection criteria and methods
3	(Stamm and Golhar)	1993	56	–	Journal paper	Supplier evaluation criteria
4	(Degraeve et al.)	2000	-	?–1999	Journal paper	Vendor selection models
5	(De Boer et al.)	2001	-	?–2000	Journal paper	Methods supporting supplier selection
6	(Aissaoui et al.)	2007	-	?–2005	Journal paper	Supplier selection and order lot sizing modeling
7	(Jain et al.)	2009	-	?–2007	Journal paper	Methods supporting supplier selection
8	(Thanaraksakul and Phruksaphanrat)	2009	76	?–2009	Conference paper	Supplier evaluation framework based on balanced scorecard with integrated corporate social responsibility perspective
9	(Ho et al.)	2010	78	2000–2008	Journal paper	MCDM approaches for supplier evaluation and selection
10	(Wu and Barnes)	2011	140	2001–2011	Journal paper	Decision-making models and approaches for partner selection
11	(Deshmukh and Chaudhari)	2011	49	1992–2007	Book Chapter	Supplier selection criteria and methods
12	(Lee et al.)	2011	20	1997–2009	Journal paper	Key criteria for green supplier selection
13	(Genovese et al.)	2013	28	1987-2010	Journal paper	Greener supplier selection: state of the art and some empirical evidence
14	(Chai et al.)	2013	123	2008–2012	Journal paper	Analyzing the process of supplier selection criteria and methods
15	(Igarashi et al.).	2013	60	1991–2011	Journal paper	Requirements on greener supplier selection and conceptual model development
16	(Govindan et al.)	2015	33	1997–2011	Journal paper	MCDM approaches for green supplier evaluation and selection
17	(Kocken and Ozkok)	2015	-	–	Book Chapter	Multi criteria decision making approaches
18	(Wetzstein et al.)	2016	221	1990–2015	Journal paper	A systematic assessment on supplier selection
19	(Zimmer et al.)	2016	143	1997–2014	Journal paper	Models supporting sustainable supplier selection
20	(Karsak and Dursun)	2016	149	2001–2013	Journal paper	Non-deterministic analytical methods for supplier selection
21	(Keshavarz Ghorabaee et al.)	2017a	339	2001–2016	Journal paper	MADM approaches in fuzzy environments on supplier evaluation and selection
22	(Simić et al.)	2017	54	1966–2016	Journal paper	Fuzzy set theory and models for supplier assessment and selection
23	(Taherdoost and Brard)	2019	-	?–2017	Conference/ Journal paper	Analyzing the process of supplier selection criteria and methods
24	(Jenssen and de Boer)	2019	39	2001–2018	Journal paper	Implementing life cycle assessment in green supplier selection
25	(Ozkok and Kececi)	2019	-	?–2019	Book Chapter	Supplier selection problem methods under uncertainty

considered individually, and the other single approaches are reviewed in a separate section. The most popular individual approach is FAHP, followed by FTOPSIS and F-MP.

2.1.1 *Fuzzy AHP (Analytic hierarchy process)*

Bottani and Rizzi (2005) defined relevant criteria added to the traditional one in an e-procurement environment for supplier selection and used FAHP to rank raw materials, maintenance, repair and operating suppliers for a major Italian company operating in the food industry. Pang (2006) and Pang (2007) provided a comprehensive evaluation method which combines both fuzzy sets and FAHP for evaluating suppliers of a process plant whose main product is graded pig iron. Four decision criteria of suppliers are selected and five potential suppliers are evaluated in the case study. Chan and Kumar (2007) presented FAHP to select the best supplier selling one of the most critical parts for a manufacturing company. To this end, they had discussed and identified some critical decision criteria, including risk factors, for the development of an effective global supplier selection system. Similarly, Chan et al. (2008) presented FAHP to effectively address both the quantitative and qualitative decision criteria involved in the selection of global suppliers for a manufacturing company.

Kong et al. (2008) presented a fuzzy decision-making approach to deal with the supplier selection problem. They determined the weighting of subjective judgments with FAHP and evaluated suppliers with the grey relation model. Yang et al. (2008) proposed an integrated fuzzy multiple criteria decision making method which consists of interpretive structural modeling, FAHP and non-additive fuzzy integral for a vendor selection problem. Lee et al. (2009a) applied the Delphi method to differentiate the criteria of traditional suppliers and green suppliers and then they exploited the fuzzy extended analytic hierarchy process to evaluate green suppliers for an anonymous TFT–LCD manufacturer in Taiwan. Similarly, Lee (2009) propose a FAHP model which incorporates the benefits, opportunities, costs and risks concept to select backlight unit's suppliers for the largest TFT-LCD manufacturers in Taiwan. Şen et al. (2010) presented a methodology, which consists of two steps for the supplier selection problem of Audio Electronics, a company in Turkey's electronics industry. Pre-defined supplier selection criteria (developed in Şen et al. (2008)) are weighted using the FAHP in the first step. 10 alternative suppliers are evaluated using a combination of a max–min approach and a non-parametric statistical test and finally, five effective suppliers are determined in the second step.

Based on the opinions of 15 Taiwanese companies that are well-known electronics manufacturers investing in China and other countries, Chiouy et al. (2011) prioritized various performance evaluation criteria for sustainable supplier selection and evaluation in the Taiwanese electronics industry by the FAHP method. Çifçi and Büyüközkan (2011) presented a decision framework based on group decision making and FAHP for a green supplier selecting problem. Kilincci and Onal (2011) investigated the supplier selection problem of a white good manufacturer in Turkey and applied FAHP to select the best supplier firm for one its critical parts used in the production of washing machines. Ertay et al. (2011) proposed an integrated decision support system methodology for supplier selection and evaluation and applied it for a real-life supplier-selection problem of a pharmaceutical company. They used Fuzzy AHP to weight criteria and ELECTRE III to rank suppliers. Punniyamoorthy et al. (2011) developed a new composite model which examines the supplier selection criteria and rank the suppliers using structural equation modeling and FAHP technique, based on the results of a survey of 151 respondents and demonstrated applicability of the model by evaluating five suppliers of a public sector company in the southern part of India whose main product of manufacture is boiler.

Azadnia et al. (2012) proposed an integrated approach for clustering and selecting suppliers. They used self-organizing map as a kind of neural network method in order to cluster suppliers. Moreover, they used FAHP in order to determine the weights of sustainable criteria and applied TOPSIS to select the best cluster of suppliers and the best of them. Rezaei et al. (2014) proposed a two-phased methodology for the supplier selection problem. In the first phase, they used a conjunctive screening method, which aims to reduce the initial set of potential suppliers and in the second phase, they used FAHP, in which suppliers are evaluated by the main and sub-criteria. Finally, they applied the methodology for choosing the best supplier of one of the largest airlines in Europe, the Royal Dutch Airlines. Gold and Awasthi (2015) proposed a two-stage fuzzy AHP approach for sustainable global supplier selection that also considers sustainability risks from sub-suppliers. They performed more than one experiment at each stage and tried to observe whether the order of the supplier changed according to the results. Kar (2015) presented the application of a hybrid approach using fuzzy AHP for prioritizing evaluation criteria and subsequently using fuzzy NN for selecting the suppliers in the supplier selection problem. He tested the group decision-supported model with an iron and steel manufacturing company based out of India.

Galankashi et al. (2016) developed an integrated Balanced Scorecard–FAHP model for the supplier selection problem in the automotive industry. They gathered measures using a literature survey and qualified them using the Nominal Group Technique. Finally, they used FAHP to select the best supplier. Büyüközkan and Göçer (2017a) proposed a new integrated methodology that consists of intuitionistic fuzzy analytic hierarchy process (IFAHP) and intuitionistic fuzzy axiomatic design (IFAD). They used IFAHP to determine the weights of supplier evaluation criteria and IFAD to rank supplier alternatives for five-alternative supplier selection process of an international sporting goods group operating in Turkey. Then they compared their own approach with IFTOPSIS, integrated IFAHP and IFTOPSIS, and integrated IFAHP and IFVIKOR. Lu et al. (2019) established a decision-making framework based on the Cloud model, possibility degree and Fuzzy AHP for green supplier selection problems. They applied the model to evaluate four alternative suppliers of a straw biomass industry in China by identifying four main and 13 sub-criteria.

Above, some of the studies implementing the FAHP approach are presented. Table 2 presents all articles in which the FAHP method as a single approach to supplier selection is investigated.

Table 2: Review summary of supplier selection studies using Fuzzy AHP method.

S. no	Author(s)	Year	Type of publication	Type of paper	Type of fuzziness
1	(Bottani and Rizzi)	2005	Journal paper	Real case study	Classical sets
2	(Li et al.)	2006	Conference paper	Theoretical study	Classical sets
3	(Pang)	2006	Conference paper	Real case study	Classical sets
4	(Pang)	2007	Conference paper	Real case study	Classical sets
5	(Chan and Kumar)	2007	Journal paper	Theoretical study	Extended classical sets
6	(Chan et al.)	2008	Journal paper	Theoretical study	Classical sets
7	(Kong et al.)	2008	Conference paper	Theoretical study	Classical sets
8	(Yang et al.)	2008	Journal paper	Real case study	Classical sets
9	(Bai)	2008	Conference paper	Theoretical study	Classical sets
10	(Pang)	2008	Conference paper	Real case study	Classical sets
11	(Xi and Jiang)	2008	Book Chapter	Real case study	Classical sets
12	(Zhao and Xu)	2008	Conference paper	Theoretical study	Classical sets
13	(Lee)	2009	Journal paper	Real case study	Classical sets
14	(Lee et al.)	2009a	Journal paper	Real case study	Extended classical sets
15	(Şen et al.)	2010	Journal paper	Real case study	Classical sets
16	(Kahraman and Kaya)	2010	Book Chapter	Real case study	Classical sets
17	(Kang et al.)	2010	Conference paper	Theoretical study	Classical sets
18	(Koul and Verma)	2011	Journal paper	Theoretical study	Classical sets
19	(Azadnia et al.)	2011	Conference paper	Real case study	Classical sets
20	(Chiouy et al.)	2011	Journal paper	Theoretical study	Classical sets
21	(Ertay et al.)	2011	Journal paper	Real case study	Classical sets
22	(Aktepe and Ersoz)	2011	Journal paper	Real case study	Classical sets
23	(Kilincci and Onal)	2011	Journal paper	Real case study	Classical sets
24	(Punniyamoorthy et al.)	2011	Journal paper	Real case study	Classical sets
25	(Çifçi and Büyüközkan)	2011	Journal paper	Real case study	Classical sets
26	(Rahman and Ahsan)	2011	Conference paper	Real case study	Classical sets
27	(Tang and Fang)	2011	Conference paper	Real case study	Classical sets
28	(Li et al.)	2012	Journal paper	Theoretical study	Axiomatic Sets
29	(Azadnia et al.)	2012	Journal paper	Real case study	Classical sets
30	(Pitchipoo et al.)	2013	Journal paper	Real case study	Classical sets
31	(Alinezad et al.)	2013	Journal paper	Real case study	Classical sets
32	(Harnisch and Buxmann)	2013	Conference paper	Theoretical study	Classical sets
33	(Rezaei et al.)	2014	Journal paper	Real case study	Classical sets
34	(Gold and Awasthi)	2015	Journal paper	Theoretical study	Classical sets
35	(Kar)	2015	Journal paper	Real case study	Classical sets
36	(Nikou and Moschuris)	2015	Book Chapter	Real case study	Classical sets
37	(Yadav and Sharma)	2015	Book Chapter	Real case study	Classical sets
38	(Liu)	2015	Conference paper	Real case study	Classical sets
39	(Yasrebdoost)	2015	Book Chapter	Real case study	Classical sets
40	(Deepika and Kannan)	2016	Conference paper	Theoretical study	Intuitionistic sets
41	(Galankashi et al.)	2016b	Journal paper	Theoretical study	Classical sets
42	(Büyüközkan and Göçer)	2017	Journal paper	Real case study	Intuitionistic sets

Table 2 contd. ...

...Table 2 contd.

S. no	Author(s)	Year	Type of publication	Type of paper	Type of fuzziness
43	(Sarwar et al.)	2017	Conference paper	Theoretical study	Classical sets
44	(Chaising and Temdee)	2017	Conference paper	Real case study	Classical sets
45	(Krishankumar et al.)	2017	Conference paper	Real case study	Intuitionistic sets
46	(Zafar et al.)	2018	Conference paper	Theoretical study	Classical sets
47	(Ayhan)	2018	Journal paper	Theoretical study	Hesitant fuzzy axiomatic design
48	(Diouf and Kwak)	2018	Journal paper	Real case study	Classical sets
49	(Lu et al.)	2019	Journal paper	Real case study	Classical sets
50	(Deshmukh and Vasudevan)	2019	Conference paper	Real case study	Classical sets
51	(Deshmukh and Sunnapwar)	2019	Conference paper	Real case study	Classical sets
52	(2019)	2019	Conference paper	Real case study	Classical sets
53	(Hu et al.)	2019	Journal paper	Real case study	Double quantitative fuzzy rough sets
54	(Mondragon et al.)	2019	Journal paper	Real case study	Classical sets
55	(Li,Sun, et al.)	2019	Conference paper	Theoretical study	Classical sets
56	(Wang et al.)	2019	Journal paper	Real case study	Classical sets
57	(Buriticá et al.)	2019	Journal paper	Real case study	Classical sets
58	(Agrawal and Kant)	2019	Book Chapter	Real case study	Classical sets

2.1.2 Fuzzy ANP (Analytic network process)

Razmi et al. (2009) developed a framework based on FAHP to evaluate and select the potential suppliers and augmented a non-linear programming model to the model to elicit the relative weights from fuzzy comparison matrices. Wei and Sun (2009) and Wei et al. (2010) showed the evaluation process of supplier selection with FANP calculating the weights of each criteria of the model. Büyüközkan and Çifçi (2011) used a new fuzzy MCDM framework based on fuzzy ANP for a sustainable supplier evaluation problem and analyzed the evaluation model in a real-life problem of a main producer of a Turkish white goods industry. Vinodh et al. (2011) used fuzzy ANP approach for the supplier selection process and carried it out in Salzer Electronics Limited, which is an Indian electronics switches manufacturing company.

Kang et al. (2012) proposed a FANP model to evaluate various aspects of suppliers and presented a case study of IC packaging company selection in Taiwan. Pang (2009) and Pang and Bai (2013) developed a supplier evaluation approach based on FANP and fuzzy synthetic evaluation for choosing a supplier in the supply chain. Sinrat and Atthirawong (2013) developed a conceptual framework model based on an integrated of FANP model and TOPSIS for evaluation procurement risk, production risk, deliver risk, and environment risk of suppliers.

Dargi et al. (2014) proposed a framework comprising of the most critical factors for the aim of a reliable supplier selection for an Iranian automotive industry. They deployed the Nominated Group Technique to summarize the most critical factors and then used FANP to weight the extracted eight criteria and determine their importance level. Galankashi et al. (2015) deployed the Nominal Group Technique to extract the most important performance measures (both classical and green) and then deployed FANP to weight these measures. Zhang et al. (2015) developed a fuzzy extended ANP methodology to deal with supplier selection problems. Ayağ and Samanlioglu (2016) proposed an intelligent fuzzy ANP-based approach to supplier selection problem and presented a case study in a leading company in the automotive sector that is in Turkey-Europe, which needs a practical evaluation system to rank supplier alternatives.

Parkouhi and Ghadikolaei (2017) proposed combination of BOCR model of Lee (2009) and the model introduced by Rajesh and Ravi (2015) and applied it for evaluating resilient suppliers of a large industrial unit in the Wood and Paper Industry. They employed FANP to weigh criteria and used grey VIKOR to identify the most resilient suppliers. Wang et al. (2018) collected data from 25 potential suppliers of the rice supply chain in Vietnam, and the four main criteria within contain 15 sub-criteria to define the most effective supplier. They used FANP to evaluate these criteria and DEA to rank suppliers. Chen et al. (2018) proposed a hybrid model that combines total interpretive structural modeling and FANP to determine the most appropriate supplier from a social responsibility perspective and illustrated an application using a case study from the Chinese food industry.

Sennaroglu and Akıcı (2019) presented FANP to select the best supplier among three alternatives in terms of ten decision criteria for the raw material requirement of a company in the chemical industry. Liao et al. (2019) proposed a model integrating the social participatory allocation network and the ANP under the hesitant fuzzy linguistic environment for low carbon supplier selection problem of a solar power company. Wang et al. (2019a) proposed a fuzzy MCDM model for the selection process of wind turbine supplier. They identified main criteria for selection of the wind turbine supplier by the SCOR metrics and literature review, applied Fuzzy ANP for identifying the weights of criteria, and then used TOPSIS to rank all potential suppliers. Finally, they considered the Wind Power Plant project in the Binh Thuan Province, Vietnam for application. Wahyuni et al. (2019) determined the criteria most influential in the choice of supplier PT Putra Gunung Kidul Company, which produces noodles and ranked three alternative suppliers using FANP.

2.1.3 Fuzzy TOPSIS (Technique for order of preference by similarity to ideal solution)

Chen et al. (2006) proposed a systematic approach based on FTOPSIS to solve the supplier-selection problem under a fuzzy environment. They considered five candidate suppliers and five criteria with three decision-makers for a high-technology manufacturing company. Boran et al. (2009) presented a multi-criteria group decision making model for supplier evaluation using intuitionistic fuzzy TOPSIS. Büyüközkan and Arsenyan (2009) proposed an axiomatic design based fuzzy group decision making approach for evaluating and selecting suppliers. They verified proposed approach through a case study of XYZ company known as a pioneering producer of the Turkish Apparel Industry. Firstly, seven main agile supplier evaluation criteria and four alternative suppliers are determined and then weights of criteria are determine using FTOPSIS. Fuzzy axiomatic design and FTOPSIS technique is applied separately for ranking the supplier alternatives. Finally, the results obtained were compared. Awasthi et al. (2010) presented a fuzzy multi-criteria approach consisting of three steps for evaluating environmental performance of suppliers. The first step involves identification of environmental criteria. The second step involves weighting selected criteria by experts and rating alternative suppliers against each of the criteria through fuzzy TOPSIS. The third step involves performing sensitivity analysis to determine the influence of criteria weights on the decision making process. Soner Kara (2011) proposed an integrated methodology based on a two stage stochastic programming model and fuzzy TOPSIS for supplier selection problems. Firstly, she evaluated 20 potential suppliers by using fuzzy TOPSIS and performed the two-stage stochastic programming method under demand uncertainty for a company whose application area is paper production.

Kilic (2013) developed a novel integrated approach including FTOPSIS and a mixed integer linear programming model to select the best supplier in a multi-item/multi-supplier environment. He used the importance value of each supplier obtained via FTOPSIS as an input in the mathematical model for determining the suppliers and the quantities of products to be provided from the related suppliers. Finally, he performed the proposed methodology in the air filter company, which is located in Istanbul. Roshandel et al. (2013) evaluated four suppliers of imported raw material “Tripolyphosphate” used to produce detergent powder for one of the largest producers of health products in Iran. They used the hierarchical FTOPSIS to rank four suppliers from South Korea, Spain, China and India based on 25 effective criteria with 10 experts. Rouyendegh and Saputro (2014) presented an integrated Fuzzy TOPSIS and Multi-Choice Goal Programming Model for supplier selection and allocation order. They suggested the best suppliers sourcing of white clay for a company producing fertilizers using the proposed method. Haldar et al. (2014) developed a quantitative approach for strategic supplier selection under a fuzzy environment and applied the FTOPSIS method to rank the suppliers of an automobile giant.

Azizi et al. (2015) discussed the selection of the best supplier in automotive industries using FTOPSIS. They considered five main criteria and 18 sub-criteria based on four alternative suppliers. Arabzad et al. (2015) developed a two-phase model for the supplier selection and order allocation problem of Gassouzan Company, which produce gas pressure regulator in Iran. In the first phase, they identified candidate suppliers and defined the evaluation criteria by considering strategic viewpoint. Then, they used SWOT to categorize criteria into two groups of external and internal. In the second phase, they utilized FTOPSIS to evaluate suppliers based on the criteria and then, they used results from FTOPSIS as an input for linear programming to allocate orders. Chatterjee and Kar (2016) developed an Interval valued FTOPSIS based method for handling supplier selection problem in uncertain Electronics Supply chain with six risk based criteria and four battery suppliers. Finally, they compared the results with some existing methods.

Fallahpour et al. (2017) determined the most important and applicable criteria and sub-criteria for sustainable supplier selection through a questionnaire-based survey and proposed a hybrid model incorporated Fuzzy Preference Programming used to weigh the criteria and FTOPSIS used for ranking the suppliers for identifying the best one with respect to the determined attributes using an Iranian textile manufacturing company as case study. Gupta and Barua (2017) proposed a novel three-phase methodology for supplier selection framework. The first phase involves the selection of criteria of green innovation through literature review and interviews with decision makers, the second phase involves ranking of selection criteria using best worst method, and third phase involves ranking of suppliers with respect to selection criteria weights using FTOPSIS. They applied the proposed approach in a leading automobile company adopting green practices. Hamdan and Cheaitou (2017) proposed a three stages approach—fuzzy TOPSIS, AHP, and a bi-objective integer linear programming model—to solve a multi-period green supplier selection and order allocation problem with all unit quantity discounts in which the availability of suppliers differs from one period to another. They solved the proposed mathematical model by MATLAB R2014, a software using the weighted comprehensive criterion method and the branch-and cut algorithm. Kumar et al. (2018) applied the FTOPSIS model to evaluate suppliers of an iron and steel manufacturing unit in eastern part of India and then performed sensitivity analysis to investigate the effect of criteria weights on selection of supplier.

Li et al. (2019) developed an extended FTOPSIS method for sustainable supplier selection, which integrates the advantage of cloud model theory in manipulating uncertainty of randomness and the merit of rough set theory in flexibly handling interpersonal uncertainty without extra information. Additionally, they proposed an integrated weighting method to determine weights of criteria. Finally, they conducted selection of a sustainable photovoltaic modules supplier for A Chinese state-owned energy company. Memari et al. (2019) presented an intuitionistic FTOPSIS to select the best sustainable supplier for an automotive spare parts manufacturer and conducted a FTOPIS with nine different scenarios to measure the sensitivity of the proposed method. Mohammed (2019) presented an integrated fuzzy TOPSIS-possibilistic multi objectives model to solving a two-stage sustainable supplier selection and order allocation problem for a meat supply chain. He determined sustainable performance of suppliers by using the FTOPSIS based on traditional, green and social criteria and then integrated into the possibilistic multi objective model for obtaining the optimal order allocation in quantity. Finally, he applied LP-metrics approach to reveal a number of Pareto solutions based on the developed model. Yu et al. (2019) proposed a group decision making sustainable a supplier selection approach using interval-valued Pythagorean FTOPSIS and conducted experiments to verify the feasibility and efficiency of the proposed approach.

Above, some of the studies implementing the FTOPSIS approach are presented. Table 3 presents all articles in which the FTOPSIS method as a single approach to supplier selection is investigated.

2.1.4 Fuzzy VIKOR (VIšeKriterijumska Optimizacija I Kompromisno Resenje—Multicriteria optimization and compromise solution)

Shemshadi et al. (2011) extended the fuzzy VIKOR method with a mechanism to extract and deploy objective weights based on the Shannon entropy concept for supplier selection processes. Zhao et al. (2013) presented a novel performance evaluation method based on combining the Interval-valued intuitionistic fuzzy VIKOR method and the cross-entropy measure for vendor selection problems and compared it with the TOPSIS method. You et al. (2015) proposed the Interval 2-tuple linguistic VIKOR method for supplier selection problems and demonstrated feasibility and practicability of the proposed approach through three realistic examples and comparisons with the existing approaches. Awasthi and Kannan (2016) presented an integrated approach for evaluating and selecting best green supplier development programs using a fuzzy Nominal Group Technique to identify criteria and FVIKOR to rank green suppliers. Wu et al. (2016) proposed an extended VIKOR based on cloud model for supplier selection in the nuclear power industry. They verified this method on supplier evaluation of a nuclear power plant in China and then compared it with fuzzy VIKOR.

Bahadori et al. (2017) provided a combined model for selecting the best supplier in a hospital using artificial neural network and FVIKOR. It was conducted for a military hospital in three phases in 2016. Zhou and Xu (2017) and Zhou and Xu (2018) proposed an integrated decision-making model for supplier selection. DEMATEL and ANP are used to find the criteria weights. Furtheron, an extended FVIKOR method is used to rank alternatives. Finally, they compared this model with FTOPSIS. Sarkar et al. (2017) proposed a multi-criteria decision making method using DEMATEL based on ANP, i.e., DANP, with FVIKOR to select the best supplier of a manufacturing company. Krishankumar et al. (2018) proposed a new two-stage decision-making framework for solving the supplier outsourcing problem. First stage; simple hesitant fuzzy-weighted geometry operator that uses hesitant fuzzy weights for better understanding the importance of each decision maker. Second stage; hesitant fuzzy statistical variance method that estimates criteria weights and three-way hesitant fuzzy VIKOR that ranks supplier outsourcings. Sharaf (2019) proposed a novel flexible multi-attribute group decision-making method for supplier selection based on interval-valued fuzzy VIKOR. Song and Wang (2019) proposed an interval intuitionistic fuzzy VIKOR method and applied this method to an automobile enterprise for selection of best auto parts supplier. Alikhani et al. (2019) proposed a multi-method approach based on the extended super-efficiency DEA model and Interval type-2 fuzzy VIKOR and illustrated this approach on the supplier selection of Shahrvand Goods & Servicing Company, which is the most advanced supermarket chain in Iran.

2.1.5 Fuzzy DEA (Data envelopment analysis)

Alem et al. (2009) presented three types of vendor selection models consisting of DEA, FDEA, and Chance Constraint DEA and a decision making scheme for choosing the appropriate method for supplier selection under certainty, uncertainty

Table 3: Review summary of supplier selection studies using Fuzzy TOPSIS method.

S. no	Author(s)	Year	Type of publication	Type of paper	Type of fuzziness
1	(Chen et al.)	2006	Journal paper	Theoretical study	Classical sets
2	(Fan et al.)	2008	Conference paper	Theoretical study	Classical sets
3	(Xiao and Wei)	2008	Conference paper	Theoretical study	Interval-Valued Intuitionistic sets
4	(Boran et al.)	2009	Journal paper	Theoretical study	Intuitionistic sets
5	(Büyüközkan and Arsenyan)	2009	Journal paper	Real case study	Axiomatic sets
6	(Yadav and Kumar)	2009	Conference paper	Theoretical study	Interval-Valued Intuitionistic sets
7	(Awasthi et al.)	2010	Journal paper	Theoretical study	Classical sets
8	(Guo et al.)	2010	Conference paper	Theoretical study	Intuitionistic sets
9	(Sevkli et al.)	2010	Conference paper	Real case study	Classical sets
10	(Soner Kara)	2011	Journal paper	Real case study	Classical sets
11	(Aghajani and Ahmadpour)	2011	Journal paper	Real case study	Classical sets
12	(Liao and Kao)	2011	Journal paper	Real case study	Classical sets
13	(Wang et al.)	2012	Conference paper	Theoretical study	Classical sets
14	(Zhang and Huang)	2012	Conference paper	Theoretical study	Intuitionistic sets
15	(Roshandel et al.)	2013	Journal paper	Real case study	Hierarchical sets
16	(Kilic)	2013	Journal paper	Real case study	Classical sets
17	(Shen et al.)	2013	Journal paper	Theoretical study	Classical sets
18	(Haldar et al.)	2014	Journal paper	Real case study	Classical sets
19	(Rouyendegh and Saputro)	2014	Journal paper	Real case study	Intuitionistic sets
20	(Kannan et al.)	2014	Journal paper	Real case study	Classical sets
21	(Kar et al.)	2014	Conference paper	Real case study	Classical sets
22	(Arabzad et al.)	2015	Journal paper	Real case study	Classical sets
23	(Azizi et al.)	2015	Journal paper	Theoretical study	Classical sets
24	(Bhayana et al.)	2015	Conference paper	Theoretical study	Classical sets
25	(Awasthi)	2015	Book Chapter	Real case study	Classical sets
26	(Igoulalene et al.)	2015	Journal paper	Real case study	Classical sets
27	(Orji and Wei)	2015	Journal paper	Real case study	Classical sets
28	(Hamdan and Cheaitou)	2015	Conference paper	Theoretical study	Classical sets
29	(Chatterjee and Kar)	2016	Journal paper	Real case study	Interval valued sets
30	(Wood)	2016	Journal paper	Theoretical study	Intuitionistic sets
31	(Wątróbski and Sałabun)	2016	Conference paper	Real case study	Classical sets
32	(Lima-Junior and Carpinetti)	2016	Journal paper	Theoretical study	Classical sets
33	(Mavi et al.)	2016	Journal paper	Theoretical study	Classical sets
34	(Nag and Helal)	2016	Conference paper	Real case study	Classical sets
35	(Solanki et al.)	2016	Conference paper	Theoretical study	Intuitionistic sets
36	(Fallahpour et al.)	2017	Journal paper	Real case study	Classical sets
37	(Gupta and Barua)	2017	Journal paper	Real case study	Classical sets
38	(Hamdan and Cheaitou)	2017	Journal paper	Theoretical study	Classical sets
39	(Mousakhani et al.)	2017	Journal paper	Real case study	Type-2 sets
40	(Mohammed et al.)	2017	Conference paper	Real case study	Classical sets

Table 3 contd. ...

...Table 3 contd.

S. no	Author(s)	Year	Type of publication	Type of paper	Type of fuzziness
41	(Wu et al.)	2017	Conference paper	Theoretical study	Hesitant sets
42	(Kumar et al.)	2018	Journal paper	Real case study	Classical sets
43	(Kumar and Singh)	2018	Book Chapter	Theoretical study	Classical sets
44	(Tian et al.)	2018	Journal paper	Real case study	Intuitionistic sets
45	(Cengiz Toklu)	2018	Journal paper	Real case study	Type-2 sets
46	(Dewi and Al Fatta)	2018	Conference paper	Real case study	Intuitionistic sets
47	(Yucesan et al.)	2019	Journal paper	Real case study	Interval Type-2 sets
48	(Liet al.)	2019	Journal paper	Real case study	Extendend rough sets
49	(Memari et al.)	2019	Journal paper	Real case study	Intuitionistic sets
50	(Mohammed)	2019	Journal paper	Real case study	Classical sets
51	(Yu et al.)	2019	Journal paper	Theoretical study	Interval-valued Pythagorean sets
52	(dos Santos et al.)	2019	Journal paper	Real case study	Classical sets
53	(Bera et al.)	2019	Journal paper	Real case study	Classical sets
54	(Rouyendegh et al.)	2019	Journal paper	Real case study	Intuitionistic sets
55	(Yadavalli et al.)	2019	Journal paper	Real case study	Z-numbers sets
56	(Chen)	2019	Journal paper	Real case study	Intuitionistic sets
57	(Abdel-Basset et al.)	2019	Journal paper	Real case study	Type-2 neutrosophic sets

and probabilistic conditions. Costantino et al. (2012) presented a novel cross efficiency FDEA technique for the supplier selection problem of an SME located in Southern Italy that provides, installs and maintains hydraulic plants. Ahmady et al. (2013) developed a novel fuzzy DEA approach with double frontiers to handle ambiguity and fuzziness for selecting suppliers. Amindoust and Saghafinia (2014) used an Affinity Diagram to obtain the criteria constituent a supplier should possess and then proposed a FDEA model based on α—cut approach to evaluate candidate suppliers according to the obtained criteria. Awasthi et al. (2014) presented a hybrid approach based on the Delphi technique, AHP and FDEA for supplier performance evaluation. First, they obtained supplier selection criteria using Delphi technique, and then determined hierarchy of criteria and relations between them using AHP, and last performed supplier performance evaluation using FDEA.

Azadi et al. (2015) developed a fuzzy DEA enhanced Russell measure model for evaluation of efficiency and effectiveness of suppliers in sustainable supply chain and presented a case study of a resin production company in Iran to exhibit the efficacy of the method. Zhou et al. (2016) developed a novel type-2 fuzzy multi-objective DEA model to evaluate and select the most appropriate sustainable suppliers and compared the model to the enhanced Russell measure DEA model and the type-1 DEA model. Azadeh, Rahimi et al. (2017) presented a decision-making scheme containing three techniques (DEA, FDEA, and stochastic DEA) for selecting an appropriate method for supplier selection under certainly, uncertainly, and stochastic conditions. Amindoust (2018) proposed a FDEA model for the supplier selection process and validated the model through its application on one of the largest suppliers of automotive parts in the Middle East and comparing it with another method. Wu, Zhang, et al. (2019) developed the DEA model in the interval-valued Pythagorean fuzzy environment for green supplier selection problems.

2.1.6 Fuzzy DEMATEL (Decision-making trial and evaluation laboratory)

Dalalah et al. (2011) presented a modified FDEMATEL model to deal with the influential relationship between the evaluation criteria and a modified TOPSIS model to evaluate the criteria against each alternative. They applied the hybrid model for the selection of four feasible cans suppliers at the Nutridar Factory in Amman-Jordan. Six experts and 17 critical criteria were suggested for the case study. Chang et al. (2011) used the FDEMATEL method to find effective criteria in selecting suppliers. Keskin (2015) proposed an integrated model composed of two steps for increasing the quality of supplier selection and evaluation for Cam Elyaf incorporate which is a part of the Şişecam Group. At the first stage, FDEMATEL is used to compute weight of criteria. At the second stage, fuzzy c-means clustering algorithm is used to assess supplier performances. Mirmousa and Dehnavi (2016) presented an integrated fuzzy Delphi, FDEMATEL and CFCS algorithm approach in order to identify factors affecting the selection of supplier in Islamic Azad University of Yazd. Firstly, they recognized 43 important criteria through literature studies and then confirmed a number of 14 criteria by using the fuzzy Delphi method. 11 of the experts and members of the universities evaluated the confirmed criteria by DEMATEL questionnaire. Ultimately, they examined the level of relationship and intensity of this relationship among factors affecting supplier selection by using the CFCS algorithm and FDEMATEL method.

Gören (2018) presented a decision framework that consists of three integrated components for sustainable supplier selection and order allocation problem of an online retailer company located in Canada, which sells different flooring and building materials. First, she used the FDEMATEL approach to calculate the weights of criteria; second, she used the Taguchi Loss Functions with these weights to rank all supplier, third she used the bi-objective optimization model by taking the ranking values as inputs to determine the optimal order quantities of each suppliers. El Mariouli and Abouabdellah (2018) developed a new mathematical model by using a hybrid approach FDEMATEL for the supplier selection problem of Moroccan company. They began with the selection of the most relevant criteria in the literature and then used the FDEMATEL to classify and calculate the weight of the selected criteria. They finished with calculating the sustainability index of each supplier using the mathematical model. Kiriş et al. (2019) proposed an integrated approach (SCOR model and FDEMATEL method) for supplier performance evaluation.

2.1.7 Fuzzy QFD (Quality function deployment)

Bevilacqua et al. (2006) suggested the fuzzy QFD method for supplier selection process of an industry that manufactures complete clutch couplings. Dursun and Karsak (2013) developed a fuzzy multi-criteria group decision-making approach that used QFD concept for the supplier selection process. They computed bounds of the weights of criteria and ratings of suppliers by using the FWA method. Similarly, Dursun and Karsak (2014) and Karsak and Dursun (2015) proposed a fuzzy multi-criteria group decision-making approach integrating fusion of fuzzy information, 2-tuple linguistic representation model, and QFD for supplier selection problems. Lima-Junior and Carpinetti (2016b) proposed a multi criteria decision-making approach based on fuzzy QFD for choosing and weighting criteria for supplier selection process. Babbar and Amin (2018) proposed a novel two-phase model based on fuzzy QFD and stochastic multi-objective techniques. In the first phase, they used a two-stage fuzzy QFD process to assess the suppliers. In the second phase, they developed a multi-objective mixed-integer linear programming to find order quantity using three methods (weighted-sums, distance, and ε-constraint methods) considering five objectives (cost, defect rate, carbon emission, weight of suppliers, on-time delivery). Liu, Gao et al. (2019) proposed a novel green supplier selection method by combining QFD with the Partitioned Bonferroni Mean operator in the context of interval type-2 fuzzy sets and used a bike-share case to illustrate the applicability of the method.

2.1.8 Fuzzy ELECTRE (ELimination Et Choix Traduisant la REalite—Elimination and choice expressing reality)

Sevkli (2010) compared crisp and fuzzy ELECTRE methods for supplier selection by applying them at Akkardan, which is a manufacturing company in Turkey. Kumar et al. (2017) evaluated the performance of suppliers based on green practices using the FELECTRE approach. Zhong and Yao (2017) extended the ELECTRE method using interval type-2 fuzzy numbers for a supplier selection application. They also conducted a sensitivity and comparative analysis to demonstrate the strength and practicality of the proposed method. Gitinavard et al. (2018) proposed an interval-valued hesitant fuzzy extended ELECTRE and verified this method using a case study of the automobile manufacturing industry to evaluate the candidate suppliers regarding environmental competencies from the recent literature. In addition, they prepared a comparative analysis, sensitivity analysis and few simulation-based experiments to sensitiveness and validation of the proposed method, respectively. Shojaie et al. (2018) analyzed green health suppliers of effective raw materials for Tehran Chemie Pharmaceutical Company by using 18 green criteria via fuzzy ELECTRE method and then, they classified suppliers via the Pareto chart using results of fuzzy ELECTRE. Komsiyah et al. (2019) proposed the fuzzy ELECTRE method for Cement Vendor Selection problem of a construction company in Indonesia.

2.1.9 Fuzzy BWM (Best-Worst Method)

Aboutorab et al. (2018) developed the ZBWM method by integrating Z-numbers into the BWM method to deal with uncertainty and applied it to a supplier development problem. According to the experimental results, ZBWM method presented a more consistent approach compared to BWM and Fuzzy BWM. Wu, Zhou et al. (2019) provided an integrated methodology based on the interval type-2 fuzzy BWM and VIKOR for green supplier selection. Similarly, Qin and Liu

(2019) presented an integrated interval type-2 fuzzy BWM and COPRAS approach for emergency material supplier selection problem. Gan et al. (2019) proposed a hybrid method based on the combination of fuzzy BWM and the modular TOPSIS in random environments for group decision-making (GMo-RTOPSIS) to solve resilient supplier selection problem. Liu, Quan et al. (2019) proposed an innovative MCDM model integrated BWM and alternative queuing method (AQM) within the interval-valued intuitionistic uncertain linguistic setting for sustainable supplier selection problems and demonstrated the applicability and effectiveness of the model with an example of a watch manufacturer.

2.1.10 Fuzzy MOORA (Multi-objective optimisation by ratio analysis)

Dey et al. (2012) presented the fuzzy MOORA in selection of alternatives in a supply chain (warehouse location selection and vendor/supplier selection). They utilized this approach to three suitable numerical examples and compared the results with those of previous research works. Pérez-Domínguez et al. (2015) presented intuitionistic fuzzy MOORA for the selection of suppliers. Büyüközkan and Göçer (2017b) presented an interval valued intuitionistic fuzzy MOORA method for supplier selection problem in a Digital Supply Chain environment. They demonstrated this method on a Turkish company, which is a global brand in airport operations. Arabsheybani et al. (2018) applied fuzzy MOORA to evaluate the overall performance of supplier, implemented failure mode and effects analysis to evaluate the risks of a supplier and also, developed a novel multi-objective mathematical model to consider supplier's order allocation. The proposed approach is implemented to a real-world case study for one of the electronic companies in Iran.

2.1.11 Fuzzy PROMETHEE (Preference ranking organisation method for enrichment of evaluations)

Wang et al. (2008) presented the fuzzy PROMETHEE method to evaluate four potential suppliers based on seven criteria and four decision makers by using a case study of a bank in Taiwan. Senvar et al. (2014) used the fuzzy PROMETHEE method for multi-criteria supplier selection problems. Krishankumar et al. (2017) presented a new two-tier decision-making framework. In the first tier, they used a linguistic based aggregation to aggregate linguistic terms directly without making any conversion. In the next tier, they used an intuitionistic fuzzy PROMETHEE to rank each alternative supplier. Finally, they tested the practicality of the framework by using supplier selection problem for an automobile factory.

2.1.12 Fuzzy Mathematical Programming

Kumar et al. (2004) formulated a fuzzy mixed integer goal programming vendor selection problem with multiple objectives; minimizing net cost, net rejections and net late deliveries subject to realistic constraints regarding demand, capacity, quota flexibility, purchase value, budget allocation, etc. (Amid et al. 2006) developed a fuzzy multi objective linear model applying an asymmetric fuzzy-decision making technique to assign different weights to various criteria. Amid et al. (2009) formulated a fuzzy multiobjective model to determine the order quantities to each supplier based on price breaks. The model minimizes the net cost, rejected items and late deliveries, and meets capacity and demand requirement. They developed a fuzzy weighted additive with mixed integer linear programming to solve the problem. Wang and Yang (2009) introduced fuzzy compromise programming for allocating order quantities among suppliers in quantity discount environments. Wu et al. (2010) proposed a fuzzy multi-objective programming model to decide supplier selection by considering risk factors. Amin et al. (2011) proposed a decisional model consists of two phases for supplier selection. In the first phase, they applied SWOT analysis for evaluating suppliers. In the second phase, they applied a fuzzy linear programming model to determine the order quantity. Finally, they utilized a case study of S.G. Company, which is a designing, engineering and supplying company of auto parts in Iran to show the efficiency of the model.

Nazari-Shirkouhi et al. (2013) introduced an extended mixed-integer linear programming model for the supplier selection and order allocation problem under multi-price level and multi-product and developed an interactive two-phase fuzzy multi-objective linear programming method for solving this problem with multiple fuzzy objectives and piecewise linear membership functions. The proposed methodology attempts to minimize the total cost and the number of defective and late delivered units simultaneously. Tiwari et al. (2013) utilized AHP to weight supplier selection criteria and modeled a multi objective program with multi-product, quantity-discounted environment to determine promising suppliers and the ordered quantities. Finally, they devised a single objective fuzzy linear program to solve the proposed model. Arikan (2013) transformed a typical multi objective (minimization of costs, maximization of quality and on-time delivery) supplier selection model into convex fuzzy programming models with a single objective function for reducing the dimension of the system and computational complexity and proposed a novel solution approach (fuzzy additive and augmented max–min model) to solve the problem. Aghai et al. (2014) outlined a fuzzy multi-objective programming model for supplier selection taking quantitative, qualitative, quantity discount and risk factors into consideration and used a real-life study from the Airplane Company to validate the proposed model. Hu and Wei (2014) proposed fuzzy multi-objective integer programming model for the supplier selection problem with multi-product purchases. Moghaddam 2015 developed a fuzzy multi-objective mathematical model for the supplier selection and order allocation in a reverse logistics system and for Pareto optimal solutions of the proposed model, they developed a Monte Carlo simulation integrated with fuzzy goal programming.

Erginel and Gecer (2016) presented a systematic approach for a calibration supplier selection problem, represented the criteria with a questionnaire and proposed the fuzzy multi-objective linear programming model for selecting the calibration supplier for a firm that manages the calibration of medical measurement devices in Ankara, Turkey. Warranties and complaint policy, Communication, Service features, Quality and Performance history are the important criteria they dealt with. Kaur and Rachana (2016) formulated a vendor selection and order allocation problem as an intuitionistic fuzzy multi objective optimization, which minimize net price, maximize quality and on time deliveries subject to supplier's constraints. Govindan et al. (2017) designed an eco-efficient CLSC and proposed a fuzzy multi-objective, multi-period model, which incorporates the firm's economic and environmental concerns for extending the existing supply chain of an Indian firm that assembles inkjet printers. They used AHP for supplier evaluation and a weighted max–min approach for generating a fuzzy, properly efficient solution. Mirzaee et al. (2018) formulated supplier selection and order allocation problem with multi-period, multi-product, multi-supplier, multi-objective, quantity discount subject to budget and capacity limitations by a mixed integer linear programming model and then solved this model by a preemptive fuzzy goal programming approach. Finally, they compared it with three other alternatives; max–min, weighted fuzzy goal programming and classical goal programming. Mari et al. (2019) proposed a possibilistic fuzzy multi-objective approach and developed an interactive fuzzy optimization solution methodology. They demonstrated effectiveness of the proposed resilient supplier selection model and solution methodology on a realistic situation of a garment-manufacturing sector. Torres-Ruiz and Ravindran (2019) presented a three-phase method for the management of suppliers of an auto parts manufacturer located in central Mexico. They proposed an interval DEA method for aggregation of environmental and economic supplier performance criteria into a single score and these scores were incorporated as goals of the three solution approaches; preemptive goal programming, non-preemptive goal programming and fuzzy goal programming for supplier order allocation.

Above, some of the studies implementing the F-MP approach are presented. Table 4 presents all articles in which the F-MP method as a single approach to supplier selection is investigated.

2.1.13 Other single fuzzy approaches

Chou and Chang (2008) applied a fuzzy SMART to evaluate the alternative suppliers/vendors for a famous manufacturer company in the Taiwanese IT industry. Aydın Keskin et al. (2010) proposed Fuzzy ART method to select the most appropriate supplier(s) and cluster all of the vendors according to chosen criteria. To test the contribution of the approach, they solved a real-life supplier evaluation and selection problem of an automotive manufacturing company. Deng and Chan (2011) developed a new fuzzy dempster MCDM method based on the main idea of the FTOPSIS to deal with the supplier selection problem. Keshavarz Ghorabaee et al. (2014) presented a new method for fuzzy multiple criteria group decision-making based on COPRAS method within the context of interval type-2 fuzzy sets for supplier selection problems. Bai et al. (2014) proposed a dynamic fuzzy multi-attribute group decision making new method for supplier evaluation and selection process. Keshavarz Ghorabaee et al. (2016) proposed an integrated approach based on the WASPAS method to deal with multi-criteria group decision-making with interval type-2 fuzzy sets for green supplier selection problems.

Qin et al. (2017) developed an extended TODIM method to solve green supplier selection problems under interval type-2 fuzzy sets. Keshavarz Ghorabaee et al. (2017b) proposed a new multi-criteria model based on EDAS method and interval type-2 fuzzy sets for evaluation of suppliers with respect to environmental criteria in a tissue paper manufacturing company. Li and Wang (2017) used an extended QUALIFLEX method with probability hesitant fuzzy information to solve green supplier selection problems. Kannan (2018) provided a decision support system for the sustainable supplier selection problem in a real world textile industry located in the emerging economy of India. He used Fuzzy Delphi Method to select critical success factors of suppliers. Chang (2019) proposed intuitionistic fuzzy weighted averaging (FWA) method and the soft set for identifying the best supplier in a supply chain. Xu et al. (2019) proposed AHPSort II method based on interval type-2 fuzzy information for sustainable supplier selection problems. Davoudabadi et al. (2019) used interval-valued intuitionistic fuzzy COPRAS method for resilient supplier selection problems. Mishra et al. (2019) developed a novel hesitant fuzzy WASPAS method for assessment of green supplier problem and found that the most significant criteria for the problem were management commitment, environmental management system and green product. In addition, they demonstrated a sensitivity analysis and compared and validated the developed method with existing approaches.

Table 4: Review summary of supplier selection studies using Fuzzy Mathematical programming.

S. no	Author(s)	Year	Type of MP	Type of solution approach	Type of publication	Type of paper
1	(Zarandi and Saghiri)	2003	F–MOP	-	Conference paper	Theoretical study
2	(Kumar et al.)	2004	F–MIGP (fuzzy mixed integer goal programming)	Max–min approach	Journal paper	Real case study
3	(Kumar et al.)	2006	F–MOIP	Zimmermann approach	Journal paper	Real case study
4	(Amid et al.)	2006	F–MOP	Asymmetric fuzzy-decision making technique	Journal paper	Theoretical study
5	(Amid et al.)	2009	F–MOP	Fuzzy weighted additive method with MILP	Journal paper	Theoretical study
6	(Wang and Yang)	2009	Fuzzy compromise programming	weighted additive- max–min approachs	Journal paper	Theoretical study
7	(Rui et al.)	2009	F–MOIP	Max–min -weighted additive approachs	Conference paper	Theoretical study
8	(Wu et al.)	2010	F–MOP	A–cut technique	Journal paper	Theoretical study
9	(Díaz-Madroñero et al.)	2010	F–MOLP	S–curve membership function	Journal paper	Real case study
10	(Amin et al.)	2011	F–LP	SWOT–Verdegay (1982) method	Journal paper	Real case study
11	(Shahrokhi et al.)	2011	Intuitionistic F–MOLP	Zimmermann approach	Journal paper	Theoretical study
12	(Ozkok and Tiryaki)	2011	F–MP	Compensatory fuzzy aggregation operator	Journal paper	Real case study
13	(Amid et al.)	2011	F–MOP	Weighted max–min approach	Journal paper	Theoretical study
14	(Haleh and Hamidi)	2011	F–LP	-	Journal paper	Theoretical study
15	(Kavitha and Vijayalakshmi)	2012	F–MOIP	Weighted additive–α –cut approachs	Journal paper	Real case study
16	(Tiwari et al.)	2013	F–LP	Weighted additive approach	Conference paper	Theoretical study
17	(Arikan)	2013	Convex fuzzy mathematical model	Fuzzy additive and augmented max–min model	Journal paper	Real case study
18	(Nazari-Shirkouhi et al.)	2013	Interactive two-phase F–MOLP	Max–min approach	Journal paper	Theoretical study
19	(Aghai et al.)	2014	F–MOP	Chance-constrained an innovative methods	Journal paper	Real case study
20	(Hu and Wei)	2014	F–MOIP (fuzzy multi-objective integer programming model)	Zimmerman approach	Conference paper	Theoretical study
21	(Sheikhalishahi and Torabi)	2014	Soft lexicographic F–GP	-	Journal paper	Real case study
22	(Moghaddam)	2015	F–MOP	Monte Carlo simulation integrated with fuzzy goal programming	Journal paper	Theoretical study
23	(Fatrias et al.)	2015	F–MOP	weighted additive aggregation function	Book Chapter	Theoretical study
24	(Arikan)	2015	F–MOLP	Fuzzy additive -augmented max–min- Chen and Tsai's fuzzy models	Journal paper	Theoretical study
25	(Darestani et al.)	2015	F–MOP	Compensatory fuzzy model	Journal paper	Theoretical study

Table 4 contd. ...

...*Table 4 contd.*

S. no	Author(s)	Year	Type of MP	Type of solution approach	Type of publication	Type of paper
26	(Memon et al.)	2015	F–MP	-	Journal paper	Theoretical study
27	(Erginel and Gecer)	2016	F–MOLP	α–cut technique	Journal paper	Real case study
28	(Kaur and Rachana)	2016	Intuitionistic F–MOP	-	Journal paper	Real case study
29	(Nasseri and Chitgar)	2016	F–MOP	Augmented weighted Tchebycheff norm	Conference paper	Theoretical study
30	(Suprasongsin and Yenradee)	2016	F–LP	Function principle – pascal triangular graded mean approachs	Conference paper	Theoretical study
31	(Afzali et al.)	2016	Interval-valued Intuitionistic F–MOLP	Zimmermann– Tiwari approachs	Journal paper	Theoretical study
32	(Gupta et al.)	2016	F–MOILP	Weighted possibilistic programming approach	Journal paper	Real case study
33	(Govindan et al.)	2017	Fuzzy multi-objective, multi-period model	Weighted max–min approach	Journal paper	Real case study
34	(Mirzaee et al.)	2018	Preemptive F–GP (fuzzy goal programming)	Additive–max–min approach	Journal paper	Theoretical study
35	(Sutrisno et al.)	2018	Expected value based fuzzy programming model	-	Journal paper	Theoretical study
36	(Mari et al.)	2019	Possibilistic F–MOP	Interactive fuzzy optimization solution methodology	Journal paper	Real case study
37	(Torres-Ruiz and Ravindran)	2019	Preemptive F–GP Non-preemptive GP F–GP	-	Journal paper	Real case study
38	(Safaeian et al.)	2019	F–MOP	Zimmermann approach– Genetic Algorithm(GA)– Non-dominated Sorting GA	Journal paper	Theoretical study

2.2 Hybrid Approaches

There are various integrated approaches for supplier selection in the literature. Based on the popularity of the approaches, we classified them into five categories: (1) FAHP-FTOPSIS; (2) Other FAHP-based approaches; (3) Other FTOPSIS-based approaches; (4) Hybrid F-MCDM and F-MP; Other fuzzy hybrid approaches (5). It was noticed that the integrated FAHP-FTOPSIS is more prevalent due to its simplicity, ease of use. In addition, F-MCDM have been studied quite a lot with F-MP.

2.2.1 Fuzzy AHP-TOPSIS

Wang et al. (2009) revised and improved the FTOPSIS and proposed fuzzy hierarchical TOPSIS as a method of analyzing the lithium-ion battery protection IC supplier selection problem. They also used FAHP to calculate the fuzzy weight of each criterion. Chen and Yang (2011) proposed a new fuzzy decision making method by comprehensively utilizing the constrained FAHP, FTOPSIS, the extent analysis technique and other transformation skills. They demonstrated advantages of method by applying to two supplier selection problems. Zeydan et al. (2011) introduced a new methodology based on FAHP-FTOPSIS-DEA and applied in the biggest car-manufacturing factory in Turkey for the selection and evaluation of quality suppliers. Jolai et al. (2011) proposed a two-phase approach for supplier selection and order allocation problem. In the first phase of the approach, they used FAHP and FTOPSIS to obtain the overall ratings of alternative suppliers and in the second phase, they constructed a multi-objective mixed integer linear programming model using the goal programming technique to determine the order quantities of each selected supplier for each product in each period. Zouggari and Benyoucef (2011) and Zouggari and Benyoucef (2012) used FAHP and FTOPSIS for evaluate supplier through four classes that are Performance strategy, Quality of service, Innovation and Risk.

Büyüközkan (2012) proposed an integrated fuzzy group decision-making framework on fuzzy AHP and fuzzy axiomatic design to evaluate green suppliers effectively, and applied the approach in a Turkish company. Then, she compared the outcome of fuzzy axiomatic design with the outcome of fuzzy TOPSIS. Ghorbani et al. (2013) proposed a three-phase approach based on the Kano model and fuzzy Multi Criteria Decision-Making for supplier selection problem of an agricultural machinery company in Iran. In the first phase, they calculated the importance weight of the criteria using a fuzzy Kano questionnaire and FAHP. In the second phase, they used FTOPSIS technique to screen out in capable suppliers. In the third phase, they evaluated filtered suppliers, once again by FTOPSIS for the final ranking. Lima Junior et al. (2014) presented a comparative analysis of FAHP and FTOPSIS methods in the context of supplier selection decision making. Beikkhakhian et al. (2015) and Lee et al. (2015) used interpretive structural model to rank and categorize the criteria of agile suppliers, FAHP to measure the weight of the evaluation criteria and FTOPSIS to evaluate agile suppliers. In addition, Lee et al. (2015) introduced approximate Pareto fronts of the resulting supplier chains for the weights of the agility criterion.

Mukherjee (2017) developed mathematical models with fuzzy analytic hierarchy process (FAHP), fuzzy TOPSIS, and multi-objective genetic algorithm for traditional supplier selection process. Görener et al. (2017) proposed a three-phase hybrid approach comprising Interval Type-2 FAHP and Interval Type-2 FTOPSIS to address the supplier performance evaluation problem in the aviation industry and presented an application at Turkish Technic Inc. Venkatesh et al. (2018) used FAHP-FTOPSIS for supply partner selection in continuous aid humanitarian supply chains. Alegoz and Yapicioglu (2019) developed a hybrid approach based on FTOPSIS, trapezoidal type-2 FAHP and goal programming for supplier selection and order allocation problems. They examined efficiency of the proposed framework in different cases and discussed the obtained results. Liu, Eckert et al. (2019) developed a fuzzy decision tool, which is a FAHP-FTOPSIS model to evaluate the sustainable performance of suppliers according to economic, environmental and social aspects and illustrated the effectiveness of the proposed tool with a sustainable agrifood value chain application. Karabayir et al. (2019) analyzed the problem of selecting the most convenient supplier for a construction company using Fuzzy AHP and Fuzzy TOPSIS.

2.2.2 Other Fuzzy AHP-based approaches

Dai et al. (2008) used a fuzzy MCDM framework based on FAHP and FVIKOR to solve the problem of supplier selection. Again, Mohammady and Amid (2011) combined FAHP method with developed FVIKOR for supplier selection in an agile and modular virtual enterprise and Awasthi et al. (2018) also presented an integrated FAHP-FVIKOR approach-based framework for multi-tier sustainable global supplier selection. Kuo et al. (2010) develop a novel performance evaluation method, which integrates both FAHP and FDEA for assisting organizations to make the supplier selection decision and then proved that the method is very suitable for practical applications by applying a case study of an internationally recognized auto lighting OEM company. Yücenur et al. (2011) used FAHP and FANP methods for selecting of the global supplier and then compared their results. Hashemian et al. (2014) proposed a hybrid fuzzy group decision-making approach for supplier evaluation integrating F-AHP and F-PROMETHEE group decision support system method and used it to evaluate the suppliers of a dairy company. PrasannaVenkatesan and Goh (2016) developed a multi-objective mixed integer linear programing model to determine the choice of suppliers and order quantity allocation under disruption risk. They evaluated and ranked suppliers using a hybrid FAHP-FPROMETHEE and then applied Multi-Objective Particle Swarm Optimization to yield a set of Pareto optimal solutions for the choice of suppliers and their order allocation. Khorasani (2018) evaluated green suppliers by integrating fuzzy AHP and fuzzy Copras.

2.2.3 Other Fuzzy TOPSIS-based approaches

Önüt et al. (2009) developed a supplier evaluation approach based on FANP and FTOPSIS methods to help a telecommunication company in the GSM sector in Turkey. Rabbani et al. (2009) applied a ranking procedure using FDEA and FTOPSIS and then used a goal programming supplier selection model to perform the trade-offs between conflicting objectives of cost, time and quality. Zhao et al. (2013) presented comparative analysis between TOPSIS and VIKOR under interval-valued intuitionistic fuzzy sets for vendor selection problems. Kuo et al. (2015) developed a framework of the supplier evaluating process for carbon management by integrating fuzzy ANP and fuzzy TOPSIS approaches and then employ an electronic company that is a pioneer in the LEDs industry to demonstrate the proposed method. Sarkar et al. (2018) proposed an integrated multi-attribute decision-making and mathematical programming-based model by combining DANP (DEMATEL-based on ANP) that evaluates the weights of the criteria, FTOPSIS and multiple segment goal programming that ranks the suppliers. They also used DANP-based FVIKOR to validate the result of proposed method, as a result, sensitivity analysis of FTOPSIS and FVIKOR are supported the findings of the best supplier. Liu, Zhang et al. (2019) presented a two-stage fuzzy MCDM model combining fuzzy QFD and improved fuzzy TOPSIS to select the optimum E-commerce supplier for aquatic product companies. Rashidi and Cullinane (2019) applied FTOPSIS and FDEA for sustainable supplier of logistics service providers in Sweden and then presented a comparative analysis of the outcomes. The results show that FTOPSIS performs better than FDEA in terms of both calculation complexity and sensitivity to changes in the number of suppliers. Sasikumar and Vimal (2019) used FVIKOR and FTOPSIS methods to select the suitable green supplier for a textile manufacturing company in southern part of India.

2.2.4 Hybrid Fuzzy MCDM approaches and Fuzzy Mathematical programming

Razmi et al. (2008) presented an integrated framework that involves FTOPSIS and F-LP for suppliers' evaluation and order allocation problem. Lee et al. (2009b) developed a fuzzy multiple goal programming (FMGP) model to select thin film transistor liquid crystal display suppliers for downstream companies. They used FAHP first to obtain the weights of the criteria and then used Multi-choice goal programming (MCGP) to find the optimal solution of allocation to suppliers. Chamodrakas et al. (2010) presented a Fuzzy Preference Programming method based on the FAHP for supplier selection processes in electronic marketplaces. Yücel and Güneri (2011) developed a new weighted additive fuzzy programming approach to capture the vagueness of the problem and preferences of decision makers. They obtained weights of factors by applying the distances of each factor between Fuzzy Positive Ideal Rating (FPIR) and Fuzzy Negative Ideal Rating (FNIR). Then, they developed a fuzzy multi-objective linear model integrating constraints, goals and weights of criteria for assign optimum order quantities to each supplier.

Yu et al. (2012) investigated a fuzzy multi-objective vendor selection program under lean procurement for a Taiwanese stereo manufacturer. They used FAHP to find the decision preferences for the objective functions and constraints and used a soft time-window mechanism to incorporate time based performance metrics for vendor evaluation. Lin (2012) proposed an integrated FANP–FMOLP model for supplier evaluation and selection. Shaw et al. (2012) presented an integrated approach for supplier selection, addressing the carbon emission issue, using FAHP and fuzzy multi-objective linear programming. Kannan et al. (2013) presented an integrated approach; fuzzy multi attribute utility theory and fuzzy multi-objective programming for selecting the best green suppliers and allocating the optimum order quantities among them and applied this approach in an Iranian automobile manufacturing company. Firstly, they used FAHP to obtain the relative importance weights of criteria. Next, they used FTOPSIS to determine the best green suppliers. Finally, they used a weighted max–min method to solve FMOLP model considering various constraints and objectives for assigning order quantities. Kar (2014) proposed an approach for the supplier selection problem by integrating FAHP and fuzzy goal programming.

Çebi and Otay (2016) developed a two-stage fuzzy approach (FMULTIMOORA to evaluate and select suppliers—fuzzy goal programming to determine the amount of order allocation) for supplier selection and order allocation problem. Govindan and Sivakumar (2016) proposed a two-phase hybrid approach for selection of the best green supplier and order allocation among the selected suppliers. They used, in the first phase, FTOPSIS to select potential suppliers and then, in the second phase, F-MOLP to determine order allocation by minimizing cost, material rejection, late delivery, recycle waste and CO_2 emissions in the production process. Bakeshlou et al. (2017) presented a hybrid algorithm including FANP, FDEMATELL, and F-MOLP to evaluate a green multi-sourcing supplier selection model. Azadeh, Siadatian et al. (2017) presented an integrated customer trust and resilience-engineering algorithm composed of standard questionnaires, fuzzy mathematical programming, statistical methods, and verification and validation mechanism to select optimum suppliers for an auto parts (bolt and nut) manufacturer in Iran. They further applied FDEA method to rank and analyze the selected suppliers. Bodaghi et al. (2018) presented a new weighted F-MOLP model for supplier selection, order allocation and customer order scheduling problem and used FANP to evaluate suppliers. Mohammed et al. (2019) proposed integrated FAHP-FTOPSIS is to assess and rank suppliers and developed a Fuzzy Multi-Objective Optimization Model (F-MOO) for allocating the optimal order quantities.

Above, some of the studies implementing the Hybrid Fuzzy MCDM-Fuzzy MP approaches are presented. Table 5 presents all the articles addressing the combination of F-MCDM and F-MP in which as a hybrid approach to supplier selection is investigated.

2.2.5 Other hybrid fuzzy approaches

Büyüközkan and Çifçi (2012) suggested a novel hybrid MCDM approach combines FDEMATEL, FANP and FTOPSIS to evaluate green suppliers and implemented it in Ford Otosan, one of the pioneering companies about environmental subjects in Turkey. Karsak and Dursun (2014) proposed a novel fuzzy multi-criteria group decision-making framework integrating FWA, FQFD and FDEA for supplier selection in a private hospital in Istanbul. Shafique (2017) developed a hybrid multi criteria decision-making approach based FDEMATEL, FANP and FTOPSIS and focused to develop the criteria for green supplier selection.

Table 5: Review summary of supplier selection studies using Hybrid F-MCDM and F-MP approaches.

S. no	Author(s)	Year	Type of Fuzzy approaches	Type of solution approach	Type of publication	Type of paper
1	(Razmi et al.)	2008	FTOPSIS F–LP	Max–min approach	Journal paper	Real case study
2	(Pang)	2009	FANP FPP	-	Conference paper	Theoretical study
3	(Lee et al.)	2009b	FAHP Fuzzy multiple goal programming (FMGP)	-	Journal paper	Real case study
4	(Lin)	2009	FANP FPP	-	Journal paper	Theoretical study
5	(Chamodrakas et al.)	2010	FAHP Fuzzy Preference Programming	-	Journal paper	Theoretical study
6	(Kaur et al.)	2010	FAHP FPP	α–cut approach	Journal paper	Theoretical study
7	(Ku et al.)	2010	FAHP FGP	Max–min method	Journal paper	Real case study
8	(Yücel and Güneri)	2011	FPIR–FNIR F-MOLP	Weighted additive model	Journal paper	Theoretical study
9	(Lin)	2012	FPP FANP F–MOLP	Zimmermann (1978) Chen and Chou (1996) approaches	Journal paper	Theoretical study
10	(Shaw et al.).	2012	FAHP F–MOLP	Max–min -weighted additive approaches	Journal paper	Real case study
11	(Yu et al.)	2012	FAHP F–MOP	Max–min approach	Journal paper	Real case study
12	(Sepehriar et al.)	2013	FELECTRE F–LP	-	Journal paper	Theoretical study
13	(Kannan et al.)	2013	FAHP, FTOPSIS F–MOLP	Max–min method	Journal paper	Real case study
14	(Kar)	2014	FAHP F–GP	-	Journal paper	Real case study
15	(Çebi and Otay)	2016	FMULTIMOORA F–GP	Augmented max–min model	Journal paper	Real case study
16	(Govindan and Sivakumar)	2016	FTOPSIS F–MOLP	Weighted additive model	Journal paper	Real case study
17	(Bhayana et al.)	2016	FAHP F–GP	-	Conference paper	Real case study
18	(Azadeh,Siadatian, et al.)	2017b	FDEA F–MP	α–cut	Journal paper	Real case study
19	(Bakeshlou et al.)	2017	FANP, FDEMATELL F–MOLP	Weighted max–min approach	Journal paper	Theoretical study
20	(Bodaghi et al.)	2018	FANP Weighted F–MOLP	Max–min operator	Journal paper	Theoretical study
21	(Gunawan et al.)	2018	FAHP, FTOPSIS F–MOLP	Weighted preservative model	Conference paper	Theoretical study
22	(Lo et al.)	2018	FTOPSIS F–MOLP	Augmented max–min model	Journal paper	Real case study

Table 5 contd. ...

...Table 5 contd.

S. no	Author(s)	Year	Type of Fuzzy approaches	Type of solution approach	Type of publication	Type of paper
23	(Mohammed et al.)	2018	FAHP, FTOPSIS F–MOP	ε-constraint method LP-metrics method	Journal paper	Theoretical study
24	(Mohammed et al.)	2019	FAHP, FTOPSIS F–MOOM	ε-constraint LP-metrics approaches	Journal paper	Real case study
25	(Çalık et al.)	2019	FAHP F–MOLP	Zimmermann, Tiwari, Fuzzy weighted additive max-min approach with group decision-making (F–WAMG)	Book Chapter	Real case study
26	(Torğul and Paksoy)	2019	FAHP, FTOPSIS F–MOLP	Weighted additive method	Book Chapter	Real case study

Singh et al. (2018) developed a novel framework based on big data cloud computing technology for eco-friendly cattle supplier selection. They applied FDEMATEL and FAHP for obtaining the importance weight of each criteria and used FTOPSIS to evaluate the available suppliers with respect to the criteria. Kafa et al. (2018) proposed a hybrid approach that combines F-AHP, F-PROMETHEE and F-TOPSIS and illustrated through a case example for sustainable partner selection problem in a real light bulbs manufacturing company located in the region of Île-de-France. Banaeian et al. (2018) provided a comparison between the performance of FTOPSIS, FVIKOR and FGRA then utilized them for a green supplier evaluation and selection study of an actual company from the agri-food industry. The comparative analysis indicated that the all three methods arrive at identical rankings, yet FGRA requires less computational complexity. Sen et al. (2018) applied three decision-making approaches; intuitionistic-FTOPSIS, intuitionistic-FMOORA and intuitionistic-FGRA to facilitate supplier selection in sustainable supply chain and obtained similar ranking order of candidate suppliers in three approaches, which proves consistency of these methods. Jahan and Panahande (2019) presented Fuzzy QFD/M-TOPSIS integrated method for supplier selection problem of Semnan Regional Power Company. Petrović et al. (2019) tried to find the appropriate method for evaluation and selection of suppliers in the case of procurement of THK Linear motion guide components by the group of specialists in the “Lagerton” company in Serbia. Firstly, they determined weight of criteria by fuzzy SWARA and then, used fuzzy TOPSIS, fuzzy WASPAS and fuzzy ARAS separately and compared results obtained from three different approaches.

3. Analysis of the Reviewed Papers

3.1 Frequency Analysis of DM Approaches

Frequency of the approaches was analysed for single approaches and hybrid approaches.

Many of the reviewed papers (78%) used a single technique and 68 papers (22%) used a hybrid approach in their analysis. Table 6 shows the frequency of all approaches. According to the Table, FAHP with 24% (58 paper) and FTOPSIS with 24% (57 paper) are the most popular approaches in the single approach category. These are followed by F-MP with 16% (38 paper), FANP with 8% (20 paper), FVIKOR with 5% (13 paper), FDEA with 4% (10 paper), FDEMATEL and FQFD with 3% (seven paper), FELECTRE (six paper), FBWM (five paper) and FMOORA (four paper) with 2%, and FPROMETHEE with 1% (three paper).

In the hybrid approach category, F-MP based approaches are the most popular with 38% (26 paper). In terms of MCDM alone, the FAHP–FTOPSIS approach is the most frequent with 25% (17 paper). This is followed by Other FAHP-based approaches and Other FTOPSIS-based approaches with 12% (eight paper) equally. 13% (nine paper) of hybrid approaches are studies using more than two methods (such as FDEMATEL-FANP-FTOPSIS or FSWARA-FTOPSIS-FWASPAS-FARAS). Therefore, FAHP and FTOPSIS can be considered the most popular approaches in hybrid approaches too.

In terms of total percentages, the ranking is as follows; FAHP with 19%, FTOPSIS with 18%, F-MP with 12%, Hybrid F-MCDM & F-MP approaches with 8%, FANP with 6%, FAHP–FTOPSIS with 5%, FVIKOR with 4%, etc. Total percentages show that using FAHP (19%) and FTOPSIS (18%) methods, as a single approach constitutes a considerable number of papers on the evaluation and selection of suppliers in the fuzzy environment, therefore we can say they are the

Table 6: Frequency of approaches in the reviewed studies.

Categories	Approaches	Journal paper	Book Chapter	Conference paper	Total	in its category	in total
Single	FAHP	30	6	22	58	24%	19%
	FANP	14	0	6	20	8%	6%
	FTOPSIS	39	2	16	57	24%	18%
	FVIKOR	11	0	2	13	5%	4%
	FMOORA	3	0	1	4	2%	1%
	FELECTRE	6	0	0	6	2%	2%
	FDEMATEL	5	0	2	7	3%	2%
	FDEA	6	2	2	10	4%	3%
	FPROMETHEE	1	1	1	3	1%	1%
	FBWM	4	1	0	5	2%	2%
	FQFD	6	1	0	7	3%	2%
	F-MP	31	1	6	38	16%	12%
	Other single fuzzy approaches	13	0	1	14	6%	5%
Hybrid	FAHP–FTOPSIS	14	1	2	17	25%	5%
	Other FAHP-based approaches	7	0	1	8	12%	3%
	Other FTOPSIS-based approaches	6	1	1	8	12%	3%
	Hybrid F-MCDM and F-MP approaches	21	2	3	26	38%	8%
	Other hybrid fuzzy approaches	8	0	1	9	13%	3%

most popular approaches in this research area. In addition, it is important to underline that F-MP is used in a significant number of papers in supplier selection problems both as a single approach and in hybrid approaches, since it is used to determine the amount of the order allocation after supplier selection.

3.2 Distribution Analysis of Publications by Years

The distribution of the 310 papers between 2003 and 2019 is shown in Table 7 and Figure 1 shows the graphical representation of the distribution. As can be seen, the general trend in the total number of papers, except for 2011, shows the increase in the number of studies to implement fuzzy MCDM approaches in the evaluation and selection of suppliers despite fluctuations in some years. In 2011, there was a noticeable increase in the number of papers compared to previous and following years. While the number of Journal papers showed the same tendency as the number of total papers, there is a continuous fluctuation of Conference papers by years and Book chapters started to reflect such studies after 2010.

Figure 2 provides the distribution of used approaches by years. As previously mentioned, the FAHP and FTOPSIS methods dominate other F-MCDM approaches in both single and hybrid approaches. FAHP is studied most in 2011 and 2019, and FTOPSIS is studied most in 2015, 2016 and 2019.

As a result, as can be seen from Figure 1 and 2, it is observed that there is a growth in the studies of the supplier evaluation and selection problem as type of both used approaches and papers (journal, chapter and conference) in recent years especially 2019. It is estimated that the number of studies will continue to increase in the coming years due to the importance and popularity of the issue.

3.3 Distribution Analysis of Papers by Journals

The reviewed papers were also analysed based on journals, books, conferences and publishers.

Figure 3 shows the journals with two and more than two papers on supplier selection and evaluation problems using F-MADM approaches. 73% of reviewed articles (164 articles) were published in 25 journals given in Figure 3.

Table 7: Distribution of papers by years.

Years	Journal paper	Book Chapter	Conference paper	Total
2005 and before	2	0	1	3
2006	4	0	2	6
2007	1	0	1	2
2008	4	1	8	13
2009	11	0	7	18
2010	10	1	4	15
2011	27	0	4	31
2012	11	0	3	14
2013	15	0	3	18
2014	12	4	3	19
2015	20	5	3	28
2016	19	0	7	26
2017	18	1	9	28
2018	30	1	4	35
2019	41	5	8	54

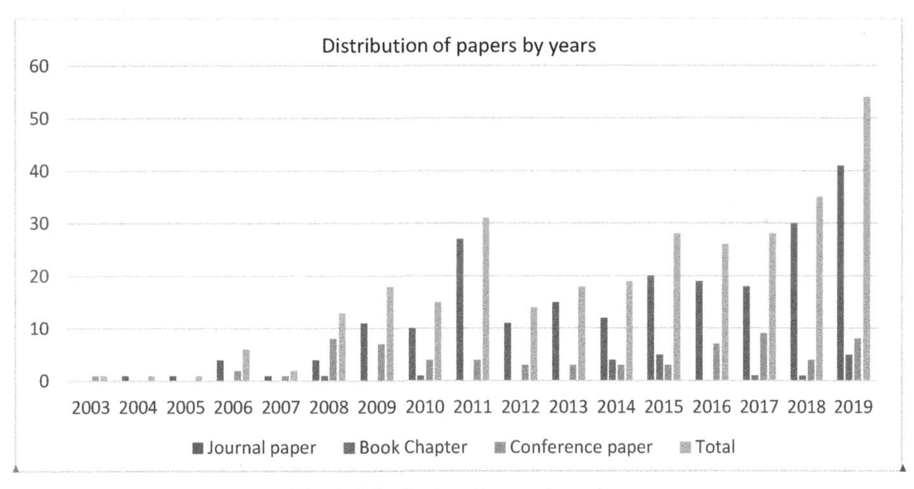

Fig. 1: Distribution of papers by years.

According to this analysis, *Expert Systems with Applications* ranked first by publishing more articles with 39 (17%) papers. The second is *Journal of Cleaner Production* with 15 (7%) articles and then respectively, *Computers & Industrial Engineering* with 13 articles (6%), *International Journal of Production Economics* with 12 articles (5%), *Applied Soft Computing and International Journal of Production Research* with nine articles (4%), *The International Journal of Advanced Manufacturing Technology eight articles* (4%), *Information Sciences* with six articles (3%), *Applied Mathematical Modeling and Journal of Intelligent Manufacturing* with five articles (2%) are listed in the top 10 journals with the highest number of articles and contained more than half of all articles (121 articles—54%).

Except for one (in *Emerging Applications in Supply Chains for Sustainable Business Development* Book by *IGI Global* Publisher) of reviewed 18 Book chapters, others are in the books published by Springer. Two Book chapters are in *Performance Measurement with Fuzzy Data Envelopment Analysis* Book and others are in separate books.

In this study, 67 Conference papers presented in 50 different conferences were examined. One conference proceeding is published by IOP, 16 conference proceedings are published by Springer and the rest are published by IEEE. Figure 4, shows the conferences whose two and more papers were reviewed in this study.

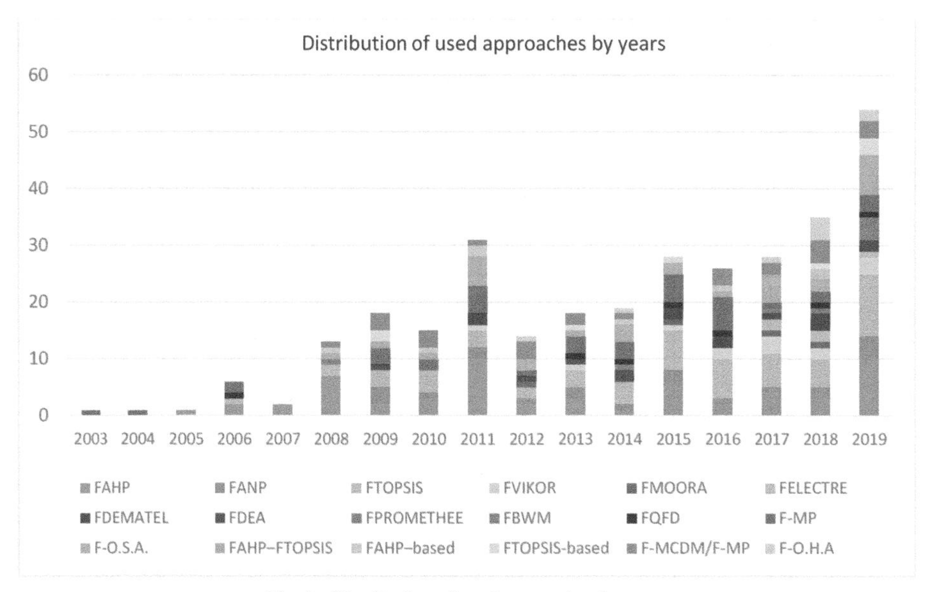

Fig. 2: Distribution of used approaches by years.

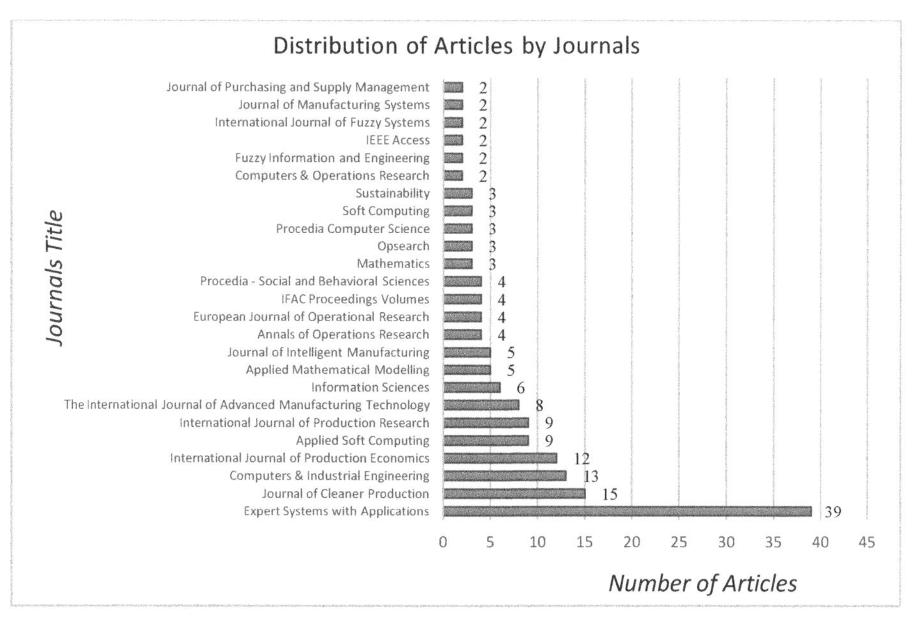

Fig. 3: Distribution of Articles by Journals.

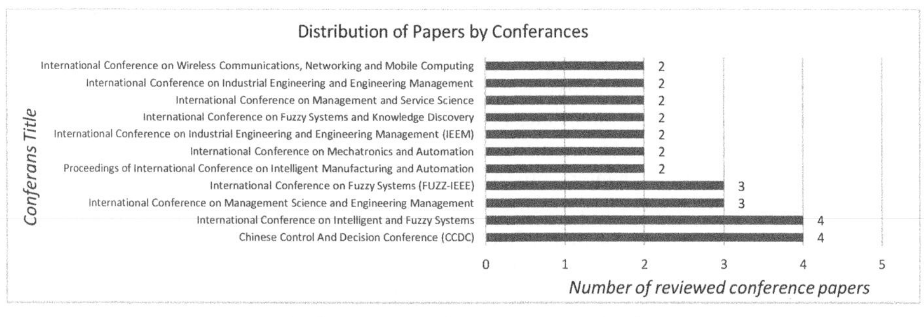

Fig. 4: Distribution of Conference papers by Conferences.

4. Observations and Discussions

According to our observations, most of the reviewed publications (176 papers–57%) examined the real case studies. The remaining papers are theoretical-assumption studies.

If we look at the most studied real cases in terms of industry: *manufacturing industry, automobile industry, white goods industry, iron and steel industry, plastic products industry, air filter industry, battery industry and electronics industry; information systems outsourcing industry and telecommunication industry; building materials industry and construction industry; textile industry, garment industry and apparel industry; health industry, hospital industry, pharmaceutical distribütör industry, emergency material industry, chemical industry and detergent production industry; food industry and agri-food industry; wind power industry, solar power industry, nuclear power industry and petroleum industry* has been mostly studied. Therefore, it may be suggested to concentrate on different sectors in the upcoming studies.

In addition, 93 (30%) of the papers examined focus on different types of suppliers with various qualifications, while the remaining studies are classical supplier selection studies. In Figure 5, the different types of supplier classes studied and the amount of papers by years are given in detail. Accordingly, *green supplier* with 41 papers and *sustainable supplier* with 34 papers have been mostly studied, especially in recent years. In the early years, it can be said that the *global suppliers* were focused on. For this reason, the authors are advised to study on *smart supplier* or hybrid supplier selection such as *smart & sustainable supplier* in future studies, since we are in the Industry 4.0 era.

We have witnessed the classic fuzzy sets of Zadeh developing rapidly over the past decade. Many studies can be found that apply fuzzy MCDM approaches to various fields of science and engineering (Keshavarz Ghorabaee, Amiri, Zavadskas and Antucheviciene 2017). Likewise, different fuzzy environments were used in these studies for supplier selection and evaluation. Reviewed 69 papers (22%) used different fuzzy sets rather than classical sets in fuzzy decision making. As can be seen in Figure 6, TOPSIS method with 27 papers, has been applied the most in different fuzzy environments. The TOPSIS method with 12 papers is overwhelmingly expanded under *Intuitionistic Sets*. Then comes AHP method with 10 papers and VIKOR method with seven papers. Therefore, evaluating suppliers by integrating other decision-making methods with different fuzzy sets can be a direction for future studies. Figure 7 shows the distribution of different fuzzy environments other than the classic sets by years. Accordingly, 23 papers reviewed in 2019 used different fuzzy sets for decision making. This is followed by 2016 with 11 papers and 2017 and 2018 with eight papers. Apart from the classic sets, the most used type of fuzziness is *Intuitionistic Sets* with 20 papers, the second is *Interval Type-2 Sets* with 12 papers and the third is *Interval-Valued Intuitionistic Sets* with nine papers. *Intuitionistic Sets* has been used the most in 2016 and 2018; *Interval Type-2 Sets* has been used the most in 2017 and 2019. It is anticipated that further studies based

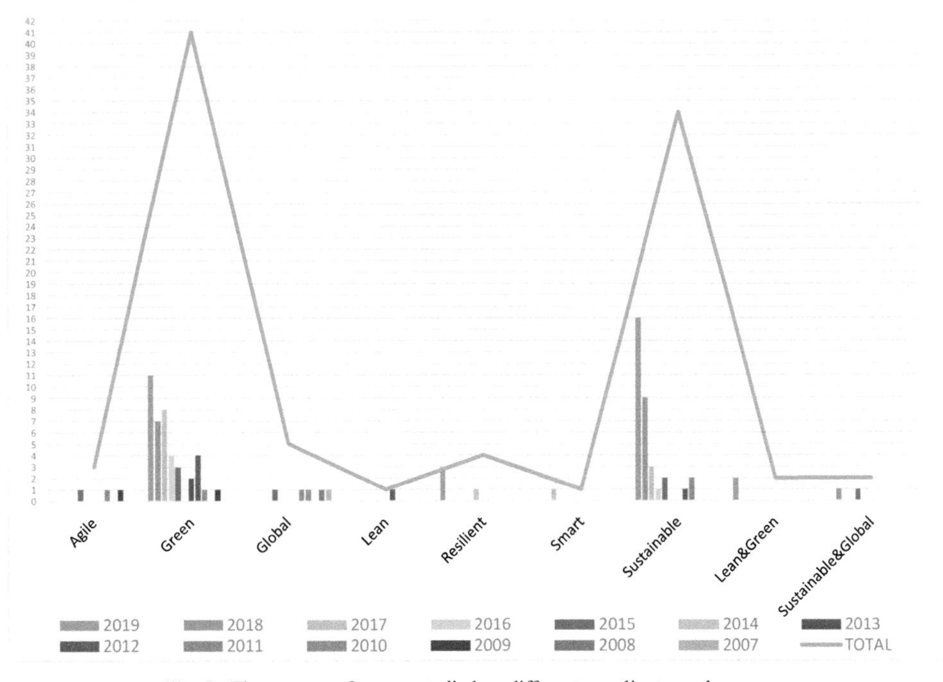

Fig. 5: The amount of papers studied on different supplier types by years.

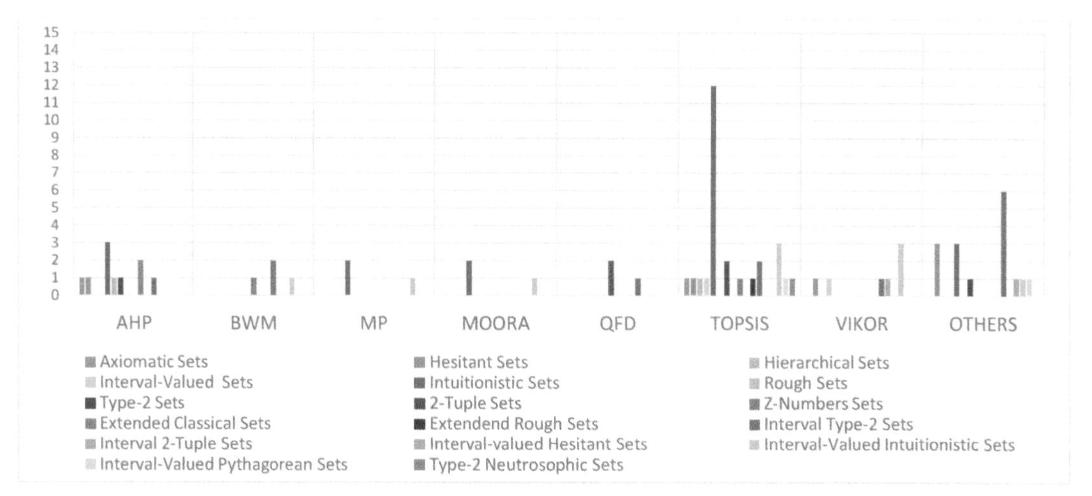

Fig. 6: The amount of papers studied by fuzzy decision-making approaches in different fuzzy environments.

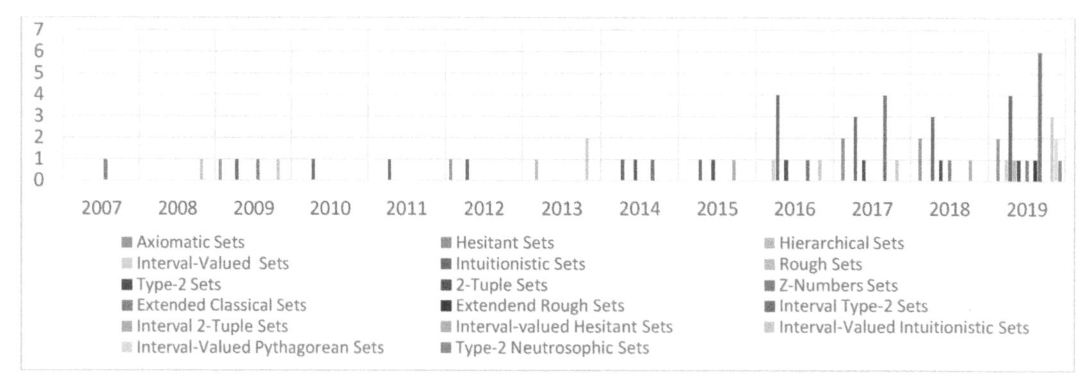

Fig. 7: The amount of papers studied on different type of fuzziness by years.

on *Neutrosophic, Pythagorean* and *Z-Numbers Sets* and new fuzzy sets that have not been studied yet will increase for supplier selection and evaluation.

5. Conclusions

Supplier selection requires to consider multiple goals and criteria. Supplier selection problem is in the class of problems that are difficult to solve because, it is very difficult to regulate the criteria in the supplier selection problem, some of these criteria are expressed qualitatively and some are quantitatively and sometimes there are contradictory or complementary criteria and also, there are a large number of alternative suppliers. For this reason, determining the method to be used in supplier evaluation is of great importance. The MCDM approaches have been the most widely used to deal with supplier selection problems. Fuzzy set theory, in conjunction with MCDM methods, has been widely used to deal with the uncertainty in the very complex real-world supplier selection decision process, because it provides a language that is suitable for processing uncertain criteria that can integrate the analysis of qualitative and quantitative factors.

This chapter presents how fuzzy set theory, fuzzy decision-making can be used in the diverse models for supplier evaluation and selection in approximately the last 20 years. We compiled the existing supplier selection literature by identifying, categorizing and examining supplier selection issues, in this context we reviewed 310 papers in this field and classified them in two categories, individual and integrated approaches, according the applied fuzzy MADM approaches.

The analyses showed that the number of studies used fuzzy MCDM approaches is gradually increasing in the evaluation and selection of suppliers and the FAHP and FTOPSIS methods dominate other F-MCDM approaches in both single and hybrid approaches. In addition, in analysis of journals, Expert Systems with Applications, Journal of Cleaner Production, Computers & Industrial Engineering and International Journal of Production Economics has been identified as the leading journals in this field. In general, the reviewed book chapters and conference papers were mostly published by Springer and IEEE respectively. According to our observations, the majority of the reviewed articles worked on real case studies and the manufacturing industry (automobile, white goods, etc.) has been mostly studied. When looking at the supplier qualifications discussed in the studies, the green and sustainable suppliers, except for traditional suppliers, have been mostly studied, especially in recent years. In addition to the classic sets in fuzzy decision-making, Intuitionistic Sets, Interval Type-2 Sets and Interval-Valued Intuitionistic Sets are the leading fuzzy environments in this field and it has been observed that these different fuzzy sets have been used in recent years, especially in 2019. As methods in the reviewed papers, the TOPSIS, then AHP and VIKOR methods were most expanded in different fuzzy environments.

Increasing interest in supplier selection due to the impact on business performance will continue both academically and practically in the future. Especially, they will focus on qualified suppliers that keep up with the needs of the age rather than traditional suppliers. We expect that more publishing will continue to increase in the coming years since both MCDM methods and fuzziness go towards new extensions.

In summary, this chapter provides a systematic literature review on papers published on the application of Fuzzy MCDM techniques for supplier selection between 2000 and 2019. In addition to its methodological value, this study contributes explicitly to this research area by providing suggestions as to what additional implementations should be done on the subject as well as the current situation, therefore it will help academics, and practitioners effectively solve the supplier selection and evaluation problem.

References

Abdel-Basset, M., M. Saleh, A. Gamal and F. Smarandache. 2019. An approach of topsis technique for developing supplier selection with group decision making under type-2 neutrosophic number. Applied Soft Computing, 77: 438–452. doi: 10.1016/j.asoc.2019.01.035.

Aboutorab, H., M. Saberi, M.R. Asadabadi, O. Hussain and E. Chang. 2018. Zbwm: The z-number extension of best worst method and its application for supplier development. Expert Systems with Applications, 107: 115–125. doi: 10.1016/j.eswa.2018.04.015.

Afzali, A., M.K. Rafsanjani and A.B. Saeid. 2016. A fuzzy multi-objective linear programming model based on interval-valued intuitionistic fuzzy sets for supplier selection. International Journal of Fuzzy Systems, 18(5): 864–874. doi: 10.1007/s40815-016-0201-1.

Aghai, S., N. Mollaverdi and M.S. Sabbagh. 2014. A fuzzy multi-objective programming model for supplier selection with volume discount and risk criteria. The International Journal of Advanced Manufacturing Technology, 71(5-8): 1483–1492. doi: 10.1007/s00170-013-5562-0.

Aghajani, H. and M. Ahmadpour. 2011. Application of fuzzy topsis for ranking suppliers of supply chain in automobile manufacturing companies in Iran. Fuzzy Information and Engineering, 3(4): 433–444.

Agrawal, N. and S. Kant. 2020. Supplier selection using fuzzy-ahp: A case study. In Trends in manufacturing processes, pp. 119–127. Published: Springer Place.

Ahmady, N., M. Azadi, S.A.H. Sadeghi and R.F. Saen. 2013. A novel fuzzy data envelopment analysis model with double frontiers for supplier selection. International Journal of Logistics Research and Applications, 16(2): 87–98. doi: 10.1080/13675567.2013.772957.

Aissaoui, N., M. Haouari and E. Hassini. 2007. Supplier selection and order lot sizing modeling: A review. Computers & Operations Research, 34(12): 3516–3540. doi: 10.1016/j.cor.2006.01.016.

Aktepe, A. and S. Ersoz. 2011. A fuzzy analytic hierarchy process model for supplier selection and a case study. Uluslararası Mühendislik Araştırma ve Geliştirme Dergisi, 3(1): 33–36.

Alegoz, M. and H. Yapicioglu. 2019. Supplier selection and order allocation decisions under quantity discount and fast service options. Sustainable Production and Consumption, 18: 179–189. doi: 10.1016/j.spc.2019.02.006.

Alem, S.M., A. Azadeh, S.N. Shirkouhi and K. Rezaie. 2009. A decision making methodology for vendor selection problem based on dea, fdea and ccdea models, 767–770 2009 Third Asia International Conference on Modelling & Simulation.

Alikhani, R., S.A. Torabi and N. Altay. 2019. Strategic supplier selection under sustainability and risk criteria. International Journal of Production Economics, 208: 69–82. doi: 10.1016/j.ijpe.2018.11.018.

Alinezad, A., A. Seif and N. Esfandiari. 2013. Supplier evaluation and selection with qfd and fahp in a pharmaceutical company. The International Journal of Advanced Manufacturing Technology, 68(1-4): 355–364. doi: 10.1007/s00170-013-4733-3.

Amid, A., S. Ghodsypour and C. O'Brien. 2006. Fuzzy multiobjective linear model for supplier selection in a supply chain. International Journal of Production Economics, 104(2): 394–407.

Amid, A., S.H. Ghodsypour and C. O'Brien. 2009. A weighted additive fuzzy multiobjective model for the supplier selection problem under price breaks in a supply chain. International Journal of Production Economics, 121(2): 323–332. doi: 10.1016/j.ijpe.2007.02.040.

Amid, A., S. Ghodsypour and C. O'Brien. 2011. A weighted max–min model for fuzzy multi-objective supplier selection in a supply chain. International Journal of Production Economics, 131(1): 139–145.

Amin, S.H., J. Razmi and G. Zhang. 2011. Supplier selection and order allocation based on fuzzy swot analysis and fuzzy linear programming. Expert Systems with Applications, 38(1): 334–342. doi: 10.1016/j.eswa.2010.06.071.

Amindoust, A. and A. Saghafinia. 2014. Supplier evaluation and selection using a fdea model. In Performance measurement with fuzzy data envelopment analysis, pp. 255–269. Published Place.

Amindoust, A. 2018. Supplier selection considering sustainability measures: An application of weight restriction fuzzy-dea approach. RAIRO-Operations Research, 52(3): 981–1001.

Arabsheybani, A., M.M. Paydar and A.S. Safaei. 2018. An integrated fuzzy moora method and fmea technique for sustainable supplier selection considering quantity discounts and supplier's risk. Journal of Cleaner Production. 190: 577–591. doi: 10.1016/j.jclepro.2018.04.167.

Arabzad, S.M., M. Ghorbani, J. Razmi and H. Shirouyehzad. 2015. Employing fuzzy topsis and swot for supplier selection and order allocation problem. The International Journal of Advanced Manufacturing Technology, 76 (5-8): 803–818.

Arikan, F. 2013. A fuzzy solution approach for multi objective supplier selection. Expert Systems with Applications. 40(3): 947–952. doi: 10.1016/j.eswa.2012.05.051.

Arikan, F. 2015. An interactive solution approach for multiple objective supplier selection problem with fuzzy parameters. Journal of Intelligent Manufacturing, 26(5): 989998. doi: 10.1007/s10845-013-0782-6.

Awasthi, A., S.S. Chauhan and S.K. Goyal. 2010. A fuzzy multicriteria approach for evaluating environmental performance of suppliers. International Journal of Production Economics, 126(2): 370–378. doi: 10.1016/j.ijpe.2010.04.029.

Awasthi, A., K. Noshad and S.S. Chauhan. 2014. Supplier performance evaluation using a hybrid fuzzy data envelopment analysis approach. In Performance measurement with fuzzy data envelopment analysis, pp. 271–285. Published Place.

Awasthi, A. 2015. Supplier quality evaluation using a fuzzy multi criteria decision making approach. In Complex system modelling and control through intelligent soft computations, pp. 195–219. Published: Springer Place.

Awasthi, A. and G. Kannan. 2016. Green supplier development program selection using ngt and vikor under fuzzy environment. Computers & Industrial Engineering, 91: 100–108. doi: 10.1016/j.cie.2015.11.011.

Awasthi, A., K. Govindan and S. Gold. 2018. Multi-tier sustainable global supplier selection using a fuzzy ahp-vikor based approach. International Journal of Production Economics, 195: 106–117. doi: 10.1016/j.ijpe.2017.10.013.

Ayağ, Z. and F. Samanlioglu. 2016. An intelligent approach to supplier evaluation in automotive sector. Journal of Intelligent Manufacturing, 27(4): 889–903.

Aydın Keskin, G., S. İlhan and C. Özkan. 2010. The fuzzy art algorithm: A categorization method for supplier evaluation and selection. Expert Systems with Applications, 37(2): 1235–1240. doi: 10.1016/j.eswa.2009.06.004.

Ayhan, M.B. 2018. A new decision making approach for supplier selection: Hesitant fuzzy axiomatic design. International Journal of Information Technology & Decision Making, 17 (04): 1085–1117.

Azadeh, A., Y. Rahimi, M. Zarrin, A. Ghaderi and N. Shabanpour. 2017. A decision-making methodology for vendor selection problem with uncertain inputs. Transportation Letters, 9(3): 123–140.

Azadeh, A., R. Siadatian, M. Rezaei-Malek and F. Rouhollah. 2017. Optimization of supplier selection problem by combined customer trust and resilience engineering under uncertainty. International Journal of System Assurance Engineering and Management, 8(S2): 1553–1566. doi: 10.1007/s13198-017-0628-2.

Azadi, M., M. Jafarian, R. Farzipoor Saen and S.M. Mirhedayatian. 2015. A new fuzzy dea model for evaluation of efficiency and effectiveness of suppliers in sustainable supply chain management context. Computers & Operations Research, 54: 274–285. doi: 10.1016/j.cor.2014.03.002.

Azadnia, A.H., P. Ghadimi, M.Z.M. Saman, K.Y. Wong and S. Sharif. 2011. Supplier selection: A hybrid approach using electre and fuzzy clustering. International Conference on Informatics Engineering and Information Science.

Azadnia, A.H., M.Z.M. Saman, K.Y. Wong, P. Ghadimi and N. Zakuan. 2012. Sustainable supplier selection based on self-organizing map neural network and multi criteria decision making approaches. Procedia - Social and Behavioral Sciences, 65: 879–884. doi: 10.1016/j.sbspro.2012.11.214.

Azizi, A., D.O. Aikhuele and F.S. Souleman. 2015. A fuzzy topsis model to rank automotive suppliers. Procedia Manufacturing, 2: 159–164. doi: 10.1016/j.promfg.2015.07.028.

Babbar, C. and S.H. Amin. 2018. A multi-objective mathematical model integrating environmental concerns for supplier selection and order allocation based on fuzzy qfd in beverages industry. Expert Systems with Applications, 92: 27–38. doi: 10.1016/j.eswa.2017.09.041.

Bahadori, M., S.M. Hosseini, E. Teymourzadeh, R. Ravangard, M. Raadabadi and K. Alimohammadzadeh. 2017. A supplier selection model for hospitals using a combination of artificial neural network and fuzzy vikor. International Journal of Healthcare Management: 1–9. doi: 10.1080/20479700.2017.1404730.

Bai, H. 2008. A fuzzy ahp based evaluation method for vendor-selection. 2008 4th IEEE International Conference on Management of Innovation and Technology.

Bai, R., F. Li and J. Yang. 2014. A dynamic fuzzy multi-attribute group decision making method for supplier evaluation and selection. The 26th Chinese Control and Decision Conference (2014 CCDC).

Bakeshlou, E.A., A.A. Khamseh, M.A.G. Asl, J. Sadeghi and M. Abbaszadeh. 2017. Evaluating a green supplier selection problem using a hybrid modm algorithm. Journal of Intelligent Manufacturing, 28(4): 913–927.

Banaeian, N., H. Mobli, B. Fahimnia, I.E. Nielsen and M. Omid. 2018. Green supplier selection using fuzzy group decision making methods: A case study from the agri-food industry. Computers & Operations Research, 89: 337–347. doi: 10.1016/j.cor.2016.02.015.

Beikkhakhian, Y., M. Javanmardi, M. Karbasian and B. Khayambashi. 2015. The application of ism model in evaluating agile suppliers selection criteria and ranking suppliers using fuzzy topsis-ahp methods. Expert Systems with Applications, 42(15-16): 6224–6236. doi: 10.1016/j.eswa.2015.02.035.

Bera, A.K., D.K. Jana, D. Banerjee and T. Nandy. 2019. Multiple-criteria fuzzy group decision-making with multi-choice goal programming for supplier selection: A case study. Discrete Mathematics, Algorithms and Applications: 1950029.

Bevilacqua, M., F.E. Ciarapica and G. Giacchetta. 2006. A fuzzy-qfd approach to supplier selection. Journal of Purchasing and Supply Management, 12(1): 14–27. doi: 10.1016/j.pursup.2006.02.001.

Bhayana, N., S. Kaur and P. Jha. 2015. A hybrid performance evaluation approach for supplier selection under fuzzy environment. Proceedings of Fourth International Conference on Soft Computing for Problem Solving.

Bhayana, N., K. Gandhi and P. Jha. 2016. Hybrid multi-criteria decision-making approach for supplier evaluation in fuzzy environment. Proceedings of Fifth International Conference on Soft Computing for Problem Solving.

Bodaghi, G., F. Jolai and M. Rabbani. 2018. An integrated weighted fuzzy multi-objective model for supplier selection and order scheduling in a supply chain. International Journal of Production Research, 56(10): 3590–3614.

Boran, F.E., S. Genç, M. Kurt and D. Akay. 2009. A multi-criteria intuitionistic fuzzy group decision making for supplier selection with topsis method. Expert Systems with Applications, 36(8): 11363–11368. doi: 10.1016/j.eswa.2009.03.039.

Bottani, E. and A. Rizzi. 2005. A fuzzy multi-attribute framework for supplier selection in an e-procurement environment. International Journal of Logistics Research and Applications, 8(3): 249–266. doi: 10.1080/13675560500240445.

Buriticá, N.C., M.Y. Matamoros, F. Castillo, E. Araya, G. Ahumada and G. Gatica. 2019. Selection of supplier management policies using clustering and fuzzy-ahp in the retail sector. International Journal of Logistics Systems and Management, 34(3): 352–374.

Büyüközkan, G. and J. Arsenyan. 2009. Supplier selection in an agile supply chain environment using fuzzy axiomatic design approach. IFAC Proceedings Volumes, 42(4): 840–845.

Büyüközkan, G. and G. Çifçi. 2011. A novel fuzzy multi-criteria decision framework for sustainable supplier selection with incomplete information. Computers in Industry, 62(2): 164–174. doi: 10.1016/j.compind.2010.10.009.

Büyüközkan, G. 2012. An integrated fuzzy multi-criteria group decision-making approach for green supplier evaluation. International Journal of Production Research, 50(11): 2892–2909. doi: 10.1080/00207543.2011.564668.

Büyüközkan, G. and G. Çifçi. 2012. A novel hybrid mcdm approach based on fuzzy dematel, fuzzy anp and fuzzy topsis to evaluate green suppliers. Expert Systems with Applications, 39(3): 3000–3011. doi: 10.1016/j.eswa.2011.08.162.

Büyüközkan, G. and F. Göçer. 2017a. Application of a new combined intuitionistic fuzzy mcdm approach based on axiomatic design methodology for the supplier selection problem. Applied Soft Computing, 52: 1222–1238. doi: 10.1016/j.asoc.2016.08.051.

Büyüközkan, G. and F. Göçer. 2017b. An extension of moora approach for group decision making based on interval valued intuitionistic fuzzy numbers in digital supply chain. 2017 Joint 17th World Congress of International Fuzzy Systems Association and 9th International Conference on Soft Computing and Intelligent Systems (IFSA-SCIS).

Çalık, A., T. Paksoy and S. Huber. 2019. Lean and green supplier selection problem: A novel multi objective linear programming model for an electronics board manufacturing company in turkey. In Multiple criteria decision making and aiding, pp. 281–309. Published: Springer Place.

Çebi, F. and İ. Otay. 2016. A two-stage fuzzy approach for supplier evaluation and order allocation problem with quantity discounts and lead time. Information Sciences, 339: 143–157. doi: 10.1016/j.ins.2015.12.032.

Cengiz Toklu, M. 2018. Interval type-2 fuzzy topsis method for calibration supplier selection problem: A case study in an automotive company. Arabian Journal of Geosciences, 11(13). doi: 10.1007/s12517-018-3707-z.

Chai, J., J.N.K. Liu and E.W.T. Ngai. 2013. Application of decision-making techniques in supplier selection: A systematic review of literature. Expert Systems with Applications, 40(10): 3872–3885. doi: 10.1016/j.eswa.2012.12.040.

Chaising, S. and P. Temdee. 2017. Application of a hybrid multi-criteria decision making approach for selecting of raw material supplier for small and medium enterprises. 2017 International Conference on Digital Arts, Media and Technology (ICDAMT).

Chamodrakas, I., D. Batis and D. Martakos. 2010. Supplier selection in electronic marketplaces using satisficing and fuzzy ahp. Expert Systems with Applications, 37(1): 490–498. doi: 10.1016/j.eswa.2009.05.043.

Chan, F.T.S. and N. Kumar. 2007. Global supplier development considering risk factors using fuzzy extended ahp-based approach. Omega, 35(4): 417–431. doi: 10.1016/j.omega.2005.08.004.

Chan, F.T.S., N. Kumar, M.K. Tiwari, H.C.W. Lau and K.L. Choy. 2008. Global supplier selection: A fuzzy-ahp approach. International Journal of Production Research, 46(14): 3825–3857. doi: 10.1080/00207540600787200.

Chang, B., C.-W. Chang and C.-H. Wu. 2011. Fuzzy dematel method for developing supplier selection criteria. Expert Systems with Applications, 38(3): 1850–1858. doi: 10.1016/j.eswa.2010.07.114.

Chang, K.-H. 2019. A novel supplier selection method that integrates the intuitionistic fuzzy weighted averaging method and a soft set with imprecise data. Annals of Operations Research, 272(1-2): 139–157.

Chatterjee, K. and S. Kar. 2016. Multi-criteria analysis of supply chain risk management using interval valued fuzzy topsis. Opsearch, 53(3): 474–499. doi: 10.1007/s12597-015-0241-6.

Chen. 2019. A new multi-criteria assessment model combining gra techniques with intuitionistic fuzzy entropy-based topsis method for sustainable building materials supplier selection. Sustainability, 11(8). doi: 10.3390/su11082265.

Chen, C.-T., C.-T. Lin and S.-F. Huang. 2006. A fuzzy approach for supplier evaluation and selection in supply chain management. International Journal of Production Economics, 102(2): 289–301. doi: 10.1016/j.ijpe.2005.03.009.

Chen, Y., S. Wang, J. Yao, Y. Li and S. Yang. 2018. Socially responsible supplier selection and sustainable supply chain development: A combined approach of total interpretive structural modeling and fuzzy analytic network process. Business Strategy and the Environment, 27(8): 1708–1719. doi: 10.1002/bse.2236.

Chen, Z. and W. Yang. 2011. An magdm based on constrained fahp and ftopsis and its application to supplier selection. Mathematical and Computer Modelling, 54(11-12): 2802–2815. doi: 10.1016/j.mcm.2011.06.068.

Chiouy, C.-Y., S.-H. Chou and C.-Y. Yeh. 2011. Using fuzzy ahp in selecting and prioritizing sustainable supplier on csr for taiwan's electronics industry. Journal of Information and Optimization Sciences, 32(5): 1135–1153. doi: 10.1080/02522667.2011.10700110.

Chou, S. and Y. Chang. 2008. A decision support system for supplier selection based on a strategy-aligned fuzzy smart approach. Expert Systems with Applications, 34(4): 2241–2253. doi: 10.1016/j.eswa.2007.03.001.

Çifçi, G. and G. Büyüközkan. 2011. A fuzzy mcdm approach to evaluate green suppliers. International Journal of Computational Intelligence Systems, 4(5): 894–909.

Costantino, N., M. Dotoli, N. Epicoco, M. Falagario and F. Sciancalepore. 2012. A cross efficiency fuzzy data envelopment analysis technique for supplier evaluation under uncertainty. Proceedings of 2012 IEEE 17th International Conference on Emerging Technologies & Factory Automation (ETFA 2012).

Dai, L., Y. Liu and Z. Zhang. 2008. Supplier selection with multiple criteria under fuzzy environment. 2008 4th International Conference on Wireless Communications, Networking and Mobile Computing.

Dalalah, D., M. Hayajneh and F. Batieha. 2011. A fuzzy multi-criteria decision making model for supplier selection. Expert Systems with Applications, 38(7): 8384–8391. doi: 10.1016/j.eswa.2011.01.031.

Darestani, S.A., M. Azizi and S. Qavami. 2015. Solving multi-objective supplier selection model using a compensatory approach. Journal of Industrial and Production Engineering, 32(6): 387–395. doi: 10.1080/21681015.2015.1065913.

Dargi, A., A. Anjomshoae, M.R. Galankashi, A. Memari and M.B.M. Tap. 2014. Supplier selection: A fuzzy-anp approach. Procedia Computer Science, 31: 691–700. doi: 10.1016/j.procs.2014.05.317.

Davoudabadi, R., S.M. Mousavi, V. Mohagheghi and B. Vahdani. 2019. Resilient supplier selection through introducing a new interval-valued intuitionistic fuzzy evaluation and decision-making framework. Arabian Journal for Science and Engineering: 1–10.

De Boer, L., E. Labro and P. Morlacchi. 2001. A review of methods supporting supplier selection. European Journal of Purchasing & Supply Management, 7(2): 75–89.

Deepika, M. and A.K. Kannan. 2016. Global supplier selection using intuitionistic fuzzy analytic hierarchy process. 2016 International Conference on Electrical, Electronics, and Optimization Techniques (ICEEOT).

Degraeve, Z., E. Labro and F. Roodhooft. 2000. An evaluation of vendor selection models from a total cost of ownership perspective. European Journal of Operational Research, 125(1): 34–58.

Deng, Y. and F.T.S. Chan. 2011. A new fuzzy dempster mcdm method and its application in supplier selection. Expert Systems with Applications, 38(8): 9854–9861. doi: 10.1016/j.eswa.2011.02.017.

Deshmukh, A.J. and A.A. Chaudhari. 2011. A review for supplier selection criteria and methods. In Technology Systems and Management, pp. 283–291. Published: Springer Place.

Deshmukh, A.J. and H. Vasudevan. 2019. Supplier selection in plastic products manufacturing msmes using a combined traditional and green criteria based on ahp and fuzzy ahp. Proceedings of International Conference on Intelligent Manufacturing and Automation.

Deshmukh, S. and V. Sunnapwar. 2019. Fuzzy analytic hierarchy process (fahp) for green supplier selection in indian industries. Proceedings of International Conference on Intelligent Manufacturing and Automation.

Dewi, M.M. and H. Al Fatta. 2018. Supplier selection using combined method of k-means and intuitionistic fuzzy topsis. 2018 International Seminar on Application for Technology of Information and Communication.

Dey, B., B. Bairagi, B. Sarkar and S. Sanyal. 2012. A moora based fuzzy multi-criteria decision making approach for supply chain strategy selection. International Journal of Industrial Engineering Computations, 3(4): 649–662.

Díaz-Madroñero, M., D. Peidro and P. Vasant. 2010. Vendor selection problem by using an interactive fuzzy multi-objective approach with modified s-curve membership functions. Computers & Mathematics with Applications, 60(4): 1038–1048.

Dickson, G.W. 1966. An analysis of vendor selection systems and decisions. Journal of Purchasing, 2(1): 5–17.

Diouf, M. and C. Kwak. 2018. Fuzzy ahp, dea, and managerial analysis for supplier selection and development; from the perspective of open innovation. Sustainability, 10(10). doi: 10.3390/su10103779.

dos Santos, B.M., L.P. Godoy and L.M.S. Campos. 2019. Performance evaluation of green suppliers using entropy-topsis-f. Journal of Cleaner Production, 207: 498–509. doi: 10.1016/j.jclepro.2018.09.235.

Dursun, M. and E.E. Karsak. 2013. A qfd-based fuzzy mcdm approach for supplier selection. Applied Mathematical Modelling, 37(8): 5864–5875. doi: 10.1016/j.apm.2012.11.014.

Dursun, M. and E.E. Karsak. 2014. An integrated approach based on 2-tuple fuzzy representation and qfd for supplier selection. In Iaeng transactions on engineering technologies, pp. 621–634. Published Place.

El Mariouli, O. and A. Abouabdellah. 2018. Development of supplier selection model using fuzzy dematel approach in a sustainable development context. International conference on the Sciences of Electronics, Technologies of Information and Telecommunications.

Erginel, N. and A. Gecer. 2016. Fuzzy multi-objective decision model for calibration supplier selection problem. Computers & Industrial Engineering, 102: 166–174. doi: 10.1016/j.cie.2016.10.017.

Ertay, T., A. Kahveci and R.M. Tabanlı. 2011. An integrated multi-criteria group decision-making approach to efficient supplier selection and clustering using fuzzy preference relations. International Journal of Computer Integrated Manufacturing, 24(12): 1152–1167. doi: 10.1080/0951192x.2011.615342.

Fallahpour, A., E. Udoncy Olugu, S. Nurmaya Musa, K. Yew Wong and S. Noori. 2017. A decision support model for sustainable supplier selection in sustainable supply chain management. Computers & Industrial Engineering, 105: 391–410. doi: 10.1016/j.cie.2017.01.005.

Fan, Z., T. Hong and Z. Liu. 2008. Selection of suppliers based on rough set theory and fuzzy topsis algorithm. 2008 IEEE International Symposium on Knowledge Acquisition and Modeling Workshop.

Fatrias, D., A.S. Indrapriyatna and D. Meilani. 2015. Fuzzy multi-objective supplier selection problem: Possibilistic programming approach. In Industrial Engineering, Management Science and Applications, pp. 521–529. Published: Springer Place.

Galankashi, M.R., A. Chegeni, A. Soleimanynanadegany, A. Memari, A. Anjomshoae, S.A. Helmi and A. Dargi. 2015. Prioritizing green supplier selection criteria using fuzzy analytical network process. Procedia CIRP. 26: 689–694. doi: 10.1016/j.procir.2014.07.044.

Galankashi, M.R., M. Hisjam and S.A. Helmi. 2016a. Lean supplier selection: A data envelopment analysis (dea) approach. Proceedings of the Sixth International Conference on Industrial Engineering and Operations Management (IEOM), Kuala Lumpur.

Galankashi, M.R., S.A. Helmi and P. Hashemzahi. 2016b. Supplier selection in automobile industry: A mixed balanced scorecard–fuzzy ahp approach. Alexandria Engineering Journal. 55(1): 93–100. doi: 10.1016/j.aej.2016.01.005.

Gan, J., S. Zhong, S. Liu and D. Yang. 2019. Resilient supplier selection based on fuzzy bwm and gmo-rtopsis under supply chain environment. Discrete Dynamics in Nature and Society, 1–14. doi: 10.1155/2019/2456260.

Genovese, A., S.C. Lenny Koh, G. Bruno and E. Esposito. 2013. Greener supplier selection: State of the art and some empirical evidence. International Journal of Production Research, 51(10): 2868–2886. doi: 10.1080/00207543.2012.748224.

Ghorbani, M., S. Mohammad Arabzad and A. Shahin. 2013. A novel approach for supplier selection based on the kano model and fuzzy mcdm. International Journal of Production Research, 51(18): 5469–5484. doi: 10.1080/00207543.2013.784403.

Gitinavard, H., H. Ghaderi and M.S. Pishvaee. 2018. Green supplier evaluation in manufacturing systems: A novel interval-valued hesitant fuzzy group outranking approach. Soft Computing, 22(19): 6441–6460.

Gold, S. and A. Awasthi. 2015. Sustainable global supplier selection extended towards sustainability risks from (1+ n) th tier suppliers using fuzzy ahp based approach. Ifac-Papersonline, 48(3): 966–971.

Gören, H.G. 2018. A decision framework for sustainable supplier selection and order allocation with lost sales. Journal of Cleaner Production, 183: 1156–1169. doi: 10.1016/j.jclepro.2018.02.211.

Görener, A., B. Ayvaz, A.O. Kuşakcı and E. Altınok. 2017. A hybrid type-2 fuzzy based supplier performance evaluation methodology: The turkish airlines technic case. Applied Soft Computing, 56: 436–445. doi: 10.1016/j.asoc.2017.03.026.

Govindan, K., S. Rajendran, J. Sarkis and P. Murugesan. 2015. Multi criteria decision making approaches for green supplier evaluation and selection: A literature review. Journal of Cleaner Production, 98: 66–83.

Govindan, K. and R. Sivakumar. 2016. Green supplier selection and order allocation in a low-carbon paper industry: Integrated multi-criteria heterogeneous decision-making and multi-objective linear programming approaches. Annals of Operations Research, 238(1-2): 243–276.

Govindan, K., J.D. Darbari, V. Agarwal and P.C. Jha. 2017. Fuzzy multi-objective approach for optimal selection of suppliers and transportation decisions in an eco-efficient closed loop supply chain network. Journal of Cleaner Production, 165: 1598–1619. doi: 10.1016/j.jclepro.2017.06.180.

Gunawan, F., G. Wang, D.N. Utama and S. Komsiyah. 2018. Decision support model for supplier selection using fuzzy logic concept. 2018 International Conference on Information Management and Technology (ICIMTech).

Guo, Z., M. Qi and X. Zhao. 2010. A new approach based on intuitionistic fuzzy set for selection of suppliers. 2010 Sixth International Conference on Natural Computation.

Gupta, H. and M.K. Barua. 2017. Supplier selection among smes on the basis of their green innovation ability using bwm and fuzzy topsis. Journal of Cleaner Production, 152: 242–258. doi: 10.1016/j.jclepro.2017.03.125.

Gupta, P., K. Govindan, M.K. Mehlawat and S. Kumar. 2016. A weighted possibilistic programming approach for sustainable vendor selection and order allocation in fuzzy environment. The International Journal of Advanced Manufacturing Technology, 86(5-8): 1785–1804. doi: 10.1007/s00170-015-8315-4.

Haldar, A., A. Ray, D. Banerjee and S. Ghosh. 2014. Resilient supplier selection under a fuzzy environment. International Journal of Management Science and Engineering Management, 9(2): 147–156. doi: 10.1080/17509653.2013.869040.

Haleh, H. and A. Hamidi. 2011. A fuzzy mcdm model for allocating orders to suppliers in a supply chain under uncertainty over a multi-period time horizon. Expert Systems with Applications, 38(8): 9076–9083. doi: 10.1016/j.eswa.2010.11.064.

Hamdan, S. and A. Cheaitou. 2015. Green supplier selection and order allocation using an integrated fuzzy topsis, ahp and ip approach. 2015 International Conference on Industrial Engineering and Operations Management (IEOM).

Hamdan, S. and A. Cheaitou. 2017. Dynamic green supplier selection and order allocation with quantity discounts and varying supplier availability. Computers & Industrial Engineering, 110: 573–589. doi: 10.1016/j.cie.2017.03.028.

Harnisch, S. and P. Buxmann. 2013. Evaluating cloud services using methods of supplier selection. International Conference on Business Information Systems.

Hashemian, S.M., M. Behzadian, R. Samizadeh and J. Ignatius. 2014. A fuzzy hybrid group decision support system approach for the supplier evaluation process. The International Journal of Advanced Manufacturing Technology. 73(5-8): 1105–1117. doi: 10.1007/s00170-014-5843-2.

Ho, W., X. Xu and P.K. Dey. 2010. Multi-criteria decision making approaches for supplier evaluation and selection: A literature review. European Journal of Operational Research, 202(1): 16–24. doi: 10.1016/j.ejor.2009.05.009.

Hu, L. and Z. Wei. 2014. Evaluation and selection of suppliers under multi-product purchase based on fuzzy multi-objective integer programming model. The 26th Chinese Control and Decision Conference (2014 CCDC).

Hu, X., B. Sun and X. Chen. 2019. Double quantitative fuzzy rough set-based improved ahp method and application to supplier selection decision making. International Journal of Machine Learning and Cybernetics, doi: 10.1007/s13042-019-00964-z.

Igarashi, M., L. de Boer and A.M. Fet. 2013. What is required for greener supplier selection? A literature review and conceptual model development. Journal of Purchasing and Supply Management, 19(4): 247–263. doi: 10.1016/j.pursup.2013.06.001.

Igoulalene, I., L. Benyoucef and M.K. Tiwari. 2015. Novel fuzzy hybrid multi-criteria group decision making approaches for the strategic supplier selection problem. Expert Systems with Applications, 42(7): 3342–3356. doi: 10.1016/j.eswa.2014.12.014.

Jahan, A. and A. Panahande. 2019. 7 supplier selection for protective relays of power transmission network with the fuzzy approach. Sustainable Procurement in Supply Chain Operations: 149.

Jain, V., S. Wadhwa and S.G. Deshmukh. 2009. Select supplier-related issues in modelling a dynamic supply chain: Potential, challenges and direction for future research. International Journal of Production Research, 47(11): 3013–3039. doi: 10.1080/00207540701769958.

Jenssen, M.M. and L. de Boer. 2019. Implementing life cycle assessment in green supplier selection: A systematic review and conceptual model. Journal of Cleaner Production, 229: 1198–1210. doi: 10.1016/j.jclepro.2019.04.335.

Jolai, F., S.A. Yazdian, K. Shahanaghi and M. Azari Khojasteh. 2011. Integrating fuzzy topsis and multi-period goal programming for purchasing multiple products from multiple suppliers. Journal of Purchasing and Supply Management, 17(1): 42–53. doi: 10.1016/j.pursup.2010.06.004.

Junior, F.R.L., L. Osiro and L.C.R. Carpinetti. 2014. A comparison between fuzzy ahp and fuzzy topsis methods to supplier selection. Applied Soft Computing, 21: 194–209.

Kafa, N., Y. Hani and A. El Mhamedi. 2018. Evaluating and selecting partners in sustainable supply chain network: A comparative analysis of combined fuzzy multi-criteria approaches. Opsearch, 55(1): 14–49.

Kahraman, C. and İ. Kaya. 2010. Supplier selection in agile manufacturing using fuzzy analytic hierarchy process. In Enterprise networks and logistics for agile manufacturing, pp. 155–190. Published: Springer Place.

Kang, H.-Y., A.H. Lee and C.-Y. Lin. 2010. A multiple-criteria supplier evaluation model. 2010 International Symposium on Computer, Communication, Control and Automation (3CA).

Kang, H.-Y., A.H. Lee and C.-Y. Yang. 2012. A fuzzy anp model for supplier selection as applied to ic packaging. Journal of Intelligent Manufacturing, 23(5): 1477–1488.

Kannan, D., R. Khodaverdi, L. Olfat, A. Jafarian and A. Diabat. 2013. Integrated fuzzy multi criteria decision making method and multi-objective programming approach for supplier selection and order allocation in a green supply chain. Journal of Cleaner Production, 47: 355–367. doi: 10.1016/j.jclepro.2013.02.010.

Kannan, D., A.B.L.d.S. Jabbour and C.J.C. Jabbour. 2014. Selecting green suppliers based on gscm practices: Using fuzzy topsis applied to a brazilian electronics company. European Journal of Operational Research, 233(2): 432–447. doi: 10.1016/j.ejor.2013.07.023.

Kannan, D. 2018. Role of multiple stakeholders and the critical success factor theory for the sustainable supplier selection process. International Journal of Production Economics, 195: 391–418. doi: 10.1016/j.ijpe.2017.02.020.

Kar, A.K. 2014. Revisiting the supplier selection problem: An integrated approach for group decision support. Expert Systems with Applications, 41(6): 2762–2771. doi: 10.1016/j.eswa.2013.10.009.

Kar, A.K. 2015. A hybrid group decision support system for supplier selection using analytic hierarchy process, fuzzy set theory and neural network. Journal of Computational Science, 6: 23–33. doi: 10.1016/j.jocs.2014.11.002.

Kar, M.B., K. Chatterjee and S. Kar. 2014. A network-topsis based fuzzy decision support system for supplier selection in risky supply chain, 288–293 2014 Seventh International Joint Conference on Computational Sciences and Optimization.

Karabayir, A.N., A.R. Botsali, Y. Kose and E. Cevikcan. 2019. Supplier selection in a construction company using fuzzy ahp and fuzzy topsis. International Conference on Intelligent and Fuzzy Systems.

Karsak, E.E. and M. Dursun. 2014. An integrated supplier selection methodology incorporating qfd and dea with imprecise data. Expert Systems with Applications, 41(16): 6995–7004. doi: 10.1016/j.eswa.2014.06.020.

Karsak, E.E. and M. Dursun. 2015. An integrated fuzzy mcdm approach for supplier evaluation and selection. Computers & Industrial Engineering, 82: 82–93. doi: 10.1016/j.cie.2015.01.019.

Karsak, E.E. and M. Dursun. 2016. Taxonomy and review of non-deterministic analytical methods for supplier selection. International Journal of Computer Integrated Manufacturing, 29(3): 263–286.

Kaur, P., R. Verma and N. Mahanti. 2010. Selection of vendor using analytical hierarchy process based on fuzzy preference programming. Opsearch, 47(1): 16–34.

Kaur, P. and K.N.L. Rachana. 2016. An intutionistic fuzzy optimization approach to vendor selection problem. Perspectives in Science, 8: 348–350. doi: 10.1016/j.pisc.2016.04.071.

Kavitha, C. and C. Vijayalakshmi. 2012. Implementation of multi objective fuzzy integer programming technique by using suppy chain management. Communications in Computer and Information Science, 330: 412–425. doi: https://doi.org/10.1007/978-3-642-35197-6_46.

Keshavarz Ghorabaee, M., M. Amiri, J. Salehi Sadaghiani and G. Hassani Goodarzi. 2014. Multiple criteria group decision-making for supplier selection based on copras method with interval type-2 fuzzy sets. The International Journal of Advanced Manufacturing Technology, 75(5-8): 1115–1130. doi: 10.1007/s00170-014-6142-7.

Keshavarz Ghorabaee, M., E.K. Zavadskas, M. Amiri and A. Esmaeili. 2016. Multi-criteria evaluation of green suppliers using an extended waspas method with interval type-2 fuzzy sets. Journal of Cleaner Production, 137: 213–229. doi: 10.1016/j.jclepro.2016.07.031.

Keshavarz Ghorabaee, M., M. Amiri, E.K. Zavadskas and J. Antucheviciene. 2017a. Supplier evaluation and selection in fuzzy environments: A review of madm approaches. Economic Research-Ekonomska Istraživanja, 30(1): 1073–1118. doi: 10.1080/1331677x.2017.1314828.

Keshavarz Ghorabaee, M., M. Amiri, E.K. Zavadskas, Z. Turskis and J. Antucheviciene. 2017b. A new multi-criteria model based on interval type-2 fuzzy sets and edas method for supplier evaluation and order allocation with environmental considerations. Computers & Industrial Engineering, 112: 156–174. doi: 10.1016/j.cie.2017.08.017.

Keskin, G.A. 2015. Using integrated fuzzy dematel and fuzzy c: Means algorithm for supplier evaluation and selection. International Journal of Production Research, 53(12): 3586–3602.

Khorasani, S.T. 2018. Green supplier evaluation by using the integrated fuzzy ahp model and fuzzy copras. Process Integration and Optimization for Sustainability, 2(1): 17–25.

Kilic, H.S. 2013. An integrated approach for supplier selection in multi-item/multi-supplier environment. Applied Mathematical Modelling, 37(14-15): 7752–7763. doi: 10.1016/j.apm.2013.03.010.

Kilincci, O. and S.A. Onal. 2011. Fuzzy ahp approach for supplier selection in a washing machine company. Expert Systems with Applications, 38(8): 9656–9664. doi: 10.1016/j.eswa.2011.01.159.

Kiriş, S.B., D.Y. Börekçi and T. Koç. 2019. A methodology proposal for supplier performance evaluation: Fuzzy dematel method with sustainability integrated scor model. International Conference on Intelligent and Fuzzy Systems.

Kocken, H.G. and B.A. Ozkok. 2015. A short review of multi criteria decision making approaches for supplier selection problem. In Encyclopedia of Information Science and Technology, third edition, pp. 4961–4969. Published Place.

Komsiyah, S., R. Wongso and S.W. Pratiwi. 2019. Applications of the fuzzy electre method for decision support systems of cement vendor selection. Procedia Computer Science, 157: 479–488.

Kong, F., Z. Zhang and Y. Liu. 2008. Selection of suppliers based on fuzzy multicriteria decision making, 198–202 2008 Fifth International Conference on Fuzzy Systems and Knowledge Discovery.

Koul, S. and R. Verma. 2011. Dynamic vendor selection based on fuzzy ahp. Journal of Manufacturing Technology Management, 22(8): 963–971.

Krishankumar, R., S. Arvinda, A. Amrutha, J. Premaladha and K. Ravichandran. 2017. A decision making framework under intuitionistic fuzzy environment for solving cloud vendor selection problem. 2017 International Conference on Networks & Advances in Computational Technologies (NetACT).

Krishankumar, R., K.S. Ravichandran and A.B. Saeid. 2017. A new extension to PROMETHEE under intuitionistic fuzzy environment for solving supplier selection problem with linguistic preferences. Applied Soft Computing, 60: 564–576.

Krishankumar, R., K.S. Ravichandran, K.K. Murthy and A.B. Saeid. 2018. A scientific decision-making framework for supplier outsourcing using hesitant fuzzy information. Soft Computing, 22(22): 7445–7461. doi: 10.1007/s00500-018-3346-z.

Ku, C.-Y., C.-T. Chang and H.-P. Ho. 2010. Global supplier selection using fuzzy analytic hierarchy process and fuzzy goal programming. Quality & Quantity, 44(4): 623–640.
Kumar, M., P. Vrat and R. Shankar. 2004. A fuzzy goal programming approach for vendor selection problem in a supply chain. Computers & Industrial Engineering, 46(1): 69–85. doi: 10.1016/j.cie.2003.09.010.
Kumar, M., P. Vrat and R. Shankar. 2006. A fuzzy programming approach for vendor selection problem in a supply chain. International Journal of Production Economics, 101(2): 273–285.
Kumar, P., R.K. Singh and A. Vaish. 2017. Suppliers' green performance evaluation using fuzzy extended electre approach. Clean Technologies and Environmental Policy, 19(3): 809–821.
Kumar, P. and R.K. Singh. 2018. Selection of sustainable suppliers. In Global value chains, flexibility and sustainability, pp. 283–300. Published: Springer Place.
Kumar, S., S. Kumar and A.G. Barman. 2018. Supplier selection using fuzzy topsis multi criteria model for a small scale steel manufacturing unit. Procedia Computer Science, 133: 905–912. doi: https://doi.org/10.1016/j.procs.2018.07.097.
Kuo, R.J., L.Y. Lee and T.-L. Hu. 2010. Developing a supplier selection system through integrating fuzzy ahp and fuzzy dea: A case study on an auto lighting system company in taiwan. Production Planning & Control, 21(5): 468–484. doi: 10.1080/09537280903458348.
Kuo, R.J., C.W. Hsu and Y.L. Chen. 2015. Integration of fuzzy anp and fuzzy topsis for evaluating carbon performance of suppliers. International Journal of Environmental Science and Technology, 12(12): 3863–3876. doi: 10.1007/s13762-015-0819-9.
Lee, A.H.I. 2009. A fuzzy supplier selection model with the consideration of benefits, opportunities, costs and risks. Expert Systems with Applications, 36(2): 2879–2893. doi: 10.1016/j.eswa.2008.01.045.
Lee, A.H.I., H.-Y. Kang, C.-F. Hsu and H.-C. Hung. 2009a. A green supplier selection model for high-tech industry. Expert Systems with Applications, 36(4): 7917–7927. doi: 10.1016/j.eswa.2008.11.052.
Lee, A.H.I., H.-Y. Kang and C.-T. Chang. 2009b. Fuzzy multiple goal programming applied to tft-lcd supplier selection by downstream manufacturers. Expert Systems with Applications, 36(3): 6318–6325. doi: 10.1016/j.eswa.2008.08.044.
Lee, J., H. Cho and Y.S. Kim. 2015. Assessing business impacts of agility criterion and order allocation strategy in multi-criteria supplier selection. Expert Systems with Applications, 42(3): 1136–1148. doi: 10.1016/j.eswa.2014.08.041.
Lee, T.R., T. Phuong Nha Le, A. Genovese and L.S.C. Koh. 2011. Using fahp to determine the criteria for partner's selection within a green supply chain. Journal of Manufacturing Technology Management, 23(1): 25–55. doi: 10.1108/17410381211196276.
Li, J. and J.-q. Wang. 2017. An extended qualiflex method under probability hesitant fuzzy environment for selecting green suppliers. International Journal of Fuzzy Systems, 19(6): 1866–1879. doi: 10.1007/s40815-017-0310-5.
Li, J., H. Fang and W. Song. 2019. Sustainable supplier selection based on sscm practices: A rough cloud topsis approach. Journal of Cleaner Production, 222: 606–621. doi: 10.1016/j.jclepro.2019.03.070.
Li, L., H. Chen and J. Wang. 2006. Research on military product supplier selection architecture based on fuzzy ahp. 2006 International Conference on Service Systems and Service Management.
Li, Y., X. Liu and Y. Chen. 2012. Supplier selection using axiomatic fuzzy set and topsis methodology in supply chain management. Fuzzy Optimization and Decision Making, 11(2): 147–176. doi: 10.1007/s10700-012-9117-x.
Li, Z., L. Sun, L. He, X. Tang, X. Zou and M. Chen. 2019. Distribution network supplier selection with fuzzy ahp and topsis. 2019 14th IEEE Conference on Industrial Electronics and Applications (ICIEA).
Liao, C.-N. and H.-P. Kao. 2011. An integrated fuzzy topsis and mcgp approach to supplier selection in supply chain management. Expert Systems with Applications, 38(9): 10803–10811.
Liao, H., Y. Long, M. Tang, A. Mardani and J. Xu. 2019. Low carbon supplier selection using a hesitant fuzzy linguistic span method integrating the analytic network process. Transformations in Business & Economics, 18(2).
Lima Junior, F.R., L. Osiro and L.C.R. Carpinetti. 2014. A comparison between fuzzy ahp and fuzzy topsis methods to supplier selection. Applied Soft Computing, 21: 194–209. doi: 10.1016/j.asoc.2014.03.014.
Lima-Junior, F.R. and L.C.R. Carpinetti. 2016a. Combining scor® model and fuzzy topsis for supplier evaluation and management. International Journal of Production Economics, 174: 128–141. doi: 10.1016/j.ijpe.2016.01.023.
Lima-Junior, F.R. and L.C.R. Carpinetti. 2016b. A multicriteria approach based on fuzzy qfd for choosing criteria for supplier selection. Computers & Industrial Engineering, 101: 269–285. doi: 10.1016/j.cie.2016.09.014.
Lin, R.-H. 2009. An integrated fanp–molp for supplier evaluation and order allocation. Applied Mathematical Modelling, 33(6): 2730–2736. doi: 10.1016/j.apm.2008.08.021.
Lin, R.-H. 2012. An integrated model for supplier selection under a fuzzy situation. International Journal of Production Economics, 138(1): 55–61. doi: 10.1016/j.ijpe.2012.02.024.
Liu, A., Y. Zhang, H. Lu, S.-B. Tsai, C.-F. Hsu and C.-H. Lee. 2019. An innovative model to choose e-commerce suppliers. IEEE Access, 7: 53956–53976. doi: 10.1109/access.2019.2908393.
Liu, H.-C., M.-Y. Quan, Z. Li and Z.-L. Wang. 2019. A new integrated mcdm model for sustainable supplier selection under interval-valued intuitionistic uncertain linguistic environment. Information Sciences, 486: 254–270. doi: 10.1016/j.ins.2019.02.056.

Liu, J. 2015. Investigation into logistics outsourcing supplier selection for automobile manufacturers. Proceedings of the Ninth International Conference on Management Science and Engineering Management.

Liu, P., H. Gao and J. Ma. 2019. Novel green supplier selection method by combining quality function deployment with partitioned bonferroni mean operator in interval type-2 fuzzy environment. Information Sciences, 490: 292–316. doi: 10.1016/j.ins.2019.03.079.

Liu, Y., C. Eckert, G. Yannou-Le Bris and G. Petit. 2019. A fuzzy decision tool to evaluate the sustainable performance of suppliers in an agrifood value chain. Computers & Industrial Engineering, 127: 196–212. doi: 10.1016/j.cie.2018.12.022.

Lo, H.-W., J.J.H. Liou, H.-S. Wang and Y.-S. Tsai. 2018. An integrated model for solving problems in green supplier selection and order allocation. Journal of Cleaner Production, 190: 339–352. doi: 10.1016/j.jclepro.2018.04.105.

Lu, Z., X. Sun, Y. Wang and C. Xu. 2019. Green supplier selection in straw biomass industry based on cloud model and possibility degree. Journal of Cleaner Production, 209: 995–1005. doi: 10.1016/j.jclepro.2018.10.130.

Mari, S., M. Memon, M. Ramzan, S. Qureshi and M. Iqbal. 2019. Interactive fuzzy multi criteria decision making approach for supplier selection and order allocation in a resilient supply chain. Mathematics, 7(2). doi: 10.3390/math7020137.

Mavi, R.K., M. Goh and N.K. Mavi. 2016. Supplier selection with shannon entropy and fuzzy topsis in the context of supply chain risk management. Procedia-Social and Behavioral Sciences, 235: 216–225.

Memari, A., A. Dargi, M.R. Akbari Jokar, R. Ahmad and A.R. Abdul Rahim. 2019. Sustainable supplier selection: A multi-criteria intuitionistic fuzzy topsis method. Journal of Manufacturing Systems, 50: 9–24. doi: 10.1016/j.jmsy.2018.11.002.

Memon, M.S., Y.H. Lee and S.I. Mari. 2015. Group multi-criteria supplier selection using combined grey systems theory and uncertainty theory. Expert Systems with Applications, 42(21): 7951–7959. doi: 10.1016/j.eswa.2015.06.018.

Mirmousa, S. and H.D. Dehnavi. 2016. Development of criteria of selecting the supplier by using the fuzzy dematel method. Procedia - Social and Behavioral Sciences, 230: 281–289. doi: 10.1016/j.sbspro.2016.09.036.

Mirzaee, H., B. Naderi and S.H.R. Pasandideh. 2018. A preemptive fuzzy goal programming model for generalized supplier selection and order allocation with incremental discount. Computers & Industrial Engineering, 122: 292–302. doi: 10.1016/j.cie.2018.05.042.

Mishra, A.R., P. Rani, K.R. Pardasani and A. Mardani. 2019. A novel hesitant fuzzy waspas method for assessment of green supplier problem based on exponential information measures. Journal of Cleaner Production, 238. doi: 10.1016/j.jclepro.2019.117901.

Moghaddam, K.S. 2015. Fuzzy multi-objective model for supplier selection and order allocation in reverse logistics systems under supply and demand uncertainty. Expert Systems with Applications, 42(15-16): 6237–6254. doi: 10.1016/j.eswa.2015.02.010.

Mohammady, P. and A. Amid. 2011. Integrated fuzzy ahp and fuzzy vikor model for supplier selection in an agile and modular virtual enterprise. Fuzzy Information and Engineering, 3(4): 411–431. doi: 10.1007/s12543-011-0095-4.

Mohammed, A., M. Filip, R. Setchi and X. Li. 2017. Drafting a fuzzy topsis-multi-objective approach for a sustainable supplier selection. 2017 23rd International Conference on Automation and Computing (ICAC).

Mohammed, A., R. Setchi, M. Filip, I. Harris and X. Li. 2018. An integrated methodology for a sustainable two-stage supplier selection and order allocation problem. Journal of Cleaner Production, 192: 99–114. doi: 10.1016/j.jclepro.2018.04.131.

Mohammed, A. 2019. Towards a sustainable assessment of suppliers: An integrated fuzzy topsis-possibilistic multi-objective approach. Annals of Operations Research. doi: 10.1007/s10479-019-03167-5.

Mohammed, A., I. Harris and K. Govindan. 2019. A hybrid mcdm-fmoo approach for sustainable supplier selection and order allocation. International Journal of Production Economics, 217: 171–184. doi: 10.1016/j.ijpe.2019.02.003.

Mondragon, A.E.C., E. Mastrocinque, J.-F. Tsai and P.J. Hogg. 2019. An ahp and fuzzy ahp multifactor decision making approach for technology and supplier selection in the high-functionality textile industry. IEEE Transactions on Engineering Management, 1–14. doi: 10.1109/tem.2019.2923286.

Mousakhani, S., S. Nazari-Shirkouhi and A. Bozorgi-Amiri. 2017. A novel interval type-2 fuzzy evaluation model based group decision analysis for green supplier selection problems: A case study of battery industry. Journal of Cleaner Production, 168: 205–218. doi: 10.1016/j.jclepro.2017.08.154.

Mukherjee, K. 2017. Modeling and optimization of traditional supplier selection. pp. 31–58. In Supplier Selection, Published: Springer Place.

Nag, K. and M. Helal. 2016. A fuzzy topsis approach in multi-criteria decision making for supplier selection in a pharmaceutical distributor. 2016 IEEE International Conference on Industrial Engineering and Engineering Management (IEEM).

Nasseri, S. and S. Chitgar. 2016. A new approach for solving fuzzy supplier selection problems under volume discount. International workshop on Mathematics and Decision Science.

Nazari-Shirkouhi, S., H. Shakouri, B. Javadi and A. Keramati. 2013. Supplier selection and order allocation problem using a two-phase fuzzy multi-objective linear programming. Applied Mathematical Modelling, 37(22): 9308–9323. doi: 10.1016/j.apm.2013.04.045.

Nikou, C. and S.J. Moschuris. 2015. Supplier selection procedure of military critical items: Mutivariate, fuzzy, analytical hierarchy procedures. In Military Logistics, pp. 19–42. Published: Springer Place.

Orji, I.J. and S. Wei. 2015. An innovative integration of fuzzy-logic and systems dynamics in sustainable supplier selection: A case on manufacturing industry. Computers & Industrial Engineering, 88: 1–12. doi: 10.1016/j.cie.2015.06.019.

Ozkok, B.A. and F. Tiryaki. 2011. A compensatory fuzzy approach to multi-objective linear supplier selection problem with multiple-item. Expert Systems with Applications, 38(9): 11363–11368.

Ozkok, B.A. and N.F. Kececi. 2019. A short review on supplier selection problem methods under uncertainty. pp. 157–168. In Handbook of Research on Transdisciplinary Knowledge Generation, Published: IGI Global Place.

Önüt, S., S.S. Kara and E. Işik. 2009. Long term supplier selection using a combined fuzzy mcdm approach: A case study for a telecommunication company. Expert Systems with Applications, 36(2): 3887–3895. doi: 10.1016/j.eswa.2008.02.045.

Pang, B. 2006. Evaluation of suppliers in supply chain based on fuzzy-ahp approach. 2006 International Conference on Mechatronics and Automation.

Pang, B. 2007. Multi-criteria supplier evaluation using fuzzy ahp. 2007 International Conference on Mechatronics and Automation.

Pang, B.-H. 2008. A method of suppliers evaluation and choice based on ahp and fuzzy theory. 2008 International Conference on Machine Learning and Cybernetics.

Pang, B. 2009. A fuzzy anp approach to supplier selection based on fuzzy preference programming. 2009 International Conference on Management and Service Science.

Pang, B. and S. Bai. 2013. An integrated fuzzy synthetic evaluation approach for supplier selection based on analytic network process. Journal of Intelligent Manufacturing, 24(1): 163–174.

Parkouhi, S.V. and A.S. Ghadikolaei. 2017. A resilience approach for supplier selection: Using fuzzy analytic network process and grey vikor techniques. Journal of Cleaner Production, 161: 431–451.

Pérez-Domínguez, L., A. Alvarado-Iniesta, I. Rodríguez-Borbón and O. Vergara-Villegas. 2015. Intuitionistic fuzzy moora for supplier selection. Dyna, 82(191): 34–41. doi: 10.15446/dyna.v82n191.51143.

Petrović, G., J. Mihajlović, Ž. Ćojbašić, M. Madić and D. Marinković. 2019. Comparison of three fuzzy mcdm methods for solving the supplier selection problem. Facta Universitatis, Series: Mechanical Engineering, 17(3): 455–469.

Pitchipoo, P., P. Venkumar and S. Rajakarunakaran. 2013. Fuzzy hybrid decision model for supplier evaluation and selection. International Journal of Production Research, 51(13): 3903–3919. doi: 10.1080/00207543.2012.756592.

Pornsing, C., P. Jomtong, J. Kanchana-anotai and T. Tonglim. 2019. Solving supplier selection problem using fuzzy-ahp for an electronic manufacturer. 2019 IEEE 6th International Conference on Industrial Engineering and Applications (ICIEA).

PrasannaVenkatesan, S. and M. Goh. 2016. Multi-objective supplier selection and order allocation under disruption risk. Transportation Research Part E: Logistics and Transportation Review, 95: 124–142. doi: 10.1016/j.tre.2016.09.005.

Punniyamoorthy, M., P. Mathiyalagan and P. Parthiban. 2011. A strategic model using structural equation modeling and fuzzy logic in supplier selection. Expert Systems with Applications, 38(1): 458–474. doi: 10.1016/j.eswa.2010.06.086.

Qin, J., X. Liu and W. Pedrycz. 2017. An extended todim multi-criteria group decision making method for green supplier selection in interval type-2 fuzzy environment. European Journal of Operational Research, 258(2): 626–638. doi: 10.1016/j.ejor.2016.09.059.

Qin, J. and X. Liu. 2019. Interval type-2 fuzzy group decision making by integrating improved best worst method with copras for emergency material supplier selection. pp. 249–271. In Type-2 Fuzzy Decision-Making Theories, Methodologies and Applications, Published: Springer Place.

R, K., R. Ks and A.B. Saeid. 2017. A new extension to promethee under intuitionistic fuzzy environment for solving supplier selection problem with linguistic preferences. Applied Soft Computing, 60: 564–576. doi: 10.1016/j.asoc.2017.07.028.

Rabbani, M., G. Ahmadi and R. Kian. 2009. A new comprehensive framework for ranking accepted orders and supplier selection in make-to-order environments. 2009 International Conference on Computers & Industrial Engineering.

Rahman, M.M. and K.B. Ahsan. 2011. Application of fuzzy-ahp extent analysis for supplier selection in an apparel manufacturing organization. 2011 IEEE International Conference on Industrial Engineering and Engineering Management.

Rajesh, R. and V. Ravi. 2015. Supplier selection in resilient supply chains: A grey relational analysis approach. Journal of Cleaner Production, 86: 343–359.

Rashidi, K. and K. Cullinane. 2019. A comparison of fuzzy dea and fuzzy topsis in sustainable supplier selection: Implications for sourcing strategy. Expert Systems with Applications, 121: 266–281. doi: 10.1016/j.eswa.2018.12.025.

Razmi, J., M.J. Songhori and M.H. Khakbaz. 2008. An integrated fuzzy group decision making/fuzzy linear programming (fgdmlp) framework for supplier evaluation and order allocation. The International Journal of Advanced Manufacturing Technology, 43(5-6): 590–607. doi: 10.1007/s00170-008-1719-7.

Razmi, J., H. Rafiei and M. Hashemi. 2009. Designing a decision support system to evaluate and select suppliers using fuzzy analytic network process. Computers & Industrial Engineering, 57(4): 1282–1290. doi: 10.1016/j.cie.2009.06.008.

Rezaei, J., P.B.M. Fahim and L. Tavasszy. 2014. Supplier selection in the airline retail industry using a funnel methodology: Conjunctive screening method and fuzzy ahp. Expert Systems with Applications, 41(18): 8165–8179. doi: 10.1016/j.eswa.2014.07.005.

Roshandel, J., S.S. Miri-Nargesi and L. Hatami-Shirkouhi. 2013. Evaluating and selecting the supplier in detergent production industry using hierarchical fuzzy topsis. Applied Mathematical Modelling, 37(24): 10170–10181. doi: 10.1016/j.apm.2013.05.043.

Rouyendegh, B.D. and T.E. Saputro. 2014. Supplier selection using integrated fuzzy topsis and mcgp: A case study. Procedia - Social and Behavioral Sciences, 116: 3957–3970. doi: 10.1016/j.sbspro.2014.01.874.

Rouyendegh, B.D., A. Yildizbasi and P. Üstünyer. 2019. Intuitionistic fuzzy topsis method for green supplier selection problem. Soft Computing, doi: 10.1007/s00500-019-04054-8.

Rui, D., C. Yan and H. Lin. 2009. Fuzzy multi-objective programming model for logistics service supplier selection. 2009 Chinese Control and Decision Conference, 17–19 June 2009.

Safaeian, M., A.M. Fathollahi-Fard, G. Tian, Z. Li and H. Ke. 2019. A multi-objective supplier selection and order allocation through incremental discount in a fuzzy environment. Journal of Intelligent & Fuzzy Systems (Preprint): 1–21.

Sarkar, S., V. Lakha, I. Ansari and J. Maiti. 2017. Supplier selection in uncertain environment: A fuzzy mcdm approach. Proceedings of the First International Conference on Intelligent Computing and Communication.

Sarkar, S., D.K. Pratihar and B. Sarkar. 2018. An integrated fuzzy multiple criteria supplier selection approach and its application in a welding company. Journal of Manufacturing Systems, 46: 163–178. doi: 10.1016/j.jmsy.2017.12.004.

Sarwar, A., Z. Zeng, R. AduAgyapong, N. ThiHoaiThuong and T. Qadeer. 2017. A fuzzy multi-criteria decision making approach for supplier selection under fuzzy environment. International Conference on Management Science and Engineering Management.

Sasikumar, P. and K. Vimal. 2019. Evaluation and selection of green suppliers using fuzzy vikor and fuzzy topsis. pp. 202–218. In Emerging Applications in Supply Chains for Sustainable Business Development, Published: IGI Global Place.

Sen, D.K., S. Datta and S.S. Mahapatra. 2018. Sustainable supplier selection in intuitionistic fuzzy environment: A decision-making perspective. Benchmarking: An International Journal, 25(2): 545–574. doi: 10.1108/bij-11-2016-0172.

Sennaroglu, B. and O. Akıcı. 2019. Supplier selection using fuzzy analytic network process. International Conference on Intelligent and Fuzzy Systems.

Senvar, O., G. Tuzkaya and C. Kahraman. 2014. Multi criteria supplier selection using fuzzy promethee method. pp. 21–34. In Supply Chain Management Under Fuzziness, Published Place.

Sepehriar, A., R. Eslamipoor and A. Nobari. 2013. A new mixed fuzzy-lp method for selecting the best supplier using fuzzy group decision making. Neural Computing and Applications, 23(S1): 345–352. doi: 10.1007/s00521-013-1458-z.

Sevkli, M. 2010. An application of the fuzzy electre method for supplier selection. International Journal of Production Research, 48(12): 3393–3405. doi: 10.1080/00207540902814355.

Sevkli, M., S. Zaim, A. Turkyilmaz and M. Satir. 2010. An application of fuzzy topsis method for supplier selection. International Conference on Fuzzy Systems.

Shafique, M.N. 2017. Developing the hybrid multi criteria decision making approach for green supplier evaluation. International Conference on Next Generation Computing Technologies.

Shahrokhi, M., A. Bernard and H. Shidpour. 2011. An integrated method using intuitionistic fuzzy set and linear programming for supplier selection problem. IFAC Proceedings Volumes, 44(1): 6391–6395. doi: https://doi.org/10.3182/20110828-6-IT-1002.01591.

Sharaf, I.M. 2019. Supplier selection using a flexible interval-valued fuzzy vikor. Granular Computing: 1–17.

Shaw, K., R. Shankar, S.S. Yadav and L.S. Thakur. 2012. Supplier selection using fuzzy ahp and fuzzy multi-objective linear programming for developing low carbon supply chain. Expert Systems with Applications, 39(9): 8182–8192. doi: 10.1016/j.eswa.2012.01.149.

Sheikhalishahi, M. and S.A. Torabi. 2014. Maintenance supplier selection considering life cycle costs and risks: A fuzzy goal programming approach. International Journal of Production Research, 52(23): 7084–7099. doi: 10.1080/00207543.2014.935826.

Shemshadi, A., H. Shirazi, M. Toreihi and M.J. Tarokh. 2011. A fuzzy vikor method for supplier selection based on entropy measure for objective weighting. Expert Systems with Applications, 38(10): 12160–12167. doi: 10.1016/j.eswa.2011.03.027.

Shen, L., L. Olfat, K. Govindan, R. Khodaverdi and A. Diabat. 2013. A fuzzy multi criteria approach for evaluating green supplier's performance in green supply chain with linguistic preferences. Resources, Conservation and Recycling, 74: 170–179. doi: 10.1016/j.resconrec.2012.09.006.

Shojaie, A.A., S. Babaie, E. Sayah and D. Mohammaditabar. 2018. Analysis and prioritization of green health suppliers using fuzzy electre method with a case study. Global Journal of Flexible Systems Management, 19(1): 39–52.

Simić, D., I. Kovačević, V. Svirčević and S. Simić. 2017. 50 years of fuzzy set theory and models for supplier assessment and selection: A literature review. Journal of Applied Logic, 24: 85–96. doi: 10.1016/j.jal.2016.11.016.

Singh, A., S. Kumari, H. Malekpoor and N. Mishra. 2018. Big data cloud computing framework for low carbon supplier selection in the beef supply chain. Journal of Cleaner Production, 202: 139–149. doi: 10.1016/j.jclepro.2018.07.236.

Sinrat, S. and W. Atthirawong. 2013. A conceptual framework of an integrated fuzzy anp and topsis for supplier selection based on supply chain risk management. 2013 IEEE International Conference on Industrial Engineering and Engineering Management.

Solanki, R., G. Gulati, A. Tiwari and Q.D. Lohani. 2016. A correlation based intuitionistic fuzzy topsis method on supplier selection problem. 2016 IEEE International Conference on Fuzzy Systems (FUZZ-IEEE).

Soner Kara, S. 2011. Supplier selection with an integrated methodology in unknown environment. Expert Systems with Applications, 38(3): 2133–2139. doi: 10.1016/j.eswa.2010.07.154.

Song, D. and J. Wang. 2019. Supplier selection problem based on interval intuitionistic fuzzy multiattribute group decision making. Open Journal of Business and Management, 07(03): 1494–1503. doi: 10.4236/ojbm.2019.73103.

Stamm, C.L. and D.Y. Golhar. 1993. Jit purchasing: Attribute classification and literature review. Production Planning & Control, 4(3): 273–282.

Suprasongsin, S. and P. Yenradee. 2016. Optimization of supplier selection and order allocation under fuzzy demand in fuzzy lead time. International Symposium on Knowledge and Systems Sciences.

Sutrisno, Widowati, Sunarsih and Kartono. 2018. Expected value based fuzzy programming approach to solve integrated supplier selection and inventory control problem with fuzzy demand. IOP Conference Series: Materials Science and Engineering, 300: 012009. doi: 10.1088/1757–899x/300/1/012009.

Şen, C.G., S. Şen and H. Başlıgil. 2010. Pre-selection of suppliers through an integrated fuzzy analytic hierarchy process and max-min methodology. International Journal of Production Research, 48(6): 1603–1625. doi: 10.1080/00207540802577946.

Taherdoost, H. and A. Brard. 2019. Analyzing the process of supplier selection criteria and methods. Procedia Manufacturing, 32: 1024–1034.

Tang, X. and S. Fang. 2011. A fuzzy ahp approach for service vendor selection under uncertainty. 2011 International Conference on Business Management and Electronic Information.

Thanaraksakul, W. and B. Phruksaphanrat. 2009. Supplier evaluation framework based on balanced scorecard with integrated corporate social responsibility perspective. Proceedings of the International MultiConference of Engineers and Computer Scientists.

Tian, Z.-P., H.-Y. Zhang, J.-Q. Wang and T.-L. Wang. 2018. Green supplier selection using improved topsis and best-worst method under intuitionistic fuzzy environment. Informatica, 29(4): 773–800. doi: 10.15388/Informatica.2018.192.

Tidwell, A. and J.S. Sutterfield. 2012. Supplier selection using qfd: A consumer products case study. International Journal of Quality & Reliability Management, 29(3): 284–294.

Tiwari, A.K., C. Samuel and A. Tiwari. 2013. Flexibility in supplier selection using fuzzy numbers with nonlinear membership functions. Proceedings of International Conference on Advances in Computing.

Torğul, B. and T. Paksoy. 2019. A new multi objective linear programming model for lean and green supplier selection with fuzzy topsis. pp. 101–141. In Lean and green supply chain management, Published: Springer Place.

Torres-Ruiz, A. and A.R. Ravindran. 2019. Use of interval data envelopment analysis, goal programming and dynamic eco-efficiency assessment for sustainable supplier management. Computers & Industrial Engineering, 131: 211–226.

Venkatesh, V.G., A. Zhang, E. Deakins, S. Luthra and S. Mangla. 2018. A fuzzy ahp-topsis approach to supply partner selection in continuous aid humanitarian supply chains. Annals of Operations Research, 283(1-2): 1517–1550. doi: 10.1007/s10479-018-2981-1.

Vinodh, S., R. Anesh Ramiya and S.G. Gautham. 2011. Application of fuzzy analytic network process for supplier selection in a manufacturing organisation. Expert Systems with Applications, 38(1): 272–280. doi: 10.1016/j.eswa.2010.06.057.

Wahyuni, R.S., P.R. Julianda and D. Wilianti. 2019. Supplier selection using fuzzy analytic network process (fanp) at pt putra gunung kidul. IOP Conference Series: Materials Science and Engineering.

Wang, C.N., V.T. Nguyen, D.H. Duong and H.T. Do. 2018. A hybrid fuzzy analytic network process (fanp) and data envelopment analysis (dea) approach for supplier evaluation and selection in the rice supply chain. Symmetry, 10(6). doi: 10.3390/sym10060221.

Wang, C.-N., C.-Y. Yang and H.-C. Cheng. 2019a. Fuzzy multi-criteria decision-making model for supplier evaluation and selection in a wind power plant project. Mathematics, 7(5). doi: 10.3390/math7050417.

Wang, C.-N., C.-Y. Yang and H.-C. Cheng. 2019b. A fuzzy multicriteria decision-making (mcdm) model for sustainable supplier evaluation and selection based on triple bottom line approaches in the garment industry. Processes, 7(7): 400.

Wang, J.-W., C.-H. Cheng and K.-C. Huang. 2009. Fuzzy hierarchical topsis for supplier selection. Applied Soft Computing, 9(1): 377–386. doi: 10.1016/j.asoc.2008.04.014.

Wang, T.-C., L.Y. Chen and Y.-H. Chen. 2008. Applying fuzzy promethee method for evaluating is outsourcing suppliers, 361–365. 2008 Fifth International Conference on Fuzzy Systems and Knowledge Discovery.

Wang, T.-Y. and Y.-H. Yang. 2009. A fuzzy model for supplier selection in quantity discount environments. Expert Systems with Applications, 36(10): 12179–12187.

Wang, Y.-J., T.-C. Han and C.-S. Kao. 2012. A fuzzy decision-making model for supplier selection of logistics, 168–171. 2012 Third International Conference on Innovations in Bio-Inspired Computing and Applications.

Wątróbski, J. and W. Sałabun. 2016. Green supplier selection framework based on multi-criteria decision-analysis approach. International Conference on Sustainable Design and Manufacturing.

Weber, C.A., J.R. Current and W. Benton. 1991. Vendor selection criteria and methods. European Journal of Operational Research, 50(1): 2–18.

Wei, J. and A. Sun. 2009. The selection of supplier based on fuzzy-anp. 2009 International Conference on Future BioMedical Information Engineering (FBIE).
Wei, J.-Y., A.-F. Sun and C.-H. Wang. 2010. The application of fuzzy-anp in the selection of supplier in supply chain management. 2010 International Conference on Logistics Systems and Intelligent Management (ICLSIM).
Wetzstein, A., E. Hartmann, W.C. Benton, jr. and N.-O. Hohenstein. 2016. A systematic assessment of supplier selection literature – state-of-the-art and future scope. International Journal of Production Economics, 182: 304–323. doi: 10.1016/j.ijpe.2016.06.022.
Wood, D.A. 2016. Supplier selection for development of petroleum industry facilities, applying multi-criteria decision making techniques including fuzzy and intuitionistic fuzzy topsis with flexible entropy weighting. Journal of Natural Gas Science and Engineering, 28: 594–612. doi: 10.1016/j.jngse.2015.12.021.
Wu, C. and D. Barnes. 2011. A literature review of decision-making models and approaches for partner selection in agile supply chains. Journal of Purchasing and Supply Management, 17(4): 256–274. doi: 10.1016/j.pursup.2011.09.002.
Wu, D.D., Y. Zhang, D. Wu and D.L. Olson. 2010. Fuzzy multi-objective programming for supplier selection and risk modeling: A possibility approach. European Journal of Operational Research, 200(3): 774–787. doi: 10.1016/j.ejor.2009.01.026.
Wu, M.-Q., C.-H. Zhang, X.-N. Liu and J.-P. Fan. 2019. Green supplier selection based on dea model in interval-valued pythagorean fuzzy environment. IEEE Access. 7: 108001–108013. doi: 10.1109/access.2019.2932770.
Wu, Q., L. Zhou, Y. Chen and H. Chen. 2019. An integrated approach to green supplier selection based on the interval type-2 fuzzy best-worst and extended vikor methods. Information Sciences, 502: 394–417. doi: 10.1016/j.ins.2019.06.049.
Wu, Y., K. Chen, B. Zeng, H. Xu and Y. Yang. 2016. Supplier selection in nuclear power industry with extended vikor method under linguistic information. Applied Soft Computing, 48: 444–457. doi: 10.1016/j.asoc.2016.07.023.
Wu, Z., X. Chen and J. Xu. 2017. Topsis-based approach for hesitant fuzzy linguistic term sets with possibility distribution information. 2017 29th Chinese Control And Decision Conference (CCDC).
Xi, X. and L. Jiang. 2008. A research on supplier selection method for cals. pp. 1473–1480. In Research and Practical Issues of Enterprise Information Systems II, Published: Springer Place.
Xiao, Z. and G. Wei. 2008. Application interval-valued intuitionistic fuzzy set to select supplier, 351–355 2008 Fifth International Conference on Fuzzy Systems and Knowledge Discovery.
Xu, Z., J. Qin, J. Liu and L. Martínez. 2019. Sustainable supplier selection based on ahpsort ii in interval type-2 fuzzy environment. Information Sciences, 483: 273–293. doi: 10.1016/j.ins.2019.01.013.
Yadav, S.P. and S. Kumar. 2009. A multi-criteria interval-valued intuitionistic fuzzy group decision making for supplier selection with topsis method. International workshop on rough sets, fuzzy sets, data mining, and granular-soft computing.
Yadav, V. and M.K. Sharma. 2015. Application of alternative multi-criteria decision making approaches to supplier selection process. pp. 723–743. In Intelligent Techniques in Engineering Management, Published: Springer Place.
Yadavalli, V.S.S., J.D. Darbari, N. Bhayana, P.C. Jha and V. Agarwal. 2019. An integrated optimization model for selection of sustainable suppliers based on customers' expectations. Operations Research Perspectives, 6. doi: 10.1016/j.orp.2019.100113.
Yang, J.L., H.N. Chiu, G.-H. Tzeng and R.H. Yeh. 2008. Vendor selection by integrated fuzzy mcdm techniques with independent and interdependent relationships. Information Sciences, 178(21): 4166–4183. doi: 10.1016/j.ins.2008.06.003.
Yasrebdoost, H. 2015. A model for evaluating of suppliers based on fuzzy ahp method at piece making firms (case study). pp. 1375–1379. In Liss 2013, Published: Springer Place.
You, X.-Y., J.-X. You, H.-C. Liu and L. Zhen. 2015. Group multi-criteria supplier selection using an extended vikor method with interval 2-tuple linguistic information. Expert Systems with Applications, 42(4): 1906–1916. doi: 10.1016/j.eswa.2014.10.004.
Yu, C., Y. Shao, K. Wang and L. Zhang. 2019. A group decision making sustainable supplier selection approach using extended topsis under interval-valued pythagorean fuzzy environment. Expert Systems with Applications, 121: 1–17. doi: 10.1016/j.eswa.2018.12.010.
Yu, M.-C., M. Goh and H.-C. Lin. 2012. Fuzzy multi-objective vendor selection under lean procurement. European Journal of Operational Research, 219(2): 305–311. doi: 10.1016/j.ejor.2011.12.028.
Yucesan, M., S. Mete, F. Serin, E. Celik and M. Gul. 2019. An integrated best-worst and interval type-2 fuzzy topsis methodology for green supplier selection. Mathematics, 7(2). doi: 10.3390/math7020182.
Yücel, A. and A.F. Güneri. 2011. A weighted additive fuzzy programming approach for multi-criteria supplier selection. Expert Systems with Applications, 38(5): 6281–6286. doi: 10.1016/j.eswa.2010.11.086.
Yücenur, G.N., Ö. Vayvay and N.Ç. Demirel. 2011. Supplier selection problem in global supply chains by ahp and anp approaches under fuzzy environment. The International Journal of Advanced Manufacturing Technology, 56(5-8): 823–833. doi: 10.1007/s00170-011-3220-y.
Zafar, A., M. Zafar, A. Sarwar, H. Raza and M.T. Khan. 2018. A fuzzy ahp method for green supplier selection and evaluation. International Conference on Management Science and Engineering Management.
Zarandi, M.H.F. and S. Saghiri. 2003. A comprehensive fuzzy multi-objective model for supplier selection process. The 12th IEEE International Conference on Fuzzy Systems, 2003. FUZZ '03., 25–28 May 2003.

Zeydan, M., C. Çolpan and C. Çobanoğlu. 2011. A combined methodology for supplier selection and performance evaluation. Expert Systems with Applications, 38(3): 2741–2751. doi: 10.1016/j.eswa.2010.08.064.

Zhang, Q. and Y. Huang. 2012. Intuitionistic fuzzy decision method for supplier selection in information technology service outsourcing. International Conference on Artificial Intelligence and Computational Intelligence.

Zhang, X., Y. Deng, F.T.S. Chan and S. Mahadevan. 2015. A fuzzy extended analytic network process-based approach for global supplier selection. Applied Intelligence, 43(4): 760–772. doi: 10.1007/s10489-015-0664-z.

Zhao, X., K. Huang and G. Li. 2013. Manufacturing vendor selection based on cross-entropy measure with fuzzy vikor method. IFAC Proceedings Volumes, 46(9): 1973–1978.

Zhao, Z. and J. Xu. 2008. Research of the application of fuzzy ahp in supplier evaluation and selection. 2008 4th International Conference on Wireless Communications, Networking and Mobile Computing.

Zhong, L. and L. Yao. 2017. An electre i-based multi-criteria group decision making method with interval type-2 fuzzy numbers and its application to supplier selection. Applied Soft Computing, 57: 556–576. doi: 10.1016/j.asoc.2017.04.001.

Zhou, X., W. Pedrycz, Y. Kuang and Z. Zhang. 2016. Type-2 fuzzy multi-objective dea model: An application to sustainable supplier evaluation. Applied Soft Computing, 46: 424–440. doi: 10.1016/j.asoc.2016.04.038.

Zhou, X. and Z. Xu. 2017. An integrated decision making model for sustainable supplier selection under uncertain environment. 2017 IEEE International Conference on Industrial Engineering and Engineering Management (IEEM).

Zhou, X. and Z. Xu. 2018. An integrated sustainable supplier selection approach based on hybrid information aggregation. Sustainability, 10(7): 2543.

Zimmer, K., M. Fröhling and F. Schultmann. 2016. Sustainable supplier management–a review of models supporting sustainable supplier selection, monitoring and development. International Journal of Production Research, 54(5): 1412–1442.

Zouggari, A. and L. Benyoucef. 2011. Simulation based fuzzy tool for supplier selection with order allocation. Proceedings of 2011 IEEE International Conference on Service Operations, Logistics and Informatics.

Zouggari, A. and L. Benyoucef. 2012. Simulation based fuzzy topsis approach for group multi-criteria supplier selection problem. Engineering Applications of Artificial Intelligence, 25(3): 507–519. doi: 10.1016/j.engappai.2011.10.012.

SECTION 4

Machine Learning in SCM

CHAPTER 7

Supplier Selection with Machine Learning Algorithms

Mustafa Servet Kıran,[1,*] *Engin Eşme*,[2] *Belkız Torğul*[3] and *Turan Paksoy*[3]

1. Introduction

Members providing input, raw materials, products or information to companies for the realization of a good or service are called suppliers. The supplier selection process consists of several steps such as identifying the purpose, determining the criteria for the purpose, pre-evaluation of the appropriate suppliers found according to the specified criteria and then making the final selection. In today's competitive conditions, effective supplier selection, management and development are crucial for companies to achieve their goals. Because, considering the effect of the material received on the product to be produced, better the factors such as quality, cost, delivery on time of the material used in production are, the higher is the value of the goods to be produced in the market and this also provides competitive advantage to the business. To support the success of the partnership, full cooperation between manufacturers, suppliers and suppliers of suppliers is required. Once businesses have identified appropriate suppliers and gathered information about them, they evaluate potential suppliers according to the determined criteria. There are three main criteria for purchasing; Quality, Cost and Delivery. However, the points to be considered when choosing suppliers have changed from past to present with the development of the supply chain concept and especially Industry 4.0 effects; nowadays supplier selection has become a process in itself and criteria for suppliers have increased while there were only a few criteria wanted before such as reasonable price, quality and close distance. The criteria may vary according to purpose and the product to be supplied, and should be defined in this direction.

The supplier selection process does not end with finding the supplier wanted, but rather, it is a continuous process that aims to follow, develop and if require, replace existing suppliers with new suppliers, which may benefit more in terms of criteria. With Industry 4.0, a lot of information is now available on supply chains. Digital technologies enable flexible decision-making by providing real-time data for all links/members of supply chains (Cavalcante et al. 2019). In addition, rapid developments in information technology make it easier to collect, transmit and store information. It is necessary to identify an effective method for evaluating suppliers in the information society, where everything is shaped according to

[1] Department of Computer Engineering, Konya Technical University, Faculty of Engineering and Natural Sciences, Konya, Turkey.
[2] Department of Computer Programming, Kulu Vocational School, Selcuk University, Konya, Turkey; Email: eesme@selcuk.edu.tr
[3] Department of Industrial Engineering, Konya Technical University, Faculty of Engineering and Natural Sciences, Konya, Turkey. Emails: belkistorgul@gmail.com; tpaksoy@yahoo.com
* Corresponding author: mskiran@ktun.edu.tr

information. In the age of Industry 4.0, in order to achieve smart results by using all these data effectively, we will apply the machine-learning method, which can analyze large, various data sets for our supplier selection problem in this chapter.

The chapter is organized as follows: Section 2 presents current relevant literature for supplier selection methods and studies on supplier selection applied machine learning. Section 3 provides fundamental content covering machine learning, learning types and learning tasks, Section 4 presents an extract of content describing the use of WEKA. Section 5 illustrates the classification of the Supplier Chain Data on the WEKA platform with four classification algorithms and finally, Section 6 presents conclusions.

2. Literature

In the current literature, multi-criteria decision making (MCDM) approaches, which support decision makers in evaluating potential alternatives according to several criteria, have been frequently used for the supplier selection problem. Such as the Analytic hierarchy process (AHP) (Chan 2003; Liu and Hai 2005), and the Analytic network process (ANP) (Sarkis and Talluri 2002; Gencer and Gurpinar 2007). In particular, fuzzy set theory with MCDM methods has been widely used to deal with uncertainty in supplier selection decision-making, such as the Fuzzy AHP (Chan and Kumar 2007; Chan et al. 2008; Lee 2009; Buyukozkan and Cifci 2011), Fuzzy ANP (Razmi et al. 2009; Kang et al. 2012; Zhang et al. 2015; Chen et al. 2018), Fuzzy technique for order of preference by similarity to ideal solution (TOPSIS) (Chen et al. 2006; Awasthi et al. 2010; Kilic 2013; Kumar et al. 2018; Yu et al. 2019), Fuzzy multi criteria optimization and compromise solution (VIKOR) (Awasthi and Kannan 2016), Fuzzy Multi objective optimization by ratio analysis (MOORA) (Dey et al. 2012), Fuzzy Elimination and choice expressing reality (ELECTRE) (Sevkli 2010), Fuzzy Decision making trial and evaluation laboratory (DEMATEL) (Keskin 2015) and combinations thereof, such as the Fuzzy AHP–TOPSIS (Chen and Yang 2011), Fuzzy ANP–TOPSIS (Kuo et al. 2015), Fuzzy AHP–VIKOR (Mohammady and Amid 2011), etc. Another commonly used methodology is mathematical programming techniques, such as linear programming (Tiwari et al. 2012), integer programming (Ding et al. 2009), mixed integer programming (Amid et al. 2009), multi objective programming (Wu et al. 2010) and goal programming (Mirzaee et al. 2018), again especially in the fuzzy environment. In addition, Stochastic Programming (Talluri and Lee 2010), Non-linear programming (Yang et al. 2007), Artificial Intelligence models (Heuristic Algorithms, Neural Networks, Gray System Theory, Rough Set Theory, Case Based Reasoning, ...) have also started to be applied for supplier selection problems (Guo et al. 2009; Guo et al. 2014).

Machine learning is a classification technique, which has been newly applied in supply chain management. Despite the remarkable improvements that Machine learning techniques have made in supply chain management, they have recently attracted researchers' attention and therefore, their researches on evaluation and selection of suppliers are few. Valluri and Croson (2005) used agent-based modeling for a supplier selection problem in literature. They modeled two techniques determining exploration reference points—auction-style focusing on probability of success and newsvendor-style focusing on profitability and studied the dynamics of high-quality and low-quality supplier interactions. Finally, they showed that it is definitely better for the buyer to take action with a few suppliers. Guo et al. (2009) introduced a new support vector machine technology combined with decision tree to address feature selection and multiclass classification on supplier selection and tested the proposed approach on the data from China. Tang (2009) proposed the support vector machine, which is kind of new machine learning technology for the assessment of the logistics suppliers in small sample case condition. Mori et al. (2012) proposed AI-based approach to find plausible candidates of business partners and used machine-learning techniques to build a prediction model of customer–supplier relationships for 30,660 manufacturing firms in the Tokyo, Japan. Omurca (2013) proposed a new solution hybridization of fuzzy c-means as a machine learning technique and rough set theory techniques for supplier evaluation, development and selection problem. The proposed method selects the best supplier(s), clusters all of the suppliers, decides the most important criteria and extracts the decision rules about data. Guo et al. (2014) suggested a model based on semi-fuzzy support vector domain description to address multi-classification problem of supplier selection. They used the semi-fuzzy kernel clustering algorithm to divided original samples into two subsets—deterministic/fuzzy and used cooperative coevolution algorithm for decision making. Finally, they tested the proposed model on the data from China. Mirkouei and Haapala (2014) suggested an integration of machine learning techniques (Support Vector Machine Method) and a mathematical programming model to select the most appropriate feedstock suppliers. Allgurin and Karlsson (2018) provided a framework for implementing the Machine Learning algorithm for a qualitative case study of the supplier selection process in Bufab Sweden AB. They identified 26 variables that are critical for supplier selection and prepared theory and empirical data and then ranked identified variables by considering Machine Learning algorithms. Cavalcante et al. (2019) developed a hybrid approach that combines machine learning and simulation and examines its applications for data-driven decision-making support in selection of resilient supplier.

3. Machine Learning

Man has struggled to invent and develop various tools to cope with the challenges of meeting his needs throughout history. Some of the inventions that were the products of intelligence expressed as creative problem-solving skills had an effect far beyond meeting the needs, and even influenced our way of life. Is it possible that intelligence is a gift that is given only to mankind? Is it possible to produce machines that can imitate cognitive skills like comprehension, application, analysis, and synthesis? The *"artificial intelligence"* concept, which John McCarthy, who was a pioneering American computer scientist in his field, described as *"science and engineering of making intelligent machines"* was used for the first time in 1956 at *"The Dartmouth College Artificial Intelligence Conference: The Next Fifty Years"*, which was organized by him; and was born as a discipline (Moor 2006). The first examples of Artificial Intelligence were able to produce problem-focused, specific solutions with classical programming approaches. In other words, machines that can react by detecting the situations around them can be said to imitate an intelligence; however, it is very difficult to develop programs in areas where we do not know exactly how the human brain works, where conditions vary and cannot be defined clearly (Hinton 2013). As an alternative to this difficulty in programming, the data mining approaches, which emerged as computers accelerated and as the Internet became more widespread, have led to significant developments in machine learning methods. Machine Learning was first used in 1959 by Arthur Samuel, who was pioneer in the field of computer gaming and artificial intelligence, and constituted a sub-field of Artificial Intelligence. It may be important to hear what the masters of this field said on machine learning to better understand it. Arthur Samuel defined machine learning as *"Machine Learning is the field of study that gives computers the ability to learn without being explicitly programmed"* (Samuel 1959). Yoshua Bengio, who is known for his works on artificial neural networks and deep learning, defines machine learning as *"Machine learning research is part of research on artificial intelligence, seeking to provide knowledge to computers through data, observations and interacting with the world. That acquired knowledge allows computers to correctly generalize to new settings"*. Tom Mitchell, American computer scientist and E. Fredkin University Professor at the Carnegie Mellon University, explained machine learning in a mathematical form as *"A computer program is said to learn from experience E with respect to some class of tasks T and performance measure P if its performance at tasks in T, as measured by P, improves with experience E"* (Mitchell 1997). Based on the definitions of the masters of this field, it may be summarized as follows; what is asked from machine learning algorithms is to discover the patterns in the data at hand, to develop a model for the solution of the problem, and generalize it, in other words, produce accurate results for new situations. In this respect, this field is closely related with computational statistics, mathematical optimization, probability theory, data mining to be able to carry out the tasks like clustering, classification, regression and estimations. Although there is no clarity and consensus in the literature, in the common sense, machine learning algorithms may be classified according to the learning type as Supervised, Unsupervised, and Reinforcement Learning. On the other hand, the problem types that are handled may be categorized as *Classification, Regression, Clustering, Association Rules, Dimensional Reduction,* and *Density Estimation* (Liao et al. 2012; Shalev-Shwartz and Ben-David 2014; Neapolitan and Jiang 2018).

3.1 Machine Learning Algorithms According to Learning Types

3.1.1 Supervised learning

The datasets in which the outputs and the inputs are known are used in the establishment of the model. The algorithm used is fed by the input vector and the output vector of the samples one by one. In time, the algorithm produces a solution space that can produce the expected output for all the samples. Example of Supervised Learning Algorithms:

1. Decision Trees
2. Naive Bayes
3. Nearest Neighbor
4. Support Vector Machine (SVM)
5. Random Forest
6. Neural Network

3.1.2 Unsupervised learning

In this approach, the targets corresponding to the inputs are not known or are not given to the algorithms. The algorithm is expected to discover the patterns in the data on its own in the construction of the model. Example of Unsupervised Learning Algorithms:

7. K-Means
8. Fuzzy C-Means
9. Soft Clustering
10. Self-Organizing Maps
11. PCA
12. Associated Rules
13. Neural Network

3.1.3 Reinforcement learning

It is a method of training used in positive and negative feedbacks like a rewarding system. The algorithm is reinforced to select the desired behaviors instead of undesirable behaviors. The algorithm that makes a lot of mistakes at first decreases its wrong responses as it is trained. Example of Reinforcement Learning Algorithms:

14. Q-Learning
15. State-Action-Reward-State-Action (SARSA)
16. Deep Q Network
17. Deep Deterministic Policy Gradient (DDPG)
18. Distributional Reinforcement Learning with Quantile Regression (QR-DQN)

3.2 Machine Learning Tasks

3.2.1 Classification

Each observation or sample in the dataset belongs to a category. The data set may consist of only two categories (Binominal, binary), or more than two categories (multinominal, multi-class). The category is often called as *class*, *label*, or *destination.* In classification problems, which mean a supervised learning task, it is expected that the algorithm that is trained with the dataset at hand learns the categories in the dataset, and then associate it with a new observation in the category it belongs to.

3.2.2 Regression

The target values are continuous in regression problems, which are supervised learning tasks. The regression approach predicts the target value by determining the linear or nonlinear relation between two (simple regression) or more variables (multiple regressions).

3.2.3 Clustering

It divides the observations or instances in the dataset into groups based on their similarities of their features. It is an unsupervised learning task. The similarity is also expressed as the linear distance, the norm in geometry, and is measured by calculating. The commonly-used distance function is the *Euclidean* distance.

3.2.4 Association rules

It is a rule-based machine learning approach targeting to rate the relations among the features of a problem observed together by identifying these features of the problem. *Apriori* is one of the most commonly known algorithms for determining relations. *The Market Basket Analysis*, which reveals the purchasing tendencies of customers, is a cliché problem. The selection of the ads that will be shown to customers on web-based shopping websites is an up-to-date application area.

3.2.5 Dimensional reduction

The real-world data has a large number of features in general. The high number of features of the observation might increase the ability to represent it; however, sometimes, it might also cause an *overfitting* problem, make it difficult to establish the model, and increase the time and resource consumption needed for the training phase. The dataset may be reduced to a more processable size by discarding the attributes with high correlations and the ones that are not representative with the *Feature Selection* and *Feature Extraction Approaches.*

3.2.6 Density estimation

The relation between the outcomes of observations and their probability is referred to as the *Probability Density*. Density Estimation is used for making estimates to a probability density function by considering the data frequency. It provides an aspect into the characteristics such as the probability distribution shape, the most likely value, the spread of the values, and thus enables the identification of anomaly or inconsistency in an observation.

It is expected from the selected machine learning approach to make a generalization to carry out the desired tasks by learning from the existing experience. There are two situations to avoid during the training process. The first one is *overfitting*, which is the start of the algorithm to memorize observations instead of learning patterns in the dataset. Although it can produce results, which are very suitable for the observations in the dataset, it produces inaccurate results for new situations. The second one is *underfitting*, which is the inability of the algorithm to capture the pattern in the observations. In general, the dataset is divided into three parts as *training*, *approval* and *testing*. In the learning phase, the memorization or little learning can be avoided by controlling the generalization performance with approval data. The final success of the established model is rated by testing it with test data. In this context, it is important that the machine learning approach that will be applied is selected according to the structure of the problem. In other words, it is possible to obtain different achievements by using different attributes and different algorithm combinations. However, since there are no methods identified in choosing the learning algorithm that is suitable for the problem, a great number of algorithms are tested in general, compared to the criteria like learning costs and accuracy success, and those with which high performance is achieved are preferred. For this reason, various machine learning platforms have been developed that can prepare and apply *cleansing, transforming, discretization, data reduction* and *attribute selection*, and that can implement a large number of learning algorithms at a fast pace. In this chapter, a concise part of *WEKA Machine Learning Software* displays that have been used from educational and academic studies to industry and commercial applications has been presented, and an application has also been provided on the supply chain problem to illustrate the use of certain algorithms within WEKA.

4. Introduction to WEKA

It is being developed by *Waikato University* in New Zealand. WEKA which stands for *"Waikato Environment for Knowledge Analysis"*, is a comprehensive collection of machine learning algorithms employed in data mining tasks. WEKA is coded in java and is open source software released under the GNU General Public License. It can be run on *Windows, Macintosh, Linux* operating systems and almost all platforms. By connecting to databases via the *Java Database Connectivity (JDBC)* driver, it can treat a query consequence and store the results of the transaction in the databases.

WEKA is kept up-to-date with commendable efforts by its developers to include even the latest algorithms in the field of data mining. The current algorithms are included in the form of plug-in packages, and users can access and install the packages through the package management system. Thanks to its diversity of algorithms, it paves the way for users to solve their problems with different and up-to-date methods and to compare the solution methods without the demand for code writing.

The GUI Chooser shown in Figure 1, which welcomes the user when WEKA starts, allows switching between five interactive interfaces. *The Explorer* is the basic section that contains the tools for the algorithms used to examine and analyze a dataset and visualization. *The Experimenter* provides performance statistics by benchmarking different classifiers or filters applied to the problem. In addition, advanced users can distribute the computational load to multiple machines by using a Java remote method invocation. *The Knowledge Flow* is an interface which serves to establish learning models in the form of a data stream by combining graphical blocks representing data sources, preprocessing tools, learning algorithms, evaluation methods and visualization modules. *The Workbench* is a stand binding all graphical interfaces within a single window in which the appearance of applications and plug-ins can be customized. Even if the interactive interfaces fulfill the need for many problems, in case they are inadequate for advanced analysis, the *SimpleCLI* which is a text based coding section completes the task. *The SimpleCLI* is also advantageous in terms of memory consumption.

4.1 Attribute-Relation File Format

Attribute-Relation File Format (ARFF) developed for WEKA is a text-based dataset file consisting of two distinct sections called *HEADER* and *DATA* as shown in the Figure 2 below. The Header Section contains information lines about the source and content of the data set, which are written after *"%"* character, in order to inform the user; however, the comment lines are not interpreted by WEKA. In addition to the description lines, *@relation* refers to the descriptive name of the data set and this information is displayed in the *Current Relation Field* in the perspective of *Preprocess* when the data set is loaded. The last part of the Header lists attributes with their types exposing the data structure. *@attribute* refers to the attribute name

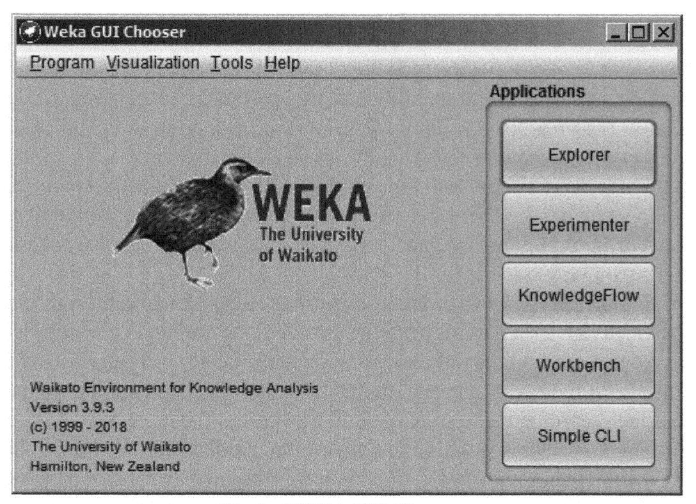

Fig. 1: WEKA GUI Chooser.

followed by its type definition or nominal values. Type definition consists of numeric, string, date and relational. Curly Brackets, commas and spaces are used composing nominal values. The data section that is started with the *@data* tag contains instances on each row where it is mandatory that the sequence corresponds to the above definition and sequence declared by *@attribute*. For unknown attribute value question mark is available instead of its value.

4.2 Explorer

It is the most commonly used interface in which tasks operated in the data mining process are provided to the user in six perspectives which are *preprocess, classify, cluster, associate, select attributes, visualize*. Initially, the data discovery to be analyzed is started on the preprocessing perspective, because the other perspectives will invisible unless a data set is properly loaded in the preprocessing section.

4.2.1 Preprocess

The data to be processed can be obtained from a file, a URL address, a database source or can be created artificially by means of *DataGenerators*. Furthermore, an interior editor is also provided for manual manipulation of data. This first step window performs two important tasks on the data set. The former provides information, such as the number of instances, the number of attributes, and statistical information for each attribute, to grasp the structure of the data set. The latter provides a variety of filters implementing processes such as cleansing, transforming, integrating, reducing and discretizing. In the *Filter Panel*, filter parameters can be assigned in the *TextField* as text-based or by selecting options on the *GenericObjectEditor*, the visual interface that appears when left-clicking. Changes made to the data can be canceled or saved for later use. *Preprocess Screen* can be viewed in Figure 3.

4.2.2 Classify

Numerous classifiers accessed by clicking the *Classifier Button* are organized according to key approaches as shown in Figure 4. The ones of the provided algorithms compatible with the dataset loaded in the previous step are visible and others are invisible in the list. The parameters of the selected algorithm can be edited with its *TextField* or *GenericObjectEditor*, as in classifiers counterpart. Four types of methods are presented to evaluate classifier performance.

1. *Use training set:* The classifier is tested with the data set used in its training.
2. *Supplied test set:* The classifier is tested with an exterior data set that is not used in its training.
3. *Cross Validation:* The data set is subdivided into groups and each one is held for testing, while others are used for training.
4. *Percentage Split:* The data set is subdivided into a training set and a test set based on a user-defined percentage.

```
%   Popular dataset related to the problem

%   of play game depending on weather conditions.


@relation weather


@attribute outlook {sunny, overcast, rainy}

@attribute temperature numeric

@attribute humidity numeric

@attribute windy {TRUE, FALSE}

@attribute play {yes, no}


@data

sunny,85,85,FALSE,no

sunny,80,90,TRUE,no

overcast,83,86,FALSE,yes

rainy,70,96,FALSE,yes

rainy,68,80,FALSE,yes

rainy,65,70,TRUE,no

overcast,64,65,TRUE,yes

sunny,72,95,FALSE,no

sunny,69,70,FALSE,yes

rainy,75,80,FALSE,yes

sunny,75,70,TRUE,yes

overcast,72,90,TRUE,yes

overcast,81,75,FALSE,yes

rainy,71,91,TRUE,no
```

Fig. 2: ARFF File.

Measurement of classifier performance can be elaborated using additional evaluation options and specific evaluation metrics. These extra options are invoked via the more options button on the test panel. *Classifier Output Panel* is the area where the results of the training and test operations are explained. The structure of the data set, the learning scheme and the test statistics are presented in detail here. As for the *Result List Panel*, it holds a list of results for each classification attempt. Through this panel, the user can compare the results of classification experiments, graphically review the results, and also store them.

4.2.3 Cluster

The Clusterer Button brings up the list of clustering schemas. Similar to classifiers perspective, the parameters of the chosen algorithm can be edited with its *TextField* or *GenericObjectEditor*. The *Ignore Attributes Button* which throws undesired attributes out is located under the cluster mode panel. *The Store Cluster* option determining whether the clustering results will be visualized is productive for data sets requiring enormous memory usage. Four methods are present for evaluating the clustering performance:

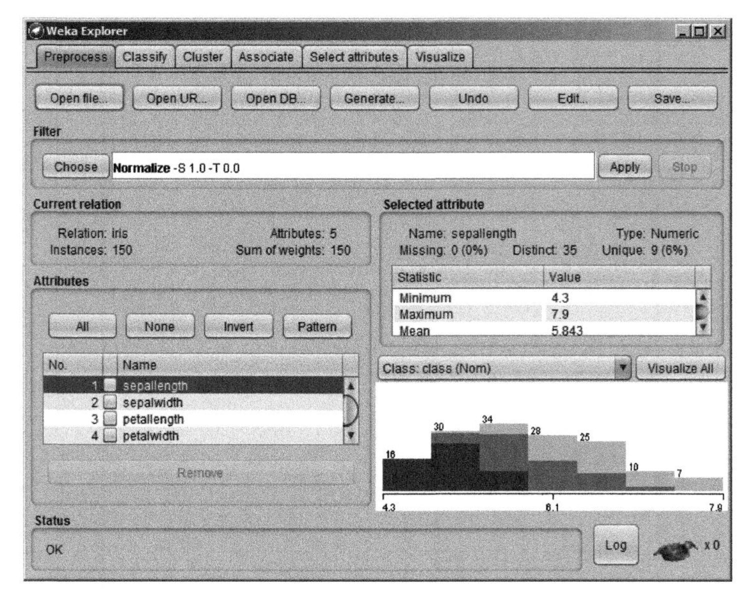

Fig. 3: Preprocess.

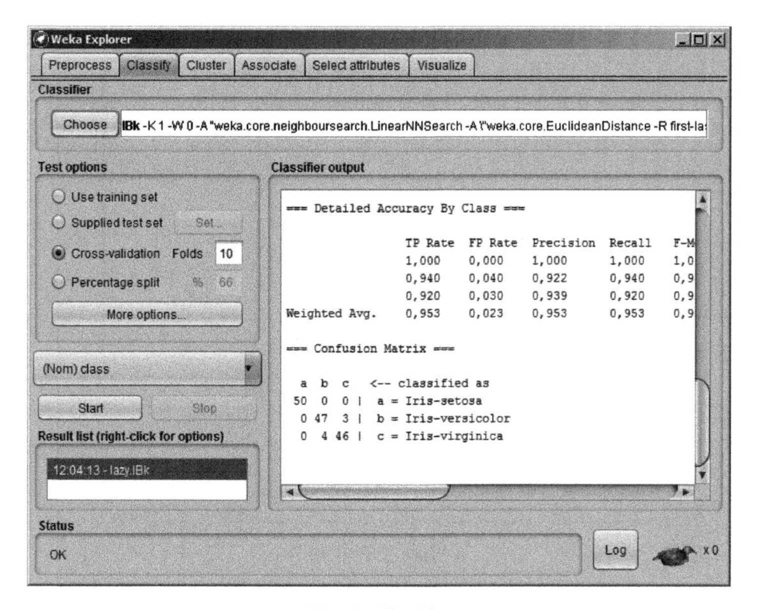

Fig. 4: Classify.

1. *Use training set:* The training set is classified according to the clusters obtained and the number of instances per cluster is calculated.
2. *Supplied test set:* The boundaries of the clusters can be evaluated on a separate test data.
3. *Percentage split:* The data set is split into two parts as the training set and the test set, considering a user-defined certain percentage. The clusters generated using the training segment are evaluated with the test segment.
4. *Classes to clusters evaluation:* Clustering is assessed by taking into account predefined classes in the data set and results are represented in the confusion matrix.

The Result List and *Clusterer Output Panels* are no different from those of the classify perspective shown in Figure 5.

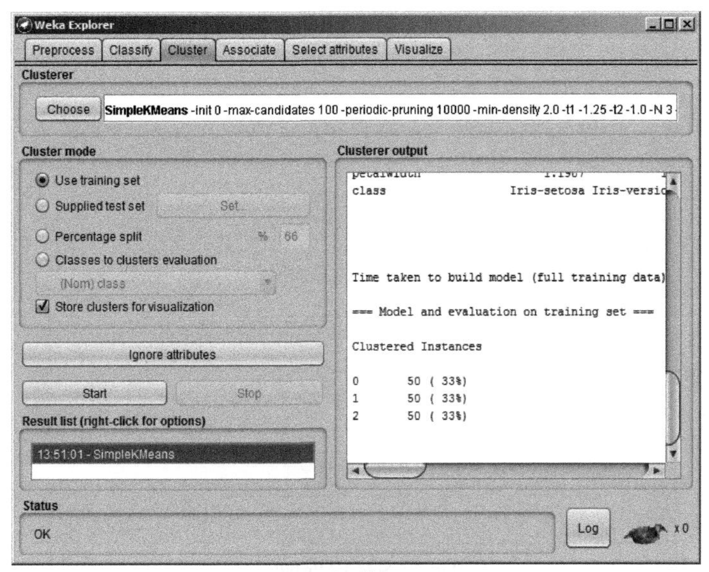

Fig. 5: Cluster.

4.2.4 Associate

The algorithms in this perspective shown in Figure 6 reveal the association rules among attributes in a data set. The algorithms are preferred from the *Associator Field.* The parameters of the algorithms can be edited with *Textfield* or *GenericObjectEditor*. Once *the Start Button* is clicked, the rules obtained are listed in *Associator Output Field.*

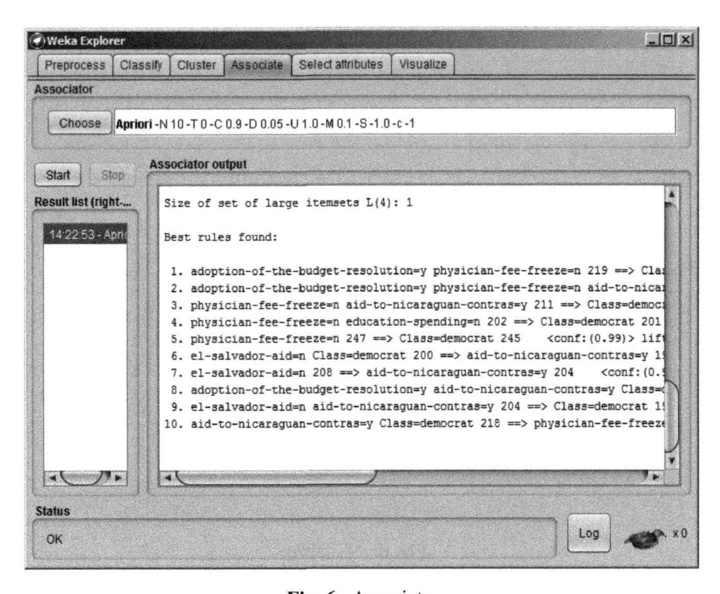

Fig. 6: Associate.

4.2.5 Select attributes

Attribute Evaluator and *Search Method* are used to establish which attributes are furthest convenient for classification or prediction. The attribute evaluator assigns which extraction method will be used, while the search method assigns what search approach will be performed. The whole data set can be handled in the evaluation, as well as cross-validation. In the *Attribute Selection Output field*, the selected attributes and their associated statistics are output. Figure 7 is a screenshot of *Select Attributes*.

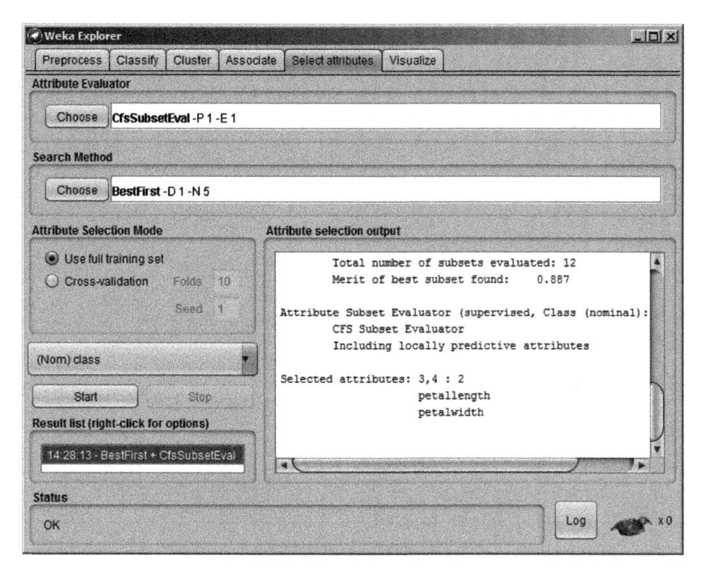

Fig. 7: Select Attributes.

4.2.6 Visualize

Two-dimensional graphs in which the distribution of attributes in the data set can be displayed are accessed through the visualize panel shown in Figure 8. Graphs can be constituted with user-defined attributes or data instances. Arrangements regarding the appearance of the graphs such as color and size can be made.

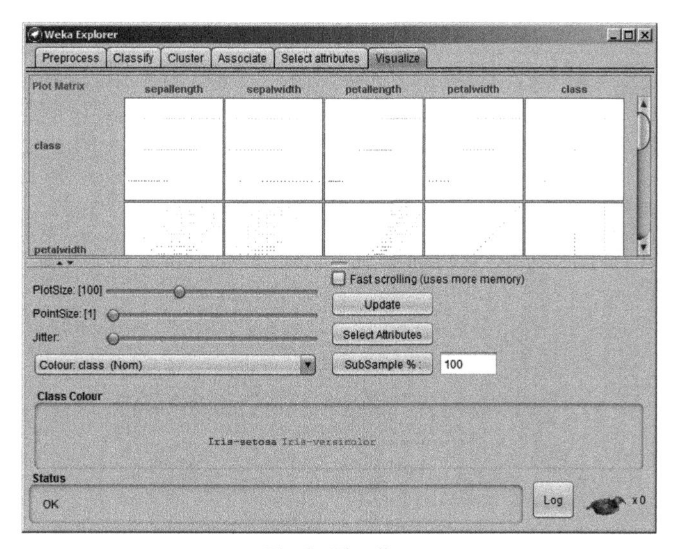

Fig. 8: Visualize.

4.3 The Experimenter

Whereas Explorer can only actuate schemes individually, more comprehensive experiments can be implemented by Experimenter facilities. As distinct from Explorer, a data set can be processed by using a number of algorithms and besides more than one data set can be processed with one or more algorithms in the application. Consequently, the user has the opportunity to analyze the performance of algorithms run on the data sets. An experiment consists of three steps entitled *Setup, Run,* and *Analyze.*

4.3.1 Setup

Setup, shown in Figure 9, has two views including simplified and advanced option. Whereas *Simple* has a simpler display for the user, *Advanced* offers access to all options. The test results can be stored in several alternative recording environments such as ARFF file, CSV file or JDBC database. Storing the results in the database is more advantageous in terms of time consumption for experiments broken or to expand them. Just as the classification counterpart, the cross-validation or percentage split techniques can be used for testing and training process. It is important to repeat the training to generate more reliable results. In the *Iteration Control Field*, the number of repetitions is set and while working multiple algorithms on multiple data sets, it is preferred whether the data sets or algorithms are handled initially. In the *Datasets and Algorithms Panels*, once the data sets and learning schemes examined are selected, they become ready to work. It is also possible to load and store the settings of algorithms that require multiple parameter settings.

As for the advanced interface, the *Result Generator Panel* has been added to allow the user to determine the result generators, which is the detailed equivalent of the experiment type in the simple view. Apart from the Result Generator, there is a *Distribute Experiment Panel* that distributes the processing load to the other nodes in the network. A database server, computers and properly generated remote engine policy file are required to perform this feature.

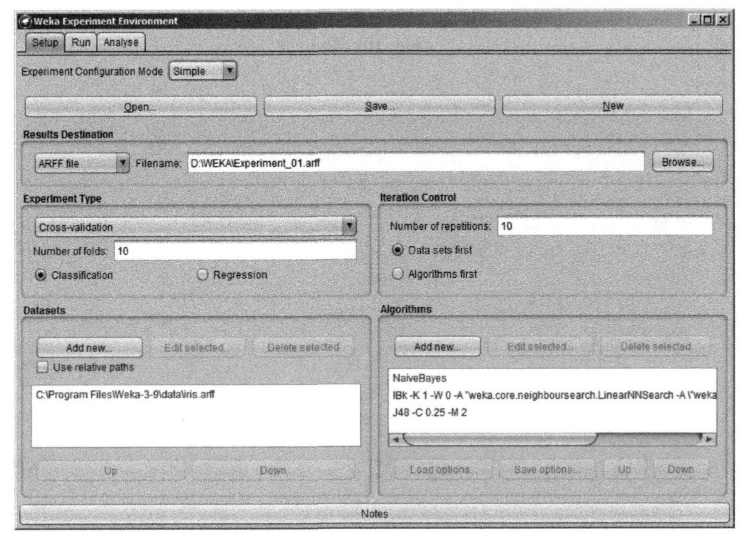

Fig. 9: Setup.

4.3.2 Run

The experiment is launched by clicking the *Start Button* in the *Run Tab* shown in Figure 10. After selected learning schemes have been employed on the data sets, a message stating that the operations were completed without error should be received. The results are stored in the specified file path in the *Result Destination Field* on the *Setup Tab.*

4.3.3 Analyse

If already saved, the experiment is accessed from a file or database source; otherwise, clicking the *Experiment Button* will bind the experiment that has just been finalized on the *Run Tab.* Perform Test is the button which generates detailed statistics, yet the test configuration must primarily be done by selecting the options in the *Configure Test Field.* Numerous criteria are proposed to assess the performance of the learning schemas. These criteria can be viewed in Figure 11.

1. *T-Test:* Measures whether there is a significant difference between the averages of the user groups.
2. *Select Rows and Cols:* Assigns the criteria to the rows and columns of the result table.
3. *Comparison Field:* Selects the type of statistics to compare.
4. *Significance:* Specifies confidence threshold
5. *Sorting (asc) by:* Sets the sorting criteria of table rows.

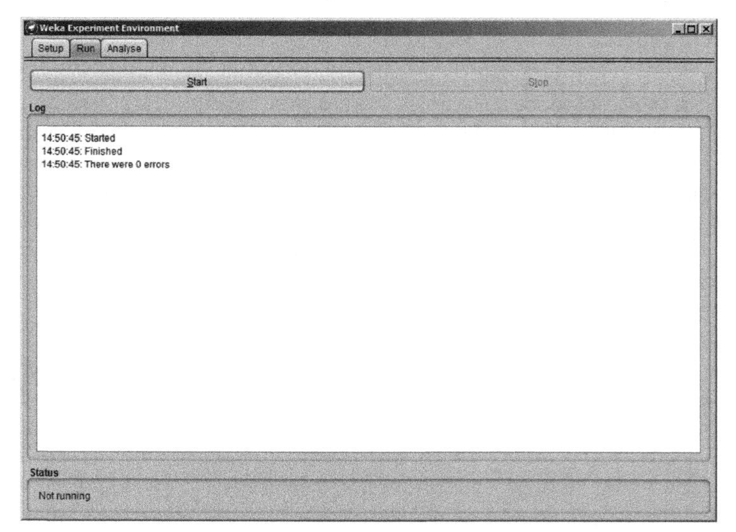

Fig. 10: Run.

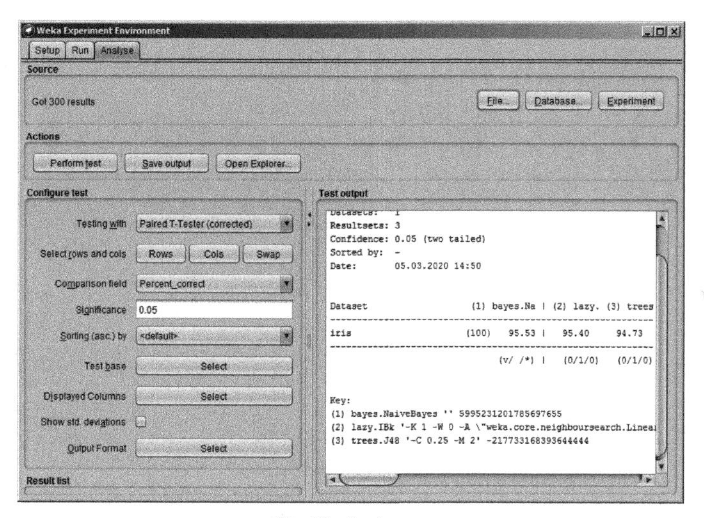

Fig. 11: Analyse.

6. *Test Base:* Uses to change the baseline scheme.
7. *Displayed Columns:* Some of the items selected in "Select rows and cols" can be taken away from the result table. However, the test base cannot be discarded from the table.
8. *Show Std. Deviations:* Adds standard deviation information to the result table.
9. *Output Format:* Provides utility tools as follows for editing the output format. Precision can be specified for Mean and std deviation. A row representing the average of each column can be added. Plain Text, CSV, HTML, LaTeX, GNUPlot and Significance Only can be specified as output formats. Using the advanced setup option, in addition to those mentioned, all adjustable properties of the output matrix can be particularized.

4.4 The KnowledgeFlow

It is an application where all the data mining methods mentioned in the *Explorer* section are served in iconic form. The block functions representing the operation processes are associated with link nodes on the edges of their symbols and thereby composing a flowchart executing the work. Unlike Explorer; In the *KnowledgeFlow* shown in Figure 12, both

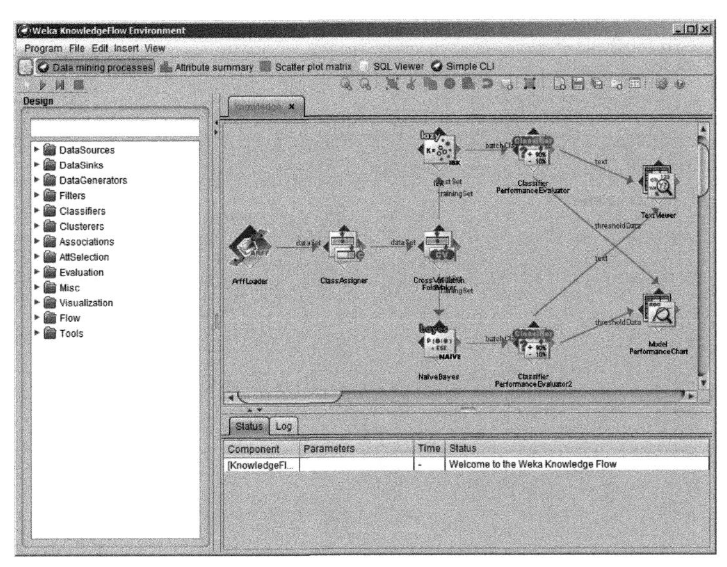

Fig. 12: KnowledgeFlow Environment.

continuous learning and components can be applied sequentially, e.g., the data can be passed through multiple filters. In addition, each of the components is executed as a separate thread.

To summarize the usage of this perspective, filters, classifiers, clusters, association rules and attribute selectors, as well as other tools, are located in folders named after them in the *Design Field* built into the left edge of the perspective. These components are placed on a layout canvas by drag and drop in the order required by the data mining tasks and thereafter the parameters of the components can be edited by double-clicking or selecting configure option from the shortcut menu.

As for general handling in a data flow, the *DataSources* tools are used to obtain the dataset from a data source or can be generated through the *DataGenerators* tools. Evaluation tools are used for the determination of the column that holds the class information in the dataset, the approach to which training and test data will be obtained, e.g., split or cross-validation and the criteria for evaluating test phase. Using the components in the visualization category, the evaluation results can be represented as text or chart. By means of *DataSinks* components, a data set subdivision, a trained model, a chart and text-based information can be recorded. Extra tools are also available to manage the data flow, in the flow and tools categories.

4.5 The Workbench

The Workbench added with WEKA version 3.8 brings together the perspectives described so far under the same roof. As viewed in Figure 13, on the Workbench, each perspective has its own tab located at the top of the layout. The user can define the settings in the perspectives, such as initial settings, default values and appearance. Apart from these adjustments, there is no difference in the functionality of the perspectives from those previously described. These settings can be accessed with the gear illustrated button located to the left of the perspectives.

As an example in the following screenshot shown in Figure 14, the settings which are some initial and default values belong to the clustering options, are shown. It is also a pleasing alternative to leave the text and background colors on the output panel to the user's preference.

4.6 SimpleCLI

Java packages running behind interactive interfaces can be activated with coding via Weka's command-line interface. Help lists the main commands of SimpleCLI.

1. *capabilities <classname> <args>:* Lists the capabilities of the specified class. If the class is a weka.core.OptionHandler then trailing options after the classname will be set as well.
2. *cls:* Clears the output area.
3. *echo msg:* Outputs a message.

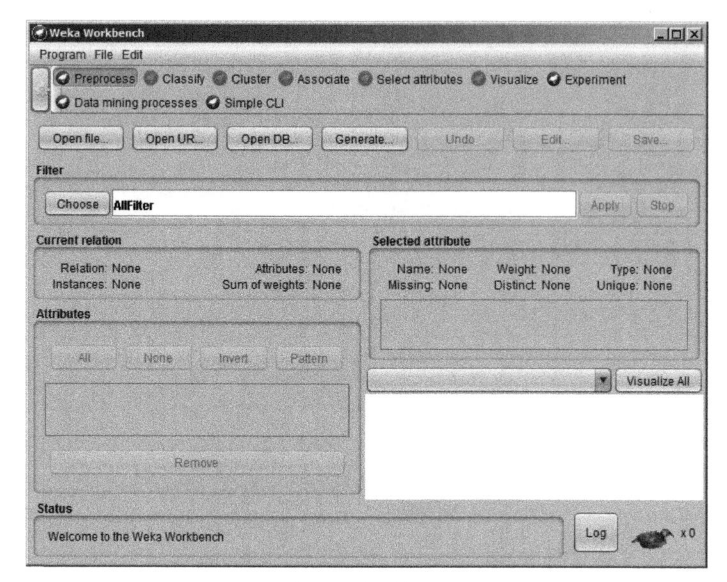

Fig. 13: Workbench.

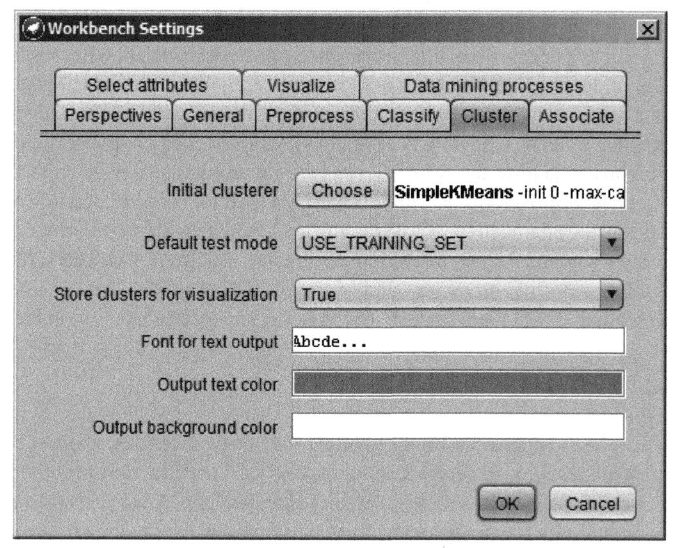

Fig. 14: Workbench Settings.

4. *exit:* Exits the SimpleCLI program.
5. *help [command1] [command2] [...]:* Outputs the help for the specified command or, if omitted, for all commands.
6. *history:* Prints all issued commands.
7. *java <classname> <args>:* Lists the capabilities of the specified class. If the class is a weka.core.OptionHandler then trailing options after the classname will be set as well.
8. *kill:* Kills the running job, if any.
9. *script <script_file>:* Executes commands from a script file.
10. *set [name=value]:* Sets a variable. If no key=value pair is given all current variables are listed.
11. *unset name:* Removes a variable.

Weka has a hierarchical Java package structure. Namely, a classifier is contained in a classifiers subpackage at the higher level, which is grouped according to the approach method; the classifiers subpackage is contained in classifiers

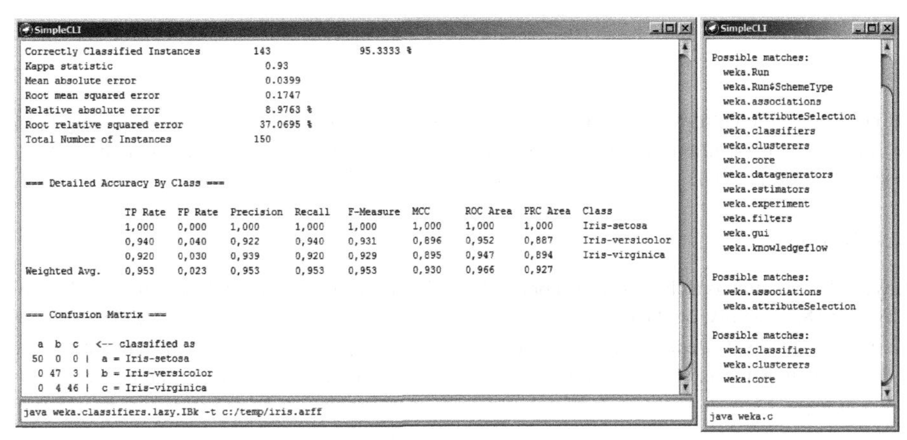

Fig. 15: SimpleCLI.

package which is at the much higher level, and the classifiers package is contained in weka package which is at the top-level. Like *IBk> Lazy> Classifiers> Weka.* Java packages are called with the java <classname> <args> command. In the example, k-Nearest Neighbour is run on iris.arff dataset with default parameters. The parameter "- t" refers to the dataset portion to be used for training.

As shown in the Figure 15, the Tab key is functional as a command complement. In the example, the Tab key lists matching packages/commands after *"java.weka.a"* and *"java.weka.c"*. If the command is composed until the classifier name, a description of both the general parameters used in each classifier and the classifier-specific parameters is displayed.

Detailed information on the schemes, algorithms and parameters of the packages can be purchased from the WEKA documentation pages (Bouckaert et al. 2018).

5. Classification of Supply Chain Data by Using WEKA

5.1 Material and Methods

The classification processes have been sampled on supply chain data by using 4 different classifiers consisting of Decision Tree, Naive Bayes (NB), K-Nearest Neighbor (k-NN), and Artificial Neural Network (ANN).

5.1.1 J48 algorithm

J48 is the application of the C4.5 decision tree algorithm, which was developed by Quinlan in Weka (Quinlan 1993). Decision Tree is the process of dividing the existing observations by using clustering approaches until each group has observations from the same class. A decision tree has a graphical representation in the tree structure, which itself includes all possible scenarios. The decision nodes in the tree are determined by calculating the Information Gain Ratio, which is given with Equation 3. The branches of the tree hold the answer to the questions, which are asked to the decision node to which it belongs, and the end-nodes-leaves represent the class labels. Each path leading from the root node to the leaves constitutes a decision rule.

$$H(D) = -\sum_{i=1}^{c} p_i \, log_2 (p_i) \tag{1}$$

Here, D refers to the observations in the dataset, c refers to the classes, p_i refers to the class i probability, and $H(D)$ refers to the entropy of the dataset in the Equation 1. Entropy means the probability of an event, and is inversely proportional to the amount of information acquired. The entropy of an attribute is calculated with Equation 2.

$$H_A (D) = -\sum_{j=1}^{s} \frac{|D_j|}{D} H(D_j) \tag{2}$$

$$\text{Information } Gain\ Ratio = \frac{H_A(D)}{H(D)} \tag{3}$$

Here, $H_A(D)$ shows the entropy of the discriminating attribute, which has a value of s dividing the dataset into s subsets. The attribute with the minimum Information Gain Ratio, the ratio of the information acquisition of the tested attribute to the total information acquisition, is determined as the decision node.

5.1.2 Naive Bayes

The Naive Bayes Classification is a classification approach that is based on the work of the English Mathematician Thomas Bayes, who lived in 18th Century, on the probability theory. It aims to determine which possible class the new observation probably belongs to. Namely, *C*, representing the c possible class, $X = \{x_1, x_2, \ldots, x_n\}$ to represents the *n* feature variables of the observation, the hypothesis of the probability of the *X* observation being of class *C* can be written as Equation 4.

$$P(C|X) = \frac{P(X|C)P(C)}{P(X)} \tag{4}$$

Here,

$P(C|X)$ is the posterior probability of *Class* given *Observation.*

$P(X|C)$ is the likelihood which is the probability of *Observation* given *Class.*

$P(C)$ is the prior probability of *Class.*

$P(X)$ is the prior probability of *Observation.*

As it is seen in Equation 5, the Naive Bayes Classifier is the product of all the conditional probabilities, and as $P(X)$ is equal to all classes, the *X* observation is considered to belong to the class that maximizes $P(X|C)\ P(C)$.

$$P(X \mid C) = \prod_{i=1}^{n} P(X_k \mid C) = P(X_1 \mid C)P(X_2 \mid C) \ldots P(X_n \mid C) \tag{5}$$

Since the Naive Bayes Classifier is not an iterative calculation method, it can work quickly classifying big data sets with high accuracy rates.

5.1.3 k-Nearest Neighbor

The k-Nearest Neighbor (k-NN), which was proposed by Fix and Hodges in 1951, is a sample-based classification algorithm estimating the class of the new observation that is based on known observations in the training set (Fix and Hodges 1989). It is also named as the Lazy Learning Method because there is no training stage in calculating the values of the variables of the method by using the training data, which is the case in some supervised learning algorithms. The k-NN algorithm examines the similarity between the new observation and other observations in the training set. The similarity is found by calculating the metric distance between the new observation, whose class is sought, and the attribute variables of previous observations. The following Figure 16 demonstrates in the two-dimensional space how the similarity is measured between the new sample whose class is sought and the neighbors whose classes are known.

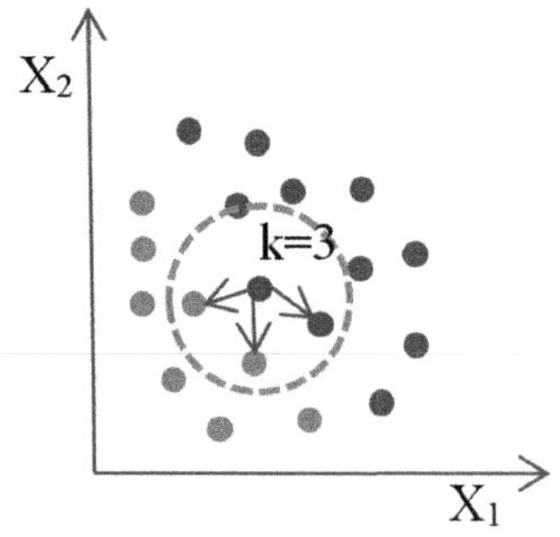

Fig. 16: Neighborhood relationship between samples.

The distance between the attribute vector of the observation whose class is sought X^u and the X_i^j, which is the attribute vectors of all observations in the training set is calculated as in the distance Equation 6.

$$distance = \|X^u - X_i^j\| \tag{6}$$

The most common metric that is used in similarity measurement is the Euclidean Distance, which is the application of Pythagorean Theorem, and is formulated in Equation 7. It is calculated by taking the square root of the sum of the squares the differences between the attribute variables of the new observation and the attribute variables of the neighboring previous observations.

$$Euclidean\ distance = \sqrt{\sum_{i=1}^{n}(x_i - y_i)^2} \tag{7}$$

The class in which the majority of the most similar k observations belongs is considered as the class of the new observation in the similarity vector calculated with Equation 6.

5.1.4 Artificial neural network

It is a supervised classification algorithm, which imitates the biological nervous system behavior. The basic functioning element is the artificial nerve cell, which is called as the neuron, and the togetherness of a large number of neurons creates an artificial neural network. The artificial neural network consists of three basic parts as the input layer, the hidden layer, and the output layer. An example of the artificial neural network model is shown in Figure 17.

The number of inputs of the artificial neural network is represented with x, which is the attribute variable of the observations, and which is equal to the number. Outputs that represent the classes are shown with y. The hidden layer can be edited once or more, and there is no definitive method to determine the number of nodes in the hidden layers. The weight coefficients that are represented with W determine the relations between the input nodes, hidden layer nodes, and output nodes. Each neuron weighs its inputs, and transfers their sum to the activation function. The activation function may be linear or some special functions like Sigmoid. The purpose of the training stage is to calculate the final values, which will produce each observation class accurately by updating the weights that were determined randomly at the initial stage. To update the weights, linear approaches like extreme machine learning (Huang et al. 2006), or iterative approaches like back propagation algorithm are used (Hinton 2007).

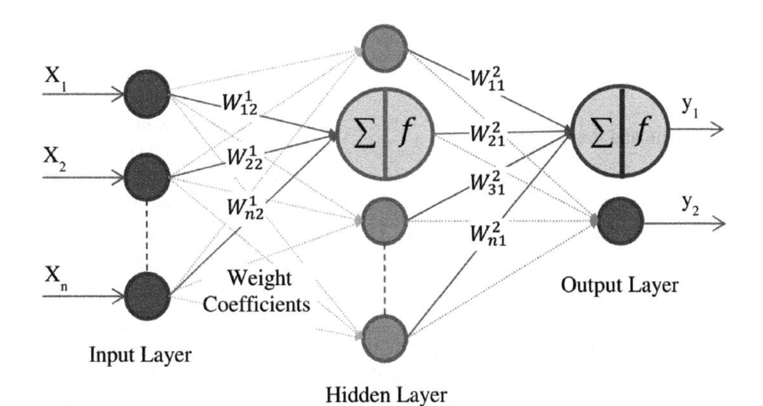

Fig. 17: ANN architecture.

5.2 Organization of the Supply Chain Data

In this exemplary application, suppliers are considered according to 10 criteria. The following Table 1 shows the criteria along with related sources and detailed descriptions. The Quality and On-Time delivery criteria are in the form of number in percentages, and other criteria consist of discrete numbers between 1–9 according to Likert Scale. The data set contains 1000 instances generated randomly, and each of them contains 10 attributes. The class distribution in the data set is that 144 samples belong to the low-grade supplier represented by class 1, 764 samples belong to the middle-class supplier

Table 1: Supplier selection criteria with their related sources and definitions.

No.	Criteria Name	Papers	Definitions
1	**Price**	(Dickson 1966; Weber et al. 1991; Sarkar and Mohapatra 2006; Jain et al. 2009; Thanaraksakul and Phruksaphanrat 2009; Deshmukh and Chttudhari 2011; Erginel and Gecer 2016; Taherdoost and Brard 2019)	Price can include unit price, transportation cost, production cost, taxes, and discount./Unit product price identified by potential suppliers.
2	**Quality**	(Dickson 1966; Weber et al. 1991; Jain et al. 2009; Thanaraksakul and Phruksaphanrat 2009; Buyukozkan and Cifci 2011; Deshmukh and Chttudhari 2011; Erginel and Gecer 2016; Taherdoost and Brard 2019; Sarkar and Mohapatra 2006; Chan and Kumar 2007; Wang et al. 2009)	Quality is defined as when the products meet customer demands and requirements and meet their specifications./ The ratio of the number of quality products to the total number of products.
3	**On Time Delivery**	(Dickson 1966; Weber et al. 1991; Sarkar and Mohapatra 2006; Jain et al. 2009; Thanaraksakul and Phruksaphanrat 2009; Wang et al. 2009; Deshmukh and Chttudhari 2011; Erginel and Gecer 2016; Taherdoost and Brard 2019)	Delivery can include lead-time, supply Ability, delivery time, location, and transportation./The ratio of the number of products delivered on time to the total number of products.
4	**Environmental responsibility**	(Chiou et al. 2008; Thanaraksakul and Phruksaphanrat 2009; Taherdoost and Brard 2019)	The supplier's responsibility to use natural resources carefully, implement recycling-reusing-refurbishing-remanufacturing operations, minimize damage, reduce energy consumption and possess environmental certificates such as ISO 14000, environmental policies./The level of fulfillment of environmental responsibility by potential suppliers.
5	**Social responsibility**	(Thanaraksakul and Phruksaphanrat 2009; Taherdoost and Brard 2019)	The supplier's responsibility to present employee benefits and rights and stakeholders' rights, disclosure of information, respect for policy and provide occupational health and safety and corporate social responsibility./The level of fulfillment of social responsibility by potential suppliers.
6	**Industry 4.0-Maturity**	(Torğul and Paksoy 2019)	Industry 4.0 incorporates internet of things, cyber physical systems, sensors, RFID technologies, robotics technologies, artificial intelligence, big data, 3D printing, cyber security, augmented reality and cloud computing./The maturity level of Industry 4.0 concepts within the organization, how well these systems are transformed, designed and functioning.
7	**Flexibility**	(Jain et al. 2009; Thanaraksakul and Phruksaphanrat 2009; Buyukozkan and Gocer 2017)	Supplier flexibility can be defined as the easy adaptation of the supplier to customer requirements./The flexibility level of potential suppliers.
8	**Warranties and claim policies**	(Dickson 1966; Jain et al. 2009; Thanaraksakul and Phruksaphanrat 2009; Deshmukh and Chttudhari 2011; Erginel and Gecer 2016)	After-sale tracking services, written warranties that promises to repair or replace the product if necessary within the specified period of time and claim policies for the scope or compensation of a loss or policy event.
9	**Mutual trust and easy communication**	(Taherdoost and Brard 2019)	Assurance on the quality of the service offered by the supplier and liabilities between the buyer and the supplier, Supplier's communication system with information on the order's progress data./The level of trust and communication with potential suppliers.
10	**Reputation and position in industry**	(Dickson 1966; Weber et al. 1991; Jain et al. 2009; Deshmukh and Chttudhari 2011; Taherdoost and Brard 2019)	The factors such as market share, status, image, past performance and reputation of potential suppliers.

represented by class 2, and 92 belong to the high-class supplier represented by class 3. The class information is in the latest column. The dataset constructed is shown in Figure 18 in the Arff format.

The dataset opened in WEKA Explorer is shown in Figure 19. The Preprocess Screen has several parts like the number of the samples that inform the user on the contents of the dataset, the number of attributes, statistical information on the values of the attributes, the attributes, and the bar-graph showing the class distributions. However, the main function of the Preprocess is its including a variety of filters to prepare the data for processing.

5.3 Classification Results

When the model is established, if the dataset is as whole as it is here, there are two approaches to create the training and test sets. The first one is the Percentage Split Method, which divides the dataset into two parts at a user-specified rate, the training and the test set. Here, the drawback is that the classification results may be very good or very bad because the training and test sets that are created with the division do not represent the overall characteristics well. The second is the Cross Validation. In this approach, the dataset is divided into specific subsets, widely to 10 subsets. Each step is considered as a subset test, and the rest is considered as a training set. The average results for each subset are considered as the final classification performance.

Figure 20 shows the classification perspective of the Supply Chain Data. The testing was done with the Cross-Validation Method. The Classifier Output Screen provides detailed statistical data for the classification process. The results of the statistics are given in Table 2 and confusion matrices of classifiers are given in Table 3.

```
@relation SupplyChain


@attribute Quality numeric

@attribute OnTimeDelivery numeric

@attribute Price {1,2,3,4,5,6,7,8,9}

@attribute EnvironmentRes {1,2,3,4,5,6,7,8,9}

@attribute SocialRes {1,2,3,4,5,6,7,8,9}

@attribute Industry40 {1,2,3,4,5,6,7,8,9}

@attribute Flexibility {1,2,3,4,5,6,7,8,9}

@attribute Warranties {1,2,3,4,5,6,7,8,9}

@attribute Trust {1,2,3,4,5,6,7,8,9}

@attribute Reputation {1,2,3,4,5,6,7,8,9}

@attribute Class {1,2,3}


@data

81,76,3,5,1,2,2,3,5,2,1

75,89,3,4,6,2,2,4,5,3,1

81,71,5,2,5,2,1,4,5,3,1

83,83,2,6,2,4,2,5,2,3,1

87,75,2,1,7,3,1,2,4,2,1

81,71,1,4,2,8,1,6,2,2,1

|||||||||||||||||||||||
```

Fig. 18: Supply Chain Data in the ARFF format.

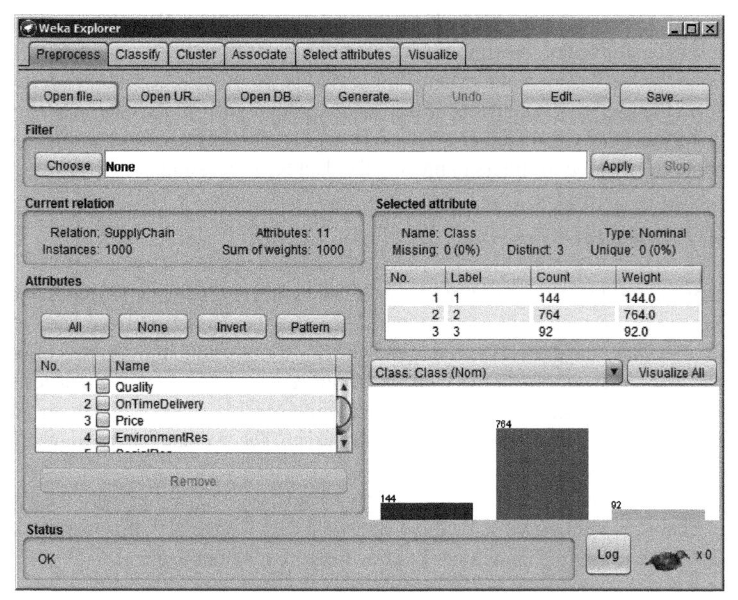

Fig. 19: Preprocess perspective of Supply Chain Data.

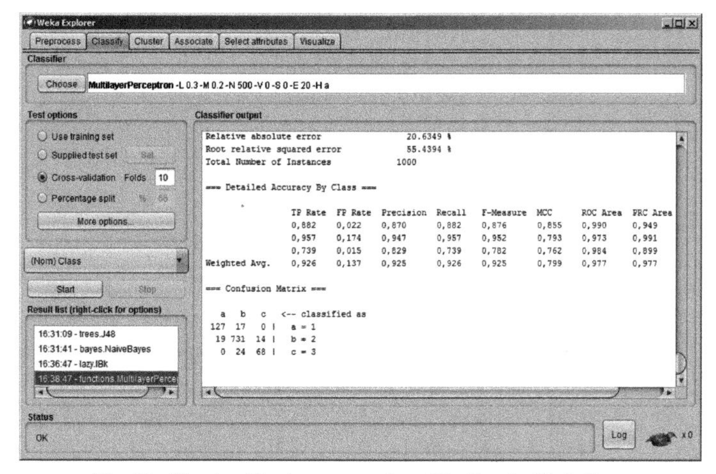

Fig. 20: The classification perspective of the Supply Chain Data.

According to Table 2, the four methods used correctly classified suppliers by more than 80%. However, the ANN method showed the best performance with 92.6%. Looking at Table 3, the ANN method also predicted classes 1 and 2 better than other methods, and predicted Class 3 worse than only the NB method. As a result, we can say that the use of ANN would be correct in terms of achieving the best results in supplier evaluation.

The suppliers who are member of Class 1 are the suppliers with high-risk level and the conditions regarding the critical criteria should be improved first. The firm recommends reducing or eliminating the high risks identified in the process of guiding suppliers of this class. In case of a negative response in which recovery cannot be achieved, these suppliers are pruned. The suppliers who are member of Class 2 are the suppliers with medium risk level. Their basic characteristics (primary criteria) are in good condition and they are potentially recommended candidates, however the findings identified as risky for the company should be corrected. The evaluation process continues until these suppliers enter in Class 3. The suppliers who are members of Class 3 are the suppliers with low risk. They can be chosen to establish a long-term relationship and do not require any action for the firm.

Table 2: Statistics of classification results.

	J48	NB	k-NN	ANN
Correctly Classified Instances	88.1 %	88.8%	84.3%	92.6%
Kappa statistic	0.684	0.726	0.5557	0.8061
Mean absolute error	0.109	0.1061	0.1503	0.0534
Root mean squared error	0.2501	0.2361	0.2788	0.1992
Relative absolute error	42.1171%	41.01%	58.1007%	20.6349%
Root relative squared error	69.6247%	65.7346%	77.6039%	55.4394%
TP-Rate	0.881	0.888	0.843	0.926
FP-Rate	0.227	0.137	0.345	0.137
Precision	0.879	0.895	0.837	0.925
Recall	0.881	0.888	0.843	0.926
F-Measure	0.880	0.890	0.835	0.925
MCC	0.673	0.717	0.548	0.799
ROC Area	0.886	0.955	0.877	0.977
PRC Area	0.883	0.956	0.862	0.977

Table 3: Confusion matrices of classifiers results.

	J48			NB			k-NN			ANN		
	1	**2**	**3**	**1**	**2**	**3**	**1**	**2**	**3**	**1**	**2**	**3**
Class 1	**103**	41	0	**124**	20	0	**92**	52	0	**127**	17	0
Class 2	23	**713**	28	43	**691**	30	41	**711**	12	19	**731**	14
Class 3	0	27	**65**	0	19	**73**	0	52	**40**	0	24	**68**

6. Conclusion

With the growth of supply chains, complex and large amounts of data have become difficult to analyze and supplier selection has begun to be influenced by many attributes with too complex effects to be determined by conventional methods.

In this chapter, we introduced a new solution approach to supplier evaluation. 10 criteria from the current literature were selected primarily to evaluate suppliers, and each was assessed for their importance in selecting an appropriate supplier. Then, hypothetic data set was created for 1000 supplier profiles evaluated according to these criteria. Artificial neural networks, decision trees, bayesian classifiers and k-nearest neighbor were applied to classify these data in the WEKA machine learning tool. Three classes were determined for the risk profiles of suppliers and the attitudes of the enterprises to their suppliers according to each classes were proposed. The results show that the use of J48, NB, k-NN and ANN algorithms with WEKA machine learning tool can support supplier selection decision-making process and may lead to improvements in suppliers' risk reduction decisions and efforts.

In the next step, different algorithms can be run for the data in the WEKA and the results can be compared. As a result, the proposed approach is flexible and so, can be used to find new partners or is easily applicable to other real case supplier selection problems however much the dataset size.

References

Allgurin, A. and F. Karlsson. 2018. Exploring Machine Learning for Supplier Selection: A case study at Bufab Sweden AB. Independent thesis Advanced level (degree of Master (One Year)) Student Thesis.

Amid, A., S.H. Ghodsypour and C. O'Brien. 2009. A weighted additive fuzzy multiobjective model for the supplier selection problem under price breaks in a supply Chain. International Journal of Production Economics, 1212: 323–332.

Awasthi, A., S.S. Chauhan and S.K. Goyal .2010. A fuzzy multicriteria approach for evaluating environmental performance of suppliers. International Journal of Production Economics, 1262: 370–378.

Awasthi, A. and G. Kannan. 2016. Green supplier development program selection using NGT and VIKOR under fuzzy environment. Computers & Industrial Engineering, 91: 100–108.

Bouckaert, R.R., E. Frank, M. Hall, R. Kirkby, P. Reutemann, A. Seewald and D. Scuse. 2018. WEKA manual for version 3-9-3. The University of Waikato, Hamilton, New Zealand.

Buyukozkan, G. and G. Cifci. 2011. A novel fuzzy multi-criteria decision framework for sustainable supplier selection with incomplete information. Computers in Industry, 622: 164–174.

Buyukozkan, G. and F. Gocer. 2017. Application of a new combined intuitionistic fuzzy MCDM approach based on axiomatic design methodology for the supplier selection problem. Applied Soft Computing, 52: 1222–1238.

Cavalcante, I.M., E.M. Frazzon, F.A. Forcellini and D. Ivanov. 2019. A supervised machine learning approach to data-driven simulation of resilient supplier selection in digital manufacturing. International Journal of Information Management, 49: 86–97.

Chan, F.T.S. 2003. Interactive selection model for supplier selection process: an analytical hierarchy process approach. International Journal of Production Research, 4115: 3549–3579.

Chan, F.T.S. and N. Kumar. 2007. Global supplier development considering risk factors using fuzzy extended AHP-based approach. Omega-International Journal of Management Science, 354: 417–431.

Chan, F.T.S., N. Kumar, M.K. Tiwari, H.C.W. Lau and K.L. Choy. 2008. Global supplier selection: a fuzzy-AHP approach. International Journal of Production Research, 4614: 3825–3857.

Chen, C.T., C.T. Lin and S.F. Huang. 2006. A fuzzy approach for supplier evaluation and selection in supply chain management. International Journal of Production Economics, 1022: 289–301.

Chen, Y.G., S. Wang, J.R. Yao, Y.X. Li and S.Q. Yang. 2018. Socially responsible supplier selection and sustainable supply chain development: A combined approach of total interpretive structural modeling and fuzzy analytic network process. Business Strategy and the Environment, 278: 1708–1719.

Chen, Z.P. and W. Yang. 2011. An MAGDM based on constrained FAHP and FTOPSIS and its application to supplier selection. Mathematical and Computer Modelling, 5411-12: 2802–2815.

Chiou, C.Y., C.W. Hsu and W.Y. Hwang. 2008. Comparative investigation on green supplier selection of the american, Japanese and Taiwanese electronics industry in China. Ieem: 2008 International Conference on Industrial Engineering and Engineering Management, 1–3: 1909–1914.

Deshmukh, A.J. and A.A. Chttudhari. 2011. A review for supplier selection criteria and methods. Technology Systems and Management, 145: 283–291.

Dey, B., B. Bairagi, B. Sarkar and S. Sanyal. 2012. A MOORA based fuzzy multi-criteria decision making approach for supply chain strategy selection. International Journal of Industrial Engineering Computations, 34: 649–662.

Dickson, G.W. 1966. An analysis of vendor selection systems and decisions. Journal of Purchasing, 21: 5–17.

Ding, R., Y. Chen and L. He. 2009. Fuzzy Multi-objective Programming Model for Logistics Service Supplier Selection. Ccdc 2009: 21st Chinese Control and Decision Conference, Vols 1-6, Proceedings: 1527–1531.

Erginel, N. and A. Gecer. 2016. Fuzzy multi-objective decision model for calibration supplier selection problem. Computers & Industrial Engineering, 102: 166–174.

Fix, E. and J.L. Hodges. 1989. Discriminatory analysis—nonparametric discrimination—consistency properties. International Statistical Review, 573: 238–247.

Gencer, C. and D. Gurpinar. 2007. Analytic network process in supplier selection: A case study in an electronic firm. Applied Mathematical Modelling, 3111: 2475–2486.

Guo, X.S., Z.P. Yuan and B.J. Tian. 2009. Supplier selection based on hierarchical potential support vector machine. Expert Systems with Applications, 363: 6978–6985.

Guo, X.S., Z.W. Zhu and J. Shi. 2014. Integration of semi-fuzzy SVDD and CC-Rule method for supplier selection. Expert Systems with Applications, 414: 2083–2097.

Hinton, G.E. 2007. Learning multiple a layers of representation. Trends in Cognitive Sciences, 1110: 428–434.

Hinton, G.E. 2013. "Lecture 1a: Why do we need machine learning?" http://www.cs.toronto.edu/~tijmen/csc321/lecture_notes.shtml#1a.

Huang, G.B., Q.Y. Zhu and C.K. Siew. 2006. Extreme learning machine: Theory and applications. Neurocomputing, 701-3: 489–501.

Jain, V., S. Wadhwa and S.G. Deshmukh. 2009. Select supplier-related issues in modelling a dynamic supply chain: potential, challenges and direction for future research. International Journal of Production Research, 4711: 3013–3039.

Kang, H.Y., A.H.I. Lee and C.Y. Yang. 2012. A fuzzy ANP model for supplier selection as applied to IC packaging. Journal of Intelligent Manufacturing, 235: 1477–1488.

Keskin, G.A. 2015. Using integrated fuzzy DEMATEL and fuzzy C: means algorithm for supplier evaluation and selection. International Journal of Production Research, 5312: 3586–3602.

Kilic, H.S. 2013. An integrated approach for supplier selection in multi-item/multi-supplier environment. Applied Mathematical Modelling, 3714-15: 7752–7763.

Kumar, S., S. Kumar and A.G. Barman. 2018. Supplier selection using fuzzy TOPSIS multi criteria model for a small scale steel manufacturing unit. Procedia Computer Science, 133: 905–912.

Kuo, R.J., C.W. Hsu and Y.L. Chen. 2015. Integration of fuzzy ANP and fuzzy TOPSIS for evaluating carbon performance of suppliers. International Journal of Environmental Science and Technology, 1212: 3863–3876.

Lee, A.H.I. 2009. A fuzzy supplier selection model with the consideration of benefits, opportunities, costs and risks. Expert Systems with Applications, 362: 2879–2893.

Liao, S.H., P.H. Chu and P.Y. Hsiao. 2012. Data mining techniques and applications—A decade review from 2000 to 2011. Expert Systems with Applications, 3912: 11303–11311.
Liu, F.H.F. and H.L. Hai. 2005. The voting analytic hierarchy process method for selecting supplier. International Journal of Production Economics, 973: 308–317.
Mirkouei, A. and K.R. Haapala. 2014. Integration of machine learning and mathematical programming methods into the biomass feedstock supplier selection process. Flexible Automation and Intelligent Manufacturing.
Mirzaee, H., B. Naderi and S.H.R. Pasandideh. 2018. A preemptive fuzzy goal programming model for generalized supplier selection and order allocation with incremental discount. Computers & Industrial Engineering, 122: 292–302.
Mitchell, T.M. 1997. Machine Learning. New York, McGraw-Hill.
Mohammady, P. and A. Amid. 2011. Integrated fuzzy AHP and fuzzy VIKOR model for supplier selection in an agile and modular virtual enterprise. Fuzzy Information and Engineering, 34: 411–431.
Moor, J. 2006. The Dartmouth College Artificial Intelligence Conference: The next fifty years. Ai Magazine, 274: 87–91.
Mori, J., Y. Kajikawa, H. Kashima and I. Sakata. 2012. Machine learning approach for finding business partners and building reciprocal relationships. Expert Systems with Applications, 3912: 10402–10407.
Neapolitan, R.E. and X. Jiang. 2018. Artificial Intelligence: With an Introduction to Machine Learning, CRC Press.
Omurca, S.I. 2013. An intelligent supplier evaluation, selection and development system. Applied Soft Computing, 131: 690–697.
Quinlan, J.R. 1993. C4.5 : programs for machine learning. San Mateo, Calif., Morgan Kaufmann Publishers.
Razmi, J., H. Rafiei and M. Hashemi. 2009. Designing a decision support system to evaluate and select suppliers using fuzzy analytic network process. Computers & Industrial Engineering, 574: 1282–1290.
Samuel, A.L. 1959. Some studies in machine learning using the game of checkers. IBM J. Res. Dev., 33: 210–229.
Sarkar, A. and P.K. Mohapatra. 2006. Evaluation of supplier capability and performance: A method for supply base reduction. Journal of Purchasing and Supply Management, 123: 148–163.
Sarkis, J. and S. Talluri. 2002. A model for strategic supplier selection. Journal of Supply Chain Management, 384: 18–28.
Sevkli, M. 2010. An application of the fuzzy ELECTRE method for supplier selection. International Journal of Production Research, 4812: 3393–3405.
Shalev-Shwartz, S. and S. Ben-David. 2014. Understanding Machine Learning : From Theory to Algorithms, Cambridge university press.
Taherdoost, H. and A. Brard. 2019. Analyzing the process of supplier selection criteria and methods. 12th International Conference Interdisciplinarity in Engineering (Inter-Eng. 2018), 32: 1024–1034.
Talluri, S. and J.Y. Lee. 2010. Optimal supply contract selection. International Journal of Production Research, 4824: 7303–7320.
Tang, X.L. 2009. Study on selection of logistics supplier based on support vector machine. Proceedings of 2009 International Conference on Machine Learning and Cybernetics, 1–6: 1231–1235.
Thanaraksakul, W. and B. Phruksaphanrat. 2009. Supplier Evaluation Framework Based on Balanced Scorecard with Integrated Corporate Social Responsibility Perspective. Imecs 2009: International Multi-Conference of Engineers and Computer Scientists, Vols. I and II: 1929–1934.
Tiwari, A.K., C. Samuel and A. Tiwari. 2012. Flexibility in Supplier Selection Using Fuzzy Numbers with Nonlinear Membership Functions, New Delhi, Springer India.
Torğul, B. and T. Paksoy. 2019. Smart and Sustainable Supplier Selection for Electric Car Manufacturers. The International Aluminium-Themed Engineering and Natural Sciences Conference (IATENS'19). Seydişehir/TURKEY pp. 1063–1067.
Valluri, A. and D.C. Croson. 2005. Agent learning in supplier selection models. Decision Support Systems, 392: 219–240.
Wang, J.W., C.H. Cheng and H. Kun-Cheng. 2009. Fuzzy hierarchical TOPSIS for supplier selection. Applied Soft Computing, 91: 377–386.
Weber, C.A., J.R. Current and W.C. Benton. 1991. Vendor selection criteria and methods. European Journal of Operational Research, 501: 2–18.
Wu, D.D., Y.D. Zhang, D.X. Wu and D.L. Olson. 2010. Fuzzy multi-objective programming for supplier selection and risk modeling: A possibility approach. European Journal of Operational Research, 2003: 774–787.
Yang, S.T., J. Yang and L. Abdel-Malek. 2007. Sourcing with random yields and stochastic demand: A newsvendor approach. Computers & Operations Research, 3412: 3682–3690.
Yu, C.X., Y.F. Shao, K. Wang and L.P. Zhang. 2019. A group decision making sustainable supplier selection approach using extended TOPSIS under interval-valued Pythagorean fuzzy environment. Expert Systems with Applications, 121: 1–17.
Zhang, X.G., Y. Deng, F.T.S. Chan and S. Mahadevan. 2015. A fuzzy extended analytic network process-based approach for global supplier selection. Applied Intelligence, 434: 760–772.

CHAPTER 8

Deep Learning for Prediction of Bus Arrival Time in Public Transportation

Faruk Serin,[1] *Suleyman Mete,*[2] *Muhammet Gul*[3] and *Erkan Celik*[4,*]

1. Introduction

Public transportation is an important issue for the city planner or decision maker. It has a direct impact on the all aspect of the community such as economy, education, health and entertainment activities. There are a lot of disadvantages during use private cars such as noise and air pollution, stress and traffic problems, excessive and unreliable travel times. Hence, most people usually prefer to use public transportation instead of their private cars. For this reason, it is gaining more and more importance day by day as the population increases. The earliest arrival and the minimum number of transfers are the most important and common preferences among public transport users even though the choices of passenger may be differ from each other. Hence, number of transfers, total travel time and cost from origin to destination are important indicators for the passenger. These indicators should be optimized by passenger preferences (Serin and Mete 2019). The bus arrival time information can decrease the passenger waiting time, make passenger informative and thus able to arrange their trip plans and choose suitable travelling routes. Many researchers and practitioners have begun to be interested in the prediction of bus arrival time. There are various methods developed in the literature for prediction of bus arrival time like artificial neural network (ANN), Kalman-filters, Non-parametric regression (NPR) model and support vector machines (SVM). Therefore, this chapter intends to apply the Long Short Term Memory (LSTM) model to predict accurate bus arrival time for public transportation system.

LTSM was used firstly by Hochreiter and Schmidhuber 1997 and improved by Flex Gers' team 3 years later. This method is mostly known for usage of natural language text recognition. Some major technology companies also use this algorithm for their speech recognition and translation applications. LSTM is a class of recurrent neural network, so its cells have activation functions that are generally logistic functions. Commonly, LSTM architecture includes a memory cell, an input gate, an output gate and a forget gate. Memory cells store their inputs for some period of time. This method is a very popular artificial intelligence technique and its applications can be seen in many different areas such as robot control, time series predictions, human action recognition, semantic parsing etc. Moreover, this chapter examines the improved methodology for real application utilization.

2. Long Short Term Memory

Hochreiter and Schmidhuber (1997) developed the LSTM network as a special kind of recurrent neural network (RNN). It has special structures of memory blocks and cells and has been successful in prediction for different application areas. Zheng et al. (2017) used LSTM for Electric load forecasting in smart grids. Zaytar and Amrani (2016) applied LSTM for

[1] Department of Computer Engineering, Munzur University, Tunceli/Turkey.
[2] Department of Industrial Engineering, Gaziantep University, Gaziantep/Turkey.
[3] Department of Industrial Engineering, Munzur University, Tunceli/Turkey.
[4] Department of Transportation and Logistics, Istanbul University, Istanbul/Turkey.
* Corresponding author: erkancelik@munzur.edu.tr

forecasting of sequence to sequence weather. Fischer and Krauss (2018) applied LSTM for financial market predictions. Especially, it is used for prediction for supply chain and transportation areas. Ma et al. (2015) applied it for predicting traffic speed using remote microwave sensor data. Toqué et al. (2016) forecasted dynamic public transport origin-destination matrices using LSTM. Wu and Tan (2016), Zhao et al. (2107), Abbas et al. (2018) aim to forecast the short-term traffic with LSTM. Chen et al. (2016) predicted traffic congestion prediction with online open data using LSTM. Song et al. (2016) used LSTM for Prediction and Simulation of Human Mobility and Transportation Mode. Liu et al. (2017) compared LSTM and DNN for Short-term travel time prediction in transportation. A literature review for deep learning in transportation is presented by Nguyen et al. (2018). They also analyzed LSTM that is used in transportation systems.

To overcome back propagation problems in RNN, the cells are replaced by gated cells. A multiplicative input gate units and multiplicative output gate units are introduced. LSTM networks use an input layer, at least one hidden layer, and an output layer. Abigogun 2005 specified the difference between the traditional NN and LSTM as the hidden layer is memory block. A memory cell is given as a more complex unit (Figure 1). The memory block encompasses memory cells with self-connections memorizing a pair of adaptive and temporal state. The memory cell of j is denoted as c_j.

While net_{cj} and cj gets input from output gate, out_j and input gate in_j. The activation of input gate at time t is presented as $y^{in_j}(t)$ and the activation of output gate at time t is $y^{out_j}(t)$.

$$y^{out_j}(t) = f_{out_j}(net_{out_j}(t)) \tag{1}$$

$$y^{in}(t) = f_{in}(net_{in_j}(t)) \tag{2}$$

Where

$$net_{out_j}(t) = \sum_u w_{out_j u} y^u (t-1) \tag{3}$$

$$net_{in_j}(t) = \sum_u w_{in_j u} y^u (t-1) \tag{4}$$

$$net_{c_j}(t) = \sum_u w_{c_j u} y^u (t-1) \tag{5}$$

u presents the summation of the memory cells, input units, gate units or hidden units. In this process, one input layer, one hidden layer, and one output layer are used in the network.
At time t, c_j's output $y^{c_j}(t)$ (t) is calculated as follows:

$$y^{c_j}(t) = y^{out_j}(t)h(s_{c_j}(t)) \tag{6}$$

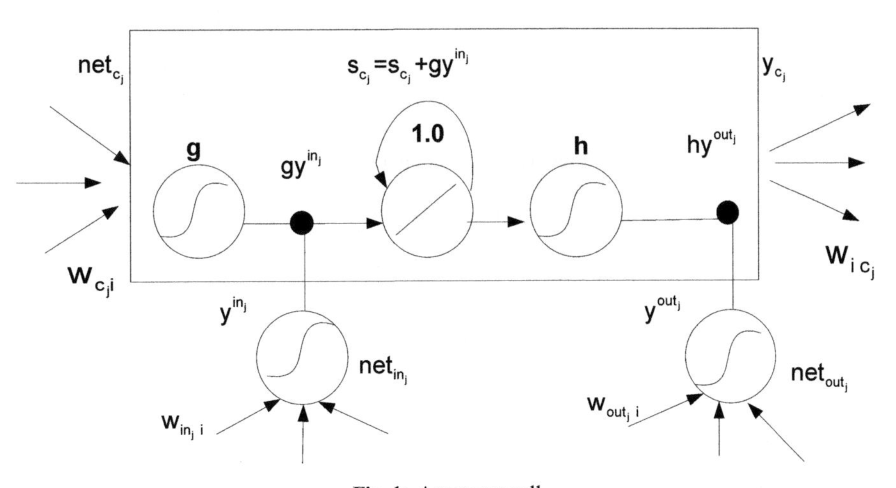

Fig. 1: A memory cell.

where the internal state $s_{c_j}(t)$

$$s_{c_j}(0) = 0,\ s_{c_j}(t) = s_{c_j}(t-1) + y^{in_j}(t)g(net_{c_j}(t)) \text{ for } t > 0 \tag{7}$$

The differentiable function g squeezes net_{c_j}; the differentiable function h scales memory cell outputs calculated from the internal state s_{cj}. In LSTM, a variant of RTRL (Robinson and Fallside 1987) is used as learning process which appropriately considers the reformed, multiplicative dynamics caused by input gates and output gates. In this chapter, the objective of LSTM is to predict of bus arrival time based on prior information.

3. Application

This section presents the developed models regarding bus arrival time prediction at bus stop using real gathered data from the Istanbul public transportation authority in Turkey. Istanbul is a highly populated city with more than 15 million population. Approximately 78% of its residents preferred road transportation (Yanık et al. 2017). Moreover, %27.26 of the residents directly use the Bus, Metrobus or private buses. Nearly 13 million passengers use public transportation in İstanbul on a daily basis (IETT 2018). The authorities of the Istanbul public transportation apply several advanced public transportation systems such as real time location tracking, transit vehicle tracking, multimodal coordination, informing passengers on stations about location of vehicles, fare collection through an electronic system. The bus stations of Istanbul city have electronic panels demonstrating the bus timelines in real time manner. Estimated arrivals of buses to the particular bus stops are possible to track. Therefore, passengers can easily plan their route even during heavy traffic. Istanbul has a technologically advanced and complex bus route network. There are more than 1000 bus routes. In our study, one of the most overcrowded and longest line called "500T: Tuzla-Cevizlibağ" was used to assess the performance of the developed prediction models. We selected this line since it consists so many bus routes and bus stops. It has a high passenger demand since there a bridge crossing in its route. The line starts from Europe and ends in Asia. The length of the route is approximately 73.6 km.

4. Implementation

Public transportation network mainly consists of route, stop, and bus. A line between two sequential stops on a route is defined as a segment. A bus travel time on a segment is calculated using automatic vehicle location data as in (8) where t_v^b is bus, v, arriving time at beginning-station, b, of segment s; t_v^e is bus, v, arriving time at end-station, e, of segment s. Travel time of all buses on segment, s, are arranged sequentially as time series as in (9). Finally, series are rearranged according to time window as in Table 1 (time window =3).

$$\Delta T_v^s = t_v^e - t_v^b \tag{8}$$

$$\Delta T^s = \{\Delta T_0^s, \Delta T_1^s, \Delta T_2^s \dots \Delta T_n^s\} = \{S_0, S_1, S_2 \dots S_n\} \tag{9}$$

Keras, the Python deep learning library, is used to apply LSTM model. The architecture of the model is given in Figure 2. The parameters of the model are as follow: number of epochs = 1000; train percentage = 70; time window = 6.

Table 1: Time series time window.

X (Input)		Y (Output)	
S_0	S_1	S_2	S_3
S_1	S_2	S_3	S_4
...	...	...	...
S_{n-3}	S_{n-2}	S_{n-1}	S_n

5. Performance Measures

In this chapter, we have applied five performance measure for evaluating the results of the proposed approach. These performance measures are mean absolute error (MAE), mean absolute percentage error (MAPE), mean square error

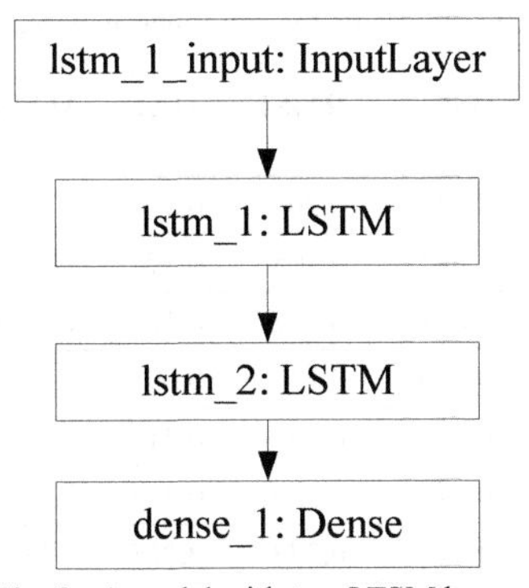

Fig. 2: A model with two LTSM layer.

(MSE), root mean square error (RMSE), and the residual sum of squares (RSS). The mathematical formulation of the performance measures are given in detail. While Y_t is the actual data, $\hat{Y}_t$ is the predicted data where n is the number of the total forecasted data.

$$MAE = \frac{1}{n} \sum_{t=1}^{n} |Y_t - \hat{Y}_t| \tag{10}$$

$$MAPE = \left(\frac{1}{n} \sum_{t=1}^{n} \frac{|Y_t - \hat{Y}_t|}{Y_t} \right) * 100 \tag{11}$$

$$MSE = \frac{1}{n} \sum_{t=1}^{n} (Y_t - \hat{Y}_t)^2 \tag{12}$$

$$RMSE = \sqrt{\frac{1}{n} \sum_{t=1}^{n} (Y_t - \hat{Y}_t)^2} \tag{13}$$

$$RSS = \sum_{t=1}^{n} |Y_t - \hat{Y}_t|^2 \tag{14}$$

6. Discussions and Results

The proposed approach is implemented for five different performance measures for predicting bus arrival time in a route in the city of Istanbul. The following data contains the following information: (1) route id, (2) segment number, (3) departure station id, (4) arrival station id, (5) sample size (cleared signal size), (6) signal size, (7) method used, and (8) performance measures (MSE, RMSE, MAE, MAPE, RSS) and elapsed time. The descriptive statistics of the data is presented in Appendix 1. The selected route includes 72 segments with data. The number of epochs is taken as 1000. While %70 of the data is used in training, the remaining data is used in testing. In addition, the lag is taken as 6 and the minimum series size is considered as 11 in LSTM. The average values of the MSE, RMSE, MAE, MAPE and RSS of each segment of the selected route are presented in Table 2.

When the MAPE is analyzed for all segment, the minimum MAPE for segment of 6, 67, and 71 is 4.439, 5.749, and 7.283, respectively. It means that the minimum MAPE is a reasonable results. On the other hand, while the minimum elapsed time is 10.51 second for segment 11, the maximum elapsed time is 70.65 for segment 45. The average elapsed time for all segment is 33.71 second. The average value of the, MAE, MAPE, MSE, RMSE and RSS with all segment for selected bus route are presented at the end of the Table 2.

7. Conclusion

This chapter proposes a method based on the deep learning approach for prediction of bus arrival time in public transportation. Therefore, the LSTM model is applied to predict accurate bus arrival time for the public transportation system. The developed model regarded bus arrival time prediction at a bus stop using a real gathered data from the Istanbul public transportation authority in Turkey. Istanbul has a technologically advanced and complex bus route network. Therefore, we examined just one of the most overcrowded and longest lines to assess the performance of the developed

Table 2: The results for performance measurement.

#	#Segment	MSE	RMSE	MAE	MAPE	RSS	Elapsed Time
0	1	0.040	0.200	0.134	20.746	0.640	25.81
1	4	0.042	0.205	0.177	15.386	0.676	27.75
2	5	0.041	0.202	0.165	15.788	0.695	25.99
3	8	0.019	0.138	0.095	13.883	0.245	15.04
4	9	0.310	0.557	0.414	18.444	4.649	25.79
5	12	0.276	0.525	0.474	27.976	4.145	26.26
6	14	**0.074**	**0.272**	**0.225**	**4.439**	**1.031**	**15.65**
7	15	0.011	0.105	0.075	16.325	0.117	10.51
8	18	0.002	0.045	0.030	9.626	0.025	16.33
9	19	0.058	0.241	0.197	23.271	0.864	27.51
10	22	0.010	0.100	0.080	7.700	0.168	27.22
11	26	0.052	0.228	0.168	24.891	0.679	16.94
12	29	0.087	0.295	0.235	12.184	1.217	29.27
13	30	0.031	0.176	0.143	17.941	0.491	29.29
14	34	0.072	0.268	0.209	15.264	1.159	29.42
15	37	0.012	0.110	0.092	11.104	0.145	18.65
16	39	0.027	0.164	0.130	13.804	0.293	19.53
17	42	0.051	0.226	0.180	18.292	0.810	32.66
18	44	0.034	0.184	0.156	23.223	0.481	30.62
19	46	0.031	0.176	0.148	14.708	0.525	31.58
20	48	0.045	0.212	0.185	28.557	0.724	33.4
21	51	0.192	0.438	0.335	11.884	2.684	31.08
22	53	0.036	0.190	0.155	21.404	0.572	32.13
23	55	0.179	0.423	0.378	17.378	1.071	21.31
24	56	0.904	0.951	0.770	16.219	8.139	20.62
25	59	0.718	0.847	0.665	9.289	11.489	32.56
26	60	0.069	0.263	0.249	33.712	1.177	32.95
27	63	0.175	0.418	0.357	32.020	2.624	33.12
28	65	0.197	0.444	0.342	20.840	3.159	33.58
29	66	0.092	0.303	0.251	12.819	1.472	34.25
30	69	0.050	0.224	0.194	15.108	0.753	34.03
31	71	0.080	0.283	0.228	13.077	1.276	34.14
32	72	0.014	0.118	0.100	13.752	0.206	34.7
33	74	0.025	0.158	0.122	13.799	0.330	23.28
34	75	0.034	0.184	0.145	20.492	0.472	34.83
35	77	*0.418*	*0.647*	*0.572*	*39.920*	*2.091*	*23.69*
36	80	0.032	0.179	0.154	21.397	0.475	36.56
37	85	0.011	0.105	0.086	12.068	0.149	24.33
38	86	0.021	0.145	0.123	16.238	0.298	24.88
39	90	0.006	0.077	0.067	10.350	0.089	36.65
40	92	*0.068*	*0.261*	*0.211*	*37.160*	*1.025*	*37.09*

Table 2 contd. ...

...Table 2 contd.

#	#Segment	MSE	RMSE	MAE	MAPE	RSS	Elapsed Time
41	95	0.067	0.259	0.209	20.577	0.866	25.6
42	97	1.522	1.234	0.988	22.274	25.869	37.96
43	100	0.014	0.118	0.097	17.959	0.209	38.11
44	102	0.037	0.192	0.153	18.063	0.484	26.96
45	104	0.053	0.230	0.191	19.821	0.794	70.65
46	106	*0.066*	*0.257*	*0.220*	*38.996*	*0.919*	*27.92*
47	108	0.067	0.259	0.203	21.863	0.941	40.39
48	109	0.000	0.000	0.016	16.424	0.005	28.07
49	110	0.026	0.161	0.146	10.566	0.344	28.52
50	115	0.045	0.212	0.190	21.879	0.759	40.76
51	118	0.112	0.335	0.288	23.725	1.684	40.73
52	122	0.130	0.361	0.291	23.822	1.685	29.3
53	124	0.012	0.110	0.087	20.272	0.127	29.51
54	128	0.032	0.179	0.149	9.136	0.485	42.18
55	130	0.114	0.338	0.287	18.418	1.825	44.43
56	132	0.051	0.226	0.173	14.842	0.759	42.57
57	133	0.057	0.239	0.197	13.877	0.967	43.04
58	135	0.050	0.224	0.183	12.378	0.751	43.39
59	136	0.057	0.239	0.199	31.173	0.796	43.51
60	138	0.091	0.302	0.254	15.598	1.453	44.1
61	141	0.248	0.498	0.381	15.378	4.213	46.32
62	143	0.095	0.308	0.249	28.365	1.425	44.58
63	145	0.048	0.219	0.178	26.238	0.821	45.11
64	146	0.227	0.476	0.354	8.666	3.628	47.78
65	148	0.055	0.235	0.206	18.609	0.929	48.01
66	149	0.107	0.327	0.266	13.332	1.925	46.34
67	152	**0.003**	**0.055**	**0.042**	**5.749**	**0.044**	**46.71**
68	155	0.265	0.515	0.407	21.012	4.502	47.99
69	156	0.446	0.668	0.452	20.331	7.128	47.44
70	158	0.141	0.375	0.255	15.773	2.402	46.86
71	161	**1.336**	**1.156**	**0.989**	**7.283**	**20.037**	**47.34**
72	162	0.051	0.226	0.147	13.769	0.758	47.77
Average		0.143	0.299	0.242	18.201	2.039	33.71

prediction models. The selected line studied has many bus routes, bus stops and a high passenger demand. Moreover, five performance measures were used to show verification of proposed model. The analysis results show that the prediction model based on LSTM method gives acceptable results according to performance measures.

References

Abbas, Z., A. Al-Shishtawy, S. Girdzijauskas and V. Vlassov. 2018. Short-term traffic prediction using long short-term memory neural networks, 57–65. International Congress on Big Data (BigData Congress), Milan, Italy, IEEE.

Abigogun, O.A. 2005. Data Mining, Fraud detection and mobile telecommunications: call pattern analysis with unsupervised neural networks. M.S. Thesis, University of the Western Cape, Cape Town, South Africa.

Chen, Y.Y., Y. Lv, Z. Li and F.Y. Wang. 2016. Long short-term memory model for traffic congestion prediction with online open data, 132–137. 19th International Conference on Intelligent Transportation Systems (ITSC), Rio de Janeiro, Brazil, IEEE.

Fischer, T. and C. Krauss. 2018. Deep learning with long short-term memory networks for financial market predictions. Eur. J. Oper. Res., 270(2): 654–669.

Hochreiter, S. and J. Schmidhuber. 1997. Long short-term memory. Neural Comput., 9(8): 1735–1780.

Liu, Y., Y. Wang, X. Yang and L. Zhang. 2017. Short-term travel time prediction by deep learning: A comparison of different LSTM-DNN models, 1–8. 20th International Conference on Intelligent Transportation Systems (ITSC), Yokohama, Japan, IEEE.

Ma, X., Z. Tao, Y. Wang, H. Yu and Y. Wang. 2015. Long short-term memory neural network for traffic speed prediction using remote microwave sensor data. Transp. Res. Part C Emer. Technol., 54: 187–197.

Narayan, A. and K.W. Hipel. 2017. Long short term memory networks for short-term electric load forecasting, 2573–2578. 2017 International Conference on Systems, Man, and Cybernetics (SMC) Banff, Canada, IEEE.

Nguyen, H., L.M. Kieu, T. Wen and C. Cai. 2018. Deep learning methods in transportation domain: a review. IET Intell. Transp. Sy., 12(9): 998–1004.

Robinson, A.J. and F. Fallside. 1987. The utility driven dynamic error propagation network. Technical Report CUED/F-INFENG/TR.1, Cambridge University Engineering Department.

Serin, F. and S. Mete. 2019. Public transportation graph: A graph theoretical model of public transportation network for efficient trip planning. Pamukkale Üniversitesi Mühendislik Bilimleri Dergisi, 25(4): 468–472.

Song, X., H. Kanasugi and R. Shibasaki. 2016. DeepTransport: Prediction and simulation of human mobility and transportation mode at a citywide level, 2618–2624. Proceedings of the Twenty-Fifth International Joint Conference on Artificial Intelligence, New York, USA.

Toqué, F., E. Côme, M.K. El Mahrsi and L. Oukhellou. 2016. Forecasting dynamic public transport origin-destination matrices with long-short term memory recurrent neural networks, 1071–1076. 19th International Conference on Intelligent Transportation Systems (ITSC), Rio de Janeiro, Brazil, IEEE.

Wu, Y. and H. Tan. 2016. Short-term traffic flow forecasting with spatial-temporal correlation in a hybrid deep learning framework. arXiv preprint arXiv: 1612.01022.

Zaytar, M.A. and C. El Amrani. 2016. Sequence to sequence weather forecasting with long short-term memory recurrent neural networks. Int. J. Comput. Appl., 143(11): 7–11.

Zhao, Z., W. Chen, X. Wu, P.C. Chen and J. Liu. 2017. LSTM network: a deep learning approach for short-term traffic forecast. IET Intell. Transp. Sy., 11(2): 68–75.

Zheng, J., C. Xu, Z. Zhang and X. Li. 2017. Electric load forecasting in smart grids using long-short-term-memory based recurrent neural network. 51st Annual Conference on Information Sciences and Systems (CISS) (pp. 1–6). IEEE.

Appendix-1: The descriptive statistics of the data

#	#segment	Count	Max	Min	Mode	Mean	Var	Std	Sum	Skew	Kurtosis	Sem	Mad	Nunique	Quantile%25	Quantile%50	Quantile%75
0	1	58	1.93	0.32	0.48	0.64	0.098	0.313	37.1	2.102	4.907	0.041	0.23	28	0.455	0.5	0.768
1	4	60	1.5	0.75	1.23	1.133	0.03	0.175	67.97	–0.272	–0.321	0.023	0.14	33	1.02	1.17	1.235
2	5	62	1.5	0.63	1.03	1.05	0.04	0.2	65.11	0.119	–0.191	0.025	0.153	31	0.93	1.03	1.13
3	8	51	0.92	0.43	0.55	0.558	0.007	0.086	28.48	1.682	4.954	0.012	0.062	17	0.51	0.55	0.58
4	9	57	3.02	1.15	1.62	1.983	0.24	0.49	113.01	0.479	–0.479	0.065	0.395	43	1.62	1.95	2.37
5	12	57	3.97	0.85	1.67	1.949	0.667	0.817	111.12	1.29	0.766	0.108	0.613	49	1.43	1.7	2.1
6	14	52	5.62	4.18	4.57	4.848	0.103	0.321	252.08	0.246	–0.326	0.044	0.259	41	4.6	4.81	5.07
7	15	44	0.57	0.28	0.37	0.37	0.004	0.059	16.28	1.384	2.521	0.009	0.043	13	0.33	0.37	0.385
8	18	52	0.47	0.25	0.3	0.328	0.002	0.042	17.07	0.846	1.557	0.006	0.032	11	0.3	0.325	0.35
9	19	57	1.1	0.4	0.6	0.676	0.026	0.16	38.55	0.731	–0.094	0.021	0.128	24	0.57	0.63	0.75
10	22	59	1.27	0.8	1.1	1.03	0.011	0.107	60.76	0.358	0.045	0.014	0.084	21	0.95	1.02	1.1
11	26	50	1.08	0.3	0.53	0.607	0.025	0.157	30.37	1.033	0.837	0.022	0.125	25	0.5	0.55	0.695
12	29	53	2.42	1.07	1.52	1.758	0.092	0.303	93.15	0.087	0.084	0.042	0.233	34	1.55	1.75	1.93
13	30	58	1.25	0.57	1.03	0.929	0.036	0.191	53.9	–0.159	–1.215	0.025	0.169	30	0.772	0.975	1.08
14	34	58	1.8	0.75	1.07	1.18	0.047	0.217	68.42	0.589	0.934	0.028	0.163	34	1.055	1.16	1.27
15	37	45	1.05	0.6	0.73	0.832	0.013	0.112	37.43	0.086	–0.623	0.017	0.092	25	0.75	0.83	0.92
16	39	44	1.25	0.58	0.87	0.82	0.023	0.151	36.1	0.611	0.258	0.023	0.118	25	0.7	0.815	0.885
17	42	59	1.13	0.6	0.7	0.858	0.023	0.152	50.6	0.189	–1.157	0.02	0.132	29	0.72	0.85	0.98
18	44	54	1	0.33	0.55	0.65	0.026	0.16	35.1	0.361	–0.567	0.022	0.132	30	0.535	0.63	0.765
19	46	63	1.4	0.67	1.17	1.073	0.034	0.186	67.59	–0.477	–0.514	0.023	0.153	35	0.96	1.1	1.2
20	48	58	1.13	0.35	0.48	0.715	0.051	0.225	41.46	0.118	–1.022	0.03	0.186	35	0.48	0.73	0.872
21	51	54	3.82	2.18	3.02	2.927	0.159	0.399	158.08	0.433	–0.29	0.054	0.313	37	2.598	2.93	3.12
22	53	58	1.13	0.48	0.65	0.757	0.022	0.15	43.9	0.339	–0.319	0.02	0.122	27	0.65	0.75	0.87
23	55	25	2.85	1.63	1.98	2.166	0.129	0.358	54.14	0.338	–0.943	0.072	0.307	24	1.87	2.1	2.45
24	56	37	5.52	2.55	3.67	4.225	0.541	0.736	156.31	0.033	–0.333	0.121	0.589	34	3.78	4.15	4.65
25	59	59	15.33	5.63	7.05	8.145	3.208	1.791	480.55	2.312	5.983	0.233	1.184	52	7.15	7.63	8.415
26	60	61	1.22	0.48	0.68	0.838	0.042	0.205	51.14	–0.034	–1.052	0.026	0.174	33	0.68	0.83	1
27	63	57	1.58	0.4	0.58	0.986	0.068	0.261	56.19	0.021	–0.265	0.035	0.202	35	0.83	0.97	1.17

Appendix 1 cont. ...

...Appendix 1 cont.

#	#segment	Count	Max	Min	Mode	Mean	Var	Std	Sum	Skew	Kurtosis	Sem	Mad	Nunique	Quantile%25	Quantile%50	Quantile%75
28	65	59	2.23	0.98	1.13	1.469	0.123	0.35	86.69	0.747	–0.563	0.046	0.294	40	1.18	1.35	1.68
29	66	59	2.65	1.48	1.83	2.062	0.095	0.308	121.67	–0.007	–0.938	0.04	0.257	40	1.83	2.08	2.3
30	69	56	1.75	0.93	1.27	1.389	0.037	0.193	77.77	–0.205	–0.098	0.026	0.152	31	1.27	1.37	1.512
31	71	60	2.27	1.05	1.78	1.637	0.076	0.275	98.24	0.018	–0.457	0.036	0.231	32	1.43	1.65	1.835
32	72	55	0.83	0.42	0.73	0.635	0.012	0.11	34.9	0.113	–0.983	0.015	0.094	23	0.55	0.63	0.73
33	74	48	1.1	0.63	0.68	0.826	0.018	0.133	39.65	0.208	–1.223	0.019	0.117	22	0.7	0.825	0.93
34	75	54	0.97	0.4	0.53	0.622	0.02	0.143	33.57	0.606	–0.458	0.019	0.12	27	0.52	0.58	0.728
35	77	24	2.38	1.32	1.85	1.875	0.113	0.336	45.01	–0.167	–0.982	0.069	0.27	20	1.688	1.85	2.135
36	80	55	1.23	0.38	0.52	0.762	0.038	0.195	41.93	0.178	–0.156	0.026	0.152	30	0.585	0.78	0.87
37	85	48	0.78	0.5	0.57	0.617	0.006	0.079	29.63	0.815	–0.284	0.011	0.062	17	0.57	0.59	0.655
38	86	52	1.02	0.53	0.62	0.739	0.017	0.132	38.41	0.523	–0.911	0.018	0.112	24	0.628	0.71	0.855
39	90	53	0.75	0.48	0.62	0.604	0.004	0.06	32.03	0.271	–0.232	0.008	0.047	15	0.57	0.6	0.63
40	92	56	0.8	0.28	0.33	0.522	0.028	0.169	29.23	–0.068	–1.526	0.023	0.153	28	0.33	0.575	0.655
41	95	48	1.32	0.55	0.9	0.876	0.023	0.151	42.04	0.419	0.708	0.022	0.117	28	0.78	0.89	0.955
42	97	63	6.32	2.15	4.48	4.084	1.032	1.016	257.31	0.368	–0.494	0.128	0.841	49	3.37	3.93	4.65
43	100	56	0.8	0.3	0.42	0.521	0.014	0.116	29.19	0.751	–0.286	0.016	0.094	21	0.445	0.48	0.58
44	102	50	1.13	0.45	0.53	0.714	0.028	0.167	35.69	0.473	–0.67	0.024	0.141	31	0.58	0.67	0.865
45	104	56	1.67	0.75	1.07	1.061	0.059	0.242	59.39	0.954	–0.024	0.032	0.193	32	0.895	0.98	1.192
46	106	52	0.83	0.23	0.32	0.473	0.024	0.155	24.59	0.733	–0.596	0.021	0.129	24	0.345	0.43	0.622
47	108	54	1.85	0.42	0.58	0.836	0.091	0.301	45.17	1.327	2.135	0.041	0.232	36	0.63	0.77	0.978
48	109	50	0.15	0.07	0.1	0.091	0	0.017	4.53	1.159	2.136	0.002	0.014	6	0.08	0.08	0.1
49	110	49	1.75	1.05	1.23	1.361	0.033	0.182	66.68	0.188	–0.773	0.026	0.153	30	1.22	1.38	1.48
50	115	62	1.15	0.48	0.9	0.803	0.032	0.18	49.77	–0.268	–0.794	0.023	0.149	30	0.683	0.84	0.92
51	118	56	1.83	0.6	1.63	1.233	0.098	0.312	69.04	0.01	–0.701	0.042	0.255	37	1.027	1.21	1.458
52	122	49	2.08	0.58	0.88	1.014	0.079	0.281	49.7	1.48	3.474	0.04	0.203	33	0.85	0.95	1.15
53	124	44	0.72	0.28	0.28	0.42	0.015	0.121	18.49	0.75	–0.406	0.018	0.1	21	0.32	0.39	0.505
54	128	55	2.3	0.85	1.53	1.585	0.089	0.298	87.18	0.171	0.734	0.04	0.216	32	1.41	1.55	1.735
55	130	60	2.13	1.02	1.75	1.657	0.071	0.267	99.42	–0.49	–0.221	0.034	0.213	36	1.515	1.73	1.83
56	132	56	1.93	0.68	1.25	1.228	0.085	0.292	68.77	0.565	–0.068	0.039	0.225	37	0.98	1.2	1.405

57	133	62	1.88	0.97	1.45	1.401	0.046	0.215	86.89	0.162	–0.749	0.027	0.179	39	1.228	1.43	1.542
58	135	56	2.13	1.03	1.33	1.512	0.06	0.245	84.69	0.491	–0.283	0.033	0.203	39	1.33	1.46	1.68
59	136	53	0.97	0.18	0.6	0.557	0.035	0.187	29.53	–0.026	–0.703	0.026	0.154	31	0.38	0.6	0.67
60	138	58	2.37	1	1.62	1.598	0.078	0.279	92.67	0.317	0.488	0.037	0.206	32	1.435	1.6	1.765
61	141	61	3.75	1.05	2.82	2.308	0.327	0.572	140.8	0.345	0.401	0.073	0.435	47	1.95	2.27	2.63
62	143	57	1.22	0.45	0.53	0.813	0.045	0.212	46.35	0.061	–1.044	0.028	0.178	39	0.63	0.82	0.98
63	145	63	1.05	0.38	0.5	0.687	0.034	0.183	43.29	0.24	–0.663	0.023	0.148	35	0.54	0.7	0.79
64	146	58	5.33	3.52	4.3	4.317	0.198	0.444	250.39	0.328	–0.327	0.058	0.353	45	4.02	4.3	4.578
65	148	62	1.48	0.72	1.18	1.073	0.036	0.19	66.53	–0.051	–0.856	0.024	0.16	34	0.92	1.1	1.2
66	149	65	2.65	0.95	1.73	1.788	0.126	0.354	116.19	0.391	0.067	0.044	0.278	48	1.5	1.73	2.02
67	152	57	1.03	0.63	0.77	0.783	0.007	0.085	44.62	1.06	1.803	0.011	0.061	17	0.73	0.77	0.82
68	155	62	2.7	0.6	1.27	1.65	0.228	0.478	102.32	0.41	–0.385	0.061	0.389	47	1.285	1.53	1.995
69	156	60	2.97	0.92	1.08	1.708	0.405	0.636	102.45	0.557	–1.091	0.082	0.56	44	1.1	1.39	2.218
70	158	64	2.63	0.95	1.43	1.64	0.113	0.337	104.94	0.685	0.547	0.042	0.265	38	1.418	1.575	1.828
71	161	55	14.48	10.95	11.62	12.735	0.767	0.876	700.44	0.012	–0.728	0.118	0.714	50	12.06	12.73	13.375
72	162	55	1.63	0.43	0.87	1.011	0.053	0.23	55.58	0.377	1.584	0.031	0.162	32	0.885	0.98	1.1

SECTION 5

Augmented Reality in SCM

Chapter 9

Augmented Reality in Supply Chain Management

Sercan Demir,[1,*] *Ibrahim Yilmaz*[2] and *Turan Paksoy*[3]

1. Introduction

Digitization is the transformation of operations, functions, models, processes, or activities by using the merits of digital technologies. Digitization is the enabler of new business models, and one of the most powerful drivers of innovation with the potential to provoke the next wave of innovation (Gürdür et al. 2019). Industry 4.0 triggers a radical change in the conventional production methods. The new wave of the Industrial Revolution is considered a global transformation of the manufacturing industry that is initiated by the introduction of digitization and the Internet. The smart factory integrates innovative digital technologies into the manufacturing and service industries, and it is considered as the future of production. These digital technologies include but are not limited to, advanced robotics, artificial intelligence (AI), hi-tech sensors, cloud computing, the Internet of Things (IoT), autonomous systems, and additive manufacturing. Smart systems aim to establish the connection between machines and human-beings within the Industry 4.0 context (Tjahjono et al. 2017).

Due to the increase in energy cost and the ongoing use of old manufacturing systems, the cost of doing business has risen dramatically. As a result, companies were motivated to lower the production cost while maintaining their quality standards within a certain level. Digital transformation age has brought many innovative technologies that have a huge impact on supply chains. AR is one of the emerging technologies that address low-cost solutions to the increasing running cost of businesses. This technology helps many players in the supply chains, such as truck drivers, warehouse workers, supervisors, and managers, by superimposing digital information into the real world. This computer-generated information assists these players to track the flow of goods from one point to another in a supply chain. Conventional, slow and paper-based logistics and supply chain processes are gradually being converted into a fast and technology-driven industry as a result of the applications of AR in businesses. Some business areas that AR technology is currently being used include, but not limited to, pick and pack services, the collaborative logistics, maintenance services, procurement, and last-mile delivery (Koul 2019).

[1] Department of Industrial Engineering, Faculty of Engineering, Harran University, Sanliurfa, Turkey.
[2] Department of Industrial Engineering, Faculty of Engineering, Yildirim Beyazit University, Ankara, Turkey.
Email: i.yilmaz@live.com
[3] Department of Industrial Engineering, Faculty of Engineering, Konya Technical University, Konya, Turkey.
Email: tpaksoy@yahoo.com
* Corresponding author: sercanxdemir@gmail.com

AR is one of the key technologies for Industry 4.0, and the field of computer science that is concerned with merging the real world with the computer-generated data. AR devices digitally process images received from the environment and augment this data by adding supplementary computer-generated graphics. AR systems combine real and virtual objects in real-time and align these objects with each other in a real environment. An AR application usually consists of a display, a camera, and a computer with application software. AR applications can be run using different devices such as camera phones, handheld computers, laptops, and head-mounted displays (Sääski et al. 2008).

AR is not a new technology since the idea of AR has been used in many applications such as virtual mirrors and mobile applications on mobile devices and tablets. Virtual mirrors are AR devices that are commonly used in the fashion retail industry. These devices film customers with the integrated cameras and superimpose the selected clothes on a customer's body displayed on the virtual mirror, hence assist a customer to judge which garment suits and fits best on him/her. Besides, mobile AR applications are widely used by many people on smartphones and tablet computers to reach information instantaneously. For instance, the Cyclopedia AR application provides information about nearby buildings or historic places, when an app user takes a picture of the mentioned building with the smart device (Ro et al. 2018).

Augmented Reality Smart Glasses (ARSGs) are another innovative AR devices that recently draw growing attention both in industry and academia, and they offer great opportunities for development in the near future. ARSGs are wearable AR devices (e.g., Google Glass and Microsoft HoloLens) that capture and digitally process the objects in an environment, and augment them with computer-generated data. Physical information is captured by the add-on technologies on ARSGs such as camera, GPS, or microphones, then the virtual information gathered from the internet and/or memory device is used to augment this physical information. The user can see the virtual information that is superimposed on the physical world on the screen of an ARSG (Ro et al. 2018).

This chapter is organized as follows. In the second section, we investigate how digitization affects the business models and reshape the organization of supply chains. The major milestone and important advancements in the AR technology are presented in the third section. The fourth section discusses the applications of AR technology in supply chains by providing real-world scenarios. Finally, the conclusion of this chapter and future research directions are presented in section five.

2. Digitization in Supply Chain Management

Supply Chain processes must adopt newly emerging technologies and transform themselves into sustainable operations to catch up with the increasing competition, rapidly changing environment and volatile markets. Failure to adapt to this fast-paced environment and harsh competition results in fatal consequences for companies.

Mechanization and harnessing mechanical power led to the transition from manual work to the first mechanical manufacturing process during the 1800s. This period was the debut of the First Industrial Revolution. The Second Industrial Revolution started as a result of electrification that led to industrialization and mass production during the late 19th century. The Third Industrial Revolution was initiated by the appearance of microelectronic devices such as transistors and microprocessors, and automated systems. In this era, flexible production was achieved by the integration of the programmable machines on flexible production lines (Rojko 2017). All industrial revolutions have brought along their unique disruptive technologies in manufacturing. Steam engine, automated electrical production line, and digital production methods were the major innovations that appeared during the first three industrial revolutions, respectively. The process of industrialization continues with the Fourth Industrial Revolution, namely Industry 4.0. The most recent industrial revolution has brought the concept of "smart products" and "smart factory". Smart products are uniquely identifiable, can be detected anytime throughout the supply chain, and their history, current status, and alternative routes to reach their destination can be easily monitored. The emerging technologies are inseparable parts of the smart factories. For instance, cyber-physical systems (CPS) take part in monitoring manufacturing processes, creating a virtual copy of the physical world, and making decentralized decisions, while they communicate and cooperate with the Internet of Things (IoT) and humans simultaneously (Carvalho et al. 2018).

The new generation of technologies such as robotics, artificial intelligence, big data, and augmented reality assist supply chains to improve and become more sustainable against growing environmental challenges. These newly emerging technologies help companies to make optimized decisions, administer automation devices, forecast demand, and plan the vital processes (Merlino and Sproge 2017). Smart manufacturing (a.k.a. intelligent manufacturing) aims to optimize production by using advanced information and manufacturing technologies. The entire life cycle of a product can be facilitated with the integration of smart technologies into the manufacturing process. Smart sensors, adaptive decision-making models, advanced materials, intelligent devices, and data analytics are some of these smart technologies that increase production efficiency, overall product quality, and customer service level. Physical processes can be easily monitored by smart manufacturing systems, and real-time optimized decision can be made by the intelligent systems that enable the interaction and cooperation between humans, machines, sensors and smart devices (Zhong et al. 2017).

Smart products and CPS are at the core of the digitization of supply chains. Products are evolving into more complex systems that require the integration of many technologies such as hardware, sensors, data storage, microprocessors, and software. This paradigm restructures many industries or discovers new ones (Klötzer and Pflaum 2017).

Industry 4.0 introduces radical changes to supply chains and business processes. Interoperability, virtualization, decentralization, real-time capability, service orientation, and modularity are the main principles of Industry 4.0. The latest industrial revolution presents more flexible manufacturing systems, reduced lead times, customized small batch sizes, and overall cost reduction. Industry 4.0 is characterized by state-of-art automation and digitization processes, and integration of information technologies (IT) into the manufacturing and service industry. The key technologies of Industry 4.0 consist of mobile computing, cloud computing, big data, and IoT. Real-time data processing feature of Industry 4.0 optimizes resource usage and brings about improved system performance. Industry 4.0 has initiated the term "smart" since factories, production lines, cities, and manufacturing equipment become "smart", as the adaptability, resource efficiency, and the integration of supply and demand processes are improved in Industry 4.0. The term "smart" is used to describe intelligence and knowledge in the applications of Industry 4.0. Smart factories, smart products, and smart cities are the main application of Industry 4.0. Factories become more intelligent and flexible by adopting sensors, actuators, and autonomous systems for their manufacturing processes. These technologies lead manufacturing systems to achieve a high level of self-optimization and automation while improving their capacity to produce more complex and better quality products (Lu 2017).

A new digital revolution arises as the Internet transformation of the digital industry takes place in manufacturing processes, together with artificial intelligence, big data, and CPS. Digitization of production, automation, and automatic data interchange are the main features of Industry 4.0 that will completely transform the industry. The current business models have been changing by the emergence of digitization that includes the Internet and mobile technologies with high-speed connectivity. Technologies such as the Internet, mobility, and sensor systems enable economic and social activities to be interconnected and networked globally. Each object has the potential to be connected and networked to each other and this leads to the development of innovative business models within the companies (Roblek et al. 2016).

Smart products, smart machine, and augmented operator characterize the Industry 4.0. Instead of treating smart products as passive work pieces, Industry 4.0 accepts them as active parts of a manufacturing system. Smart products can store operational data and request the required resources while coordinating the necessary production processes. Smart machines are the decentralized and self-operating devices that utilize the CPS technology. These intelligent systems can communicate with each other and smart products, leading the production line to become more flexible and modular. Augmented operator refers to the automation of knowledge in a manufacturing system in which an employee is supported by the mobile, context-sensitive user interfaces, and user-focused assistance systems. These systems allow an employee to manually interfere with the autonomous manufacturing systems and be in the role of strategic decision-maker while facing a large variety of jobs (Mrugalska and Wyrwicka 2017).

3. Development of Augmented Reality

3.1 Augmented Reality (AR)

AR is the technology that integrates computer-generated information with the real-world environment. Existing AR applications integrate computer graphics into the user's view of his current surroundings and provide him an improved experience of working conditions in which he can access and interact with information directly related to their immediate surrounding (Paelke 2014).

AR is a variation of Virtual Reality (VR), but with slight differences. The user of a VR device completely involves in an artificial environment and he has no interaction with the real world surrounding him. However, AR allows the user to see and interact with the real world and the virtual objects that are combined with it. While VR substitutes reality with artificial environment, AR enhances the real environment rather than completely replacing it. AR allows the coexistence of virtual and real objects in the same place, and the users of this technology can interact with both kinds of objects (Azuma 1997).

Three key characteristics of an AR system are (Azuma et al. 2001):

1. It combines real and virtual objects in a real environment,
2. It runs interactively, and in real-time,
3. It aligns real and virtual objects with each other.

Milgram and Kishino (1994) define a reality—virtuality continuum to present the mixture of classes of objects, as illustrated in Figure 1. Real environments are placed at the left end of the continuum, while virtual environments are placed at the right end of the continuum. The left side of the continuum, real environments, defines environments consisting

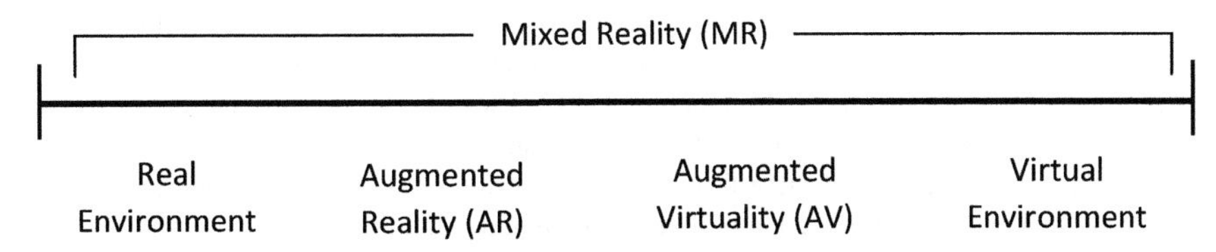

Fig. 1: Virtuality Continuum (Milgram and Kishino 1994).

of only real objects. These objects can be viewed directly without any particular electronic devices, or via conventional video display. The right side of the continuum, virtual environments, refers to the environments consisting of only virtual objects. Computer graphic simulation is an example of a virtual environment. A mixed reality environment is the one in which real and virtual environment objects coexist within a single display. Mixed Reality (MR) refers to the environments in which real and virtual worlds merge into each other to generate a new environment or visualization. MR consists of any objects located between the extreme points of Milgram's reality—virtuality continuum (Milgram and Kishino 1994).

According to this continuum, AR is located within the mixed reality area. In an AR application, the surrounding environment is real, and virtual objects are added to this environment. On the other hand, in Augmented Virtuality (AV) and Virtual Environment (Virtual Reality), the surrounding environment is virtual. In an AV application, real objects are added to the virtual surrounding (Azuma et al. 2001).

3.2 History and Development of AR

AR has recently gained popularity, however it is not a new concept. An optical illusion technique called "pepper's ghost" was used by the theatres and museums at the beginning of the 17th century. This technique involves placing a large piece of glass in order to merge the reflection of the hidden objects with the real world. Even though mirrors, lenses, and light sources were used to create virtual images for centuries, the first genuine AR system was not developed until 1968. Ivan Sutherland and Bob Sproull created the first head-mounted display (HMD) system, Sword of Damocles, at Harvard University in 1968. It was the first prototype of the AR system. While, display, tracking, and computing components were brought together in this very earliest application of AR, it was able to create three-dimensional components and superimpose them on the real environment. Boeing researchers Dave Mizell and Tom Caudell developed an AR application that assists workers in creating wire harness bundles efficiently. The researchers coined the term "Augmented Reality" in a paper they published in 1992. Many AR projects were initiated after the first successful application in Boeing (Billinghurst et al. 2015).

AR-related research made a breakthrough in the areas of communication and medical applications in the middle of the 1990s, focusing on key enabling technologies such as tracking, display, and interaction. Enhancing collaboration of people sharing the same place, computer-assisted surgery, visualization of surgical operations and X-Ray were some of the key research areas during the 1990s. HMDs were developed at this decade and they were the first examples of vision-based tracking systems on wearable computers (Billinghurst et al. 2015). The Global Position System, officially named "NAVSTAR-GPS" started it is operations in 1993. Even though this satellite-based radio-navigation system was intended for military use, today millions of people use it for navigation, geocaching, and AR (Arth et al. 2015).

Julier et al. (2001) introduced the Battle Field Augmented Reality System (BARS). This system was initially presented to assist soldiers to deal with challenges in a battle field. It provides information about the battlefield environment, locations of the team members, possible enemy ambushes, and assists soldiers to plan and coordinate the military operations. The system is composed of a wearable computer, GPS unit and antenna, wireless network receiver, sensors, and a see-through HMD.

The world's first outdoor AR game, AR-Quake, was launched by Thomas et al. in 2000. The players of the game were equipped with a wearable computer, HMD, and a simple input device. The first mobile AR game with high-quality content at the level of commercial games, ARhrrrr!, was developed by Kimberly Spreen at the Georgia Institute of Technology in 2009. iPad was released in April 2010 by Apple and has sold millions shortly after. The device had essential features that enable to create AR applications on a tablet computer. Some of these features were an assisted GPS, accelerometer, magnetometer, and advanced graphics chipset. In 2012, Google launched an optical HMD, Google Glass, which can be controlled with an integrated touch-sensitive sensor or natural language commands, allowing users to remain hands-free. Google Glass was a beneficial product not only for the ongoing research on AR and MR but also for clarifying the public perception of MR technology (Arth et al. 2015).

In July 2016, the mobile application game "Pokemon GO" was released by Niantic. The game uses geolocation to create AR gaming scenarios for players. The gaming components are incorporated into real-life surroundings, and players have to move to capture free-roaming Pokemon. The game was downloaded more than 550 million times and has made more than $ 470 million in sales within the three months of its release date (Wagner-Greene et al. 2017). ARKit was announced by Apple and ARCore was launched by Google in 2017. These AR frameworks use the smartphone's camera to add interactive elements to an existing environment. Developers can add advanced AR features,such as advanced motion tracking, to their AR apps with the help of these frameworks. According to Google, ARCore does two things: tracking the position of the mobile device as it moves, and building its own understanding of the real world (How-To Geek 2019). The history and development phases of AR are depicted in Figure 2 below.

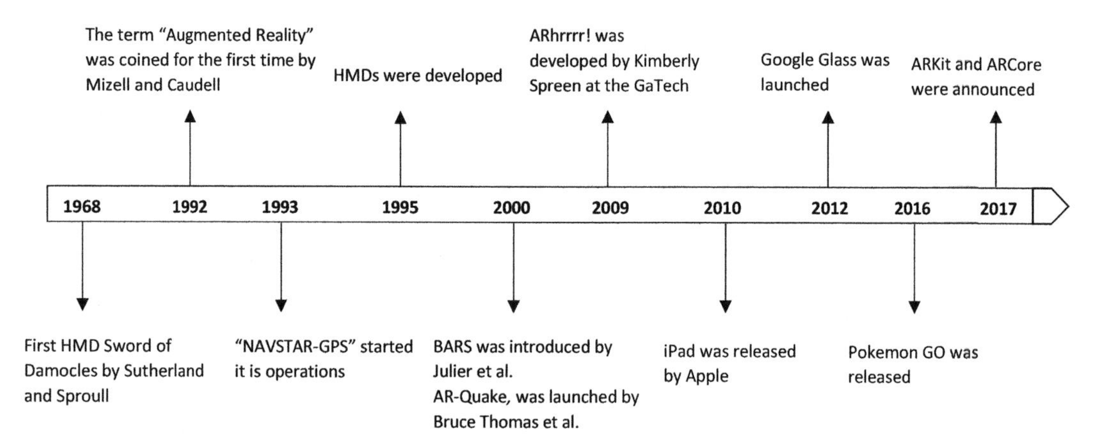

Fig. 2: History and Development of Phases Augmented Reality.

4. AR applications in Supply Chain Management

Digital technology has been transforming most of the industries and changing the way businesses operate. While this new trend brings about new opportunities for companies, it has a huge impact on supply chain management. Businesses cannot reveal their full potential of digitization unless they reorganize and adapt their supply chain strategy. This reinvented supply chain should be more connected, scalable, intelligent, and faster than the traditional supply chains (Merlino and Sproge 2017).

Due to the rapid increase in e-commerce transactions, the need for reduced inventory levels and increased customer responsiveness, the role of warehousing strategy became more crucial for the companies. Undoubtedly, warehouse operation strategy is one of the key drivers of the supply chain performance of a company. Computer systems together with information and communication technologies offer warehouse management solutions for many decades. AR is one of the newly emerging technologies that will take an important part in warehouse management. The potential use of AR in warehouse operations can be classified into four main areas: receiving, storing, order picking, and shipping. Among these areas, order picking accounts for more than 50% of the warehousing cost (Stoltz et al. 2017). Table 1, below shows some of the potential uses of AR in warehouse operations.

Order picking is one of the logistics operations that AR technology can be effectively used. Since order picking operation requires flexibility, workers cannot be replaced by machines. However, a worker equipped with an HMD can improve his information visualization. These pick-by-vision systems allow workers to act faster and work with fewer errors (Reif and Günthner 2009).

The development and implementation of AR software solutions rely on robust AR hardware platforms. These AR platforms appear in many forms such as handheld devices, stationary AR systems, spatial augmented reality (SAR) systems, HMDs, smart glasses, and smart lenses (Figure 3). AR applications allow logistics providers to access significant information easily and rapidly. This information is crucial for planning and executing delivery and load optimization operations. DHL Logistics Company explores many use of AR in various supply chain functions such as warehousing operations, transportation optimization, last-mile delivery, and enhanced value-added services (Glockner et al. 2014).

Smart glasses are wearable computers which offer human-computer interface solution between CPS and factory workers. These devices are capable of displaying information proactively and enable workers to interact with the information hands-free during work because of its capability to communicate with other information systems using wireless communication technology such as Wi-Fi or Bluetooth (Hobert and Schumann 2016).

Google's Glass Enterprise Edition 2 is currently one of the most popular smart glasses in the market. Glass is a small, lightweight wearable computer with a transparent display for hands-free work (Figure 4). Glass Enterprise Edition 2 is a

Table 1: Some potential uses of augmented reality in warehouse operations (adopted from Stoltz et al. 2017).

Operation	Potential Use
Receiving	- Indicate the unloading dock to the incoming truck driver - Check received goods against the delivery note - Show where to put the items/how to arrange them in the waiting zone
Storing	- Inform an operator about a new allocated task - Display the storage location of incoming items - Display picture and details of the item to be stored
Order Picking	- Display picture and details of the item to be picked - Display the storage location of the item to be picked - Scan the item's barcode to assign to picking cart or to see more information
Shipping	- Show the type of cardboard to be used - Indicate the right location/pallet for the shipment - Indicate appropriate loading area

Fig. 3: Handheld device and head-mounted display (Glockner et al. 2014).

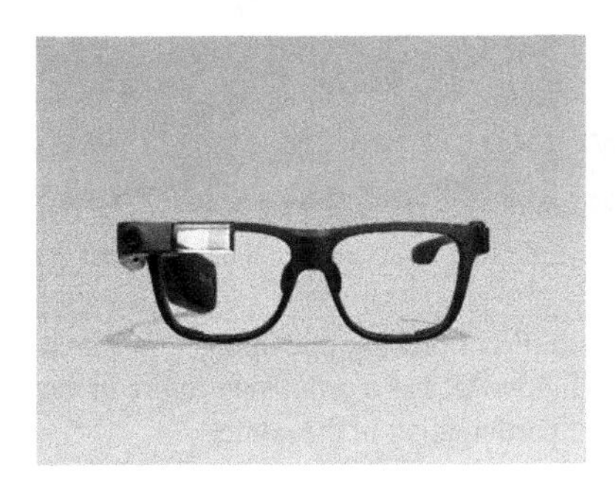
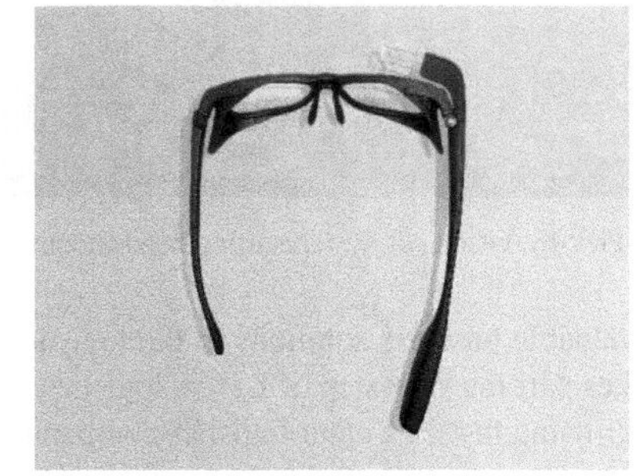

Fig. 4: Glass Enterprise Edition 2 (Glass 2019a).

wearable device that helps businesses improve the quality of their output, and help their employees work smarter, faster and safer. It provides hands-on workers and professionals with glanceable, voice-activated assistance that is designed to be worn all day with its comfortable, lightweight profile (Glass 2019a).

Smart glasses currently gain popularity and they are intensely used in various manufacturing and service sectors. Table 2 below shows some companies that use Glass Enterprise Edition 2 in their operations and the benefits they gain.

Table 2: Companies use Glass Enterprise Edition 2 based AR (Glass 2019b).

Company	Industry Field	Benefits of Using Smart Glass
AGCO	Agricultural Machines	25% reduction in production time on complex assembly operations
DHL	Logistics	15% more operational efficiency
GE	Energy	34% increased efficiency in top box wiring process
Sutter Health	Health Service	2 hours of doctor time saved per day on average

4.1 DHL Case

AR applications in the logistics sector make it possible to access anticipatory information quickly. From warehouse operations to last-mile delivery, DHL uses this technology for various functional areas of its business operations. Picking process optimization is the most popular AR application in the logistics sector. Most warehouses still use the pick-by paper approach. This method is slow and open to many errors. AR systems provide workers a digital picking list in their field of vision and reduce their travel time by indicating the shortest path to the items. Automated barcode scanning of an AR device indicates whether the worker is at the right location and pick the right item (Glockner et al. 2014).

Table 3 demonstrates DHL's logistics operations where AR technology actively in use.

Table 3: DHL's use of AR in business operations (adopted from Glockner et al. 2014).

Business Category	**Logistics Operations**
Warehousing Operations	Pick-by-Vision: Optimized Picking Warehouse Planning
Transportation Optimization	Completeness Checks International Trade Dynamic Traffic Support Freight Loading
Last-mile Delivery	Parcel Loading and Drop-off Last-meter Navigation AR-secured Delivery
Enhanced Value-added Services	Assembly and Repair Customer Services

Fig. 5: Pick-by-vision AR systems using head-mounted devices (Reif and Günthner 2009; Glockner et al. 2014).

AR offers valuable business solutions in the logistics field. From picking operations in a warehouse to after-sales activities, AR takes part in every step of DHL's logistics operations. AR has a promising future in the logistics industry, and the trend is growing faster as other logistics companies are participating in the game.

4.2 Boeing and Airbus Case

AR has been widely utilized by the important aircraft manufacturers in the global market. Boeing is the world's largest aerospace company and leading manufacturer of commercial jetliners and defense, space and security systems. The company exports to nearly 150 countries to commercial and government customers (Frigo et al. 2016). Boeing is currently using AR technology for electrical wiring applications in the aircraft fuselage. Installing electrical wiring on an aircraft is a complex task, and requires working with zero error. Boeing Company utilizes the benefits of AR to provide technicians real-time, hands-free, interactive 3D wiring diagrams. This system allows technicians to easily see and follow where the electrical wiring goes in the aircraft fuselage (Boeing.com 2019).

Airbus is an aircraft manufacturer with facilities mainly in France, Germany, Spain, and the United Kingdom. The company develops its product family in response to market needs and close consultation with airlines and operators, suppliers and aviation authorities. Currently, the Airbus Company produces the world's largest passenger airliner, the Airbus A380. The company uses MiRA (Mixed Reality Application) to increase productivity by scanning parts and detecting errors in their production line. By using this system, brackets on the fuselage can be rapidly checked, and missing or displaced brackets can be easily determined and replaced (Frigo et al. 2016).

4.3 IKEA Case

IKEA Place is one of the first major commercial AR applications developed with Apple's ARKit. This tool allows users to visualize how a virtual IKEA item would look in any space (Figure 6). Thousands of IKEA items can be placed in living rooms, bedrooms, and offices by using a customer mobile device's camera. This system allows a customer to visualize how an item would look like in a place without measuring the room space or purchasing and assembling the item. So, customers save time and money before their purchase decision. At the same time, this system helps IKEA to cut down on returns and increase customer satisfaction by allowing users to test and preview a major purchase via AR technology (Medium 2019).

Fig. 6: IKEA Place AR Tool (Medium 2019).

4.4 Coca Cola Case

The Coca Cola Company uses AR technology (designed by AR developer Augment) to solve a typical problem for the company's business to business (B2B) sales department. AR system visualizes how beverage coolers would look and fit in retail stores without any need to physically placing them (Figure 7). Indeed, the store managers can see how a cooler would look like on an aisle instead of just checking various types of coolers on catalogs or websites. Coca Cola's AR system allows potential B2B customers to browse different shapes, sizes, and designs of coolers, hence it assists them to make better product decisions (Medium 2019).

Fig. 7: Coca Cola's AR Tool (Medium 2019).

5. Conclusion and Future Directions

Industry 4.0 paradigm is rapidly converting the conventional production methods into the technology-driven smart manufacturing systems. This shift has not only been impacting the way businesses manage their key functions, but also forcing all supply chain players to adapt to the future of industrial value creation. The Fourth Industrial Revolution was shaped by physical and digital trends and technological innovations. Disruptive technologies such as CPS, IoT, AI, advanced robotics, cloud technology, additive manufacturing, autonomous vehicles, VR, and AR constitute the framework of Industry 4.0.

Smart factory is an important outcome of the Industry 4.0 concept and it works by employing the main drivers of Industry 4.0, such as CPS, IoT, AI, cloud computing, big data, and advanced autonomous robotics systems. The components of a smart factory are visible, connected and autonomous, thus these systems are able to run without much human intervention, make decentralized decisions, and learn, adapt, and respond in real-time. Smart factories present highly flexible and adaptive manufacturing solutions and they are considered as the future of manufacturing systems.

AR is the integration of computer-generated data into the real-world environment. AR devices capture the images from the real world and merge these graphics with the computer-generated information. Hence, AR applications enhance a person's perception and awareness of the surroundings by superimposing useful information on the screen of a device. AR has a wide range of applications in supply chains, especially in warehousing and transportation operations. AR devices allow users to interact with real-time information related to their immediate environment. For instance, a worker holding an AR device can navigate, locate, and perform barcode reading and item data synchronization in a large warehouse. Considering the share of the warehousing cost in total logistics cost, AR technology and its extensive use in warehouses can help businesses to minimize their cost dramatically. In addition, AR applications can optimize the efficiency of transportation operations by providing smart solutions for delivery and loading tasks. AR devices can assist loading/unloading workers and truck drivers by calculating precise truck routing, ensuring safety guidelines, and identifying unseen risks and problems of inbound and outbound parcel delivery.

AR has great potential to make supply chain operations more efficient, responsive, and cost-conscious. As the AR technology develops, it will bestow new and effective ways of presenting information, hence supply chains will become more robust and sustainable.

References

Arth, C., R. Grasset, L. Gruber, T. Langlotz, A. Mulloni and D. Wagner. 2015. The history of mobile augmented reality. arXiv preprint arXiv: 1505.01319.

Azuma, R.T. 1997 'A survey of augmented reality', Presence: Teleoperators and Virtual Environments, 6(4): 355–385. doi: 10.1162/pres.1997.6.4.355.

Azuma, R., Y. Baillot, R. Behringer, S. Feiner, S. Julier and B. MacIntyre. 2001. Recent advances in augmented reality. IEEE Computer Graphics and Applications, 21(6): 34–47.

Billinghurst, M., A. Clark and G. Lee. 2015. A survey of augmented reality. Foundations and Trends® in Human–Computer Interaction, 8(2-3): 73–272.

Boeing.com. 2019. Boeing: Boeing Tests Augmented Reality in the Factory. [online] Available at: https://www.boeing.com/features/2018/01/augmented-reality-01-18.page [Accessed 25 Oct. 2019].

Carvalho, N., O. Chaim, E. Cazarini and M. Gerolamo. 2018. Manufacturing in the fourth industrial revolution: A positive prospect in sustainable manufacturing. Procedia Manufacturing, 21: 671–678.

Frigo, M.A., E.C. da Silva and G.F. Barbosa. 2016. Augmented reality in aerospace manufacturing: a review. Journal of Industrial and Intelligent Information, 4(2).

Glass. 2019a. Tech Specs – Glass. [online] Available at: https://www.google.com/glass/tech-specs/ [Accessed 25 Oct. 2019].

Glass. 2019b. Glass – Glass. [online] Available at: https://www.google.com/glass/start/ [Accessed 9 Nov. 2019].

Glockner, H., K. Jannek, J. Mahn and B. Theis. 2014. Augmented Reality in Logistics: Changing the way we see logistics–a DHL perspective. DHL Customer Solutions & Innovation, 28.

Gürdür, D., J. El-khoury and M. Törngren. 2019. Digitalizing Swedish industry: What is next? Data analytics readiness assessment of Swedish industry, according to survey results. Computers in Industry, 105: 153–163.

Hobert, S. and M. Schumann. 2016. Application Scenarios of Smart Glasses in the Industrial Sector. i-com, 15(2): 133–143.

How-To Geek. 2019. What Are the ARCore and ARKit Augmented Reality Frameworks? [online] Available at: https://www.howtogeek.com/348445/what-are-the-arcore-and-arkit-augmented-reality-frameworks/ [Accessed 15 Jul. 2019].

Julier, S., Y. Baillot, M. Lanzagorta, D. Brown and L. Rosenblum. 2001. Bars: Battlefield augmented reality system. Naval Research Lab Washington DC Advanced Information Technology.

Klötzer, C. and A. Pflaum. 2017. Toward the development of a maturity model for digitalization within the manufacturing industrys supply chain. Proceedings of the 50th Hawaii International Conference on System Sciences, pp. 4210–4219. doi: 10.24251/hicss.2017.509.

Koul, S. 2019. Augmented reality in supply chain management and logistics. Int. J. Recent. Sci. Res., 10(02): 30732–30734. DOI: http://dx.doi.org/10.24327/ijrsr.2019.1002.3113.

Lu, Y. 2017. Industry 4.0: A survey on technologies, applications and open research issues. Journal of Industrial Information Integration, 6: 1–10.

Medium. 2019. 5 Use Cases Of Augmented Reality That Boosted Businesses' Sales. [online] Available at: https://medium.com/swlh/5-use-cases-of-augmented-reality-that-boosted-businesses-sales-2114ac35bf5a [Accessed 25 Oct. 2019].

Merlino, M. and I. Sproge. 2017. The augmented supply chain. Procedia Engineering, 178: 308–318. doi: 10.1016/j.proeng.2017.01.053.

Milgram, P. and F. Kishino. 1994. A taxonomy of mixed reality visual displays. IEICE TRANSACTIONS on Information and Systems, 77(12): 1321–1329.

Mrugalska, B. and M.K. Wyrwicka. 2017. Towards lean production in industry 4.0. Procedia Engineering. The Author(s), 182: 466–473. doi: 10.1016/j.proeng.2017.03.135.

Paelke, V. 2014. Augmented reality in the smart factory: Supporting workers in an industry 4.0. environment, 19th IEEE International Conference on Emerging Technologies and Factory Automation, ETFA 2014. IEEE, pp. 1–4. doi: 10.1109/ETFA.2014.7005252.

Reif, R. and W.A. Günthner. 2009. Pick-by-vision: augmented reality supported order picking. The Visual Computer, 25(5-7): 461–467.

Ro, Y.K., A. Brem and P.A. Rauschnabel. 2018. Augmented reality smart glasses: Definition, concepts and impact on firm value creation. In Augmented reality and virtual reality (pp. 169–181). Springer, Cham.

Roblek, V., M. Meško and A. Krapež. 2016. A Complex View of Industry 4.0. SAGE Open, 6(2). doi: 10.1177/2158244016653987.

Rojko, A. 2017. Industry 4.0 concept: background and overview. International Journal of Interactive Mobile Technologies (iJIM), 11(5): 77–90.

Sääski, J., T. Salonen, M. Liinasuo, J. Pakkanen, M. Vanhatalo and A. Riitahuhta. 2008. Augmented reality efficiency in manufacturing industry: a case study. In DS 50: Proceedings of Nord Design 2008 Conference, Tallinn, Estonia, 21.-23.08. 2008 (pp. 99–109).

Stoltz, M.H., V. Giannikas, D. McFarlane, J. Strachan, J. Um and R. Srinivasan. 2017. Augmented reality in warehouse operations: opportunities and barriers. IFAC-PapersOnLine, 50(1): 12979–12984.

Thomas, B., B. Close, J. Donoghue, J. Squires, P. De Bondi, M. Morris and W. Piekarski. 2000. October. ARQuake: An outdoor/indoor augmented reality first person application. In Digest of Papers. Fourth International Symposium on Wearable Computers (pp. 139–146). IEEE.

Tjahjono, B., E. Esplugues, E. Ares and G. Pelaez. 2017. What does industry 4.0 mean to supply chain? Procedia Manufacturing, 13: 1175–1182.

Wagner-Greene, V.R., A.J. Wotring, T. Castor, J.K. MSHE and S. Mortemore. 2017. Pokémon GO: Healthy or harmful? American Journal of Public Health, 107(1): 35.

Zhong, R.Y., X. Xu, E. Klotz and S.T. Newman. 2017. Intelligent Manufacturing in the Context of Industry 4.0: A Review. Engineering. Elsevier LTD on behalf of Chinese Academy of Engineering and Higher Education Press Limited Company, 3(5): 616–630. doi: 10.1016/J.ENG.2017.05.015.

SECTION 6

Blockchain in SCM: The Impact of Block Chain Technology for SCM-Potentials, Promises, and Future Directions

Chapter 10

Blockchain Driven Supply Chain Management

The Application Potential of Blockchain Technology in Supply Chain and Logistics

Yaşanur Kayıkcı

1. Introduction

As the today's business environment continues to become increasingly connected and transparent, the development of new emerging technologies such as internet of things, big data analytics, artificial intelligence, and blockchain revolutionizes the way of existing business and industrial processes and it enables the creation of new business models. At the same time, organisations have to struggle with challenges such as limited asset management, empowered customers, high transaction fees, counterfeit products and the lack of end-to-end visibility. Moreover, today's record keeping systems in supply chain are centralized, trust-based and require immediately third-party enforcements which can lead to bottlenecks, miscommunication and even slowdowns to optimal transaction time. Companies can greatly benefit and address these challenges notably by using blockchain applications. Blockchain technology creates unprecedented visibility and accountability through peer-to-peer, distributed and time stamping transactions in the supply chain. In essence, blockchain is a decentralized and distributed ledger technology to provide transparency, data security and integrity. Blockchain can record each sequence of

Turkish-German University, Department of Industrial Engineering, Sahinkaya Cad. No: 86 - 34820 Beykoz, Istanbul, Turkey.
Email: yasanur@tau.edu.tr

transactions from raw material to finished product along the supply chain on a series of blocks or ledgers which are organized in chronological order and are linked through cryptographic proof. The records are accessible to all authorized participants involved, but cannot be modified or manipulated. Since introducing the first cryptocurrency, blockchain technology has drastically expanded with potential use-cases being identified across a myriad of sectors including finance, healthcare, energy, government and manufacturing (Al-Jaroodi and Mohamed 2019). In addition, it is expected that as a technology solution, blockchain will be widely adopted in different sectors including supply chain and logistics in the next two to five years (Ahlmann 2018; i-scoop2018). Blockchain technology with the integration of other emerging technologies provides not only the capability of capturing both mapping data (transparency) and operational data (traceability) throughout the supply chain ecosystem, it also optimizes business transactions and fosters trading relationships with ecosystem participants. Therefore, it is important to understand the application potentials of blockchain technology in supply chain and logistics.

This chapter explores basically the following points:

- The basics of blockchain technology, its attributes and public, private and hybrid blockchain
- The uses and benefits of blockchain in supply chain and logistics context
- Integration of blockchain technology with other emerging technologies
- SWOT analysis for blockchain technology and adoption strategies in the supply chain
- Finally, future directions of this technology in the supply chain and logistics industry

2. Basics of Blockchain Technology

Blockchain technology was presented for the first time in 2008 in the document written by Satoshi Nakamoto (2008) as peer-to-peer electronic cash system to develop the first fully distributed cryptocurrency, namely Bitcoin, which radically changes existing payment systems. Since the beginning of this initiative, the popular awareness and implementation of the context of blockchain technology has grown and evolved greatly in different sectors, especially in the last decade. Blockchain technology is essentially an encrypted protocol (blockchain-based registration protocol) and filing system that ensures that a data block in a network is simultaneously monitored, authenticated, and permanently recorded in a single decentralized distributed database by all users allowed to enter that network (Leng et al. 2018). Blockchain simply denotes a type of digital ledger with specific characteristics that stores all transaction data or digital interactions permanently and securely. The ledger data is organized in a form of a chain of blocks which are linked one after another based on cryptographic protocols. Every transaction log is stored in the digital ledger which is replicated and distributed to all partners across a network. The blockchain is operated by a consensus mechanism, which is the most important part of blockchain system (Viriyasitavata and Hoonsopon 2019). In the consensus operations, the network partners must come to an agreement to create the next chain block (Christidis and Devetsikiotis 2016). In this technology, the security depends on the advanced encryption techniques called cryptography that validates each transaction block and links them together. Thus, none of the participating members can change, delete, tamper or modify the approved activity, namely data block subject to the blockchain (Biswas et al. 2017; Al-Jaroodi and Mohamed 2019). It provides proof-of-asset ownership and allows easy and secure transfer of assets. Bitcoin is the name of cryptocurrency or digital money, while blockchain is the name of the technology used to transfer digital money from one place to another. Theoretically,the implementation of a blockchain provides better transparency, traceability, integrity, efficacy and interoperability, enhanced security, reduces data replication, speeds up processing times and eliminates data errors, resulting in increased productivity and efficiency and reduces costs for all interested parties in a network (Christidis and Devetsikiotis 2016; Niranjanamurthy et al. 2018).

Blockchain uses smart contract in order to obtain blockchain benefits for process executions, where a business process can be encoded into smart contract transactions (Viriyasitavata and Hoonsopon 2019). Smart contract is a software application designed to disseminate, verify and enforce consensus contracts agreed on by parties. Beside this, blockchain uses private keys and digital wallets for data security requirements (Neuburger and Choy 2019). A private key is a sophisticated form of cryptography and it allows which participant can encrypt and decrypt data, whereas digital wallet refers to a utility to store blockchain-based digital assets and effectuate transactions.

There are basically three types of blockchain systems, namely public blockchain, private blockchain and hybrid blockchain (Niranjanamurthy et al. 2018). Each type has both advantages and disadvantages, which allow them to meet the needs of various applications:

(1) Permissionless (public); is the mostly used blockchain, anyone can participate in the network without authorization by a third party and has access to full data transparency for all participants. Blockchain uses an open ledger or so-called "distributed open ledger", which can enable all network participants to authenticate and submit data. Examples included: Bitcoin, Ethereum, Litecoin, Lisk, Stratis.

(2) Permissioned (private); blockchain system uses a private ledger or so-called "distributed ledger" and limits participations by central authority (mostly a company) exercising the power to control as to who can view, read or validate transactions on the blockchain. System is only accessible to selected members. Examples included: Hyperledger Fabric, R3, Corda, Ripple, MultiChain.
(3) Hybrid (consortium); blockchain system combines both permissionless and permissioned ledgers in a solution. Companies can secure background transactions with business partners on a private ledger, while also sharing product information with customers on an open ledger. It also allows flexibility to invite more players into the blockchain. The blockchain is managed by a group of participants. A hybrid blockchain is secure and helps to protect privacy. Examples included: XinFin, DragonChain.

There are several open-source blockchain platforms that allows customers to create and run their own public or private blockchain networks. This service is called as Blockchain-as-a-service (BaaS) and it has standard templates based on the cloud to develop three types of blockchain-based solutions. Several larger cloud computing providers (i.e., Microsoft, IBM, HP, Oracle) have responded by launching BaaS offerings. The BaaS services could differ in terms of functionality, infrastructure and scalability; therefore, users should make trade-off and decide on which would be preferable for them. For instance, one of the BaaS providers, Microsoft Azure offers industry leading frameworks (i.e., Ethereum and Hyperledger Fabric) to allow users to quickly create private, public and consortium blockchain environments. Ethereum is general-purpose, permissionless and "public blockchain" that is more suitable to describe business logic through smart contracts. All participants manage a shared open ledger without a trusted party. Hyperledger Fabric is a permissioned and "private blockchain" with limited number of participants. It can be used to improve performance and reliability with a blockchain-based distributed ledger (Lin and Liao 2017). Apart from public and private blockchain applications, Facebook launches its cryptocurrency Libra in 2020, the concept is based on "hybrid blockchain", where one is public or consumer-facing for customers who purchase items using digital wallets and the other is permissioned or "private blockchain" for corporate transactions.

In a nutshell, blockchain technology has four main characteristics combining with aforementioned features: (i) *immutable*, blockchain records transactions that are permanent and tamper-proof. Once a block is added to a blockchain, it cannot be changed, modified or altered. (ii) *decentralized*, blockchain is stored in a ledger that can be replicated, distributed and accessed by any party on the network. (iii) *consensus*, every block in the ledger is verified by consensus models (i.e., proof-of-work, proof-of-stake, delegated byzantine fault tolerance), that provide set of rules to validate a block. Consensus mechanism works without the existence of a central authority or intermediary responsible. (iv) *transparent*: since the blockchain is a decentralized and distributed ledger which can be accessed by any authorized party in the blockchain network, this allows that assets in transparent blockchain can be traced and tracked throughout their lifetime from manufacturing to delivery.

3. Blockchain in Supply Chain and Logistics

Although blockchain is primarily used in the financial sector as a technology to develop crypto-asset products and services, it continues to rise rapidly and develop new solutions in other sectors such as retail, insurance, healthcare, energy and real estate for asset ownership, accelerating transaction times, reducing cost, eliminating recall, counterfeiting and fraud risks (Pilkington 2016; Banerjee 2019). In addition, Blockchain technology is a demonstrable successful solution for information communication, control and management of the supply chain that enables monitoring of the entire product life cycle (Korpela et al. 2017) and it is an ideal tool to revolutionize supply chain management (Saberi et al. 2018). Blockchain accelerates the Key Performance Indicators (KPIs) monitoring efficiency of key processes of supply chain management and achieves valid and effective outcomes. For example, by using blockchain technology, companies can build productive relationships among business partners, make trade more transparent to customers, avoid making mistakes, quickly assess data and track quality problems that may occur across the supply chain. In a trustless environment, supply chain costs can be reduced by eliminating intermediary auditors (Kshetri 2018). Blockchain is able to allocate trust among partners due to unchanged tracking data (Christidis and Devetsikiotis 2016). The unique technological qualities of a blockchain such as immutability, autonomy, pseudonymity and undeniability (irreversibility), can contribute to supply chain reliability, transparency and efficiency (Treiblmaier 2018).Furthermore, the technology potential allows companies to reduce the amount of waste, product degradation and defect. Blockchain also has the capacity to transform natural resources and manage waste and recycle (Saberi et al. 2018).

Blockchain has greater impact on the supply chain performance, in terms of cost, quality, speed, reliability, risk reduction and flexibility (Bigliardi and Bottani 2010) as well as sustainability (Kshetri 2018; Kouhizadeh and Sarkis 2018; Helo and Hao 2019). While Blockchain leads to these mutually beneficial results, transparency in the supply chain brings competitive advantage (Tian 2016; Francisco and Swanson 2018).

The supply chain is the industry to benefit from both the private blockchain as well as the consortium blockchain system as they are more suitable for business-to-business (B2B) applications where privacy is concerned (Chang et al. 2019). Different parties from the entire supply chain ecosystem come together to build a consortium to move finished product from producer to end customer efficiently and cost effectively. Implementing a private or consortium blockchain solution helps to maintain the privacy and accuracy of the network, at the same time it allows to invite partners to the blockchain as needed. From this perspective, blockchain in the supply chain for transactions requires both a private ledger to communicate with the partners in the consortium and an open ledger to communicate with end consumers in order to provide a secure way to track and transfer assets through the supply chains.

In the supply chain and logistics industry, companies establish consortiums in their ecosystems by aiming at developing a number of blockchain pilots to showcase how this technology can be applied in their businesses and bring benefits to them. A good example of blockchain applications is the TradeLens platform (TradeLens 2019) in the transport supply chain. Maersk and IBM developed a Hyperledger Fabric based open, neutral and distributed blockchain platform to enable digital collaboration across multiple partners involved in international trade. Ecosystem participants such as shippers, shipping lines, freight forwarders, port and terminal operators, inland transport companies and customs authorities can interact more efficiently through real-time access to shipping data and documents. With the support of digital products and integration services such as smart containers, the platform aims to reduce transportation cost, eliminate inefficiencies burdened by paper-based processes and increase traceability in container transportation from South America and Africa to Europe (DHL 2018). This developed blockchain concept has been tested for the sending of flowers from Kenya to Royal Flora Holland, also mandarin oranges from California and pineapple from Colombia to the port of Rotterdam (Louppova 2017). In the food supply chain many retailers adapt the blockchain technology to trace the authenticity of food products. The giant retailer, Walmart collaborated with IBM and others prominent food producers in the food industry, like Dole Food, Nestlé, Unilever and Tyson Foods to set up a food traceability system based on Hyperledger Fabric, called the IBM Food Trust, (Hyperledger 2019). Walmart uses blockchain technology to improve transparency, standardization and efficiency throughout the food supply chain. The process of tracking information for food safety could take less than 3 seconds by using blockchain (Browne 2017). Here, the manufacturer-origin data from the field or farm, lot numbers, factory and process data, expiration dates, storage temperatures and transport details are stored securely and unchanged on the Blockchain. The IBM Food Trust (https://www.ibm.com/blockchain/solutions/food-trust) program expanded with the cooperation of Carrefour in France and launched Europe's first food blockchain to digitally track the movement of Carrefour's Quality Line products (Carrefour 2018). Originally used for free-range Auvergne chickens, this technology is also being developed for eggs, cheese, milk, beef steak, salmon, oranges and tomatoes. Consumers can get information about the origin of the product by scanning a Quick Response Code (QR code) on the chicken package and installing the blockchain application on their smart phones: where and how each animal was reared, the name of the farmer, what feed was used (locally grown cereals and soya beans, on GMO-free products, etc.), what treatments were used (antibiotic-free), any quality labels (organic, free-range, etc.) and where they were slaughtered. Carrefour is also planning to track additional 100 non-food items to the system by this year. Apart from this, the Switzerland's supermarket chains, Migros has also recently launched a Hyperledger Fabric based blockchain solution with the cooperation of TE-FOOD (https://www.te-food.com/) in order to offer safety and transparency for its fresh fruits and vegetables supply chains. Migros implemented blockchain technology to optimize food processes, to enhance faster distribution and to reduce food waste (TE-FOOD 2019). Implementing blockchain technology for retailers cannot only bring marketing advantages to customers by providing transparent food information, but also food traceability can provide greater value from easier product recalls to improved supply chain control. In addition, technology speeds up the processes and allows farmers to get paid faster for the products they sell. It also prevents price coercion and retroactive payments, common in the food industry. For example, Louis Dreyfus company conducted the first blockchain-based agricultural commodity trade (LDC 2019) over easy trade connect platform, selling 60,000 tons of American soybeans to the Chinese government. The entire transaction took a week, reducing total logistics time by 80%.

4. Blockchain and other Emerging Technology Applications

As a game-changer, blockchain alone does not solve visibility and traceability challenges in supply chains, implementing emerging technologies with blockchain effectively can facilitate connection and enhance efficiency, transparency and accountability from origin to completion among participated partners. Blockchain is clearly used with both Internet of Things (IoT), B2B and machine-to-machine (M2M) integrations (Korpela et al. 2017). Notably, IoT and blockchain technology have been rapidly approaching each other, in the very near future, blockchain systems will work with data generated from both near-edge or far-edge physical IoT devices (i.e., sensors, actuators, embedded devices) used in logistics and transport (Banker 2018; Ioti 2018; Provenance 2018). Combining blockchain's distributed ledger framework with these applications and other emerging technologies such as smart mobile devices, artificial intelligence, augmented reality/virtual reality, cloud computing, edge computing, 5G, Radio Frequency IDentification (RFID), etc., can improve real-time process monitoring and tracking capability of (AR/VR) supply chain and logistics systems.

In this area, industrial applications have begun to be seen even at the pilot stage. In particular, those with established blockchain platforms—logistics companies and customers doing business in specific areas automate their commercial transactions during freight shipment by employing smart contracts based on blockchain technology. Smart contracts improve traditional contracts by implementing rules that control the transfer of currencies or assets under certain conditions (Christidis and Devetsikiotis 2016).

Figure 1 denotes asset tracking examples for blockchain driven supply chain management using smart contracts and emerging technologies with supply chain ecosystem participants such as supplier, producer, transport provider, distributor, retailer and customer. Supplier A supplies the raw materials in bulk trucks, the transaction with information about raw materials, their origins and properties addressing environmental issues are recorded. Barcode and IoT applications can be used to generate data. After manufacturing factory B received the raw materials from supplier A, the quality and quantity of the materials are checked. Here, a smart contract is established and electronic entries are generated about this transaction. If the properties of raw materials are matched with the requirements of manufacturer B, then the goods are accepted, else sent back to suppliers. This rejection is also recorded to the blockchain ledger. Factory B produces goods which leave the factory in containers on wagons by rail transport and reach the shipping terminal C. Each product gets its own QR code and also containers are tracked by implementing RFID tags and IoT sensors.

The containers are carried by sea transport in vessels from terminal C to terminal D. All road conditions are tracked by using Global Positioning System (GPS), General Packet Radio Service (GPRS) technologies, also the ambient temperature is tracked via time temperature sensors with sensor devices connected to a wireless sensor network (WSN) (Rejeb et al. 2019). Then, containers are carried with trucks by road transport to warehouse/distribution center E. The transportation data is also recorded with temperature and localisation sensors. The containers are handled at distribution center E, the quality and quantity of goods are checked and recorded by using indoor localisation sensors, room temperature and humidity sensors and RFID tags. Smart contract is also established to check whether the products meet the requirement and then these are sent to retail point F by city logistics. Along the city distribution process, all temperature, humidity, localisation data are recorded. After reaching the retail point F, room temperature, localisation sensors are used to track the selling goods at the retail stores. Smart contract is used to check whether the goods are in required conditions. At the end, product item is bought from customer G by using Near-Field Communication (NFC) technology and its quality checked

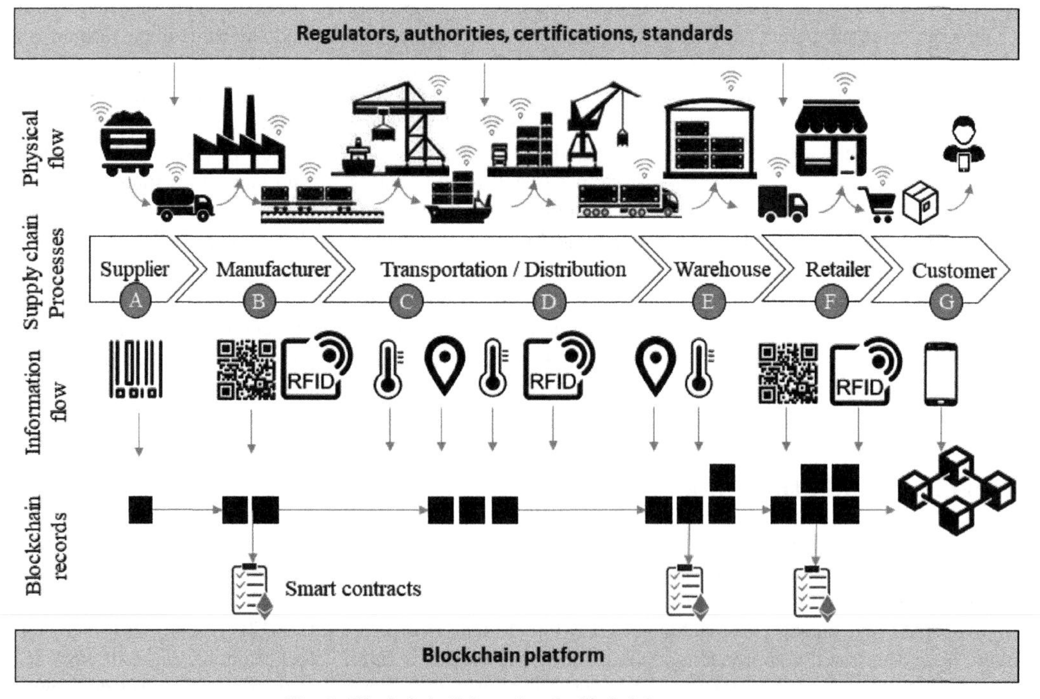

Fig. 1: Blockchain Driven Supply Chain Management.

through the QR code or RFID on the packaging by using smart phone application. Cloud computing and edge computing to operate big data and instant data (advanced) analytics are also used as complementary technologies with blockchain to increase security or quality of the data. Cloud computing operates on "bigdata" to identify risky transactions along supply chain which gives alerts and enables to make better decisions in the blockchain platform (Rejeb et al. 2019), while edge computing operates on "instant data" that is on-site real-time data generated by sensors or users to assist instant decision making. For example, defective products are detected through big data analytics and returned to the factory before arrived at retail shelves which minimize recall costs of the defected product. Using emerging technologies with blockchain help to connect the participants of the supply chain to each other, so that, all transactions throughout the supply chain system are recorded on the blockchain platform.

Ecosystem participants access the blockchain platform via open Application Programming Interfaces (APIs), which allow communication with protocols and smart contracts and their integration to blockchain services. The blockchain platforms in the supply chain industry are mostly established by using standard templates of BaaS based business models.

5. SWOT Analysis of Blockchain Technology in Supply Chain and Logistics

SWOT analysis stands for strengths, weaknesses, opportunities, and threats. It is a method designed to support the strategic planning of projects that are often related to the adoption of new technologies. *Strengths*; It relates to the characteristics of a project that reflects an advantage over other similar initiatives to help achieve its core objectives. *Weaknesses*; what makes a project less successful than others. *Opportunities*; external factors in the environment leading to the improvement of performance. *Threats*; external factors that can challenge the success of a project. Strengths and weaknesses point to the *internal factors* of a system, while opportunities and threats indicate *external factors*. These four categories force companies to understand the current status of the blockchain technology to ensure whether it can be adoptable in the future. The purpose of a SWOT analysis is to systematically formulate recommendations to help determine the adoption of a particular technology, especially in the absence of empirical evidence.

In this chapter, the application potential of blockchain technology as one of the developing new technologies to the supply chain has been investigated by using SWOT analysis, where it has been demonstrated whether it is possible to make any investment in the blockchain technology. By reviewing the existing literature, SWOT analysis of blockchain technology is performed as shown in the Table 1 and the strengths, weaknesses, threats and opportunities are briefly outlined. Within this framework, the application of blockchain technology to any supply chain can be decided according to this analysis.

The SWOT analysis shows that blockchain technology may be a good solution for some supply chains, but is not yet ready for mass acceptance (Niranjanamurthy et al. 2018; Carson et al. 2018). First of all, the contribution of technology to the supply chain has not been widely proven with blockchain pilots and real-case deployments. Blockchain technology cannot fully capture data from all stakeholders in the supply chain due to its uncommon and other infrastructure problems (Korpela et al. 2017). Full visibility and traceability in the supply chain can be achieved not only by using blockchain technology but also by using other technologies, which is not the only solution to ensure transparency (Francisco and Swanson 2018). In addition, the cost of developing and operating the blockchain-based supply chain is not fully known (Tian 2016). Currently there are only a few standards (Banerjee 2019; Rejeb et al. 2019).

The gap between the current capacity of blockchain technology and the capacity required by the supply chain is huge. In this sense, as can be seen in Table 2, blockchain technology needs to be integrated into the supply chain and logistics industry theoretically according to four different strategies. These strategies demonstrate: SO, involves making good use of opportunities by using the existing strengths. ST, the strategies associated with using the strengths to remove or reduce the effects of threats. WO, seeks to gain benefit from the opportunities presented by the external environmental factors by taking into account the weaknesses. WT, in which the organization tries to reduce the effects of its threats by taking its weaknesses into account.

The adoption strategies of blockchain technology must meet the expectations of participants to exploit potential market positions (Carson et al. 2018). The adoption strategies according to SWOT analysis are explained as follows:

(1) *SO - High Attention:* If blockchain technology holds up to expectations, then the blockchain consortium is built and the high-impact blockchain solution is employed with the first strategy. Blockchain technology should be integrated into the supply chain to increase brand awareness and consumer confidence, especially in new segments. The consortium shapes new standards that will disrupt the current businesses. Despite the huge potential, blockchain adoption cannot be done overnight, as the consortium could face regulatory and standardization barriers.

(2) *ST - Special Attention:* If expectations due to lack of technological infrastructure are not met properly, investments in research and development (R&D) are pursued to support best practices and to constitute blockchain-based industry

Table 1: SWOT Analysis for Adoption of Blockchain Technology.

Strengths (S)	**Threats (T)**
• Provides transparency throughout the supply chain network • No need of authority/intermediary for transaction • All commercial transactions are verified • Provides efficiency in business processes • Decentralized consensus approach • High quality and proof-of-data • Provides higher productivity • Low cost and risk • Fast and secure ledger system • High security and data privacy • Trustless, no need to trust a third party • Enable trust unreliable networks • Durability (no single point of failure)	• New technology, further research is required and application potential needs to be proved • Eliminated existing bank functions • Requires government agencies to be willing to implement • Applications require high investments • Requires large regulatory impact • Requires compliance with laws and regulations • Confidentiality and security concerns • Time - lost during negotiation • Uncertainty about its impact • Distributed systems • Insufficient human resources
Weaknesses (W)	**Opportunities (O)**
• Technology maturity • Access difficulties among ecosystem participants • Integration with legacy system applications (such as ERP, SaaS and supply chain applications) • Lack of industry standards • Low capacity and processing speed • Property issue • Latest digital technology but not advanced • Scalability and compatibility • Security against cyber criminals • Storage issues • Change management	• Automation • Optimization of business processes • Eliminating the need for trust • Faster secure (international) payment transfers • Accelerates transactions • Improved real-time customer experience • Increased product and service quality • Brings innovation in every sector • Instant agreements can be made • Streamlining know-your-customer processes • Triggers new collaborations • Does not require trust from rating agencies • Capturing new opportunities with IoT and other emerging technologies • Programmable consensus mechanisms • Smart contracts in blockchain ecosystem

Table 2: Adoption Strategies of Blockchain Technology in the Supply Chain.

Blockchain technology adoption in supply chain and logistics		**External factors**	
		Threats (T)	**Opportunities (O)**
Internal Factors	**Strengths (S)**	**(1)SO - High Attention** – Build consortium to employ blockchain solution – Invest in it for brand awareness, reputation building and consumer confidence, especially in **new segments** – Focus on **high-impact use-cases** which requires shared standards *Strengths–Opportunities Strategies*	**(2) ST - Special Attention** – Constitute industry standards for blockchain – Invest in **R&D** to eliminate the problems that prevent blockchain integration to the supply chain – Focus on **use-cases with highest business value and network impact** *Strengths–Threats Strategies*
	Weaknesses (W)	**(3) WO - Regular Attention** – Prepare to act fast to adopt emerging standards – Invest in more **customer-oriented** services in supply chain – Focus on **use-cases with high internal benefits** *Weaknesses-Opportunities Strategies*	**(4) WT - Low Attention** – Not seeking further cooperation in this area – Focus on disruptive **peer-to-peer use-cases** with other new and emerging technologies *Weaknesses–Threats Strategies*

standards with the second strategy. The consortium adopts use-cases with highest application value and network impact; however, it has a high risk that leads to losing the competitive advantages compared to competitors.

(3) *WO - Regular Attention:* If expectations are pitched at the right level, to be both challenging and realistic, then this is the time to invest in blockchain technology to offer more customer-oriented services with consortium partners in the supply chain, with the third strategy. Blockchain technology should be invested in more customer-oriented services in the supply chain. The application requires standardization and regulations; therefore, consortium would not have high capability to influence the other supply chain parties.

(4) *WT - Low Attention:* If expectations to invest in blockchain technology are not realized and the risks are high, not seeking further cooperation in this area and other ways can be searched to implement alternative digital technology solution to enable end-to-end supply chain traceability with the fourth strategy. Especially, the new entrants without existing market share can focus on this strategy to adopt an innovative business model with highest disruptive use-cases.

6. Conclusions and Future Directions of Blockchain Technology

More and more consumer goods are produced, transformed and distributed by an ever-increasing number of players across the world, where visibility and traceability gain deeper insight into assets at every stage of the life cycle. Meanwhile, new and emerging technologies are introducing faster, safer and more intelligent ways to design, optimize and manage the supply chain. One of these new technologies is blockchain, which is mostly known as the underlying technology behind cryptocurrency. However, he potential of the technology has been deployed widely and in many applications built to improve business operations across various industries. But they are still scarce in supply chain and logistics, and larger investments being made are expected in this industry. Blockchain enables a holistic view of product's lifecycle from origin to retail outlet with real-time permanent record keeping and provides consumers with end-to-end confidence in the supply chain. In nutshell, what the Internet does for communication, blockchain technology can do for transparency and traceability. Although, there are doubts about its applicability in the real-world, as it is still in a nascent stage, blockchain technology continues to increase its potential especially in the supply chain domain. However, only a few pilot implementations are available to show proof-of-concept, as some leading retailers launch blockchain use-case projects to develop open-source blockchain platforms to track and trace food in the supply chain, larger-scale testing is needed to determine the potential of blockchain. One of the most important handicaps related to the implementation of the blockchain is the technology cost uncertainty, which may be costly for small companies with limited margins. Furthermore, small enterprises may experience problems about capturing and transmitting of data from origin to their recipients, especially in integration with IoT. In addition, there is a need for a number of trained people in the supply network to capture and process data, and the technical difficulties of facilities in rural areas could be an obstacle to its implementation. However, blockchain technology is not owned or operated by a single authority, the ownership of the technology might be more beneficial for the success, if it is led by, for example, large retail companies or perhaps larger consortium rather than small businesses.

Blockchain technology can support consumers to prevent fraud and counterfeiting and reduce waste and losses as well as companies to operate efficiently, cut costs and reduce environmental impact throughout the supply chain. Implementing adequate adoption strategy is important to make a decision such as what kind of use-cases need to be selected for blockchain application, so that the both sides could be benefited through blockchain technology. But first of all, in order to apply this technology to the supply chain, the consumer awareness about product safety issues needs to be increased. It is a fact that investments in Blockchain technology can increase traceability and transparency in business processes as long as there is a demand for quality, production process and origin of the products purchased from the consumers' point of view. In addition, considering the costs of fraud and food-borne illnesses in the food sector, traceability and transparency in food products can be made a legal obligation for manufacturers and retail companies with the amendments of laws and regulations that states can enact. In this sense, blockchain technology will become widespread and generally adopted in the supply chain and logistics industry.

Acknowledgments

This study was supported by Turkish–German University Scientific Research Projects Commission under the grant no: 2019BM0013.

References

Ahlmann, J. 2018. Farm-to-fork transparency: food supply chain traceability. Cutter Business Technology Journal, 31(4): 3–9.

Al-Jaroodi, J. and N. Mohamed. 2019. Blockchain in industries: A survey. IEEE Access, 7: 36500–36515.

Banerjee, A. 2019. Blockchain with IoT: Applications and use-cases for a new paradigm of supply chain driving efficiency and cost. vol. 115, pp. 259–292. *In:* Kim, S., G. Chandra and D. Peng Zhang [eds.]. Advances in Computers. Elsevier, New York, USA.

Banker, S. 2018. Will Blockchain technology revolutionize supply chain applications. Available at: https://logisticsviewpoints.com/2016/06/20/will-blockchain-technology-revolutionize-supply-chain-applications/(accessed 20 September 2019).

Bigliardi, B. and E. Bottani. 2010. Performance measurement in the food supply chain: a balanced scorecard approach. Facilities, 28(5/6): 249–260.

Biswas, K., V. Muthukkumarasamy and W.L. Tan. 2017. Blockchain based wine supply chain traceability system. Proceedings of future technologies conference. Available at: https://saiconference.com/Downloads/FTC2017/Proceedings/6_Paper_425-Blockchain_based_Wine_ Supply_Chain_Traceability.pdf(accessed 20 September 2019).

Browne, R. 2017. IBM partners with Nestle, Unilever and other food giants to trace food contamination with blockchain. CNBC. Available at: https://www.cnbc.com/2017/08/22/ibm-nestle-unilever-walmart-blockchain-food-contamination.html (accessed 20 September 2019).

Carrefour. 2018. Carrefour launches Europe's first food blockchain. Available at: http://www.carrefour.com/current-news/carrefour-launches-europes-first-food-blockchain (accessed 20 September 2019).

Carson, B., G. Romanelli, P. Walsh and A. Zhumaev. 2018. Blockchain beyond the hype: What is the strategic business value?, McKinsey & Company.

Chang, S.E., Y.-C. Chen and M.-F. Lu. 2019. Supply chain re-engineering using blockchain technology: A case of smart contract based tracking process. Technol. Forecast Soc., 144: 1–11.

Christidis, K. and M. Devetsikiotis. 2016. Blockchains and smart contracts for the Internet of Things. IEEE Access, 4: 2292–2303.

DHL. 2018. Blockchain in logistics: Perspectives on the upcoming impact of blockchain technology and use-cases for the logistics industry. White Paper, DHL.

Francisco, K. and D. Swanson. 2018. The supply chain has no clothes: technology adoption of blockchain for supply chain transparency, Logistics, 2(2).

Helo, P. and Y. Hao. 2019. Blockchains in operations and supply chains: A model and reference implementation. Computers & Industrial Engineering, 136: 242–251.

Hyperledger. 2019. How Walmart brought unprecedented transparency to the food supply chain with Hyperledger Fabric. Hyperledger. Case Study. Available at: https://www.hyperledger.org/resources/ publications/walmart-case-study(accessed 20 September 2019).

Ioti. 2018. Blockchain in logistics and transportation: transformation ahead. Internet of Things Institute. Available at:http://www.ioti.com/transportation-and-logistics/blockchain-logistics-and-transportation-transformation-ahead(accessed 20 September 2019).

i-scoop. 2018. Digital transformation in transportation and logistics. Available at: https://www.i-scoop.eu/digital-transformation/transportation-logistics-supply-chain-management/(accessed 20 September 2019).

Korpela, K., J. Hallikas and T. Dahlberg. 2017. Digital supply chain transformation toward blockchain integration. Proceedings of the 50th Hawaii international conference on system sciences.

Kouhizadeh, M. and J. Sarkis. 2018. Blockchain practices, potentials, and perspectives in greening supply chains. Sustainability, 10: 3652.

Kshetri, N. 2018. 1 Blockchain's roles in meeting key supply chain management objectives. Int. J. Inform Manage, 39: 80–89.

LDC. 2019. Agricultural commodity blockchain: A soybean shipment transaction from the united states to china. Louis Dreyfus Company. Available at: https://www.ldc.com/blog/in-field/blockchain-buzzword-or-future-commodities-transactions/. (accessed 20 September 2019).

Leng, K., Y. Bi, L. Jing, H.C. Fu and I. Van Nieuwenhuyse. 2018. Research on agricultural supply chain system with double chain architecture based on blockchain technology. Future Gener. Comp. Sy., 86: 641–649.

Lin, I.C. and T.C. Liao. 2017. A Survey of blockchain security issues and challenges. International Journal of Network Security, 19(5): 653–659.

Louppova, J. 2017. Maersk and IBM introduce blockchain in shipping. Available at:https://port.today/maersk-and-ibm-introduce-blockchain-in-shipping/(accessed 20 September 2019).

Nakamoto, S. 2008. Bitcoin: a peer-to-peer electronic cash system. Available at: https://bitcoin.org/bitcoin.pdf, (accessed 20 September 2019).

Neuburger, J.D. and W.L. Choy. 2019. Supply chain management implementing blockchain technology. Practical Law., 27–34.

Niranjanamurthy, M., B.N. Nithya and S. Jagannatha. 2018. Analysis of Blockchain technology: pros, cons and SWOT. Cluster Computing, 1–15.

Pilkington, M. 2016. 11 Blockchain technology: principles and applications, Research handbook on digital transformations. 225.
Provenance. 2018. Blockchain: The Solution for Transparency in Product Supply Chains. Available at: https://www.provenance.org/whitepaper(accessed 20 September 2019).
Rejeb, A., J.G. Keogh and H. Treiblmaier. 2019. Leveraging the internet of things and blockchain technology in supply chain management. Future Internet, 11(7): 161.
Saberi, S., M., Kouhizadeh and J. Sarkis. 2018. Blockchain technology: A panacea or pariah for resources conservation and recycling? Resour. Conserv. Recy., 130: 80–81.
TE-FOOD. 2019. Migros, Switzerland's largest supermarket chain implements TE-FOOD for blockchain based food traceability. Available at: https://www.prweb.com/releases/migros_switzerlands _largest_supermarket_chain_implements_te_food_for_blockchain_based_food_traceability/prweb16530943.htm(accessed 20 September 2019).
Tian, F. 2016. An agri-food supply chain traceability system for China based on RFID & blockchain technology. 13th International Conference on Service Systems and Service Management (ICSSSM), IEEE, 1–6.
TradeLens. 2019. Digitizing the global supply chain. Available at: https://www.tradelens.com/(accessed 20 September 2019).
Treiblmaier, H. 2018. The impact of the blockchain on the supply chain: a theory-based research framework and a call for action. Supply Chain Manag., 23(6): 545–559.
Viriyasitavata, W. and D. Hoonsopon. 2019. Blockchain characteristics and consensus in modern business processes. J. Ind. Inf. Integr., 13: 32–39.

SECTION 7

AI, Robotics and Autonomous Systems in SCM

Chapter 11

Artificial Intelligence, Robotics and Autonomous Systems in SCM

Sercan Demir[1,*] and *Turan Paksoy*[2]

1. Introduction

In his 1968 paper "Memo Functions and Machine Learning", the British Computer Scientist, Donald Michie quotes the following: "If computers could learn from experience, their usefulness would be increased. When I write a clumsy program for a contemporary computer, a thousand runs on the machine do not re-educate my handiwork. On every execution, each time-wasting blemish and crudity, each needless test and redundant evaluation, is meticulously reproduced" (Michie 1968). Since then, one of the ongoing goals in the field of computer science has been to build machines that can automatically learn and improve themselves based on their experience.

Artificial Intelligence (AI) and autonomous technology have yielded the most remarkable applications that impact our lives since the beginning of the 2000s. Some of the most prominent applications include, but are not limited to, self-driving cars, drones, weapon systems, unmanned spacecraft exploring space, underwater robots exploring the deep ocean, software agents used in financial decision making, and deep learning methods used in medical diagnosis. This fast-paced development can be attributed to the recent improvements in the field of machine learning that is a subset of AI, and accessibility to the large data sets from various domains in daily life. These innovative digital technologies become more efficient and productive as they are together incorporated into new products and services. AI integrated services modify the job done, improves productivity and work conditions while minimizing human intervention during the operation (EGE 2018).

The concept of the smart machine, which can communicate with other machines and surroundings is the driving force behind the Fourth Industrial Revolution. The new industrial transformation takes advantage of several innovations that

[1] Department of Industrial Engineering, Faculty of Engineering, Harran University, Sanliurfa, Turkey.
[2] Department of Industrial Engineering, Faculty of Engineering, Konya Technical University, Konya, Turkey.
Email: tpaksoy@yahoo.com
* Corresponding author: sercanxdemir@gmail.com

constitute the formation of smart systems that can operate autonomously. Together with these innovations, AI shares the same background theory that suggests analyzing and filtrating a large amount of data coming from different sources will assist to interpret and propose the most valuable course of action. AI aims to build smart systems that can perceive and understand their surroundings and make better decisions to increase the odds of success (Dopico et al. 2016).

2. Artificial Intelligence and its Development

The beginning of AI dates back to the invention of computers around the 1940s and 1950s. During the development stage of AI, the very first applications focused on programming the computers to make them do things intelligently just like a human. Programming computers to imitate human behavior started a discussion of the differences between a computer and a human brain, and how close a computer to a human brain could be in the 1960s and 1970s. In the 1980s and 1990s, the research took a different path with the emergence of artificial brains. From that point, AI was not limited to purely imitating human intelligence, rather it could be intelligent in its own way since its potential has been substantially grown. This has brought the idea that argues an artificial brain had the potential to be faster and more efficient than a human brain, and it could potentially outperform it. Recently, research efforts on AI are on the rise, and the real-world applications of AI in finance, manufacturing, and military fields have greater results than a human brain can achieve. Today, artificial brains are being designed to learn, adapt, and carry out their needs. They perceive their surroundings, move around in their own body, and make their decision independently (Warwick 2013).

Smart factories, that are the unification of software and hardware devices, constitute the main framework of Industry 4.0. Smart factories are the manufacturing environments where workers, machines, and resources communicate and cooperate in complex manufacturing networks. The integration of AI into the manufacturing systems will facilitate to set up these networks, and enable them to learn, infer, and act autonomously based on the collected data during the industrial process (Dopico et al. 2016). AI focuses on computer programs that are capable of making their own decision to solve a problem of interest, and the systems that are created with AI are intended to mimic the intelligent behavior of expertise (Kumar 2017).

The history of AI comprises of imaginations, fictions, possibilities, demonstrations, and promise. From the ancient philosophers to today's scientists, the possibility of non-human intelligent machines made humankind cast about this subject. Fiction writers, who used intelligent machines in their novels, such as Jules Verne, Isaac Asimov, and L. Frank Baum were ahead of their time and have inspired many AI researchers (Buchanan 2005).

In the 18th and 19th centuries, chess-playing machines were exhibited as intelligent machines, however, these machines were not playing autonomously. The most notable one named "the Turk" that was invented by Hungarian inventor Wolfgang von Kempelen in 1770. A human chess player hid inside the machine and decided the moves on behalf of the machine. Although it was a fake chess-playing machine, it brought about the idea that a machine could perform intelligent assignment like a human (Stanford.edu 2019). Even though chess was used as an instrument for studying inference and representation mechanisms in the early AI research efforts, the first major improvement was the time when IBM's chess-playing computer, named Deep Blue, defeated the world chess champion, Gary Kasparov, in 1997 (Buchanan 2005).

During its development stage, AI has been affected by many disciplines such as engineering, biology, experimental psychology, communication theory, game theory, mathematics, statistics, logic and philosophy, and linguistics. Computers and programming languages were capable of conducting experiments and generating results about AI research starting from the last half of the 20th century. The advances in electronics and the increase in the computing power of modern computers in the 20th century made computers "giant brains", and paved the way for the quick development of AI. Today, robots are also powerful devices that take part in AI research efforts. Robots are being given common knowledge about the objects that we come across every day in a human environment and tested whether they make intelligent decisions that we expect (Buchanan 2005).

Vannevar Bush's 1945 paper, "As We May Think", sheds light on the possible advancements in technology, with the growth of computer science, during and in the future of the information age (Bush 1945). In his 1950 paper, "Computing Machinery and Intelligence", A.M. Turing proposed a question: "Can machines think?" His paper was the starting point in the history of AI. He described the imitation game, also known as the Turing Test, and presented his ideas about the possibility of creating intelligent machines that can think and behave rationally (Turing 1950).

O.G. Selfridge defines the term "Pattern Recognition" in his 1955 paper, "Pattern Recognition and Modern Computers". In his paper, Selfridge defines pattern recognition as: "the extraction of the significant features of data from a background of irrelevant detail". Today, pattern recognition is one of the main research branch in the field of AI (Selfridge 1955).

The Dartmouth Summer Research Project of 1956 is accepted as the event that initiated AI as a research discipline. The proposal of the event was written by computer scientists John McCarthy, cognitive scientist Marvin Minsky, mathematician Claude Shannon, and computer scientist Nathaniel Rochester. John McCarthy is credited with coining the term "artificial intelligence" and shaping the future of the field (Moor 2006). The major milestones in the history of AI is given in Table 1 below.

Table 1: A Quick Look to the History of AI (adopted from Buchanan 2005; Bosch Global 2018).

Year	Event
1945	Bush's paper "As We May Think" published in the "Atlantic Monthly"
1950	Turing's seminal paper "Computing Machinery and Intelligence" published in the philosophy journal "Mind"
1955	Oliver Selfridge's paper, "Pattern Recognition and Modern Computers", was published in Proceedings of the Western Joint Computer Conference
1956	Arthur Samuel's checker-playing program was able to learn from experience by playing against human and computer opponents and improve its playing ability
1956	John McCarthy coined the term "Artificial Intelligence" in a conference at the Dartmouth College
1966	The first chatbot was developed by MIT professor Joseph Weizenbaum. The program was able to simulate real conversation like a conversation partner.
1972	MYCIN, an expert system that used artificial intelligence to treat illnesses, was invented by Ted Shortliffe at Stanford University.
1986	Terrence J. Sejnowski and Charles Rosenberg developed a program called "NETtalk" that was able to speak, read words and pronounce them correctly.
1997	IBM's AI chess computer "Deep Blue" defeated the incumbent chess world champion, Garry Kasparov.
2011–2015	Powerful processors and graphics cards in computers, smartphones, and tablets led AI programs to spread in everyday life. Apple's "Siri", Microsoft's "Cortana", and Amazon's "Echo" and "Alexa" were introduced to the market.
2018	IBM's "Project Debater" debated complex topics with master debaters, and performed unusually well. Google's "Duplex" called a hairdresser and made an appointment on behalf of a customer while the lady on the other end of the line did not notice that she was talking to a machine.

2.1 Artificial Intelligence

AI and artificial neural networks (ANN) are interconnected domains in computer science. AI can be successfully executed through ANN. There are still key differences between AI and ANN. First and foremost, neural networks are the basis for research in the field of AI. The primary aim of AI is to build intelligent machines that can achieve a specific task, such as playing chess, without crossing the boundaries set by the computer scientist. ANN are being used to surpass the limitations of the task-orientated AI. ANN create computer programs that can receive feedback and respond to a problem through adaptive learning. These programs can optimize their response by solving the same problem many times and adjusting the response based on the feedback received each time (Techopedia.com 2017).

Nowadays, AI refers to any machine that can simulate human cognitive skills, such as problem-solving. In other words, it is an attribute of machines that conceptualize a form of intelligence rather than merely perform the orders programmed by the users. Checker and chess-playing machines, and language analysis software were among the early applications of AI. Machine learning is an AI technique that allows machines to learn from input data without having particularly programmed (The Scientist Magazine 2019). The goal of AI is the development of algorithms that lead machines to perform cognitive tasks. An AI system must be capable of storing knowledge, applying this knowledge to solve problems, and acquiring new knowledge through experience. An AI system consists of three key components as shown in Table 2 below (Haykin 1998):

Table 2: Three Key Component of an AI System (adopted from Haykin 1998).

Component	Definition
1. Representation	It is the pervasive use of the language of symbol structures to represent both general knowledge about a problem of interest and specific knowledge about the solution to the problem. Clarity of symbolic representation of AI paves the way for easier human-machine communication.
2. Reasoning	Reasoning refers to the ability to solve problems.
3. Learning	The learning element acquires some information from the environment and uses this information to improve the knowledge base component of the machine learning process.

2.2 Machine Learning

Machine learning (ML) deals with the building process of computers that can improve themselves through experience. The field of ML is the foundation of AI and data science, and it is situated at the intersection of computer science and statistics. Even though it is relatively young, ML is one of the most swiftly growing technical fields today. The rapid development of the new learning algorithms, rapid growth in the availability of accessible data, and constantly lowering computation cost are among the main reasons for the recent progress in ML. The field of ML involves in our everyday life by penetrating across major sectors of the economy including health care, manufacturing, education, financial modeling, policing, and marketing (Jordan and Mitchell 2015).

ML applications have become an important part of our daily life routine over the last few decades. Some of the top ML applications in our everyday life are given in Table 3.

The main goal of ML is to develop computer systems that can automatically improve themselves through experience. ML also focusses on the area of the fundamental computational statistics methods that involve in all learning systems. The discipline of ML has emerged as a branch of AI and it is used as a practical software for computer vision, speech recognition, natural language processing, and robot control. AI researchers all agree that training a system by presenting the examples of desired input-output behavior is much easier than programming the system manually by predicting the desired outcome for all possible inputs (Jordan and Mitchell 2015).

A simple model of ML consists of four elements depicted in Figure 1. The environment supplies information to the learning element. This information is processed and used by the learning element to ensure progress in the knowledge base. At the final stage, the performance element performs its task by using the knowledge base. The information received by the machine is usually imperfect, hence the learning element does not know the details about information in advance. For this reason, the machine operates by guessing while receiving "feedback" from the performance element. As the machine receives feedback, it evaluates its decisions and revises them if needed (Haykin 1998).

Table 3: Top applications of Machine Learning (adopted from Mitchell 2006 and Datasciencecentral.com 2017).

Application	Description
1 Computer Vision	Many current vision systems, e.g., face recognition systems, and systems that automatically classify microscope images of cells, are developed using ML.
2 Speech Recognition	Speech recognition (SR) is the translation of spoken words into text, where a software application recognizes spoken words. Currently available commercial systems for speech recognition all use ML to train the system to recognize speech.
3 Medical Diagnosis	Methods, techniques, and tools that are built by using ML can assist in solving diagnostic and prognostic problems.
4 Statistical Arbitrage	In finance, statistical arbitrage refers to automated trading strategies that are typical of the short term and involve a large number of securities. ML methods such as linear regression and support vector regression (SVR) are applied to build an index arbitrage strategy.
5 Learning Associations	The process of generating insights into various associations between products refers to learning association. ML techniques help to understand and analyze customer buying behaviors.
6 Classification	Classification is a process of placing each individual from the population under study in many classes. In other words, it separated observations into groups based on their characteristics. This ML technique helps the analyst to identify the category of an object in a large set.
7 Prediction	Prediction is one of the most used ML algorithms. It helps businesses to take required decisions (based on historical data) on time.
8 Extraction	Information Extraction is the process of extracting structured information from unstructured data, e.g., web pages, articles, reports, and e-mails. The input is taken as a set of documents and the output is produced as a structured data, e.g., excel sheet, table in a relational database.
9 Regression	The principles of ML can be applied to optimize parameters in regression, e.g., to cut the approximation error and calculate the closest possible outcome.
10 Robot Control	ML methods have been successfully implemented in many robotics applications such as stable helicopter flight, and self-driving vehicles.

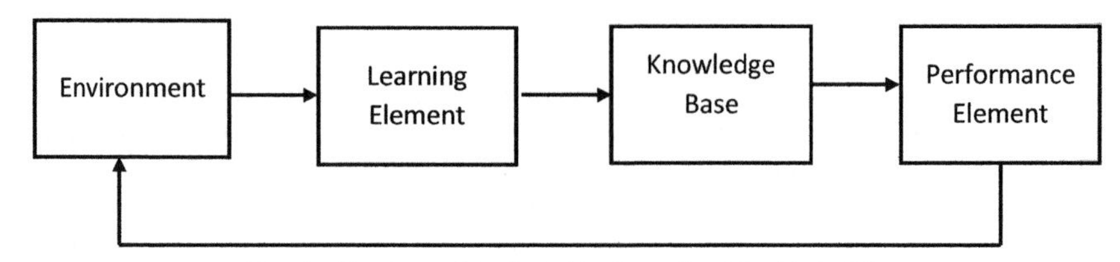

Fig. 1: Simple Model of Machine Learning (Haykin 1998).

2.3 Two Methods of ML: Artificial Neural Network and Deep Learning

The neural network is one of the techniques used in ML research that allows processing algorithms via interconnected nodes called artificial neurons. The first artificial neural network was created by Marvin Minsky in 1951. The structure of the human brain was the inspiration to the researchers for the design of the ANN. In its early years, the ANN approach was very limited, and other approaches in the field of AI has drawn more attention. In the last few years, ANN came back to the stage as a form of an ML approach called deep learning (The Scientist Magazine 2019).

A neuron is an information processing unit of a neural network, and it is necessary for the neural network operation. Figure 2 below depicts the model of a neuron that is the basic building block of an ANN. A neuronal model consists of three basic elements that are explained in Table 4 (Haykin 1998).

Deep learning (DL) is an AI technique and a type of ML approach that leads computer systems to improve with experience and data. ML is the only feasible way to construct successful AI systems that can adapt and operate in complex real-world environments. DL is a specific kind of ML technique whose power and flexibility comes from its ability to continuously learn from the physical world and it describes this world as a nested hierarchy of concepts. While DL is considered as an emerging technology, its history dates back to the 1940s. Before it gained its current popularity, it was a relatively unpopular field for several decades. It was also influenced by many researchers and perspectives, and recently named "deep learning" after it had been called in many different names. DL was used to be known as **cybernetics** in the 1940s–1960s, and it became to be known as **connectionism** in the 1980s–1990s. Then it was coined as **deep learning**

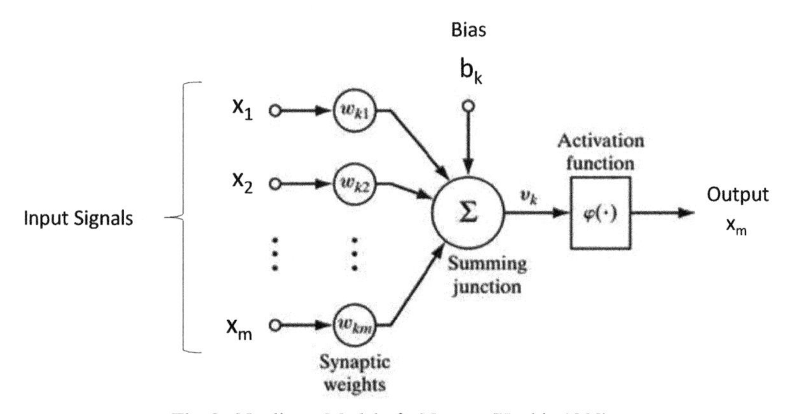

Fig. 2: Nonlinear Model of a Neuron (Haykin 1998).

Table 4: Three Elements of a Neuronal Model (adopted from Haykin 1998).

Element	Attribute
A set of synapses or connecting links	Each of them is identified by a weight or strength of its own. The synaptic weight of an artificial neuron may lie in a range that includes both negative and positive values. In the figure given below, a signal Xj at the input of synapse j connected to neuron k is multiplied by the synaptic weight wkj.
An adder	It sums the input signals that are weighted by the respective synapses of the neuron.
An activation function	It limits the amplitude of the output of a neuron. Since the activation function squashes (limits) the permissible amplitude range of the output signal to a finite value, it referred to as a squashing function.

Table 5: Three-way Categorization of Deep Learning Networks (adopted from Deng and Yu 2014).

Category	Goal
1. Deep networks for unsupervised or generative learning	These networks are designed to analyze the high-order correlation of the observed data for pattern analysis or synthesis purposes when no information about target class labels is available.
2. Deep networks for supervised learning	This type of network is designed to provide solutions for pattern classification problems by aiming to characterize the distributions of classes behind the visible data. These networks are also named as discriminative deep networks.
3. Hybrid deep networks	Hybrid refers to the deep architecture that comprises of both generative and discriminative model components. The outcomes of the generative component are mostly used for discrimination which is the final goal of hybrid deep networks.

around 2006 (Goodfellow et al. 2016). DL is considered an ML approach that can use many layers of hierarchical non-linear information processing. DL architecture or technique can be constructed based on its intention to use and it is categorized into three major domains (Deng and Yu 2014). Table 5 explains three major categorization domains of DL networks.

Figure 3 below represents a Venn diagram showing the AI technology and its relations with ML, ANN, and DL.

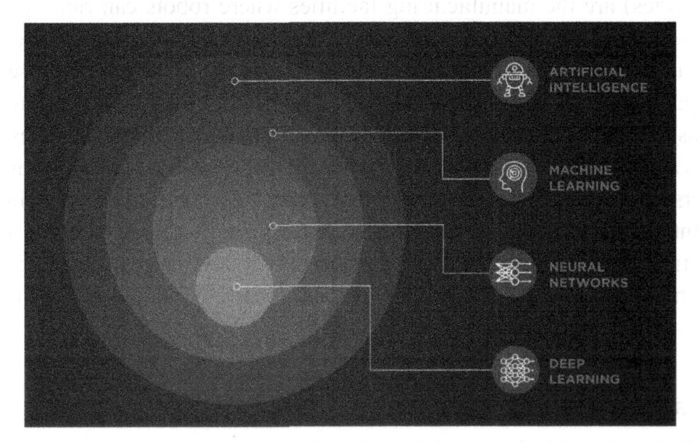

Fig. 3: Artificial Intelligence and its Subsets (The Scientist Magazine 2019).

3. Robotics and Autonomous Systems

The term robotics was defined by Brady (1985) as the intelligent connection of perception to action. Since this connection is intelligent, AI's role in robotics is undeniable. AI addresses the problem of dealing with real objects in the real world in robotics research. Robotics incorporates mechanical effectors, sensors, and computers. The typical sensing of modalities of robots include vision, force and tactile sensing, and proprioceptive sensing of the robot's internal state. Arms, wheels, grippers, and legs are the mechanical parts that allow robots to act (Brady 1985).

The industrial revolution is an ongoing process that consists of four main stages. The First Industrial Revolution started with the integration of steam and water power-based mechanization into the manufacturing process at the end of the 18th century. While the factory electrification had been sprawling through the beginning of the 20th century, conveyor belts and mass production techniques started the Second Industrial Revolution. The last half of the 20th century witnessed the Third Industrial Revolution as electronics and information technology (IT) led to the rise of digital automation in production. Today, newly emerging technologies such as autonomous robots, cyber-physical systems (CPS), the internet of things (IoT), the internet of services (IoS), and virtual reality are not only shaping the future of the manufacturing processes, but also urging the industrial economies to transform themselves towards to the Fourth Industrial Revolution. Industrial robots had improved remarkably and became more productive, flexible, versatile, safer, and collaborative since the last quarter of the 20th century. Industrial robots are one of the main components of smart factories, and they are accepted as one of the key drivers of Industry 4.0 (Bahrin et al. 2016).

Unlike the past, advance robots are currently equipped with improved senses, agility, dexterity, and intelligence, and a growing number of manufacturing and service jobs are being replaced by robots each day. Although the increased use of robotics in manufacturing and service is creating a threat for the career prospect of white and blue collars by aiming to diminish the human quantities on the shop floor, this rise still requires the need for more human qualities in assembly lines (Pfeiffer 2016).

Unintelligent robots are used in simple tasks that do not require any intelligent action. These tasks do not require higher-level skills or dexterity during their execution. Transferring parts, welding, spray painting, packing are examples of these tasks. Major industrial plants have automated their highly repetitive manufacturing processes by using robots (Brady 1985). The big breakthrough in robotics has boosted the robotics industry and the companies in this sector, while it has brought up a concern regarding the future of industrial production jobs, especially assembly work, at the same time. Less costly and more effective robotics technology has been threatening assemblers' job that still constitutes the main part of the mass production (Pfeiffer 2016). Smart factories will bring about opportunities, as well as challenges. The transition from the current manufacturing systems to smart manufacturing could be the greatest challenge for the companies. The next generation of smart, flexible, and low-cost robotics technology will be the backbone of the new age of factory automation (Kusiak 2018).

The role of robots in modern manufacturing industries is undeniable. Since 2004, the number of multipurpose industrial robots build-up by the members of these industries in Europe has almost doubled. Autonomous robots are one of the main pillars of Industry 4.0. Autonomous robots in smart manufacturing plants function intelligently while providing safe, flexible, versatile and collaborative working opportunities. Industry 4.0 concept requires workers and robots to work collaboratively using human-machine interfaces, and robots can be controlled remotely while taking part in a variety of business functions such as production, logistics, and office management. Innovative manufacturing methodologies appear as automated robots have been integrating into the production. For instance, "lights out factories" (aka. lights out manufacturing or dark factories) are the manufacturing facilities where robots can run the production without light or heat, or when workers are not physically at the facility (Bahrin et al. 2016). Intelligent robotics technologies have brought many benefits to daily life in a wide range of environments such as land, space, and undersea. Industrial robots involve in manufacturing by manipulating parts and manufacturing tools. Assembly, material handling, welding, and painting are some of the basic operations that can be done by using these robots. Automated guided vehicles (AGVs) transport materials and inventories in factories and warehouses. Telerobots can be controlled at a distance by a human operator, and they can operate in space and undersea. Walking robots (legged robots) can move with the help of limbs, and they are useful in hazardous environments. Intelligent robots possess advanced sensory feedback mechanisms, and they can make decisions autonomously while operating in partially structured and unstructured environments (Gruver 1994).

In the next section of this chapter, we will investigate the industrial applications of intelligent robots and autonomous systems in Industry 4.0 perspective.

4. Industry Applications

Industry 4.0 has brought the idea of smart factories consist of smart objects that offer integrated processing and communication capabilities. Employee communication and the interaction level of humans and technology were improved since the smart objects had emerged in factories. Autonomous systems have brought new problems in factories; however, their autonomous and self-organizing characteristics made the complex manufacturing systems more controllable and sustainable (Gorecky et al. 2014). The new generation of manufacturing technologies and smart factory concepts are changing the production systems in many ways. Additive manufacturing (3D printing), hybrid machines that are capable of performing multiple jobs, new materials and components, factory automation supported by low-cost robots are transforming the conventional manufacturing systems into the smart manufacturing systems. This new paradigm will also impact transportation which is a vital resource for companies. Transportation can be categorized into two groups, internal and external. While internal transportation includes material handling within a manufacturing system, external transportation covers the supply and distribution network of a company. Advances in robotics and autonomous vehicles expose huge improvement opportunities to internal and external transportation by offering a high level of autonomy and sharing, on a regional and global scale (Kusiak 2018).

Mobile robots are capable of moving around via wheels, tracks or legs. AGVs are wheeled mobile robots that usually operate in a factory. AGVs can operate both in an office environment and heavy industrial surroundings. While most AGVs use sensors to follow guide wires attached on the floor, some types can be programmed to pursue a trajectory and to make decisions on the way using the signals they receive (Gruver 1994). The use of AGVs is a widespread phenomenon among the flexible manufacturing systems that involve transport robots in their manufacturing processes. The paths of AGVs are restricted to predetermined routes by incorporating magnetic stripe navigation or guide wires, and they require the workplace to be restructured for them to work efficiently (Arkin and Murphy 1990). Automobile manufacturer SEAT is one of the companies that intensively utilizes AGVs on its shop floor (Figure). The company reports that 125 AGVs are in use in their Martorell facility in Spain. These robots convey nearly 24,000 parts daily, participating in manufacturing with 7000 employees. AGVs facilitate and optimize the workers' jobs and lead to an almost 25% reduction in production time (Volkswagenag.com 2019).

Fig. 4: AGVs used in Seat Martorell Facility Spain (Volkswagenag.com 2019).

The birth of Industry 4.0 paradigm led to the development of the Smart Logistics concept. Many warehousing and shipping companies have been taking advantage of information technologies, robotics, and automated systems since they integrated AI into their business models. The rapid development in AI and robotics technologies has offered ground breaking systems for the logistics industry such as unmanned warehouse and delivery drones. Companies expedited shipping time and improved the quality of customer service significantly as a result of adopting the intelligent warehousing and delivery service. Unlike the manufacturing industry, the logistics sector copes with adaptation problem that stems from the variety of orders from many customers. Since the orders in the delivery sector are unique in terms of sorting, packing, and delivering, the technological machines and equipment should be equipped with intelligent features. This technology helps the firms to focus on customization by taking each customer's requirement into account, and improve the customer service level by delivering the right good in the right place at the right time. Also, large order quantities urge the delivery companies to employ automated and intelligent systems to avoid delay in lead times (Wen et al. 2018).

Many technological devices that ease our everyday life already exist in the human environment. When various kinds of robots will be designed to cooperate with each other to perform our daily tasks in the near future, they will be an indispensable part of human life. Wheeled robots, legged robots, humanoid robots, and network sensors will provide various services to humans by either working autonomously or working together. This cooperation among various robots is beneficial to many human activities such as warehouse management, industrial assembling, military applications, and daily-life tasks. The logistics sector will benefit from the coordination of several mobile robots. Heterogeneous multi robot systems, composed of different types and sizes of robots, already became a vital part of warehouse management systems. These multi-robot systems consist of many autonomous robots that are capable of communicating with each other via wireless networks, and they are used to transport different objects in warehouses (Wang et al. 2012).

Robots have involved many activities in manufacturing processes since the 1950s. They have been intensively used for repetitive tasks such as cutting, welding, and assembling in the automotive industry. In addition to these repetitive activities, the optimization of the internal material flow of a company can be accomplished using the robotics system in the logistics activities. Some of these activities completed by robotics systems are loading/unloading and palletizing/depalletizing of goods and materials (Echelmeyer et al. 2008).

E-commerce (aka. electronic commerce or internet commerce) giants such as Alibaba and JD.com receive millions of orders every single day, while these online orders constitute a major problem of delivery. E-commerce companies have to deal with the problems of slow/wrong deliveries, lost packages, damaged goods, and incorrect packing while fulfilling millions of orders placed online each day. This challenge encourages e-commerce companies to integrate automated systems into their distribution network. E-commerce logistics activities comprise of three main stages. Replenishment of the goods from the suppliers to the warehouse or distribution center is the first stage. The fulfillment of the customers' orders at distribution centers is the second stage, and this stage usually consists of picking, sorting, and packing operations. Finally, the third stage is the delivery of the orders from the distribution centers to customers. E-commerce companies usually collaborate with 3PL service providers to carry out the first and third stages. The second stage is the source of the bottleneck for e-commerce logistics operations, especially during the peak season. Order picking is an extremely labor-intensive task and it requires human operators to move long distances in a highly limited space for storage and order processing. Companies invest in automated systems and robots to reduce this bottleneck in their warehouse operations. This endeavor covers both the automation of the flow of materials and the flow of information (Huang et al. 2015).

Drones, commonly known as unmanned aerial vehicles (UAVs), are electronic devices that are capable of sustained flight without any human operator on board. Drones perform useful actions under sufficient control such as the delivery of small items that are urgently needed in areas that are not easily accessible. Drone delivery has been applied to healthcare and humanitarian logistics areas in recent years. For instance, delivery of urgently needed medications, blood, and vaccines at the right time when land transport is challenging due to the poor transportation infrastructure, traffic congestion, or severe

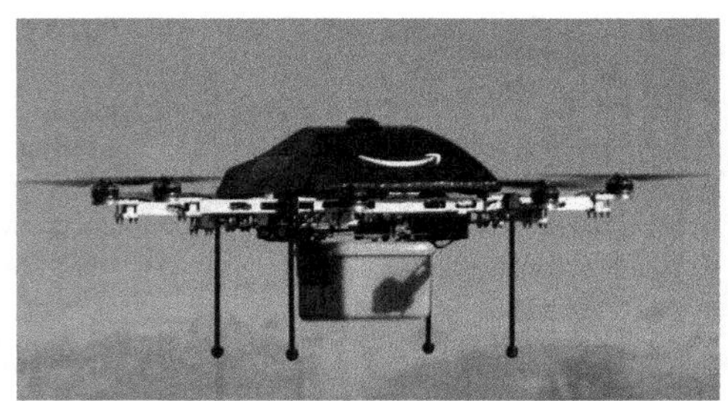

Fig. 5: Amazon's Octocopter Drone (left) and The Horsefly Drone Intended to Use by UPS (Bamburry 2015).

natural conditions (e.g., weather or disasters). Drones are useful when human lives are in danger. For instance, drones helped rescue teams to pinpoint the survivors after the Nepal earthquake in 2015 (Scott and Scott 2017).

Cost-saving and high delivery speed are the two main drivers behind the spread of drones used in supply chains. Commercial drones are still in their early stages of gaining attention; however, they already are considered as disruptive technology that will impact the future of product delivery service. Giant companies such as Amazon, Google, and UPS started to invest in this innovative product delivery method. Amazon named her drone project for its online shopping portal as Amazon Prime Air. The company aims to deliver customer orders in less 30 minutes using UAVs when this technology becomes fully functional. UPS also attempts to adopt drone delivery to improve productivity, reduce fuel costs and accidents at work. Google stepped into the drone business by starting the Project Wing drone program in 2014. The company mainly focuses on delivering first-aid kits, defibrillators, and medical products to a scene of a crisis promptly (Bamburry 2015).

Google's Project Wing is a drone delivery service aiming to increase access to goods, reduce traffic congestion in cities, and help to reduce the CO_2 emissions resulting from the transportation of goods. The project also includes the development of an unmanned traffic management system that will allow UAVs to navigate around other drones, manned aircraft, and other obstacles like trees, buildings and power lines (X – Wing 2019).

5. Conclusion and Future Research

The 21st century has become the era of the digital transformation accompanied by newly emerging technologies. It is not possible to disjoin human life from emerging technologies since these technologies impact every aspect of our lives. Smart factories and smart manufacturing processes have been converting the traditional way of manufacturing into a technology-driven manufacturing approach that utilizes the merits of technology. Digitization of manufacturing changed the way goods are made and delivered while improving the operational efficiency of the manufacturers and making them more profitable. As companies shifted from the linearly organized supply chains to the interconnected supply chain operations, the manufacturing process became more dynamic and controllable.

AI together with the robotics technology are two of the crucial drivers of digital transformation. As the players of manufacturing systems become autonomous and self-driven, the manufacturing efficiency and employee productivity greatly improve. Industrial automation led to the birth of the intelligent warehouse and delivery systems. Many interconnected warehousing technologies that are capable of working together form intelligent warehouse systems. The goods are received, identified, sorted, organized, and prepared for shipment automatically without the need of any human operator. These systems automated the entire operations (from suppliers to the end customer) with minimal cost and error while providing companies a strong market position and competitive edge by increasing customer responsiveness and quality of their service.

References

Arkin, R.C. and R.R. Murphy. 1990. Autonomous navigation in a manufacturing environment. IEEE Transactions on Robotics and Automation, 6(4): 445–454.

Bahrin, M.A.K., M.F. Othman, N.N. Azli and M.F. Talib. 2016. Industry 4.0: A review on industrial automation and robotic. JurnalTeknologi, 78(6-13): 137–143.

Bamburry, D. 2015. Drones: Designed for product delivery. Design Management Review, 26(1): 40–48.

Bosch Global. 2018. The history of artificial intelligence. [online] Available at: https://www.bosch.com/stories/ history-of-artificial-intelligence/ [Accessed 12 Aug. 2019].

Brady, M. 1985. Artificial intelligence and robotics. Artificial Intelligence, 26(1): 79–121.

Buchanan, B.G. 2005. A (very) brief history of artificial intelligence. Ai Magazine, 26(4): 53–53.

Bush, V. 1945. As we may think. The Atlantic Monthly, 176(1): 101–108.
Datasciencecentral.com. 2017. Top 9 Machine Learning Applications in Real World. [online] Available at: https://www.datasciencecentral.com/profiles/blogs/top-9-machine-learning-applications-in-real-world [Accessed 15 Sep. 2019].
Deng, L. and D. Yu. 2014. Deep learning: methods and applications. Foundations and Trends® in Signal Processing, 7(3-4): 197–387.
Dopico, M., A. Gomez, D. De la Fuente, N.García, R. Rosillo and J. Puche. 2016. A vision of industry 4.0 from an artificial intelligence point of view. In Proceedings on the International Conference on Artificial Intelligence (ICAI) (p. 407). The Steering Committee of The World Congress in Computer Science, Computer Engineering and Applied Computing (WorldComp).
Echelmeyer, W., A. Kirchheim and E. Wellbrock. 2008. September. Robotics-logistics: Challenges for automation of logistic processes. In 2008 IEEE International Conference on Automation and Logistics (pp. 2099–2103). IEEE.
European Group on Ethics in Science and New Technologies (EGE). 2018. Statement on Artificial Intelligence, Robotics and 'Autonomous' Systems. Available at: http://ec.europa.eu/research/ege/pdf/ege_ai_statement_2018.pdf.
Goodfellow, I., Y. Bengio and A. Courville. 2016. Deep learning. MIT press.
Gorecky, D., M. Schmitt, M. Loskyll and D. Zühlke. 2014. July. Human-machine-interaction in the industry 4.0 era. In 2014 12th IEEE international conference on industrial informatics (INDIN) (pp. 289–294). Ieee.
Gruver, W.A. 1994. Intelligent robotics in manufacturing, service, and rehabilitation: An overview. IEEE Transactions on Industrial Electronics, 41(1): 4–11.
Haykin, S. 1998. Neural Networks: A Comprehensive Foundation, Prentice Hall PTR, Upper Saddle River, NJ.
Huang, G.Q., M.Z. Chen and J. Pan. 2015. Robotics in ecommerce logistics. HKIE Transactions, 22(2): 68–77.
Jordan, M.I. and T.M. Mitchell. 2015. Machine learning: Trends, perspectives, and prospects. Science, 349(6245): 255–260.
Kumar, S.L. 2017. State of the art-intense review on artificial intelligence systems application in process planning and manufacturing. Engineering Applications of Artificial Intelligence, 65: 294–329.
Kusiak, A. 2018. Smart manufacturing. International Journal of Production Research, 56(1-2): 508–517.
Michie, D. 1968. "Memo" functions and machine learning. Nature, 218(5136): 19.
Mitchell, T.M. 2006. The discipline of machine learning (Vol. 9). Pittsburgh, PA: Carnegie Mellon University, School of Computer Science, Machine Learning Department.
Moor, J. 2006. The Dartmouth College artificial intelligence conference: The next fifty years. Ai Magazine, 27(4): 87–91.
Pfeiffer, S. 2016. Robots, Industry 4.0 and humans, or why assembly work is more than routine work. Societies, 6(2): 16.
Scott, J. and C. Scott. 2017, January. Drone delivery models for healthcare. In Proceedings of the 50th Hawaii international conference on system sciences.
Selfridge, O.G. 1955. March. Pattern recognition and modern computers. In Proceedings of the March 1–3, 1955, western joint computer conference (pp. 91–93). ACM.
Stanford.edu. 2019. CS221. [online] Available at: https://stanford.edu/~cpiech/cs221/apps/deepBlue.html [Accessed 5 Aug. 2019].
Techopedia.com. 2017. What is the difference between artificial intelligence and neural networks? [online] Available at: https://www.techopedia.com/2/27888/programming/what-is-the-difference-between-artificial-intelligence-and-neural-networks [Accessed 9 Sep. 2019].
The Scientist Magazine®. 2019. A Primer: Artificial Intelligence Versus Neural Networks. [online] Available at: https://www.the-scientist.com/magazine-issue/artificial-intelligence-versus-neural-networks-65802 [Accessed 9 Sep. 2019].
Turing, A.M. 1950. Computing machinery and intelligence. Mind 59, 236, pp. 433–460
Volkswagenag.com. 2019. The daily routine of autonomous robots. [online] Available at: https://www.volkswagenag.com/en/news/stories/2018/01/the-daily-routine-of-autonomous-robots.html [Accessed 24 Oct. 2019].
Wang, T., D.M. Ramik, C. Sabourin and K. Madani. 2012. Intelligent systems for industrial robotics: application in logistic field. Industrial Robot: An International Journal, 39(3): 251–259.
Warwick, K. 2013. Artificial intelligence: the basics. Routledge.
Wen, J., L. He, and F. Zhu. 2018. Swarm robotics control and communications: Imminent challenges for next generation smart logistics. IEEE Communications Magazine, 56(7): 102–107.
X - Wing. 2019. Retrieved October 21, 2019, from https://x.company/projects/wing/.

SECTION 8

Smart Factories: Transformation of Production and Inventory Management

CHAPTER 12

Smart Factories

Integrated Disassembly Line Balancing and Routing Problem with 3D Printers

Zülal Diri Kenger,[1] *Çağrı Koç*[2] and *Eren Özceylan*[1,*]

1. Introduction

As consumption is rapidly increasing worldwide, discarded devices produce a huge amount of electronic waste. Disassembly of EOL products helps to reduce this waste which promotes recycling. Disassembly phase generates the desired components via separation of EOL products into its units. The disassembly line balancing (DLB) is a crucial member of the reverse supply chain and is to a great extent, effective to successfully deal with EOL products (Deniz and Ozcelik 2019). Balancing the disassembly line is critical in terms of recycling and (re)manufacturing.

The 3DP technology is one of the current supply chain trends which is also considered as additive manufacturing (AM) (Khajavi et al. 2014). This technology provides new opportunities in supply chain operations. In recent years, many manufacturers heavily invested in 3D printers. For example, General Electric manufactures nozzles obtained by assembling 20 different parts in one step using 3D printer technology (Catts 2013). This helps General Electric to take the advantage of creating prototype part designs in a faster and more efficient way than before. Airbus produces aircraft components for its jetliners. They are working on spare parts production on demand with 3D printers (Airbus 2014). In 2017, Siemens achieved to produce gas turbine wings using completely AM (Siemens 2017). The common reasons why all these companies use 3DP technology are that less material and more economical parts will be produced in shorter time. NASA has announced that 70 parts produced by additive manufacturing are used on Mars Rover test vehicles (Küçükkoç 2019).In addition, it is possible for the manufacturing system to be functional during the transportation, allowing the products to be produced

[1] Department of Industrial Engineering, Gaziantep University, Gaziantep, Turkey, Email: zulal-88@hotmail.com
[2] Department of Business Administration, Social Sciences University of Ankara, Ankara, Turkey, Email: cagri.koc@asbu.edu.tr
* Corresponding author: erenozceylan@gmail.com

en-route. Although this form of production is rare in traditional manufacturing, Amazon has received for a patent for a system that performs 3DP during en-route (Ryan et al. 2017).

In logistics and supply chain management, determining the position of the delivery points and constructing routes is crucial. Optimization of distribution operations saves large amounts of cost and time. The vehicle routing problem (VRP), one of the most well-known and studied optimization problems, aims to distribute goods between depots and customers (Toth and Vigo 2014). Integration of disassembly and distribution planning problems gains interest by researchers because of the significant importance in practice (see Özceylan and Paksoy 2013, 2014; Özceylan et al. 2014; Koç 2017; Habibi et al. 2017a, 2017b, 2018; Kannan et al. 2017).

Both the DLB and 3DP technology are current subject areas that have many significant implications on supply chain. However, while the DLB is generally effective on the reverse supply chain, 3DP technology is commonly effective on the forward supply chain. The DLB and 3DP technology have common advantages such as sustainable clean environment, cost and time savings, and raw material saving. Furthermore, less labor and inventory cost, process simplifications, high efficiency and flexibility are among potential advantages of 3DP technology (Li et al. 2016; Chan et al. 2018). For example, instead of assembling several components, using 3DP technology simplifies the process by producing the required part. Integration of the supply chain and DLB is now a popular field of study. The biggest challenge in supply chain management is the delivery of precious products and services to customers efficiently and effectively (Holmström et al. 2010).

The VRP is a very critical issue for the delivery of components formed after disassembly process. In addition, AM technology has a significant effect on producing more precious and solid products. Hence, co-operation of the VRP and AM technology gains importance in this manner. 3DP technology is now more accessible since the cost of processing has reached the level everyone can afford (Chan et al. 2018). The 3DP technology saves time up to 75% of the old development time in component production time and consumes up to 65% less resources, reducing gas emissions by up to 30%, and results in longer-life components (Siemens 2017). The presence of 3DP technology speeds up the progress of industries such as spare parts manufacturing (Khajavi et al. 2014). Although 3DP is a popular study field and an affordable technology, utilization in the literature and industry is still very limited. In the VRP integrated problems, vehicles are used to collect EOL products. However, we use vehicles for the distribution of the components formed after disassembly process. To our knowledge, the current literature has not investigated in the 3D printers within the studied integrated problems.

This chapter makes three main scientific contributions. The first is the introduction of the integrated disassembly line balancing and routing problem with 3DP printers (I-DLB-RP-3DP) within the context of smart factories. The second is to propose a mixed integer nonlinear mathematical formulation for the I-DLB-RP-3DP. The third contribution is to conduct extensive computational experiments to investigate the I-DLB-RP-3DP, and to provide several policy and managerial implications.

The rest of the chapter is structured as follows. Section 2 reviews the related literature. Section 3 formally defines the I-DLB-RP-3DP and presents the mathematical formulation. Computational experiments are presented in Section 4. Conclusions are presented in Section 5.

2. Literature Review

In this section, we first review the DLB and the VRP in Section 2.1, we then review the 3D printers in supply chain management in Section 2.2, we finally review the integrated disassembly and distribution planning problem in Section 2.3.

2.1 A Brief Review on the DLB and the VRP

The DLB problem introduced by Gungor and Gupta (2001). Özceylan et al. (2019) and Deniz and Ozcelik (2019) reviewed the related literature on the DLB in detail. Özceylan et al. (2019) classified the literature in nine main aspects: Models and solution approaches, objectives, product types, the parameter structure, disassembly levels, complications in disassembly lines, line types, disassembly process, and disassembled product. Deniz and Ozcelik (2019) applied bibliometric and social network analysis to systematically define the trend and key direction. The authors also implemented future study realization analysis to observe if the future work promises of published studies were accomplished or not.

For about sixty years, the VRP and its rich variations have been intensively studied in the literature by researchers. For more details, we refer the readers to the following survey papers of Braekers et al. (2016); Koç et al. (2016); Koç and Laporte (2018), and to the book of Toth and Vigo (2014).

2.2 A Brief Review on the 3D Printers in Supply Chain Management

The history of AM technology dates back to 1980s (Khajavi et al. 2014). Huang et al. (2013) surveyed the AM literature and analyzed the impact on population health and wellbeing, energy consumption and environmental impact, impact on manufacturing supply chain, and potential health and occupational hazards. Rogers et al. (2016) reviewed the studies related with implications of 3DP technology on the supply chain, and assessed the firms provided by 3DP services in selected European markets. They emphasize that the market can change as 3DP technology matures. Sasson and Johnson (2016) presented an alternative that with traditional production and completes traditional serial production. The authors discuss the economic reasons for modern manufacturing and for 3DP. A process for 3DP super centers that focus on low volume, customized, and high emergency production is proposed, and implications of 3DP technology on supply chain are examined. Jia et al. (2016) suggested two business models for the 3D printed chocolate supply chain and evaluate the financial viability of these models by using modeling and simulation. Ford and Despeisse (2016) focused on advantages and challenges of AM and their implications on sustainability. Li et al. (2016) considered the impact of AM technology on the spare parts supply chain. Conventional, distributed AM-based and centralized AM-based supply chains are compared and the superiority of AM technology is detected for spare parts supply chain. Chan et al. (2018) discussed the impacts of 3DP technology on the supply chain and evaluate in terms of manufacturing and legal perspective. The authors focused on challenges that resist mass-scale applications of 3DP.

2.3 A Brief Review on the Integrated Disassembly and Distribution Planning Problem

This section presents the literature on the integrated disassembly problem and distribution planning. These studies and solution methods are summarized in Table 1.

Table 1: The literature on the integrated disassembly problem and distribution planning.

Study	Disassembly	DLB	VRP	CLSC	RSC	Solution Method
Özceylan and Paksoy (2013)		✓			✓	NLP
Özceylan et al. (2014)		✓		✓		NLP
Özceylan and Paksoy (2014)		✓			✓	Fuzzy mathematical programming
Habibi et al. (2014)		✓	✓		✓	LP
Habibi et al. (2017a)	✓		✓		✓	LP
Habibi et al. (2017b)	✓		✓		✓	Two-phase iterative heuristic
Kannan et al. (2017)		✓			✓	NLP
Habibi et al. (2018)	✓		✓		✓	Stochastic mathematical programming, Two-phase iterative heuristic
The current study		✓	✓			NLP

CLSC: Closed loop supply chain, RSC: Reverse supply chain, LP: Linear programming; NLP: Nonlinear programming

3. Problem Definition and Modeling

In our problem, a supply chain network consists of disassembly centers with 3DP machines and (re)manufacturing centers. In disassembly centers, EOL products are disassembled by balancing disassembly lines and missing components are produced by 3DP machines. Then vehicles with fixed capacities deliver components to (re)manufacturing centers. A 3DP machine in a disassembly center, working simultaneously with disassembly lines, is used in the production of missing components. Since the current problem provides spare parts to (re)manufacturing centers through disassembly process, the cooperation of 3DP technology and disassembly lines gains highly importance. Figure 1 illustrates the I-DLB-RP-3DP.

We now present a nonlinear mathematical formulation for the I-DLB-RP-3DP which considers a single disassembly center, a single 3DP machine, single-type of product, and multi-component. Mathematical model is developed by extending the formulation of Koç et al. (2009) and integrated with formulation of the classical capacitated VRP. Our assumptions are as follows.

- The demand of each (re)manufacturing center for a component is known and must be fully satisfied.
- Distances between disassembly center and (re)manufacturing centers and between (re)manufacturing centers are known.
- A single-type of product are completely disassembled.

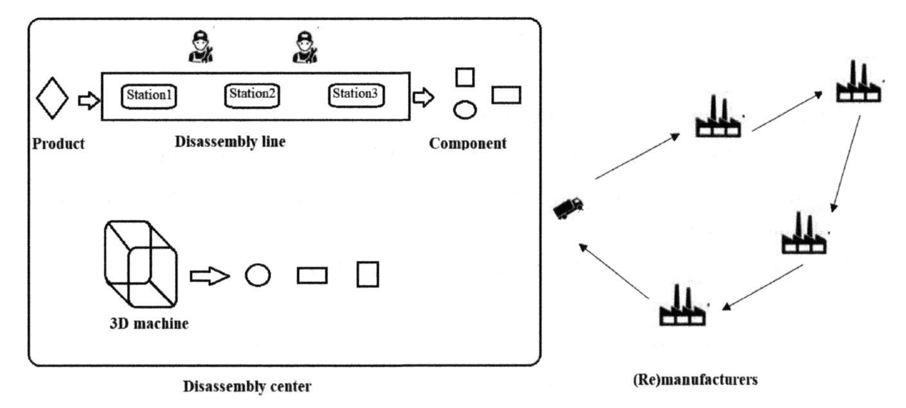

Fig. 1: Illustration of the I-DLB-RP-3DP.

- Setup times are ignored.
- No work-in-process inventory is allowed.
- A task cannot be split among two or more workstations.
- The precedence relations of the problem are known.
- All workstations can process any of the tasks and all have the same associated costs.
- The task process times are not sequence dependent.
- One disassembly center responsible for disassembling of EOL products.
- Each disassembly takes apart the product or sub-assembly into exactly two or more new sub-assemblies.
- Any task can be processed at any workstation.
- Waste is ignored.
- All (re)manufacturing centers demand more than one type of component.
- Capacity of vehicles are known and are ready for service to customers in the(re)manufacturing center.
- Inventory is ignored.
- 3DP machine has the ability of producing different components.
- Each component is different from each other.

Notations

Indices

c,g	remanufacturing center, $c,g = \{1,2,\ldots,N\}$
i	task
j,h	workstation
a	artificial task
k	component

Sets and Parameters

A_a	artificial nodes of a
B_i	nodes of i
$P(A_a)$, $P(B_i)$	immediate predecessor set of A_a and B_i, respectively
$S(A_a)$, $S(B_i)$	immediate successor set of artificial node set of A_a and B_i respectively
d_{B_i}	task time (normal node) of B_i

J	number of workstations (upper bound)
W_{time}	working time
De_{gk}	demand of (re)manufacturing center g
Ct	fixed cost of shipping from disassembly center to (re)manufacturing center g
D_{cg}	distance between disassembly center and (re)manufacturing center or between (re) manufacturing centers
O	fixed cost to open a workstation in the disassembly line
Cap	vehicle capacity
St	fixed cost to disassembly a product
Tt	fixed cost to operate 3DP machine
t_k	operating time of part k on 3DP machine

Variables

CT	cycle time
U_g	load of vehicle just after leaving cluster
Dp	the number of disassembled products on disassembly line
TD_k	the number of component k produced by a 3DP machine
x_{ij}	1, if task is assigned to workstation j; 0 otherwise
y_{cg}	1, if vehicle travels from disassembly center to (re)manufacturing center or between (re)manufacturing centers; 0 otherwise
z_i	1, if task is performed; 0 otherwise
F_j	1, if workstation j is opened; 0 otherwise

The mathematical formulation of the I-DLB-RP-3DP is then:

$$\min Z = O * \sum_{j}^{J} F_j + Ct * \sum_{c}^{N} \sum_{g}^{N} D_{cg} * y_{cg} + St * Dp + Tt * \sum_{k}^{K} t_k * TD_k \tag{1}$$

$$\sum_{i:B_i \in S(A_0)} z_i = 1 \tag{2}$$

$$\sum_{i:B_i \in S(A_a)} z_i = \sum_{i:B_i \in P(A_a)} z_i \qquad \forall a, a \neq 0 \tag{3}$$

$$\sum_{j=1}^{J} x_{ij} = z_i \qquad \forall i \tag{4}$$

$$\sum_{i:B_i \in S(A_a)} x_{ih} \leq \sum_{i:B_i \in P(A_a)} \sum_{j=1}^{h} x_{ij} \quad \forall a, a \neq 0, h \in J \tag{5}$$

$$\sum_{i=1}^{I} x_{ij} * d_{B_i} \leq CT * F_j \qquad \forall j, j = 1, \ldots, J \tag{6}$$

$$CT * Dp = W_{time} \tag{7}$$

$$F_{j+1} \leq F_j \qquad \forall j, j = 1, \ldots, J-1 \tag{8}$$

$$\sum_{g=1,g\neq c}^{N} y_{cg} = 1 \quad \forall c, c = 2,\ldots,N \tag{9}$$

$$\sum_{c=1,c\neq g}^{N} y_{cg} - \sum_{c=1,c\neq g}^{N} y_{gc} = 0 \quad \forall g, g = 1,\ldots,N \tag{10}$$

$$U_c - U_g + Cap * y_{cg} \leq Cap - \sum_{k=1}^{K} De_{gk} \quad \forall c,g,c,g = 2,\ldots,N, c \neq g \tag{11}$$

$$\sum_{k=1}^{K} t_k * TD_k \leq W_{time} \tag{12}$$

$$Dp + TD_k \geq \sum_{g=2}^{N} De_{gk} \quad \forall k \tag{13}$$

$$CT,\ Dp,\ TD_k \geq 0 \quad \forall g,k \tag{14}$$

$$\sum_{k=1}^{K} De_{gk} \leq U_g \leq Cap \quad \forall g \tag{15}$$

$$x_{ij},\ z_i,\ y_{cg},\ F_j \in \{0,1\}\ \forall i,j,c,g \tag{16}$$

The objective function (1) minimizes the total traveling cost of vehicles, the total fixed cost of operating disassembly workstations, total disassembly cost of products, and total fixed cost of operating a 3DP machine. Constraints (2) and (3) guarantee that exactly one of the OR successors is selected. Constraints (4) ensure that each selected task is assigned to at most one workstation in the disassembly center. Constraints (5) tackle with the precedence relations between the nodes. Constraints (6) prevent the cycle time being exceeded by a disassembly workstation. Constraints (7) show that the cycle time in the disassembly center is equal to the total working time divided by the number of disassembled products. Constraints (8) ensures that workstations open sequentially. Constraints (9) guarantee that each customer must be visited once. Constraints (10) define the flow. Constraints (11) are the capacity constraints. Constraints (12) ensure that the time required for components produced by the 3DP machine does not exceed the total working time. Constraints (13) ensure that sum of the number of disassembled products and the component k produced by the 3DP machine cannot be less than the total demand of the component k. Constraints (14)–(16) define the domain of the decision variables.

4. Computational Experiments

This section presents the results of our computational experiments. We first present the details of the benchmark instance, and then the results obtained on a numerical example by applying the proposed mathematical model. We finally present the detailed results of the analyses of the effect of changing working time and fixed cost for 3DP machine.

4.1 Benchmark Instance

We generated the benchmark instance for the I-DLB-RP-3DP by considering the well-known test problems from literature. For the DLB part, we used a set of a sample product from the study of Koç et al. (2009). The details about this sample product are given in the Appendix. For the VRP part, we modified the classical VRP instances of Augerat et al. (1995). In particular, we selected the smallest data set (P_n16_k8) of Set P, and for simplicity we selected 9 nodes (1 disassembly center and 8 remanufacturing center) without changing the coordinate of nodes and vehicle capacity. All demand characteristics were generated by using the discrete uniform distributions. Demands of (re)manufacturing centers are in the range [1, 6]. Operating time of component k on 3DP machine is in the range [5–20]. The details about the benchmark instance are presented in the Appendix.

Unit shipping cost (*Ct*) and cost of opening a workstation (*O*) were set as in Özceylan and Paksoy (2013). *Ct* is fixed to 5.23 cents per tonne-km for a general freight truck, and *O* is fixed to 100 USD per each workstation in the disassembly line. W_{time} is set to 1000 min. Number of maximum workstation (*J*) is set to 5. Disassembling cost of a product (*St*) is fixed to 50USD per product, and fixed cost of operating 3DP machine (*Tt*) is fixed to 2USD per component.

4.2 Results Obtained on a Numerical Example

Since the presented model is nonlinear, GAMS/SCIP with its default settings as the optimizer is used to solve the model and experiments are conducted on a computer Intel Corei5 1.60 GHz processor with 8 GB RAM.

Table 2 and Figure 2 present the obtained results. In Table 2, C1, C2, C3, C4, C5, C6, and C7 denote component 1, 2, 3, 4, 5, 6, and 7, respectively. 3D-NC denotes the number of components produced by 3DP machine and DL-NC denotes

Table 2: The number of components obtained from 3DP machine and disassembly line.

	3D-NC	DL-NC
C1	-	17
C2	-	17
C3	9	17
C4	-	17
C5	5	17
C6	-	17
C7	2	17

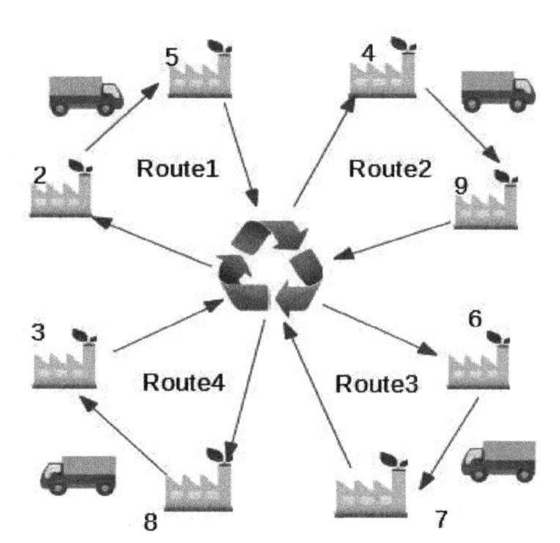

Fig. 2: Optimal distribution for the numerical example.

the number of components obtained via the disassembly line. The optimal objective function value is obtained as 1327. 5USD for the numerical example. According to the optimal result, the number of disassembled products is 17, the number of opened workstations is 2, and the cycle time is determined as follows:

$$CT = W_{time}/Dp = 1000/17 = 58.82 \text{ min}$$

4.3 Scenario Analysis

In order to investigate the impact of several parameters on the solution, we now analyze two different scenarios: (i) the effect of changing working time on the performance measures (total cost, number of workstations, cycle time, the number of disassembled products and the number of components produced by a 3DP machine), (ii) the effect of changing fixed cost of operating a 3DP machine without changing the disassembling cost of a product on performance measures (total cost, number of workstations, cycle time, the number of disassembled products and the number of components produced by a 3DP machine).

We first analyzed the effect of changing working time. Working time of initial problem increased by +%25 and +%50, and decreased by –%25 and –%50. Table 3 presents the details of all solutions. WT, FC-3D, NS, TN denote the working time, fixed cost to operate 3DP machine, the number of workstations and the number of vehicle tour, respectively. Results show that increasing working time increases the number of components produced by 3DP machine (3DP in Table 3), and decreases the number of disassembled products (DP in Table 3). Furthermore, as working time is decreased objective function value is increased, and the number of workstations is tented to increase.

Figure 3 presents the changing objective function value and cycle time according to the variation of working time. When the working time is increased, objective function value decreases and cycle time increases. The reason behind the decreasing of the objective function value as the working time increases probably depends on the assumed costs. This is because, fixed disassembling cost of a product may be much more than fixed cost of operating the 3DP machine due to several cost such as labor and opening workstations in disassembly line.

Figure 4 presents the relationship between the number of disassembled products and components produced by a 3DP machine. In this case, as working time is increased, the number of disassembled products decreases and the number

Table 3: Results of the effect of changing working time.

WT	FC-3D	*Obj*	NS	*CT*	TN	CPU	DP	3DP
500	0.5	1140	3	83.3	4	2.160	12	47
750	0.5	882.02	1	125	4	0.920	8	75
1000	0.5	838.5	1	200	4	1.540	5	96
1250	0.5	780.5	1	1000	4	0.870	1	124
1500	0.5	666	-	-	4	0.560	-	131

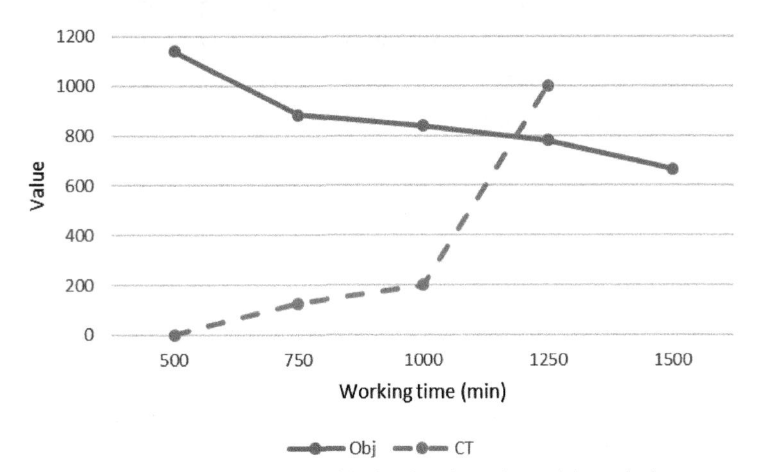

Fig. 3: Relationship between objective function value and the cycle time.

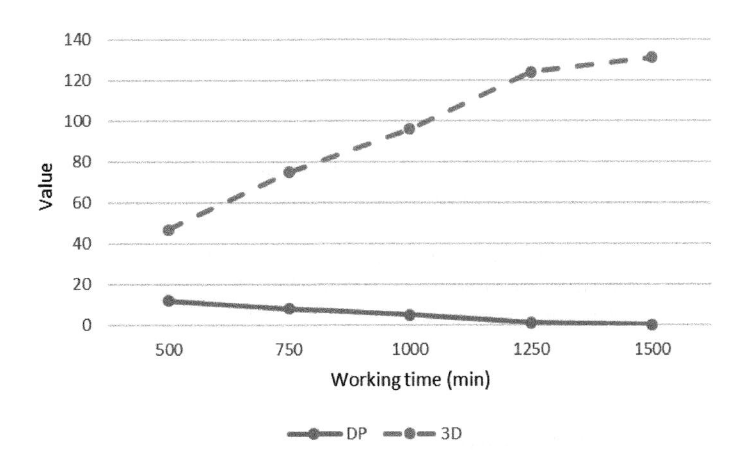

Fig. 4: Relationship between the number of disassembled products and components produced by 3DP machine.

of components produced by the 3DP machine increases. Because of the fixed cost of operating a 3DP machine is much lower, the model tends to increase the number of components produced through the 3DP machine to minimize the value of the objective function. Most of the demanded components cannot be produced with the 3DP machine due to the limited working time for the numerical example, although the fixed cost of operating the 3DP machine is lower. Increasing the working time makes it possible to produce components in 3DP machine and decreases the value of the objective function.

In the second scenario, we investigate the effect of changing the fixed cost of operating a 3DP machine without changing the disassembling cost of a product. Fixed cost of operating the 3DP machine of initial problem is increased up to 16USD/component and decreased down to 0.5USD/component for scenario analysis. Table 4 presents the obtained results. In the frame of this scenario, as the fixed cost of operating the 3DP machine increases, objective function value, the number of workstations and the number of disassembled products also increases. However, cycle time and the number of components produced by the 3DP machine decrease.

Table 4 shows that as the fixed cost of the 3DP machine increases, the model tends to increase the number of disassembled products in the disassembly lines. This also causes an increase in the value of the objective function. In these conditions, the total cost is higher due to obtaining components via the disassembly line.

Figure 5 shows the variation of the number of disassembled products and the number of components produced by a 3DP machine according to changing fixed cost of the 3DP machine. The scenario analysis results show that when we

Table 4: Results of fixed cost change for a 3DP machine.

FC-3D	*Obj*	NS	*CT*	CPU	DP	3DP
0.5	838.5	1	200	1.540	5	96
1	1168.5	1	83.3	1.530	12	47
2	1327.5	2	58.82	0.840	17	16
4	1471.5	2	45.45	1.140	22	4
8	1601.5	2	43.47	1.240	23	3
16	1611.5	3	38.46	1.060	26	-

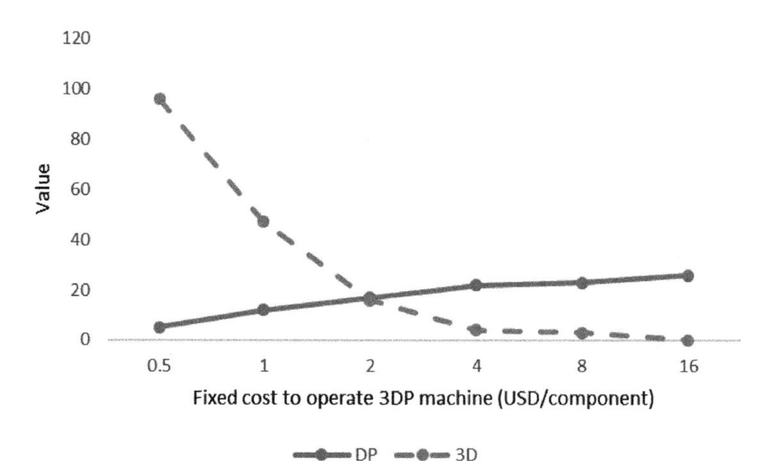

Fig. 5: Relationship between the number of disassembled products and components produced by 3DP machine according to fixed cost of operating 3DP machine.

increase the fixed cost of operating a 3DP machine, the number of disassembled products through the disassembly line increases and the number of components produced by the 3DP machine decreases. This is because the disassembly line and the 3DP machine run in parallel at the same working time, and affect each other in opposite directions.

5. Conclusion

This chapter focused on the integration of 3DP technology with the disassembly line balancing and vehicle routing within the context of smart factories. The disassembly process sustains a clean environment as well as time and cost saving as it promotes the recycling process. The effective distribution of components formed after the disassembly process is a challenging operation. The 3DP technology is an essential part of smart factories since it reduces the cost of inventory, saves time and provides quicker response to demands. It is also a highly effective process for creating new designs and producing better quality products when compared with the traditional production techniques.

We have proposed a mixed integer nonlinear mathematical formulation for the I-DLB-RP-3DP. We have conducted extensive computational experiments to investigate the I-DLB-RP-3DP, and have provided several policy and managerial insights. Scenario analysis show that costs and working time are an important factor on the results. When the fixed cost of the 3DP machine increases, the number of disassembled products and the value of the objective function also increase for the current test problem and vice versa. In addition, increasing the working time causes the increase in the number of components produced by 3DP machine and cycle time and decrease the number of disassembled products and objective function value.

For future work, uncertainty or linearization may be considered for the same problem. In addition, the researchers are expected to concentrate on mixed and multi products, multiple disassembly center or multiple 3DP machine and to apply effective metaheuristics for solving the problem.

Acknowledgement

We thank YÖK (Higher Education Council of Turkey) for 100/2000 Ph.D. scholarship and TÜBİTAK (The Scientific and Technological Research Council of Turkey) for 2211/C Ph.D. research scholarship.

Appendix

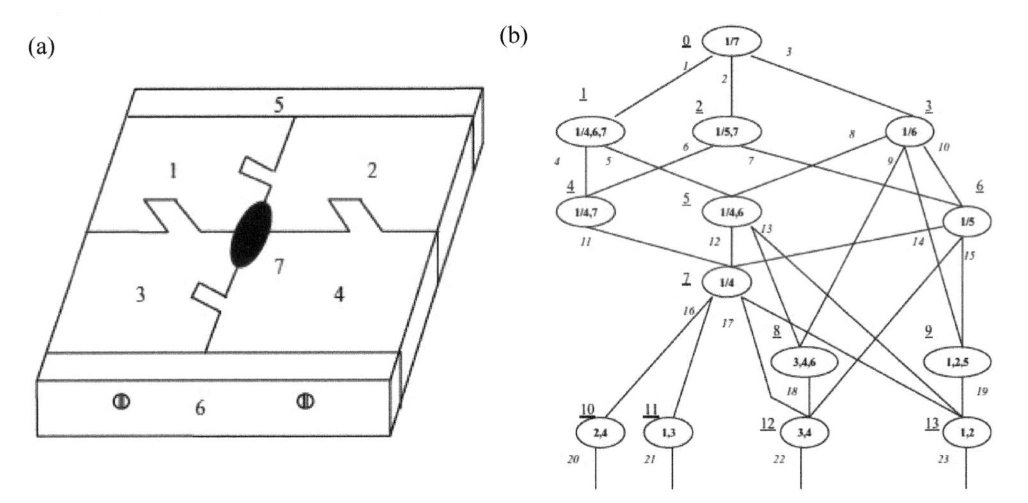

Fig. A.1: A sample product: (a) overview and (b) AND/OR graph (AOG) of the sample product (Koç et al. 2009).

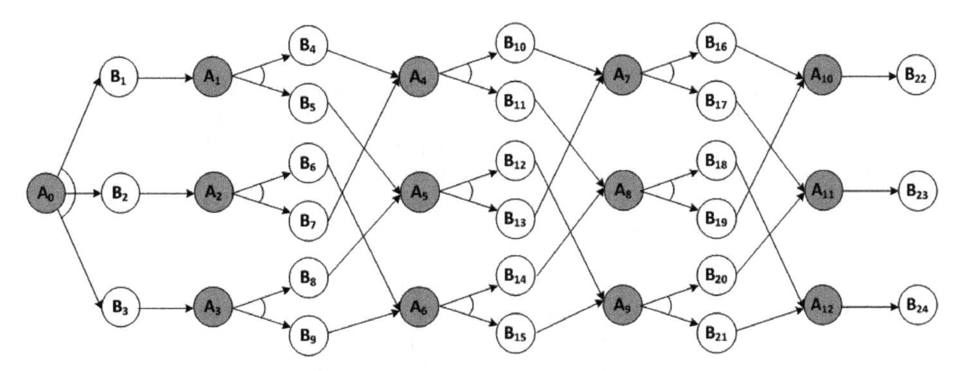

Fig. A.2: TAOG of the sample product (Koç et al. 2009).

Table A.1: Times of disassembly tasks for sample product.

Task	Task time	Task	Task time	Task	Task time
1	22	9	16	17	14
2	22	10	21	18	18
3	14	11	12	19	18
4	21	12	20	20	7
5	13	13	15	21	7
6	21	14	20	22	7
7	13	15	15	23	7
8	21	16	14		

Table A.2: Demands of (re)manufacturing centers.

RC	C1	C2	C3	C4	C5	C6	C7
2	3	1	4	3	2	1	2
3	2	4	3	2	3	1	2
4	1	1	6	2	2	1	3
5	2	1	2	3	4	3	4
6	1	1	5	2	1	2	1
7	5	2	1	1	3	3	4
8	2	1	3	3	4	2	1
9	1	3	2	1	3	3	2

Table A.3: Operating time of components.

Component	Operating time
1	12
2	8
3	10
4	20
5	5
6	7
7	9

References

Airbus, 2014. Printing the future: Airbus expands its applications of the revolutionary additive layer manufacturing process. [online] Available at:<https: //www.airbus. com/ newsroom/news/en/2014/03/printing-the-future-airbus-expands-its-applications-of-the-revolutionary-additive-layer-manufacturing-process.html> [Accessed 26October 2019].

Augerat, P., J.M. Belenguer, E. Benavent, A. Corberan, D. Naddef and G. Rinaldi. 1995. Computational Results with a Branch and Cut Code for the Capacitated Vehicle Routing Problem. Rapport de recherche-IMAG.

Braekers, K., K. Ramaekers and I.V. Nieuwenhuyse. 2016. The vehicle routing problem: state of the art classification and review. Computers and Industrial Engineering, 99: 300–313.

Catts, T. 2013. GE Turns to 3D Printers for Plane Parts. [online] Bloomberg View. Available at:<https://www.bloomberg.com/news/articles/2013-11-27/general-electric-turns-to-3d-printers-for-plane-parts> [Accessed 26 October 2019].

Chan, H.K., J. Griffin, J.J. Lim, F. Zeng and A.S.F. Chiu. 2018. The impact of 3D Printing Technology on the supply chain: Manufacturing and legal perspectives. International Journal of Production Economics, 205: 156–162.

Deniz, N. and F. Ozcelik. 2019. An extended review on disassembly line balancing with bibliometric and social network and future study realization analysis. Journal of Cleaner Production, 225: 697–715.

Ford, S.J. and M. Despeisse. 2016. Additive manufacturing and sustainability: an exploratory study of the advantages and challenges. Journal of Cleaner Production, 137: 1573–1587.

Güngör, A. and S.M. Gupta. 2001. A solution approach to the disassembly line balancing problem in the presence of task failures. International Journal of Production Research, 39(7): 1427–1467.

Habibi, M.K.K., O. Battaïa, V.D. Cung and A. Dolgui. 2014. Integrated procurement-disassembly problem. pp. 382–390. *In:* Grabot, B., B. Vallespir, S. Gomes, A. Bouras and D. Kiritsis [eds.]. Proceedings of the IFIP Advances in Information and Communication Technology. Ajaccio, Corsica, France. Springer.

Habibi, M.K.K., O. Battaïa, V.-D. Cung and A. Dolgui. 2017a. Collection-disassembly problem in reverse supply chain. International Journal of Production Economics, 183: 334–344.

Habibi, M.K.K., O. Battaïa, V.-D. Cung and A. Dolgui. 2017b. An efficient two-phase iterative heuristic for collection-disassembly problem. Computers and Industrial Engineering, 110: 505–514.

Habibi, M.K.K., O. Battaïa, V.-D. Cung, A. Dolgui and M.K. Tiwari. 2018. Sample average approximation for multi-vehicle collection–disassembly problem under uncertainty. International Journal of Production Research, 57(8): 1–20.

Holmström, J., J. Partanen, J. Tuomi and M. Walter. 2010. Rapid manufacturing in the spare parts supply chain. Journal of Manufacturing Technology Management, 21(6): 687–697.

Huang, S.H., P. Liu, A. Mokasdar and L. Hou. 2013. Additive manufacturing and its societal impact: a literature review. The International Journal of Advanced Manufacturing Technology, 67: 1191–1203.

Jia, F., X.F. Wang, N. Mustafee and L. Hao. 2016. Investigating the feasibility of supply chain-centric business models in 3D chocolate printing: a simulation study. Technological Forecasting and Social Change, 102: 202–213.

Kannan, D., K. Garg, P.C. Jha and A. Diabat. 2017. Integrating disassembly line balancing in the planning of a reverse logistics network from the perspective of a third party provider. Annals of Operations Research, 253(1): 353–376.

Khajavi, S.H., J. Partanen and J. Holmström. 2014. Additive manufacturing in the spare parts supply chain. Computers in Industry, 65(1): 50–63.

Koç, A., I. Sabuncuoglu and E. Erel. 2009. Two exact formulations for disassembly linebalancing problems with task precedence diagram construction using an AND/OR Graph. IIE Transactions, 41(10): 866–881.

Koç, Ç., T. Bektaş, O. Jabali and G. Laporte. 2016. Thirty years of heterogeneous vehicle routing. European Journal of Operational Research, 249(1): 1–21.

Koç, Ç. 2017. An evolutionary algorithm for supply chain network design with assembly line balancing. Neural Computing and Applications, 28(11): 3183–3195.

Koç, Ç. and G. Laporte. 2018. Vehicle routing with backhauls: review and research perspectives. Computers and Operations Research, 91: 79–91.

Küçükkoç, I. 2019. MILP models to minimise makespan in additive manufacturing machine scheduling problems. Computers and Operations Research, 105: 58–67.

Li, Y. G. Jia, Y. Cheng and Y. Hu. 2016. Additive manufacturing technology in spare parts supply chain: A comparative study. International Journal of Production Research, 55(5): 1498–1515.

Özceylan, E. and T. Paksoy. 2013. Reverse supply chain optimisation with disassembly line balancing. International Journal of Production Research, 51(20): 5985–6001.

Özceylan, E. and T. Paksoy. 2014. Fuzzy mathematical programming approaches for reverse supply chain optimization with disassembly line balancing problem. Journal of Intelligent and Fuzzy Systems, 26: 1969–1985.

Özceylan, E., T. Paksoy and T. Bektaş. 2014. Modeling and optimizing the integrated problem of closed-loop supply chain network design and disassembly line balancing. Transportation Research Part E: Logistics and Transportation Review, 61: 142–164.

Özceylan, E., C.B. Kalayci, A. Güngör and S.M. Gupta. 2019. Disassembly line balancing problem: a review of the state of the art and future directions. International Journal of Production Research, 57(15-16): 4805–4827.

Rogers, H., N. Baricz and K.S. Pawar. 2016. 3D printing services: classification, supply chain implications and research agenda. International Journal of Physical Distribution and Logistics Management, 46(10): 886–907.

Ryan, M., D. Eyers, A. Potter, L. Purvis and J. Gosling. 2017. 3D printing the future: scenarios for supply chains reviewed. International Journal of Physical Distribution and Logistics Management, 47(10): 992–1014.

Sasson, A. and J.C. Johnson. 2016. The 3D printing order: variability, supercenters and supply chain reconfigurations. International Journal of Physical Distribution and Logistics Management, 6(1): 82–94.

Siemens. 2017. From design to repair—AM changes everything. [online] Available at:<https://new.siemens.com/global/en/products/energy/services/maintenance/parts/additive-manufacturing.html> [Accessed 26 October 2019].

Toth, P. and D. Vigo [eds.]. 2014. Vehicle routing: Problems, methods, and applications. MOS-SIAM Series on Optimization, Philadelphia.

CHAPTER 13

Enterprise Resource Planning in the Age of Industry 4.0

A General Overview

İbrahim Zeki Akyurt,[1] *Yusuf Kuvvetli*[2] and *Muhammet Deveci*[3,*]

1. Enterprise Resource Planning

Nowadays, information technology has become an essential tool for the operation and management of all activities ranging from production scheduling to supply chain management. Accordingly, the general name given to an integrated software based management system including basic business functions such as production, finance and marketing as well as side functions such as cost accounting, purchasing, distribution, customer relations, cash flows, warehouse management, human resources, material management, electronic banking, and quality control is called Enterprise Resource Planning (ERP). ERP provides efficient use of all resources in enterprises that produce both products and services. For this purpose, the company collects all the data in one place, consolidates and uses this data to realize the activity in case of need.

According to The 16th Edition of the APICS Dictionary ERP (enterprise resource planning) is defined as a "Framework for organizing, defining, and standardizing the business processes necessary to effectively plan and control an organization so the organization can use its internal knowledge to seek external advantage. An ERP system provides extensive databanks of information including master file records, repositories of cost and sales, financial detail, analysis of product and customer hierarchies, and historic and current transactional data." ERP software integrates all departments and functions in the business into a single computer system that can meet the specific needs of all these departments and functions. Normally, each department, such as human resources, production, distribution and finance, has a computer system that operates its own department. However, ERP aggregates them all into a single integrated software program running in a single database; so various departments can share information more easily and communicate with each other more easily (Koch et al. 1999). This integrated software facilitates the flow of information between the internal and external supply chain processes in the organization (Al-Mashari and Zairi 2000). In this respect, ERP has attracted the attention of both academic and industrial communities in recent years (Shehab et al. 2004).

ERP renews the old standalone computer systems such as finance, human resource, production and warehouse functions and breaks them down into modules running under a single program. Thus, an employee in production or finance can easily see whether a raw material ordered has been brought to the warehouse. At this point, one of the biggest benefits that ERP brings to businesses is that it prevents unnecessary search and information traffic. Each department has its own story and figures. When the top manager wants to evaluate them in general, they have to learn the method of each department. However, since ERP operates on a single system, it does not provide separate methods to each department, but integrates the necessary financial information through a single story. Again, ERP can follow orders more easily and simultaneously coordinate manufacturing, stock and transportation in many different locations. ERP saves time, increases productivity and reduces the number of employees by standardizing processes such as manufacturing and assembly with

[1] Department of Production Management, Faculty of Business, Istanbul University, Istanbul, Turkey.
Emails: akyurt@istanbul.edu.tr; akyurt@istanbul.edu.tr

[2] Department of Industrial Engineering, Cukurova University, Adana, Turkey, Email: ykuvvetli@cu.edu.tr

[3] Department of Industrial Engineering, Naval Academy, National Defense University, 34940, Istanbul, Turkey.

* Corresponding author: muhammetdeveci@gmail.com,

a single, integrated computer system. With the same system, semi-finished stocks are reduced, material orders are issued and products to be delivered come out of production on time.

ERP systems have made great improvements over the years as aforementioned before. Integrating business functions together provides more efficient production environments for the firms. Furthermore, ERP systems have more capabilities by using new technologies. Industry 4.0 technologies have made some mile-stone improvements to the firms in context with the ERP systems such as data acquiring, analysing, storing and decision-making capabilities. This chapter discusses the future ERP systems on the new production systems that are influenced by industry 4.0 technologies.

2. Literature Review of ERP

In the manufacturing processes, computers were used in the early 1960s. The first application in manufacturing is limited to stock control and order process. This was generally designed to allow accountants to calculate the value of the stock (Gumaer 1996). In the 1970s, manufacturing computer systems were known as MRP (material requirements planning). Even though the roots of MRP are fairly old, most of the MRP functionality is still available in today's ERP systems (Kurbel 2016). MRP is a time-phased ordering system that plans production work orders and purchase orders, so that subassemblies and components reach the assembly station as required. Some of the benefits of MRP include inventory reduction, improved customer service, efficiency and productivity (Rao Siriginidi 2000). The main disadvantage of the MRP approach is that it does not take into account the production capacities. Since it works with endless source logic, it is not clear whether the job and purchase orders can be fulfilled or not. To create an applicable plan, MRP must be supported by capacity planning. Accordingly, in the 1980s, an MRP version known as MRP II (manufacturing resource planning) was introduced. MRP II has specific modules such as rough cut capacity planning and capacity requirements planning for production planning (Rao Siriginidi 2000). ERP is the latest developed phase of MRP II, which takes into account the additional functions of an organization such as finance, distribution and human resources through an integrated network. While planning a resource, the concept of resource planning is essential (Kurbel 2016). ERP expanded in the mid-1990s to include more functions such as order management, financial management, warehousing, distribution production, quality control, asset management and human resource management, sales force and marketing, electronic commerce and supply chain systems. All these operations are shown in Figure 1 (Adapted from (Shehab et al. 2004)).

The fact that ERP is included in all processes of an enterprise also shows the fact that ERP is a process for the enterprise. Since the mid-1990s, ERP systems have been installed in thousands of companies worldwide (Mabert et al. 2003). Implementation of the ERP system is difficult for the enterprise and requires the waste of corporate resources and time. It is also a costly process. Many ERP implementations are classified as failures because they do not meet predetermined corporate objectives (Umble et al. 2003). However, implementing ERP systems can be quite easy when organizations are simply configured and run in one or more locations (Markus et al. 2000). Accordingly, the ERP process and taxonomy of an enterprise is shown in Figure 2 (Al-Mashari et al. 2003).

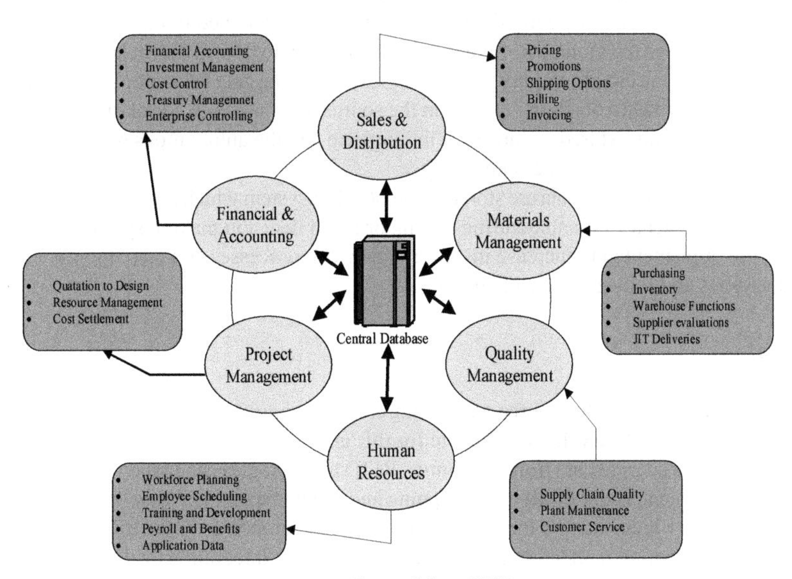

Fig. 1: The modules of ERP.

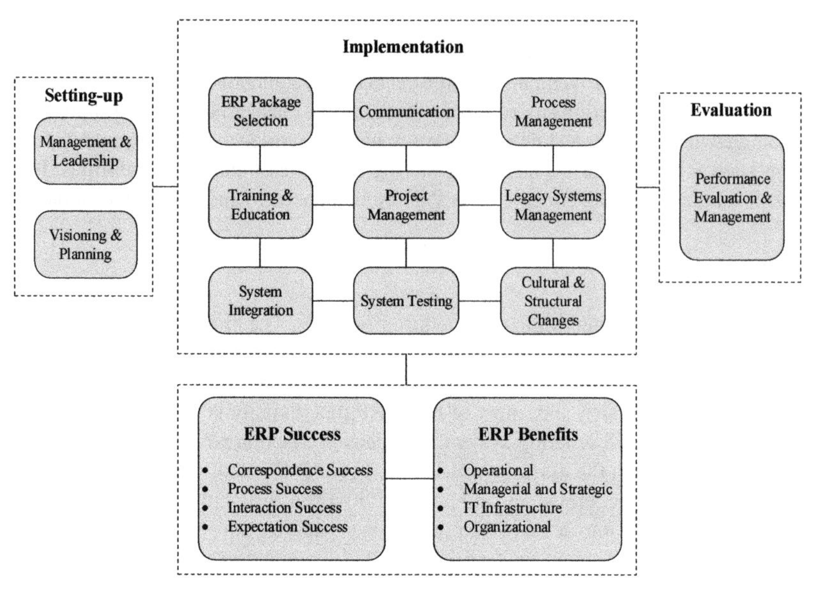

Fig. 2: The process and taxonomy of ERP.

3. Effect of the Industry 4.0 on ERP Systems

ERP systems have a complex software environment which includes different layers. An ERP software includes a database design and a database layer in order to keep different numbers of data from the system. The database can only access by core business logics such as security applications, firewalls that is provided by server level. The business functions such as finance, marketing, production etc. are modeled as applications at the business applications layer. Finally, end-users are accessed to the system by using user interfaces at the end-user layer. The whole architecture is summarized in the Figure 3.

Industrialization is the paradigm that covers progress on producing materials in better conditions. The changes in industrialization has been called "industrial revolutions" and the first industrial revolution is the development on mechanization, the second industrial revolution is the development on electrical energy, the third industrial revolution is the development on automation and digitalization and the fourth industrial revolution is the development on internet and smart objects (Lasi et al. 2014). The production environment is changing day by day in order to achieve a more productive manufacturing environment. The 4th industrial revolution provides more efficient production systems by connecting devices and equipment together via internet. The evolutions of Industry 4.0 are depicted in Figure 4.

Most of Industry 4.0 related technologies influence the new ERP systems at different levels of obtaining, analyzing and mining the data. For example, the cloud computing technologies and the big data technologies make a great improvement on the ERP software capabilities due to obtaining data from the source automatically. Similarly IoT devices make easier data integration with the ERP systems. Moreover, artificial intelligence and the autonomous robots provide to gain inferences from the ERP systems and achieving business intelligence.

In classical approach the corporate data are stored in the SCADA system which is used as data acquisition systems. The data are analyzed in order to achieve manufacturing tracing or better decision making systems. Finally, in the knowledge level, ERP systems are implemented to increase integration between processes and data. Figure 5 shows the relationship between classical systems and Industry 4.0 technologies.

3.1 Data Level

In the data level, obtaining data from the source and storing it is an important extension point of the future ERP systems. IoT and cloud technologies of industry 4.0 are suitable for this aim. RFID (Radio Frequency Identification) systems are one of the equipment that can acquire data from the source. RFID technologies can make variety of processes more visible such as receiving, replenishment, order fulfillment, shipping and product tracking (Angeles 2005). RFID technologies are incorporated by IoT devices which changes the information acquiring processes of the ERP systems. This will lead to the IoT and ERP integration becoming mandatory in future factories. These integrations will reinforce the ERP systems into

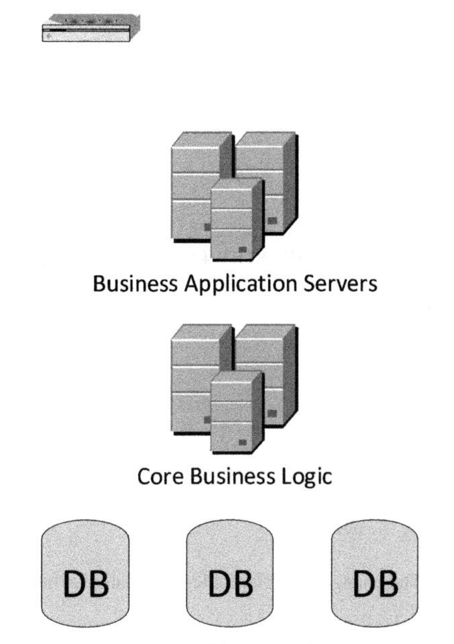

Fig. 3: Software Architecture of ERP systems.

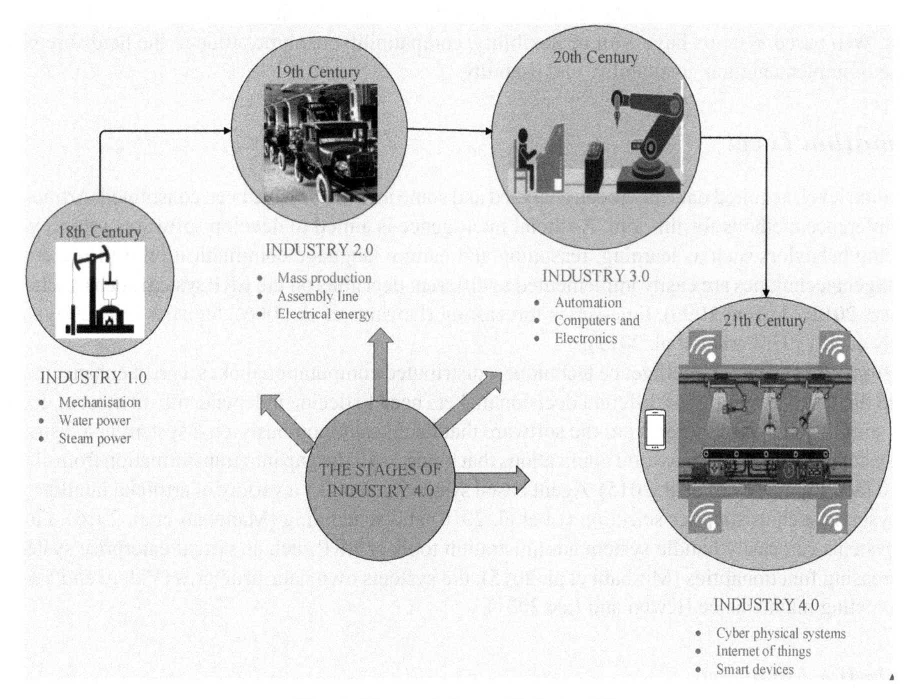

Fig. 4: The evolutions of Industry 4.0.

interfaces of customer, human-machine, partner, and the employee with different manners such as mobile apps, predictive models, automatic data exchange to achieve total cost ownership (Ranjan et al. 2017).

Cloud ERP systems are the new trends in the ERP industry. Cloud computing may define the hardware and software systems that ensure the services for using applications over Internet (Armbrust et al. 2010). Cloud computing provide to use the applications as platform free and it enables ERP systems more capability. For enhanced capability of cloud ERP systems strategic decisions on how firms could effectively respond to market dynamism are required (Gupta et al. 2019).

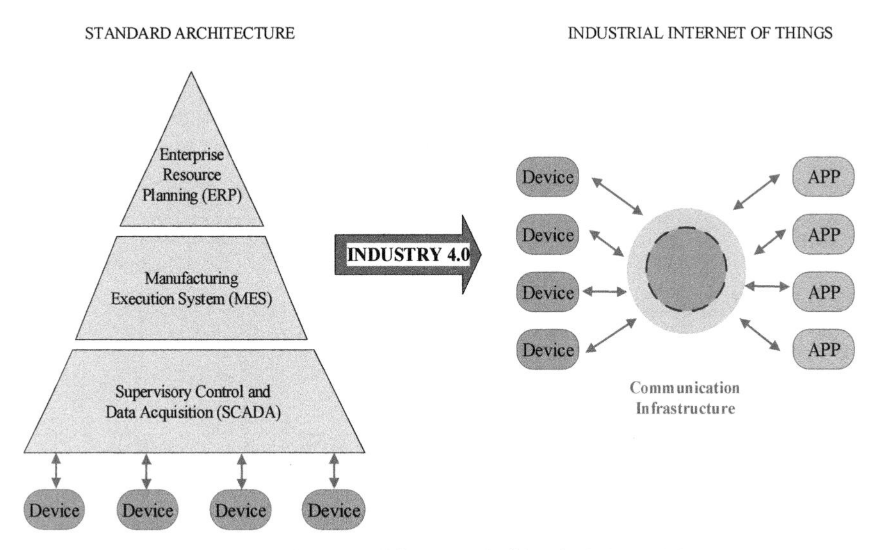

Fig. 5: The change on different level of data by Industry 4.0.

Cloud systems changes ERP software to ERP service by using Software As A Service (SAAS) architecture. There are lots of important factors making decision on the SAAS architecture and the important ones are given as follows (Johansson and Ruivo 2013): costs, security, availability, usability, implementation, ubiquity, flexibility, compatibility, analytics and best-practices. Web based systems have a lot of flexibility, compatibility, analytics (due to the hardware power of cloud systems), ease of implementation, availability and usability.

3.2 Information Level

In the information level, acquired data have been analyzed and some inferences have been constituted. Artificial intelligence is one of the inference methods for this aim. Artificial intelligence is aimed to develop software systems which simulate the human being behaviors such as learning, reasoning and natural language communication (O'Hare et al. 1996). The artificial intelligent techniques are easily implemented to different decisions on the ERP system such as scheduling (Rojek and Jagodziński 2012; De Toni 1996), time-series forecasting (Doganis et al. 2006), logistics (Tan 2001), and inventory decisions (Roy et al. 1997; Farhat et al. 2017).

Among from other artificial intelligence techniques, distributed computation makes a great enhancement on the ERP systems due to the fact that number of different decision makers need to decide independently in most of decisions related with ERP systems. Agent based systems are the software that decide autonomously on a system in a distributed manner. Agent based systems have lots of real-world applications that guide to fill the gap into transformation from classical systems to Industry 4.0 facilities (Adeyeri et al. 2015). Agent based systems can make a variety of artificial intelligent applications on the ERP systems such as supplier selection (Li et al. 2018) and scheduling (Manupati et al. 2016). Furthermore, the agent based systems can easily handle system administration tools of ERP such as virtual enterprise systems (Sadigh et al. 2017), increasing functionalities (Mesbahi et al. 2015), the systems own data structures (Vidoni and Vecchietti 2015), and even conducting maintenance (Kwon and Lee 2001).

3.3 Knowledge Level

Intelligent robots are new manufacturing models for the new technological era. Robots need to make acceleration, flow, and some operations due to the fact that they should process and analyze lots of raw data obtained from the different sources such as gyro, force, image and sound sensors (Sakagami et al. 2002). Intelligent robots need cloud systems and big data applications for this amount of raw data (Anton et al. 2020). Robots are important to achieve a sustainable manufacturing environment which includes redesigning, reusing, remanufacturing, recovering, recycling and reducing operations (Bi et al. 2015).

Smart facilities are the future factory concept that connects intelligent manufacturing, planning and autonomous decision-making. Therefore, the smart factory model is primarily aimed at facilitating and ensuring the availability of all relevant information for real-time storage, which will be possible through the integration between all elements in the value chain (Majeed and Rupasinghe 2017).

4. Future Trends of ERP in Industry 4.0

Recently, most of the companies are actively working for digital transformation. As technology develops in impressive and unexpected ways, ERP will continue to be a very important business tool. Future trends of ERP in Industry 4.0 are as follows (Cre8tive Technology 2019):

4.1 Additive Manufacturing

We call Additive Manufacturing to the integration of technologies that are rapidly adopted by people and widely used in our lives and used effectively as rapid prototyping and 3D printing methods. With layer production, parts can be produced in a short time according to the requirements and no cost or time is required for the change of design. In layer production, complex geometries and difficult-to-make parts can be produced by removing design boundaries.

Layer production will have an important place in the 4th Industrial Revolution concept in line with the added value and benefits. With Industry 4.0, the industry needs machines that can produce the desired components more flexibly and precisely than ever before. Less prototype construction, mold, process additive production will be realized thanks to production (Turkey's Industry 4.0 Platform 2019). Thus, it is aimed to reduce costs and increase productivity with larger and faster 3D printers.

The biggest challenge for manufacturers is to manage the data stack that comes with 3D printing. Additional production increases the data volume at each step of the production process, thereby revealing the necessity of the ERP system. Manufacturers will need to carefully review their existing ERP capabilities to receive the rewards of this trend.

4.2 Data Analytics

Previous ERP systems may have been good at collecting and organizing data, but were more limited in analytics and reporting. Now data-driven decision-making has become the priority of every manufacturer. Therefore, ERP solutions are developing analytical capabilities to meet the needs of manufacturers. Modern ERP solutions enable users to run temporary reports, visualize data, or embed analytical tools into existing applications(Babu and Sastry 2014).

4.3 Cloud ERP

Cloud ERP is not a new term, it is an enterprise resource planning solution where all data, applications and programs are stored on a virtual server instead of physical machines and accessed via internet. Cloud-ERP emerges as a new trend in the ERP market owing to cloud-ERP. The world is changing so fast that in this new internet-oriented world, organizations are thinking of moving IT services to the cloud every day. It is rapidly becoming the industry standard for many manufacturing sectors. The restructuring of the competitive environment with the many advantages provided by Cloud ERP raises concerns for many companies.

Benefits of Cloud ERP are as follows: (i) eliminating unnecessary expenses, (ii) having a flexible and agile solution, (iii) overcoming remote access barriers, (iv)the reduction of computing costs,(v) making data more secure, (vi) increasing productivity, (vii) provides new business opportunities for organizations(Demi and Haddara 2018).

4.4 Machine Integration

The Internet of Things (IoT), with Industry 4.0, makes the manufacturer more productive and increases transparency. It also has the potential to improve data availability and accuracy and thus has a significant impact on the manufacturing sector. The sensors with internet connection form a direct link between a working machine part and ERP system. The use of Industry 4.0 provides communication and integration between all systems (Majeed and Rupasinghe 2017).

5. Conclusion

Industry 4.0 ensures many new research and implementation areas for the firms such as investigating and adapting new technologies to classic production environment, changing jobs and skills and training the organization regarding new technologies.

Enterprise resource planning and the future of this concept will be well affected by this new transformation. It is obvious that some concerns are eliminated by implementation of the Industry 4.0 and related activities such as: (i) gathering data from the real production environment will be quite easier than manual data entering, (ii) more reliable data is acquired from the source at once, (iii) more amount of data can be handled and analyzed by cloud and big data approached, (iv) decision capabilities of ERP software are increased by applying artificial intelligence techniques, and (v) future ERP systems are one of the key concepts that controls the bottom and intermediate data storage and analysis in the smart facilities.

References

Al-Mashari, M. and M. Zairi. 2000. Supply-chain re-engineering using enterprise resource planning (ERP) systems: an analysis of a SAP R/3 implementation case [Journal Article]. International Journal of Physical Distribution & Logistics Management, 30(3/4): 296–313.

Al-Mashari, M., A. Al-Mudimigh and M. Zairi. 2003. Enterprise resource planning: A taxonomy of critical factors [journal article]. European Journal of Operational Research, 146(2): 352–364.

Angeles, R. 2005. RFID technologies: supply-chain applications and implementation issues. Information Systems Management, 22(1): 51–65.

Anton F., T. Borangiu, S. Raileanu, S. Anton, N. Ivănescu and I. Iacob. 2020. Secure sharing of robot and manufacturing resources in the cloud for research and development. *In*: Berns, K., D. Gorges [eds.]. Advances in Service and Industrial Robotics. Advances in Intelligent Systems and Computing. Vol. 980.

Cham: Springer International Publishing Ag. pp. 535–543.

Armbrust, M., A. Fox, R. Griffith, A.D. Joseph, R. Katz, A. Konwinski, G. Lee, D. Patterson, A. Rabkin, I. Stocia and M. Zaharia. 2010. A view of cloud computing. Commun ACM, 53(4): 50–58.

Adeyeri, M.K., K. Mpofu and T.A. Olukorede. 2015. Integration of Agent Technology into Manufacturing Enterprise: A Review and Platform for Industry 4.0. New York: Ieee. English (2015 International Conference on Industrial Engineering and Operations Management).

Babu, M.P. and S.H. Sastry. 2014. Big data and predictive analytics in ERP systems for automating decision making process. In 2014 IEEE 5th International Conference on Software Engineering and Service Science. 2014 June (pp. 259–262). IEEE.

Bi, Z.M., Y.F. Liu, B. Baumgartner, E. Culver, J. Sorokin, A. Peters, B. Cox, J. Hunnicutt, J. Yurek and S. O'Shaughnessey. 2015. Reusing industrial robots to achieve sustainability in small and medium-sized enterprises (SMEs) [Article]. Ind. Robot., 42(3): 264–273.

Cre8tive Technology [Internet]. 2019. US: Cre8tive Technology; [cited 30.10.2019]. Available from: https://www.ctnd.com/the-future-of-erp-top-5-erp-trends-for-2019/.

De Toni, A., G. Nassimbeni and S. Tonchia. 1996. An artificial, intelligence-based production scheduler. Integrated Manufacturing Systems, 7(3): 17–25.

Demi, S. and M. Haddara. 2018. Do Cloud ERP Systems Retire? An ERP Lifecycle Perspective [Article]. Procedia Computer Science, 138: 587–594.

Doganis, P., A. Alexandridis, P. Patrinos and H. Sarimveis. 2006. Time series sales forecasting for short shelf-life food products based on artificial neural networks and evolutionary computing. Journal of Food Engineering, 2006 2006/07/01/;75(2): 196–204.

Farhat, J. and M. Owayjan. 2017. ERP neural network inventory control. pp. 288–295. *In*: Dagli, C.H. [ed.]. Complex Adaptive Systems Conference with Theme: Engineering Cyber Physical Systems, Cas. Procedia Computer Science. Vol. 114. Amsterdam: Elsevier Science Bv.

Gumaer, R. 1996. Beyond ERP and MRP II: Optimized planning and synchronized manufacturing [journal article]. IIE Solutions, 28(9): 32–36.

Gupta, S., X.Y. Qian, B. Bhushan and Z. Luo. 2019. Role of cloud ERP and big data on firm performance: a dynamic capability view theory perspective [Article]. Manag Decis. 2019 Sep; 57(8): 1857–1882.

Johansson, B. and P. Ruivo. 2013. Exploring Factors for Adopting ERP as SaaS. Procedia Technology, 2013/01/01/;9:94–99.

Koch, C., D. Slater and E. Baatz. 1999. The ABCs of ERP [Article]. CIO Magazine, pg. 22.

Kurbel, K.E. 2016. Enterprise Resource Planning and Supply Chain Management. Springer-Verlag Berlin An.

Kwon, O.B. and J.J. Lee. 2001. A multi-agent intelligent system for efficient ERP maintenance. Expert Systems with Applications. 2001/11/01/; 21(4): 191–202.

Lasi, H., P. Fettke, H.G. Kemper, T. Feld and M. Hoffamn. 2014. Industry 4.0 [journal article]. Business & Information Systems Engineering. 2014 August 01; 6(4): 239–242.

Li, J.H., M.M. Sun, D.F. Han, X. Wu, B.Yang, X. Mao and Q. Zhou. 2018. Semantic multi-agent system to assist business integration: An application on supplier selection for shipbuilding yards [Article]. Comput. Ind.Apr., 96: 10–26.

Mabert, V.A., A. Soni and M.A Venkataramanan. 2003. Enterprise resource planning: Managing the implementation process [journal article]. European Journal of Operational Research, 146(2): 302–314.

Manupati, V.K., G.D. Putnik, M.K. Tiwari, P. Ávila and M.M. Cruz-Cunha. 2016. Integration of process planning and scheduling using mobile-agent based approach in a networked manufacturing environment [Article]. Comput. Ind. Eng. Apr., 94: 63–73.

Majeed, A.A. and T.D. Rupasinghe. 2017. Internet of things (IoT) embedded future supply chains for industry 4.0: An assessment from an ERP-based fashion apparel and footwear industry [Article]. International Journal of Supply Chain Management, 6(1): 25–40.

Markus, M.L., C. Tanis and P.C. Van Fenema. 2000. Enterprise resource planning: multisite ERP implementations [journal article]. Communications of the ACM, 43(4): 42–46.

O'Hare, G.M., N.R. Jennings and N. Jennings. 1996. Foundations of distributed artificial intelligence. Vol. 9. John Wiley & Sons.

Mesbahi, N., O. Kazar, S. Benharzallah, M. Zoubeidi and S. Bourekkache. 2015. Multi-agents approach for data mining based k-means for improving the decision process in the ERP systems [Article]. Int. AJ Decis. Support Syst. Technol. Jun–Apr., 7(2): 1–14.

Ranjan, S., V.K. Jha and P. Pal. 2017. Application of emerging technologies in ERP implementation in Indian manufacturing enterprises: an exploratory analysis of strategic benefits [journal article]. The International Journal of Advanced Manufacturing Technology. January 01, 88(1): 369–380.

Rao Siriginidi, S. 2000. Enterprise resource planning in reengineering business [journal article]. Business Process Management Journal, 6(5): 376–391.

Rojek, I. and M. Jagodziński [ed.]. 2012. Hybrid Artificial Intelligence System in Constraint Based Scheduling of Integrated Manufacturing ERP Systems2012; Berlin, Heidelberg: Springer Berlin Heidelberg; (Hybrid Artificial Intelligent Systems.

Roy, B.V., D.P. Bertsekas, Y. Lee and J.N. Tsitsiklis [ed.]. 1997. A neuro-dynamic programming approach to retailer inventory management. Proceedings of the 36th IEEE Conference on Decision and Control; 12–12 Dec. 1997.

Sadigh, B.L., H.O. Unver, S. Nikghadam, E. Dogdu, A. Murat Ozbayoglu and S. Engin Kilic. 2017. An ontology-based multi-agent virtual enterprise system (OMAVE): part 1: domain modeling and rule management [Article]. Int. J. Comput. Integr. Manuf., 30(2-3): 320–343.

Sakagami, Y., R. Watanabe, C. Aoyama, S. Matsunaga, N. Higaki and K. Fujimara [ed.]. 2002. The intelligent ASIMO: system overview and integration. IEEE/RSJ International Conference on Intelligent Robots and Systems; 30 Sept.–4 Oct. 2002.

Shehab, E.M., M.W. Sharp, L. Supramaniam and T.A. Spedding. 2004. Enterprise resource planning: An integrative review. Business Process Management Journal [Journal Article], 10(4): 359–386.

Tan, K.C., L.H. Lee and K. Ou. 2001. Artificial intelligence heuristics in solving vehicle routing problems with time window constraints. Engineering Applications of Artificial Intelligence. 2001/12/01/; 14(6): 825–837.

Turkey's Industry 4.0 Platform. 2019. [Internet]. Turkey: ElektrikPort; [cited 30.10.2019]. Available from: https://www.endustri40.com/endustri-4-0-ile-katmanli-uretim/.

Umble, E.J., R.R. Haft and M.M. Umble. 2003. Enterprise resource planning: Implementation procedures and critical success factors [journal article]. European Journal of Operational Research, 146(2): 241–257.

Vidoni, M.C. and A.R. Vecchietti. An intelligent agent for ERP's data structure analysis based on ANSI/ISA-95 standard [Article]. Comput. Ind. 2015 Oct; 73: 39–50.

Chapter 14
Smart Warehouses in Logistics 4.0

Muzaffer Alım[1] and *Saadettin Erhan Kesen*[2,*]

1. Introduction

In recent years, the world has been witnessing dizzying changes in many areas in the light of developing technologies. All these changes have led to the beginning of a new age to the industrial revolutions journey. The journey started with the first industrial period as Industry 1.0 which had been started by the use of steam as a power to the machinery and this was followed by Industry 2.0 in which electricity was used as energy source and mass production began. By the establishment of computer and electronic systems in the production which results in automated systems, a new industrial period has rapidly permeated the industry. The history of industrial revolutions is presented in Figure 1. Following these breakthroughs, we have entered in to a new era which is triggered by the developments in information and communication technologies.

The concept of a new industrial age was first initiated in Hannover Fair in Germany in 2011 and named Industry 4.0 (Rojko 2017). In the same context, similar technological programs were announced and examples to these programs such as "Made in China 2025" by China, "Advanced Manufacturing Partnership" by the United States, "La Nouvelle France Industrielle" by France and Brazil's "Towards Industry 4.0" are such initiatives,which aim to understand and spread the advances in the context of Industry 4.0 to local companies (Dalenogare et al. 2018; Liao et al. 2018). The interest about Industry 4.0 is not only at the scale of governments but also from academia and industry as well.

Various features of Industry 4.0 distinguish it from the other three industrial periods. First, for the first time in history, an industrial revolution is predicted prior to its existence unlike others which are evaluated as revolutions posteriori (Rainer and Alexander 2014). This will allow to shape its structure by foreseeing and controlling the implications and its effects. Despite the great interest from the market, Industry 4.0 is said to encompass the future to great extent. Second, the

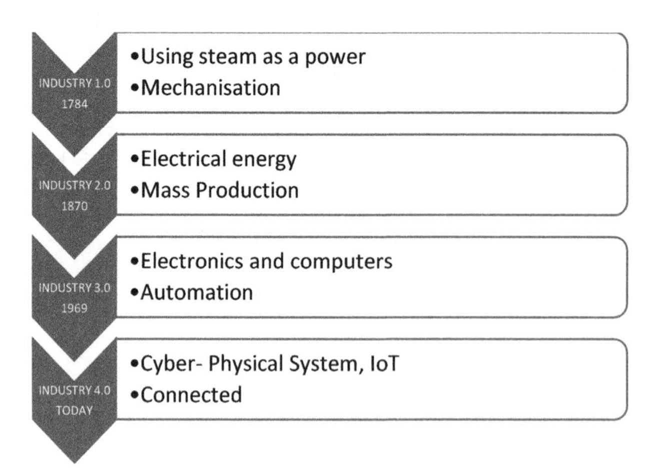

Fig. 1: Historical developments of industrial revolutions.

[1] Batman University, Technology Faculty, Batman/Turkey.
[2] Konya Technical University, Dept. of Industrial Engineering, Konya/Turkey.
* Corresponding Author: sekesen@ktun.edu.tr

impact of Industry 4.0 is expected to be huge on the economic scale due to its substantial improvements on effectiveness of operations as compared to the other periods (Hermann et al. 2016). The connectivity enabled by technologies changes not only the industry but also the society and the speed of this change and impact size have made it so unique from other periods (Schwab 2017).

Despite all its popularity, a common and comprehensive definition of Industry 4.0 has yet to be made as its boundaries are still not fully predictable. Thus, instead of a complete definition, studies have usually identified the structure and purposes of it. The promoters of the Industry 4.0 stated its main purpose as to make fundamental improvements in industrial processes including manufacturing, engineering, supply chain systems, usage of materials and life cycle management (Hermann et al. 2016). Under this purpose the main components of Industry 4.0 have been pointed out by Hermann et al. 2016 as following;

1. Cyber-physical systems (CPS), combination of physical and digital systems by the integration of electronic and physical processes.
2. The Internet of things (IoT), where all the parts of the process are connected to each other in the network.
3. Smart Factory, refers to the manufacturing which use the technological advances such as sensors, actuators and adopted the autonomous systems.

The idea lying behind the emergence and implications of these technologies has created great expectations for benefits due its huge potential. Smart factories enable companies to meet the customer requirements in a more profitable way and the systems become more flexible with the changing working environments in Industry 4.0 (Kagermann et al. 2013; Rojko 2017). The connectivity between the smart machines will make them automate the production systems, and also analyse and solve some of the production issues without human intervention (Tjahjono et al. 2017). In addition, monitoring the systems and the detection of failures could be made easier with Industry 4.0. It also offers some solutions to the environmental issues such as effective use of resources and energy (Frank et al. 2019; Kagermann et al. 2013). All these changes in the business systems lead to introduction of new and innovated business models (Gilchrist 2016; Hofmann and Rüsch 2017). All these benefits brought by Industry 4.0 are expected to have major influences not only in manufacturing but also in logistic systems, leading to the observation of revolutionary changes in classical logistic systems which drive through Logistic 4.0 (Strandhagen et al. 2017).

2. Logistics and Industry 4.0

Logistics can basically be defined as management of the movements of semi or finished goods and information between the partners within the business. This movement includes production, storage and distribution phases starting from supply to the end-user or vice-versa for reverse logistics. While considering a logistic strategy and planning, the main aim is to deliver the finished products to the final customers at a reasonable service level while keeping the capital and operational costs at minimum (Ghiani et al. 2004). The increasing competitive business environment makes it difficult to achieve these goals which entails companies undertaking search of better alternatives/improvements for their logistic systems.

A recent survey by Deloitte Insights 2018 which is conducted with 1600 C-level executives in over 19 countries shows that only 6 percent give importance to the implications of technological advances on logistic systems whereas the main focus is manufacturing. This seems that the strategic significance and impact of logistics are underestimated or ignored. However, the logistic systems must be integrated to the manufacturing supported by Industry 4.0 in order to achieve the most effective results. Otherwise, the benefits will be diminished due to the bullwhip effect in the supply chain. Another motivation behind why logistics should be taken into consideration is that the logistic sector has a share of 10–11 percent of GDP in developed countries and this share tends to increase in the future (Arvis et al. 2018).

Having a more responsive, collaborative, efficient, sustainable and traceable logistic system can be obtained by transforming the classical logistic to Logistic 4.0 (Strandhagen et al. 2017). Logistic 4.0 or smart logistics is a new concept that results from the implications of smart technologies which can automatically control all processes and coordinate with each other into the traditional logistics systems. As an economic impact of these features, Rojko (2017) note that by exploiting technological advances to transform into a smart system, logistics costs can be reduced by about 10–30 percent. Tang and Veelenturf (2019) demonstrate the strategic importance of logistics by explaining real-life applications of technological advances and their effects on logistic systems. The flexibility, connectivity and the effective use of resources as the benefits of Industry 4.0 have also been highly effective in logistics and have altered the perception that logistics is a cost centre.

3. Warehouse Management and its Functions

Warehousing plays a vital role in logistics and along with inventory activities it accounts for about 50 percent of the overall logistic costs in EU countries (Ghiani et al. 2004). The increasing trend on e-commerce and great number of fulfilment centres in recent years have been responsible in increasing the importance of warehousing.

Warehousing can be defined as the process of receiving the products/loads coming from an upstream supplier(s), storing for some certain period of time and shipping to the downstream receiver(s). The main purpose is to make the movements of products between the parts of the supply chain so as to satisfy the customers' demand in an effective fashion.

The warehouses can be classified into three types, namely distribution, production and contract warehouses, based upon their operational functions (Berg and Zijm 1999). Distribution warehouses receive products from several suppliers and ship them to customers. These products could be either finished products which are ready for sale to the customers or intermediate products which are customized or assembled at the warehouse. Production warehouses are those in which raw materials or semi-finished products are stored and are located within the manufacturing facilities or in extractions points in close proximity (Richards 2014). This type of warehousing is essential for maintaining continuous production. Contract warehouses operate on behalf of the multiple customers and meet their storage needs.

Warehousing is vital for the efficient operation of the supply chain and to deliver the right products with right quantities at the right time to the receivers. There are some other reasons that enable an effective warehousing essentials for companies (Ten Hompel and Schmidt 2007).

- ✓ Dealing with the uncertain and varying demand
- ✓ Having a seasonal production
- ✓ Getting benefits due to the bulk order discounts
- ✓ Having an economic transportation and shipment
- ✓ Maintaining a continuous production
- ✓ The distance between supplier and consumer
- ✓ Needs for buffer stock for production shutdowns
- ✓ Support against price fluctuations on raw material, spare parts

4. Historical Developments towards Smart Warehouses

The concept of warehousing began almost with the human history and continues to be an important part of the human life. With the changings in social, commercial life, needs and developing technology, warehouses have developed incessantly.

In prehistoric ages, people felt the need to store some of their surplus food due to secure themselves against food shortages that may arise in the future. For this reason, they used storage pits and thus kept the excess food in their hands as a buffer stock in case of an emergency or famine. With the transformation to settled life and the beginning of regular agricultural life, as warehouses in form of granaries began to store some of the excessive food for the uncertainty in harvest and seed for the next sowing period. In the Roman period, the developments of trade led to the use of buildings called *Horreum* to stock the trade products in port cities. Transportation of goods between the port cities and others via railways began in the 1800s. The monopolistic structure of the railways led to control of the transportation and warehousing as an additional service at the time of moving products (Tompkins and Smith 1998).

The industrial revolution and the resulting mass production resulted in further developments and specialisation of warehouses. The warehouses which were once seen solely as a building, soon began to conducting different activities and became an indispensable part of the supply chain. They usually are located close to the transportation hubs such as, ports, canals or railways. The storage of raw material coming from different parts of the world to meet the increasing mass production requirements, receiving, storing, packing, labelling, dispatching of the goods and even in some cases displaying the good for customers as a commercial were among the activities done in warehouses (Tompkins and Smith 1998).

During the World War I, hand trucks were started to be used in the material handling and the stocking was still done by hand. The introduction of forklift trucks and wooden pallet after the World War II has transformed the space use in warehouses and led to an increase in the efficiency and effectiveness. The wider availability and usage of electricity and the invention of internal combustion engines have also contributed to observing great developments in warehouses (Tompkins and Smith 1998).

As it can be seen from the historical flow as is illustrated in Figure 2, warehousing has undergone enormous development to become million-dollar facilities from the pits originally built to store food against famine and drought. It should be

Fig. 2: Storage pit, Horreum and 21st Century warehouse.

admitted that increasing needs and new technologies coming within the concept of Industry 4.0 will also contribute to this development and the warehouses are prone to further changes in the future.

5. Warehouse Functions

Warehousing's role is to act as a bridge between the supplier and end-user. Although there are different types, the flow of the materials in a warehouse usually follows three processes including receiving, storage and shipping to the customers (Ten Hompel and Schmidt 2007; Berg and Zijm 1999). The functions of a typical warehouse is presented in Figure 3.

The first important step is to receive the order. The trucks approach to the unloading docks to deliver the products coming from suppliers or manufacturing. When the products are unloaded, the quantity and quality of order are checked. If the control is positive, the products are labelled with a barcode/tag and moved to the storage areas and placed to their predefined locations.

When a customer's order arrives, the required products are picked up from their storage places and this is named *order picking*. Different products with varying quantities are retrieved from their storage locations and collected based on the customer's demand and prepared for sending to the shipping area. After the final checks, the products are loaded into trucks for distribution.

6. Warehouse Management and Performance Indicators

Warehouse management can simply be stated as planning, controlling and optimising of the complex activities of warehouse and distribution systems (Ten Hompel and Schmidt 2007). The issues to be considered by the warehouse manager involve inventory management, storage location assignment, sequencing problems and capacity planning (Berg and Zijm 1999; Faber et al. 2013). The order time and the order quantity to replenish the stock level are the main decisions to be made for the inventory management. For the effective usage of warehouse capacity and reduction in the travel times between the products for storage and picking activities, determining the storage location of the products is a critical decision. The sequencing problems include the decisions of routings when storing and order picking of the multiple products.

In order to evaluate the effectiveness of a warehouse management and customer satisfaction, the performance and the productivity of the warehouse need to be measured. The main focus should not be only the customer satisfaction, which might lead to high operational costs. Thus, the performance metrics need to be determined while considering all side of the system. These metric may vary depending on the types of warehouse. Chen et al. (2017) summarized the warehouse performance indicators as keeping less inventory, increasing the accuracy, effective picking, increasing the efficiency of time, customer service, rapid shipment, space saving and reducing the costs. The performance measures by the functions of warehousing is presented in tabular form in Table 1.

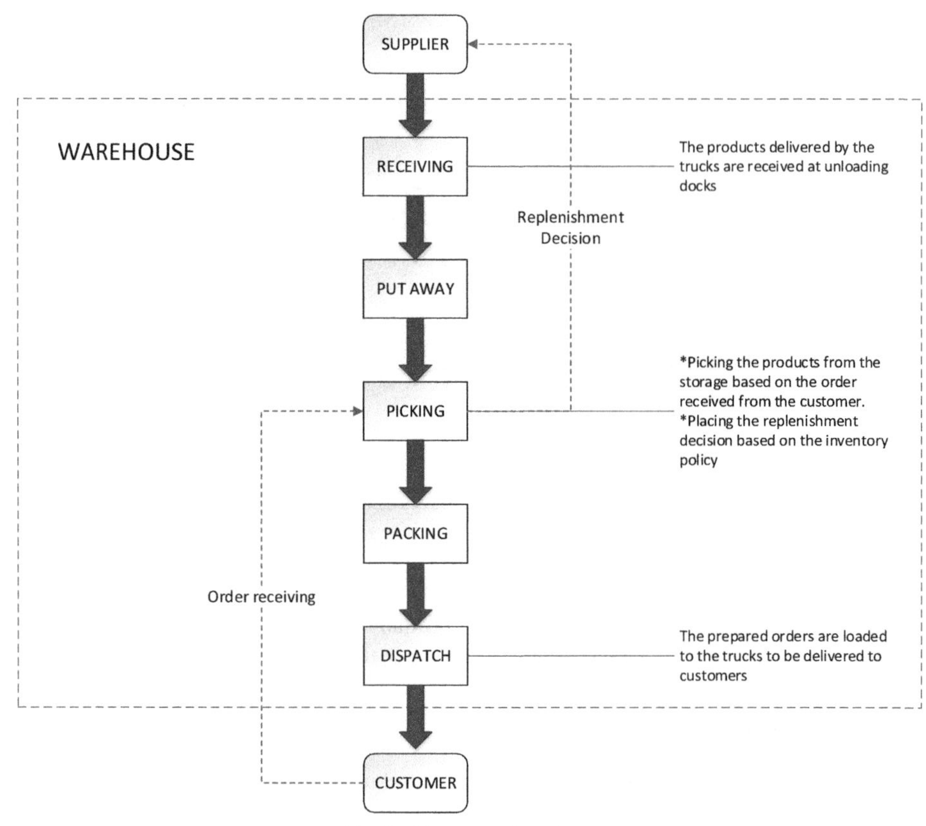

Fig. 3: Warehouse Functions.

Table 1: The performance measure of warehouse operations (Bayraktar et al. 2011).

	Economic	**Time**	**Efficiency**	**Accuracy**
Receiving	The cost of unloading item	Time between dock and the storage	Receiving product per man-hour	The percentage of correct checks
Storage	The cost of storage per item per space	Time to keep on hand inventory	The amount of inventory per space	The percentage of storing at the right place
Order picking	The cost of picking an order list	Time to collect an order list	Picking rate per man-hour	The percentage of right picking
Shipment	Shipping cost per shipment	Time to prepare a shipment	Shipment rate per man-hour	The percentage of right shipments
Overall	The operational cost of an order per each activity	The total time spent for an order per each activity	Overall number of orders per man-hour	The overall percentage of right orders

7. New Technologies used in Warehouses

The increasing importance of logistics and the strategic value of the warehousing in logistics encourage companies to seek improvements in warehousing activities. In addition, volatile conditions in warehousing make it difficult to perform the warehousing operations with the desired performance criteria. These changing conditions include (Frazelle 2002; Renko and Ficko 2010).

- ➢ The increase of the amount and variety of the products in warehouses,
- ➢ The increase of value-added activities in warehouses,

- Shorter life cycles of products and fast delivery times,
- The high volume of product returns,
- Expansion of international orders due to globalisation,
- The increasing need to complete all warehousing activities in less time and with less error due to the high business competition.

As a result of all these requirements, it has become mandatory to utilise the latest technologies in the warehouses to reduce the costs, optimise operations and improve the quality of services. The main technologies used in the warehouse and the references are summarized in Table 2.

Table 2: Literature papers for warehouse technologies.

The technology	References
Warehouse Management System (WMS)	Wang et al. 2010; Atieh et al. 2016; Ten Hompel and Schmidt 2007; Burke and Ewing Jr. 2014; A et al. 2012
Radio Frequency RFID	Wang et al. 2010; Chen et al. 2013; Zhou et al. 2017
AS/RS systems	De Koster et al. 2008; Lerher et al. 2010; Gagliardi et al. 2012
Internet of Things (IoT)	Fleisch 2010; Gilchrist 2016; Lee et al. 2017; Witkowski 2017; Macaulay et al. 2015
Robotics	Burke and Ewing Jr. 2014; Niku 2001; Bonkenburg 2016
Automated Guided Vehicles (AGVs)	Schwab 2017; Schulze and Zhao 2007; Schulze and Wullner 2006
Drones	Schwab 2017; Kückelhaus and Niezgoda 2014

7.1 Warehouse Management Systems

Warehouse Management System (WMS) is a software that performs the management of warehousing and its operations. It guides planning, optimising, and controlling of the daily warehouse operations starting from receiving the products to shipment. Addition to the efficient use of storage areas, another purpose of WMS is to make the most efficient use of other resources such as warehouse elements, handheld terminals and transport vehicles and operators. The benefits of having an effective WMS can be listed as;

- Increasing the efficiency
- Monitoring the inventory and whole operations in real time
- Speeding up the process of order preparing and shipment
- Better management for the use of resources
- Increasing the accuracy of satisfying customers' orders
- Maintaining the integration between the units
- Reducing the operational costs
- Enabling to easily report and online management as paperless
- Keeping record of all activities, help to monitor the performance of the warehouse

The WMS can be integrated with the connected devices and sensors which help increase the responsiveness and flexibility of the system. The automated WMS emerges as another benefit of this integration. Some human errors that may occur due to manual handling can be minimised by the automated WMS (Atieh et al. 2016). In an automated WMS, the transportation devices can communicate the WMS regarding their locations and expected arrival times to another position, to which the WMS in turn, can determine appropriate slot and route for the device and optimise the system in real-time (Barreto et al. 2017). Simultaneously, an RFID system can inform the WMS about the status of products shipped, in transit, on process and request a transportation device to move the goods. WMS can assign the most appropriate transport device to the job and optimise the routing.

Effective management and optimisation of the processes in a warehouse require to have a WMS. A company should be careful when choosing a WMS and determine the system based on the needs and the characteristic of the warehouse.

7.2 RFID Technologies

Radio frequency identification (RFID) is a technology for recognizing tags attached to the objects individually and automatically via radio waves. The two components of RFID are tag and reader. The best feature of RFID technology is that it does not need to read on the line of sight unlike the barcoding systems, which needs to scan the tag to read.

RFID tags can be of three different types such as passive, semi-passive and active tags. Passive tags, the cheapest and the simplest ones, do not have their own energy power and their energy needs are supplied by the reader. Semi-passive tags, on the other hand, have a small battery and do not need to get power from reader. These are more reliable and can respond quickly to the customer as they have a wide range of readability. Active tags have their own power supplies which allow them for running their circuits and generating a response signal. Although their performance is higher compared with other types of tags, they also have higher costs.

The implementation of RFID technology into warehouses and distribution centres bring about several benefits to the system. RFID technology improves the accuracy of inventory and order picking since it enables managers to monitor the products in real-time. Increasing the accuracy of the system will result in reducing errors which allows for keeping up with the customer expectations. As a result of being processes easier and faster, the labour costs will be reduced. As an another benefit, RFID tags are more durable, resulting in less errors such as falling off or damaged tags which are common for barcodes labels.

Despite all the benefits, there are some critical points to be considered for the implementation of RFID. The cost of having an RFID system can be very higher compared to the other barcoding systems. Thus, there should be a trade-off between the benefits and the costs as all the equipment and facilities have to be upgraded in compliance with RFID in order to obtain the desired benefit. Also, RFID systems can keep a very large dataset which requires to be managed and this is a challenging issue.

7.3 AS/RS Systems

An Automated Storage and Retrieval System (AS/RS) is a part of an automated warehouse technologies which is capable of processing the storing and picking of the products activities in an automated way. In spite of its applicability to a wide range of warehouses, the implementation of an AS/RS system in a high-altitude storage areas where shelves reach up to 40 meters is more essential.

Main component of an AS/RS system comprises of a crane or storage/retrieval machine, storage racks, input and output locations, and picking positions. The racks are the stationary places positioned on both sides of an aisle and the goods are stored within these racks. A crane or a storage/retrieval machine,which is capable of moving in the both horizontal and vertical directions, operates in the aisle to put the items into the shelves and retrieve them. The goods are taken from input location by the crane to store in the racks or the retrieved goods from the shelves are dropped to the output location. The input/output location is placed at the end of the aisle. The goods are picked up from the retrieval boxes from the shelves and placed in the picking zone. An example of an AS/RS system is presented in Figure 4.

The integration of an AS/RS system with a warehouse has several potential benefits that are listed below.

- ✓ Capacity usage; AS/RS systems can utilize the space in warehouses and the automated system can reach to the places where manual access cannot be done to greater extent. It can also operate in a narrow aisles on which forklifts cannot work.
- ✓ Reduction on labour cost; one time establishment of an AS/RS system can significantly reduce the labour cost since a single AS/RS can perform all order picking activities.
- ✓ Increase in the accuracy of order picking and storage; in any case, there is an error factor for human who might retrieve wrong products or put in the wrong places. This causes inefficiencies of the functions of warehouse and these can be minimized by an AS/RS system.
- ✓ Safety; occupational accidents caused by instant distraction, stress, hunger or other emotional reasons of the people using forklifts or other heavy machinery might result in a great tragedy. Replacing human factor with an automated system might reduce these unpredictable incidents.

As an example of the benefits of AS/RS, there could be a saving on aisles space up to 50%, reducing the waste space in a warehouse up to 85% and improving the accuracy of material handling up to 99.99% (Lerher et al. 2010; De Koster et al. 2008).

In addition to all the benefits of AS/RS systems, there exist some issues that should not be ignored. The cost of having an AS/RS system including initial setup cost and maintenance cost can be more expensive than traditional system in the

Fig. 4: An example of AS/RS system.

short term. Once the system is established, it is very difficult to make changes. Thus, it should be more suitable for the facilities in which regular and repetitive operations are carried out. Everything comes at a cost and a trade-off between the benefits and these issues should be taken into account when considering the implementation of an AS/RS system.

7.4 Internet of Things (IoT)

Internet of Things (IoT) is a system that can establish the connection between the interrelated computer systems, mechanical and digital machines and human computer interaction. Although propounded in 1999 for the first time, IoT has made great strides with the developments of hardware in recent years.

Understanding the structure of IoT requires the knowledge of what makes it different from the other available technologies. The first difference is the context of IoT which provides an advance level of interaction between the objects and the existing system and the ability to react to the changes in the system. Omnipresence of IoT which distinguishes it from other technologies can be expressed in a fashion that the IoT is more than just an interaction of human-machines and it is everywhere. Optimization as another feature can be expressed as improving the functions of the connected objects. One of the biggest misconceptions is to describe the IoT as the connection of devices via internet. Because, IoT includes the connection and data generation of sensors like RFIDs and identifiers. Thus, when comparing the IoT with the internet, the number of connection is much higher for IoT and Fleisch 2010 use the statement "*trillions versus billions of network nodes*".

The objects equipped with sensors and electronic circuits gain an access to communicate with the people. By doing so, they are capable of updating their status and sharing this information with the people. IoT devices such as smart bracelets, watches, glasses, rackets, home automation systems, smart vehicles, etc., have already been in our lives and benefitted us in many ways. For example, Shyp as a shipping company has announced a new development of their processes called "address-free shipping". Unlike the traditional distribution systems, the packages sent via Shy do not have address labels on it. Instead, there is only a username written on the packages and during the preparation of the shipment, the customer can update the address with a smartphone application. In this way, customers will be able to be more flexible with their address preferences and to process their transactions faster. Postybell is another device as a smart mailbox which alerts the phone of the owner when there is a post or mail inside. Even if you are a long way away, you will be aware of the mail coming to your mailbox (see Figure 5).

These devices, which are indispensable part of the smart logistics system, offer many convenience on optimization, controlling, monitoring and planning stages. Overall, the smart systems based on IoT devices work more efficiently, effectively and sustainable with the help of data and the technology in order to facilitate the work of the people.

Fig. 5: Postybell smart mailbox.

7.5 Robotics

Robots can perform many different tasks and operations accurately and do not require safety and working conditions that people need. The design, application and operation of robots into the human works are evaluated within the scope of robotics. Transforming the business models which welcomes robots and the technology advances have been causing an increase on the robotic technologies which is expected to reach a total sale of 17.5 billion USD within 2019–2021.

There are some components integrated with each other to form a robot. These are manipulator, end effector, actuators, sensors, controller, processor and software. Manipulator as the main body of the robot includes joints, links and the structure. End effector is the part that can hold products, perform operations and connect with other machines. Actuators as the power of robots enable them to move and run by receiving the signal from the controller. Sensors collect information about the external environment and the internal state. Controller enables actuators to act or move based on information it receives from the computer. Processor as the brain calculates the motions of the robot and measures the speed requirements to fulfil the given task. Software is the program designed for the movements and working principles of the robot. It could also be seen as the operating system of a robot.

The fundamental difference between a crane and a robot is based on the controlling unit. In spite of its similarity, a crane is not a robot since it has to be run by a human operator who controls the actuator. On the other hand, the control of a robot has been done by a computer which is running a software. This is what distinguishes a simple crane or manipulators from a robot.

The participation of robots in the warehousing activities provides many benefits. It is a known fact that robots work more efficiently, effectively and accurately and increase the productivity. Based on MIT Technology Review, robots are able to process the material handling tasks four times more efficiently than non-automated systems. As a result of this, there will a reduction on operational costs. Another benefit is that robots do not need the working and environmental conditions that human needs. They can operate in hazardous working environments which require safety regulations for human. Robots do not experience physical problems such as fatigue, hunger, insomnia and emotional problems such as stress, boredom and anger. Because of all these reasons, taking part of the robots in warehousing operations minimizes the work accidents.

Despite the benefits mentioned above, the use of robots brings about some disadvantages. The robots can only do what they are programmed for. They can work very well for standard jobs. However, in case of an emergency, there will be a lack of capability to intervene the problem. Being costly is another disadvantages of robots. They require high initial installation cost which might be higher than operating as manual for the short term.

7.6 Automated Guided Vehicles (AGVs)

Automated guided vehicles (AGVs) are unmanned vehicles that can operate without the need for an operator. They can autonomously move on aluminium or magnetic tapes, laser reflectors along the specified route and have the ability to select the path and position. The AGVs are generally used in production, logistics, warehousing at which transportation of goods occur between receiving and storage, loading and unloading of the goods.

The concepts of automatic and autonomous are often used interchangeably. But, they are quite different from each other. Automatic is meant that machines perform predetermined tasks and are controlled by a computer. On the other hand, automation enables the machines to decide if they encounter new and unexpected situations. This in fact reveals the difference between AGVs and driverless vehicles. While driverless vehicles are capable of manoeuvring against any situation they may encounter on the road, the AGVs can only perform the given tasks on the specified routes.

The advantages of using AGVs in a warehouse can be as;

- Working 7/24
- Reducing the possibility of work accidents
- Decreasing the operational cost
- Increasing the efficiency
- Working in harmony with the human in the same environment by the help of sensor,
- Flexible Changing or rearranging the path of AGVs

7.7 Drones

Drones, known as flying objects, are used increasingly and safely for the inventory related activities in warehousing. During these operations, drones are either controlled by an operator or can also operate autonomously by the help of a navigation software. In the event of a danger or a collision, the sensors assist the drones and in case of any unexpected situation or disconnection, the drones automatically stop working.

Drones can perform a wide range of operations in warehousing and thus several benefits may be achieved. As can be used in inventory counting, drones speed up the process with a higher accuracy. They facilitate access to products on higher shelves and therefore reduce the possible workplace incidents. All the operations which they are involved can be completed more economically. In conclusion, they reduce the human labour costs.

8. The Benefits Gained From Emerging Technologies: Real Cases

In the above sections, we introduced what the warehousing is, what it covers, and the next generation technologies used in the concept of Industry 4.0 in warehousing. Now, we present how these technologies are applied into the warehousing systems of the leading companies in their sectors and what effects they create.

8.1 Amazon

Amazon has been serving as the largest online retailer in the last decades. The most important reason for this is that it has been a front runner in the transition to automated warehousing systems by making continuous improvements by leveraging the developing technologies, resulting in the successful operation of one of the most complex supply chains. The same day delivery and the next day delivery options offered by Amazon takes the competition to a higher level, and is quite effective in terms of attracting customers.

The main objectives of Amazon is to deliver a wide range of products to the customers with a reasonable prices and fast deliveries. To achieve these goals, a large scale of products at the right quantity must be kept in inventory and distributed at low costs. In Amazon products where the profit margin is as low as 1.7%, warehouse and distributions costs stand for 13.4% of the total cost. Thus, a slight improvement on warehousing and distribution system can make a big difference on total profit. For this reason, it is essential for Amazon to maintain its profitability by using automation technologies to facilitate warehousing operations and to complete them at a lower cost.

Amazon has made great strides in the development of automation in its warehouses with the robotic company Kiva Systems which it purchased in 2012. Although initially selling robots to different companies, Kiva System started to develop robots for only Amazon after some time. By the help of robots developed in Kiva, most of the material handling activities such as order picking, packaging, etc., have been automated. These robots (see Figure 6), 40 cm and 145 kg can carry approximately 300 kg of load. Since 2016, 45,000 robots have been serving in Amazon's 20 logistic centres. All these implications result in 20% saving of warehousing costs. Amazon is expected to save $22 million from each automated warehouse. Addition to all benefits on performance, these robots also offer the most environmentally friendly solution to the warehouse operations. They can work with low energy and no need for light, resulting in 30% saving in total energy consumption in a warehouse.

The increasing number of robots and the need for human beings in the system make the human-robot compatibility obligatory. In particular, the fact that robots do not recognize people has made human-robot working difficult and this causes robots to work only in certain areas. To solve this problem, Amazon has provided its employees with smart vests. These vests send signal which are able to prevent possible accidents by warning the robots to move more slowly. This makes it possible for robots to work more comfortably in all areas of the warehouse, resulting in their ability to work more compatibly with people.

Fig. 6: Amazon's Kiva robots.

Fig. 7: Amazon flying warehouse.

Another innovative project of Amazon is the flying warehouses called Amazon Prime Air as presented in Figure 7. A warehouse system is installed in a flying airship and the delivery to the customers who place orders in special events like concerts, football matches, festivals, etc., will be made instantly from the sky with the aid of drones. The flying warehouse whose patent has already been taken by Amazon, operates as fully autonomous warehouse and can reach up to 14 km high and carry tons of cargo. The drone delivery has been tested by Amazon in UK and in the future, the idea of seeing these flying warehouses and getting service from them will not be a dream.

Amazon considers the integration of automated guided vehicles like forklifts, truck, etc., into the logistic system to reduce the distribution costs. With the automated and driverless forklifts to be used in warehouse operations and the driverless trucks in distribution, both fuel consumption and labour costs can be reduced.

8.2 DHL

DHL is one of the largest logistics companies in the world, specializing in international courier, parcel deliveries and express posts. DHL operates in 220 countries around the world and distributes approximately 1.3 billion parcels annually.

The profound effects of Industry 4.0 have been observed on logistics which is an important part of our daily life and trade. It is a common mistake for a company not to follow technological developments or even not to be a pioneer while nowadays many customers want a faster, technology friendly and connected delivery services. That is why, DHL applies many technologies advances in its logistics networks, from the usage of electrical delivery vehicles to augmented reality glasses and robotic assisted warehouse systems. The cooperation with the Chinese giant, Huawei, has contributed to these technological implementations. The sensors developed by Huawei and placed in the mailboxes alert the nearest DHL driver if there is a product in there for picking which causes to reduce the waiting times of drivers by half.

The Internet of Things (IoT) which focuses on the connections of objects is a perfect match with the logistics sector, requiring millions of shipments and tracking, storing the products by multiple vehicles and human to be carried out in harmony. The DHL has already published a report which shows the implementation of IoT in their warehouses and is presented in Figure 8.

In addition to IoT, DHL also examines robotics, unmanned aerial vehicles and other technological advances to create an application map on their system for the future. Based on the robotic report they broadcasted, a shortage of labour in the next two decades is expected and DHL would like to solve this problem by leveraging the robotics technologies into their logistics involving material handling, warehouses and even in last mile delivery. The application areas of robotics in a distribution centre is illustrated in Figure 9.

Companies that keep up with current technological applications and potentials in future can continue their existing position in the logistics field by getting above of their position. As one of the exemplary companies, DHL, not only uses the existing technologies in its warehouses, distribution and all logistics activities but also allocates resources to investigate the further potentials. For this reason, it will be one of the leading companies in which the applications and effects of technological developments coming with the concept of Industry 4.0 can be observed.

8.3 Alibaba

Alibaba is a large Chinese company serving on e-commerce, retail, technology and internet. With a market value of approximately $56 billion, the company achieved a new sales record of $27 billion in just twelve hours on a single day on 11th November. Meeting such a demand and arranging the right products at the right time necessitate to have an effective logistics and warehouse management.

Similar to its biggest competitor Amazon, Alibaba also pays attention to technological developments and the implications in their systems. The smart warehouse in Huiyan, China is operated by 60 robots. These robots (see Figure 10) which can carry up to 500 kg, are assigned with Wi-Fi signals and carry the products from storage to human workers. These workers pack the products and post them. Through their laser sensors, the robots have the ability to quickly scan the surroundings, recognize the objects and avoid collisions by recognizing each other. The robots can rotate 360 degrees and automatically plug themselves to the charger when they are at low battery. They are highly energy efficient and can operate 4–5 hours with a 5 minutes charging. Their usage triples the process volume of the warehouse and the labour has been reduced by 70%. Another smart warehouse powered by internet of things has been opened in Wuxi in China that employs around 700 automated robots.

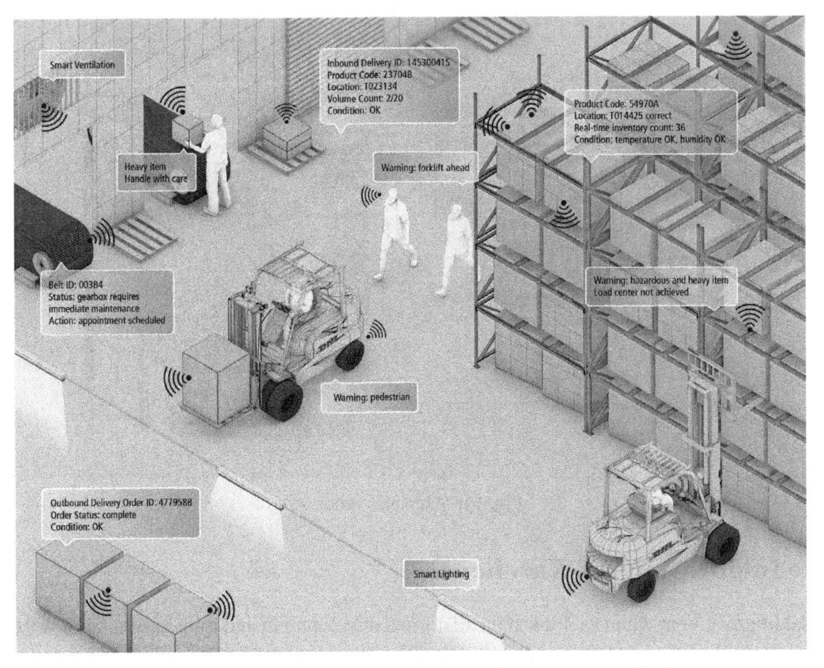

Fig. 8: IoT applications in a warehouse (Macaulay et al. 2015).

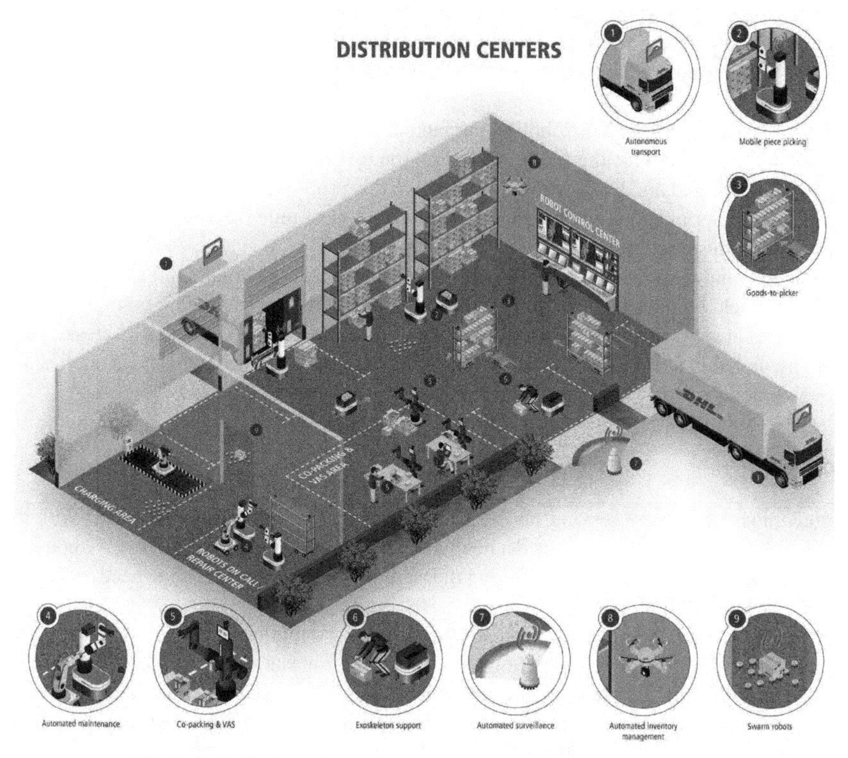

Fig. 9: Robotics in a distribution centre of DHL (Bonkenburg 2016).

Fig. 10: Alibaba's robot.

9. The Future Implications of Warehouses

There is a substantial improvement when we look at the historical development and transformation of warehouses. Initially, warehouses that only store food against famine and whose sole purpose is to maintain food intact have evolved into million dollar facilities which are now the most important part of the logistic systems and can be operated by automation. Seeing these big changes have made it difficult to estimate how the warehouses will be transformed in the future.

One of the biggest debates on the future warehouses is made with respect to their numbers and sizes. It is important to be as close as possible to the end-buyer due to the shortening of delivery times. However, the increasingly wider range of products also increases the size of warehouses. From our point of view, the more important focal point than the number and size of the warehouses will be the technologies for easier access to the customers and more effective demand forecasting methods. The "predictive sales" project which Amazon is currently taking into full consideration, will be another stage of making the order ready for shipment before customer places the order by using the big data and artificial intelligence. This means that the company is capable of knowing the demand even before the customer. As long as the company accurately forecasts the demand, there will be enough time to prepare without needing more and larger warehouses.

In the near future, the use of 3D printers will increase and the accessibility of these systems will be easier. When we think about what can be produced by 3D printing today (camera lens, jet engines, medical models, boats, etc.), it would not be wrong to say that each house with these devices will be a small factory. People can produce some of the products that they need at their home based on their specific requirements. Since the producing is getting easier and transportation is getting faster, the need for warehouses might be reduced in the future.

Another prediction for the future that we may expect is to have an increase on third party logistics (3PL). Since the automation requires high level of investments, the 3PL companies offer warehousing/logistics services to multiple companies with their automated facilities in order to return their investments. The companies which are not capable of making such investment can benefit the latest technologies by these 3PLs.

Research shows that 30% of the energy consumption of a warehouses consists of lighting. The implications of robotic technologies into the warehousing and production, there will be less need for the lighting and heating/cooling systems. Therefore, it will not be a surprise that we will encounter "*dark facilities*" in the future. Some of these facilities, in which many of the human needs are not met, can only be operated by autonomous systems. We believe that there still be a need for human factor but they could monitor, control and plan the processes with their computers at their homes.

Some expect that the usage of robots and other technologies in manufacturing and services will displace the human factor. There will be a further unemployment due to the robotics. Since the first invention of machines followed by computer and others, there has been always a discussion of human displacement which never occurred. We believe that the same will happen for the Industry 4.0 as well. It is true that some jobs will not exist after this period but definitely there will new jobs and people will adapt their work to these new standards and requirements.

Using more high techs in logistics systems and recording such as big data will make the process easier but creates more complex systems at the same time. As the complexity increases, there will be more challenges such as security, reliability and integrity, etc. The interconnection of all devices jeopardizes the security of the entire system under a cyber-attack. For this reason, we think that cyber security problem will be one of the most important problems of the coming days. The protection of the autonomous systems in the warehouses against external interferences will also lead to the emergence of new business lines. All kinds of personal information have been recorded with the smart devices we use and the internet. The issues of how these information will be used and with whom it will be shared emerges as an ethical problem.

10. Conclusions

Warehouses, which have been of great importance throughout the history of humanity, have come to the present day by a continuous development with the increasing needs and improving trade. It has become an obligation for companies to have an effective and efficient logistics and more specifically warehouses in line with the challenging competition conditions and continuously increasing customer needs. Especially the increase in e-commerce and the necessity to respond to thousands or even sometimes millions of orders within minutes have meant that traditional warehouse systems have been ineffective. Thus, there is a need for warehouse systems that can meet all the required needs.

The next generation of technological developments with Industry 4.0 has a great impact on warehouses as well as in all other areas. The successful implications of these technologies into the warehouses transform the traditional ones into Smart Warehouses. The usage of robotics, internet of things, RFID, wireless devices, driverless vehicles and even drones has changed the way how warehouses work completely. Consequently, more economical, productive, accurate and automated warehouses operate with minimum human intervention.

Today's warehouses are quite different from the systems in the past and undoubtedly the future systems will be very different from todays. In particular, we can foresee that the difference between a manufacturing factory and a warehouse disappears since the smart systems allow to produce on demand and make the shipment to the customer effectively. The dominance of China's low cost production is therefore losing its effectiveness. Although local production is more expensive, it is able to cope with cheaper global production thanks to a more efficient and cheaper smart logistics systems. The fact that these advanced systems require less manpower reduces the labour costs, too.

Besides all the benefits, some argue that the excessive use of technology in logistics systems is expected to have some disadvantages. The human displacement, lack of cyber security against connected systems, grappling with big data that all smart devices produce and store, the ethical issues regarding the use and share of personal information are such concerns raised by opponents. It is true that some jobs will disappear and some will transform by the effect of technology. We believe that some of these new jobs will be able to deal with these concerns. Although more robots and automated systems will be in process, we think that there will always be a need for people in these systems. Only the companies and people who can keep up with the advances will survive in this competitive environment.

In summary, although the fundamental functions of warehouses do not change much, the requirements that the warehouses have to meet have been increasing. For this reason, technological developments have started to be widely used in this field,resulting in Smart Warehouses. The future of logistics and warehouses will be based on automation, IoT and Artificial Intelligent (AI) systems. By having such systems, the workload of labour will be more manageable and productive and perhaps there will be no-human workers facilities. The sector is continuously evolving and this means we observe smart and even smarter systems.

References

Ramaa, A., K.N. Subramanya and T.M. Rangaswamy. 2012. Impact of warehouse management system in a supply chain. International Journal of Computer Applications, 54: 14–20.

Arvis, J.-F., L. Ojala, C. Wiederer, B. Shepherd, A. Raj, K. Dairabayeva and T. Kiiski. 2018. Connecting to compete 2018: trade logistics in the global economy. World Bank.

Atieh, A.M., H. Kaylani, Y. Al-Abdallat, A. Qaderi, L. Ghoul, L. Jaradat and I. Hdairis. 2016. Performance improvement of inventory management system processes by an automated warehouse management system. Procedia CIRP, 41: 568–572.

Barreto, L., A. Amaral and T. Pereira. 2017. Industry 4.0 implications in logistics: an overview. Procedia Manufacturing, 13: 1245–1252.

Bayraktar, D., H.B. Bolat, B.M. Fakı and S.G. Çelikkol. 2011. Depo Süreçlerinde Performans Ölçümü Ve Değerlendirmesi İçin Bir Model Önerisi.

Berg, J.P.V.D. and W.H.M. Zijm. 1999. Models for warehouse management: Classification and examples. International Journal of Production Economics, 59: 519–528.

Bonkenburg, T. 2016. Robotics in logistics, A DPDHL perspective on implications and use cases for the logistics industry. 1.

Burke, E.M. and D.L. Ewing, Jr. 2014. Improving warehouse inventory management through rfid, barcoding and robotics technologies. Naval Postgraduate School Monterey Graduate School of Business and Public Policy.

Chen, J.C., C.H. Cheng, P.B. Huang, K.J. Wang, C.J. Huang and T.C. Ting. 2013. Warehouse management with lean and RFID application: a case study. The International Journal of Advanced Manufacturing Technology, 69: 531–542.

Chen, P.-S., C.Y. Huang, C.C. Yu and C.C. Hung. 2017. The examination of key performance indicators of warehouse operation systems based on detailed case studies. Journal of Information Optimization Sciences, 38: 367–389.

Dalenogare, L.S., G.B. Benitez, N.F. Ayala and A.G. Frank. 2018. The expected contribution of Industry 4.0 technologies for industrial performance. International Journal of Production Economics, 204: 383–394.

De Koster, R.B., T. Le-Duc and Y. Yugang. 2008. Optimal storage rack design for a 3-dimensional compact AS/RS. International Journal Of Production Research, 46: 1495–1514.

Deloitte Insights. 2018. The Fourth Industrial Revolution is here—are you ready? UK: Deloitte Insight, 20.

Faber, N., M. De Koster and A. Smidts. 2013. Organizing warehouse management. International Journal of Operations Production Management, 33: 1230–1256.

Fleisch, E. 2010. What is the Internet of things? An Economic Perspective. Electronic text at http://www. autoidlabs. org/ uploads/media/AUTOIDLABS-WP-BIZAPP-53. pdf.

Frank, A.G., L.S. Dalenogare and N.F. Ayala. 2019. Industry 4.0 technologies: Implementation patterns in manufacturing companies. International Journal of Production Economics, 210: 15–26.

Frazelle, E. 2002. World-class Warehousing and Material Handling. McGraw-Hill: New York.

Gagliardi, J.-P., J. Renaud and A. Ruiz. 2012. Models for automated storage and retrieval systems: a literature review. International Journal of Production Research, 50: 7110–7125.

Ghiani, G., G. Laporte and R. Musmanno. 2004. Introduction to logistics systems planning and control, John Wiley & Sons.

Gilchrist, A. 2016. Industry 4.0: the Industrial Internet of Things, Apress.

Hermann, M., T. Pentek and B. Otto. 2016. Design principles for industrie 4.0 scenarios. 2016 49th Hawaii international conference on system sciences (HICSS), IEEE, 3928–3937.

Hofmann, E. and M. Rüsch. 2017. Industry 4.0 and the current status as well as future prospects on logistics. Computers in Industry, 89: 23–34.

Kagermann, H., J. Helbig, A. Hellinger and W. Wahlster. 2013. Recommendations for implementing the strategic initiative INDUSTRIE 4.0: Securing the future of German manufacturing industry; final report of the Industrie 4.0 Working Group, Forschungsunion.

Kückelhaus, M. and D. Niezgoda. 2014. Unmanned aerial vehicle in logistics, A DHL perspective on implications and use cases for the logistics industry.

Lee, C.K.M., Y. Lv, K.K.H. Ng, W. Ho and K.L. Choy. 2017. Design and application of Internet of things based warehouse management system for smart logistics. International Journal of Production Research, 56: 2753–2768.

Lerher, T., M. Sraml, I. Potrc and T. Tollazzi. 2010. Travel time models for double-deep automated storage and retrieval systems. International Journal of Production Research, 48: 3151–3172.

Liao, Y., E.R. Loures, F. Deschamps, G. Brezinski and A. Venâncio. 2018. The Impact of the fourth Industrial Revolution: A cross-country/region Comparison. Production, 28.

Macaulay, J., L. Buckalew and G. Chung. 2015. Internet of things in logistics DHL. DHL Trend Research and Cisco Consulting Services.

Niku, S.B. 2001. Introduction to robotics, Prentice Hall Professional Technical Reference.

Rainer, D. and H. Alexander. 2014. Industrie 4.0: hit or hype? Industrial Electronics Magazine, 8: 56–58.

Renko, S. and D. Ficko. 2010. New logistics technologies in improving customer value in retailing service. Journal of Retailing Consumer Services, 17: 216–223.

Richards, G. 2014. Warehouse Management, Kogan Page.

Rojko, A. 2017. Industry 4.0 Concept: Background and Overview. International Journal of Interactive Mobile Technologies (IJIM), 11: 77.

Schulze, L. and A. Wullner. 2006. The approach of automated guided vehicle systems. IEEE international conference on service operations and logistics, and informatics. IEEE, 522–527.

Schulze, L. and L. Zhao. 2007. Worldwide development and application of automated guided vehicle systems. International Journal of Services Operations Informatics, 2: 164–176.

Schwab, K. 2017. The fourth industrial revolution, Currency.

Strandhagen, J.O., L.R. Vallandingham, G. Fragapane, J.W. Strandhagen, A.B.H. Stangeland and N. Sharma. 2017. Logistics 4.0 and emerging sustainable business models. Advances in Manufacturing, 5: 359–369.

Tang, C.S. and L.P. Veelenturf. 2019. The strategic role of logistics in the industry 4.0 Era. Transportation Research Part E, 129: 1–11.

Ten Hompel, M. and T. Schmidt. 2007. Warehouse management, Springer.

Tjahjono, B., C. Esplugues, E. Ares and G. Pelaez. 2017. What does industry 4.0 mean to supply chain? Procedia Manufacturing, 13: 1175–1182.

Tompkins, J.A. and J.D.Smith. 1998. The warehouse management handbook, Tompkins press.

Wang, H., S. Chen and Y. Xie. 2010. An RFID-based digital warehouse management system in the tobacco industry: a case study. International Journal of Production Research, 48: 2513–2548.

Witkowski, K. 2017. Internet of things, big data, industry 4.0–innovative solutions in logistics and supply chains management. Procedia Engineering, 182: 763–769.

Zhou, W., S. Piramuthu, F. Chu and C. Chu. 2017. RFID-enabled flexible warehousing. Decision Support Systems, 98: 99–112.

SECTION 9

Smart Operations Management

Chapter 15

Comparison of Integrated and Sequential Decisions on Production and Distribution Activities

New Mathematical Models

Ece Yağmur and *Saadettin Erhan Kesen**

1. Introduction

Economic difficulties coupled with shortening product life cycles with smaller quantities necessitate companies to synchronize and coordinate their supply chain operations to continue their sustainability. Thomas and Griffin (1996), indicated that production and distribution phases are two main activities in a supply chain. Therefore, companies are increasingly acknowledging that these activities must be inter-linked for higher customer satisfaction level and lower system cost.

Conventionally, production and distribution activities are managed separately, leading to sub optimal solution as optimality of one activity does not necessarily guarantee the global optimality. Reimann et al. (2014), claimed that extensive coordination among the levels in the supply chain is a must in order to achieve a high performance of the overall system and to satisfy customers' expectations. Although this coordination can be examined at three decision levels (i.e., strategic, tactical, and operational) it is seen from the literature that most of the studies concentrate on the strategic and tactical level. Chen (2010), emphasized the lack of integration at the operational level and reviewed the studies considering the joint decisions of production and distribution plans, whereby showing the application areas of integration. For instance, in order to make business companies implement the JIT policy, production of an item begins only after a customer order is received and inventory levels are, therefore, kept at minimum. Additionally, delivery of some products have to be made just after they are produced due to their time sensitive structure, such as newspapers, catering, ready mixed concrete, nuclear medicine and industrial adhesive materials (Hurter and Van Buer 1996; Van Buer et al. 1999; Devapriya et al. 2006; Russell et al. 2008; Armstrong et al. 2008; Geismar et al. 2008; Chen et al. 2009; Farahani et al. 2012; Lee et al. 2014; Viergutz and Knust 2014).

Konya Technical University, Faculty of Engineering and Natural Sciences, Department of Industrial Engineering, Selçuklu/Konya, Turkey.

* Corresponding author: sekesen@ktun.edu.tr

Chen (2010), divided existing operational level models into several classes based on their delivery method. While early studies focus, to a great extent, on relatively simple delivery methods such as individual and immediate delivery, batch delivery to a single or multiple customers, recent studies have paid more attention to more complex delivery methods involving routing decisions. The studied problem includes both machine scheduling and vehicle routing decisions. We define the integrated production and distribution problem where jobs are first processed in a single manufacturing facility with permutation flow shop environment and subsequently orders are delivered to customers by a single vehicle. The objective is to find a joint production and distribution schedule so as to determine the minimum time needed to produce and deliver orders to all customers subject to vehicle capacity and product lifespan constraints.

From a practical point of view, jobs undergo a series of operations and these operations are conducted in the same order in many production facilities and assembly lines. Only after the operation of a job on preceding machine is completed, can the next operation of the same job start on a succeeding machine. Therefore, each job has to follow the same route (or processing sequence). The machines are then setup in series. This machining configuration is called the flow shop environment. One special case of the flow shop environment emerges when sequence of jobs in front of each machine is the same for all machines, this is referred to as the permutation flow shop environment.

This chapter studies an Integrated Production and Distribution Scheduling (IPDS) problem, which includes determining joint production and distribution schedules at an individual order in a two-stage supply chain. More specifically, a set of given orders, each with a destined customer has to undergo a series of operations on flow shop machines in a single facility and subsequently distributed to related customer by a single vehicle with limited capacity before a predetermined lifespan, after which the order perishes. From the view of operational effectiveness, it would be ideal to distribute all orders in one shipment but this may lead to the violation of the vehicle capacity constraint and order expiry date. Since the number of machines is limited and a single vehicle is available, the number of trips expectedly increases. Therefore, the objective is to determine the minimum returning time of the vehicle to the depot before which all order deliveries are made to customers. The joint schedule involves production starting and completion times on each machine, and delivery time of order to each customer. For the sake of simplicity, we assume that sequence of production and distribution for any batch including various number of orders are the same. We further assume that deterioration (or expiration) starts only after the production of the last order belonging to the associated batch completes. Some orders, therefore, have to wait other jobs in the same batch to start its delivery.

As the purpose of the study is to provide the potential economies of integrating the production and distribution decisions over sequential (or hierarchical) decision, we divide the integrated problem into two subproblems, namely the permutation flow-shop scheduling problem and the vehicle routing problem with release dates. First, orders are sequenced in the same order on all machines in a flow-shop with the objective of makespan minimization and then completion times of the orders on the last machine are given as a release dates for the vehicle for a particular trip. With given release dates, orders are consolidated and customer sequence of visitation is determined so as to minimize there turning time of the vehicle to the depot.

Vehicle capacity and lifespan of the orders are two main constraints for the integrated problem. Although maximum demand is simply assigned for vehicle capacity in order not to violate feasibility, finding a lower bound for lifespan is hard. Thus, we investigate the impact of different lifespan levels on the integrated problem. Three scenarios that include lifespan values (tight, medium and loose) are chosen to see how varying lifespan lead to a change on problem in terms of objective value and computational time.

The contribution of the study as stated in this chapter is as follows. This study, for what we believe to be the first time, compares the integrated and sequential decisions in a two-stage supply chain at an individual order level. In particular, we define our problem in which batch of orders are processed on machines in series and the associated orders in the same batch are distributed to customers by a single vehicle. Since transportation times are not negligible and vehicle capacity is limited, routing decisions are taken into consideration. Since the problem mainly composes of production scheduling and vehicle routing problems, the generation of test instances are very sensitive to processing and travelling times. If processing times are relatively large as compared to travelling times, machine scheduling problem becomes trivial and the problem boils down to the vehicle routing problem or vice versa. During the generation of test instances, we fine tune both parameters. In practice, it is quite common that processing and travelling times are well-balanced. So, we report idle times for both machines and vehicle to show that both sub problems are in equilibrium. We extend our analysis on the integrated model by investigating the impact of product lifespans.

The remainder of the chapter is organized as follows: Subsequent section reviews the relevant studies on the IPDS problem with routing decisions and limited lifespan. Section 3 describes the definitions of integrated and sequential problems with an illustrative example. The proposed Mixed Integer Linear Programming (MILP) formulations for both models are explained in Section 4. Section 5 is fully devoted to investigating the lifespan effect on integrated model. Section 6 presents the generation of random test instances. In Section 7, experimental studies are presented,in which the comparative results of the models are discussed and effects of varying levels of lifespan values on integrated model are given. Concluding remarks and future directions are stated in Section 8.

2. Overview of The Literature

It should be admitted that literature relating to the IPDS problems has been rapidly growing in recent years. In particular, IPDS problems attract more attention of researchers in the years following 2010. The existing literature covering IPDS can, on the one hand, be categorized into two areas based on the type of delivery method. In the first area, researchers study simple delivery methods, such as (i) immediate and individual shipping of orders upon its completion, (ii) batch delivery to a single customer by direct shipping, and (iii) batch delivery to multiple customers by direct shipping. In the second area, however, researchers consider delivery methods in which orders belonging to different customers can be shipped together by the routing method. On the other hand, literature can also be divided into two categories with respect to the durability of orders. While some authors assume that time has no effect on deteriorating the orders, others take the time into consideration and define orders with limited lifespan, after which orders decay or perish. Hence, we limit our literature review to the studies in which routing decisions are involved and orders have limited lifespan as shown in Table 1.

Table 1: A chronological list of studies on IPDS literature considering routing decisions and limited lifespan of orders.

Paper	Order size	Vehicle type	Vehicle number	Machine Config.	Objective	Solution method
Hurter and Van Buer 1996	General	Homogenous	Limited	Single	Cost	Heuristic
Van Buer, Woodruff and Olson 1999	General	Homogeneous	Sufficient	Single	Cost	Exact/Heuristic
Devapriya et al. 2006	General	Homogenous	Sufficient	Single	Cost	Exact/Heuristic
Armstrong, Gao and Lei 2008	General	-	Limited (single)	Single	Service	Exact
Geismar et al. 2008	General	-	Limited(single)	Single	Service	Exact/Heuristic
Russell, Chiang and Zepeda 2008	General	Heterogeneous	Limited	Parallel	Cost/ Service	Heuristic
Chen, Hsueh and Chang 2009	General	Homogenous	Limited	Single	Profit	Exact
Park and Hong 2009	General	Homogeneous	Limited	Single	Cost	Exact/Heuristic
Geismar, Dawande and Sriskandarajah 2011	General	Homogeneous	Limited	Single	Cost	Heuristic
Farahani, Grunow and Günther 2012	General	Homogeneous	Limited	Parallel	Cost/ Quality	Exact/Heuristic
Amorim et al. 2013	General	Homogeneous	Sufficient	Parallel	Cost	Exact
Lee et al. 2014	General	Heterogeneous	Limited	Parallel	Cost	Exact/Heuristic
Viergutz and Knust 2014	General	-	Limited (single)	Single	Service	Exact/Heuristic
Belo-Filho, Amorim and Lobo 2015	General	Homogeneous	Sufficient	Parallel	Cost	Exact/Heuristic
Devapriya, Ferrell and Geismar 2017	General	Homogeneous	Sufficient	Single	Cost	Exact/Heuristic
Karaoglan and Kesen 2017	General	-	Limited (single)	Single	Service	Exact
Kergosien, Gendreau and Billaut 2017	Equal	-	Limited (single)	Parallel	Service	Exact/Heuristic
Marandi and Zegordi 2017	General	Heterogeneous	Limited	Flow shop	Cost/ service	Exact/Heuristic
Lacomme et al. 2018	General	-	Limited (single)	Single	Service	Exact/Heuristic

We classify the relevant studies according to order size, vehicle type, vehicle number, machine configuration, objective function and solution method. It is obvious from Table 1 that while the most of the studies concerns with single and parallel machine production environments, only one article deals with the flow shop production environment.

Printing and distribution of newspapers is one of the main important application areas where integrated decisions should be made. Upon printing, the daily newspapers have to be delivered to news agents by early morning (let's say 04:00 am). Hurter and Van Buer (1996), proposed the joint decision of production and distribution in a newspaper company and made sensitivity analysis for various parameter levels, where by using a greedy algorithm with forward looking strategy to construct routes. In a more recent study, Van Buer, Woodruff and Olson (1999), extended the work of Hurter and Van Buer (1996), by allowing trucks to be re-used in order to reduce the number of trucks required to distribute the newspaper. They compare the performance of the variants of tabu search and simulated annealing algorithms on the problem. As another example, Russell, Chiang and Zepeda (2008) used real world data for the newspaper production and distribution problem which is modeled as an open vehicle routing problem with time windows and zoning constraints. First, they synchronized the loading of vehicles with a heuristic approach, after that they improved the initial routes by using tabu search. Devapriya et al. (2006) formulated two mixed integer programming models to solve the single plant and two-plant variants and provided heuristics based on evolutionary algorithms so as to find good quality solutions in a more reasonable time. Armstrong, Gao and Lei (2008) made a use of routes with a fixed customer sequence with delivery time window requirements. They compared the branch and bound procedure with a heuristic approach which determines a lower bound on the maximum amount of demand satisfied on randomly generated test instances. Geismar et al. (2008) proposed a two-phase heuristic and subsequently developed a lower bound on the problem with the objective of determining the minimum time required to produce and deliver the products to customers. While the former phase used either a genetic or a memetic algorithm to select a locally optimal permutation of the given set of customers; the latter phase used the Gilmore-Gomory algorithm to order the subsequences of customers. For the same problem setting, Karaoğlan and Kesen (2017) developed a branch and cut algorithm to further improve the results of Geismar et al. (2008). Lacomme et al. (2018) extended the same problem by permitting the use of multiple vehicles and demonstrated the effectiveness of the proposed algorithm against the results of Geismar et al. (2008) and Karaoğlan and Kesen (2017) in some problem instances. A greedy randomized adaptive search procedure with an evolutionary local search was proposed to solve the instances. Their results demonstrated that valid inequalities have great impact in increasing lower bounds. In the food industry, Chen, Hsueh and Chang (2009) formulated a mixed-integer nonlinear programming model for fresh food products under stochastic demands and proposed a solution algorithm based on the decomposition concept. The objective of the model was to maximize the expected total profit of the supplier. In a more recent study conducted in food industry, Farahani, Grunow and Günther (2012) proposed a mixed integer linear programming model for real settings of a food catering company. Park and Hong (2009) proposed a hybrid genetic algorithm with several local optimization techniques for single-period inventory products and compared the integrated model with the uncoordinated one. For pool-point distribution of perishable products, Geismar, Dawande and Sriskandarajah (2011) presented two real world examples which include the production and distribution of ready concrete for construction of many venues for the 2004 Summer Olympics and the cross-docking distribution of home movie and game entertainment. For the integrated problem they explored the longest processing time sequence and then developed a genetic algorithm for minimizing makespan and maximum lateness. In contrast to the previous studies with a parallel machine environment, sequence-dependent setup costs and times are taken into account in this study. The study of Amorim et al. (2013) distinguishes from the other studies as it investigated the advantages of the job splitting into sub groups processed on different machines. In a quite different application area, Lee et al. (2014) applied an integrated model to radioactive materials for nuclear medicine so as to minimize the total cost including production costs, fixed vehicle costs and travel costs. They proposed a route reduction procedure in the case that start and end time of each trip is fixed and the fleet is homogeneous. Viergutz and Knust (2014) extended the study of Armstrong (2008) by assuming that the production and distribution sequence are the same with the objective of maximizing the total satisfied demand subject to lifespan and time window restrictions. Belo-Filho, Amorim and Lobo (2015) proposed an adaptive large neighborhood search to tackle large size instances for the problem presented in Amorim et al. (2013) where job splitting decisions are taken into consideration in a parallel machine environment. In the study of Devapriya, Ferrell and Geismar (2016) three heuristics based on genetic and memetic algorithm were developed for the model which determines the fleet size as well as trucks' routes. Kergosien, Gendreau and Billaut (2017) addressed the chemotherapy production and delivery problem where independent jobs are prepared by pharmacy technicians working in parallel and used Benders decomposition-based heuristic to find feasible solutions and lower bounds. Marandi and Zegordi (2017) developed a mixed integer nonlinear model for the problem which composes of permutation flow shop scheduling with due date and vehicle routing with pickup and delivery. An improved particle swarm optimization was improved to deal with the complexity of the problem in this research.

This chapter differs from the literature that integrated and sequential decisions are compared for the first time in the case that machine environment is a permutation flow-shop and a single vehicle is available. We also break away from the existing studies by examining the lifespan effect on the integrated model.

3. Problem Descriptions with An Illustrative Example

This section is committed to explaining both integrated and sequential problems along with an illustrative example, which helps readers to better understand the distinction between the two. We now proceed to describe the integrated problem.

3.1 Description of the Integrated Problem

The integrated problem can be formally stated as follows: We consider a single plant which consists of a set of machines $\{1,\ldots, M\}$in series. A set of orders $\{1,\ldots, N\}$ arrive at the plant where each order has to visit each of the machines exactly once in the same sequence. The processing time of order $i \in \{1,\ldots, N\}$ on machine $m \in \{1,\ldots, M\}$ is denoted by p_{im}. Each order belongs to a unique customer, for which reason we use the terms orders and customers interchangeably as there is a one-to-one correspondence between them. We assume that each order i is non-splittable, ready at time 0 and once its operation on machine m starts, it cannot be interrupted until the completion (i.e., preemption is not allowed). Each customer locating in different region has a demand size of d_i, for which p_{im} units of time is required to complete on machine m. We also assume that orders are produced in batches and each batch composes of several different orders. Once all orders in a particular batch complete their processing, they must be delivered to the associated customers before the lifespan of B by a single vehicle with a capacity of Q. Due to capacity limitation of the vehicle, some batches may have to temporarily wait at the depot for delivery at the next trips. For the sake of simplicity, we further assume that orders constituting a particular production batch are delivered in the same trip. The routing part of the problem is modelled on a graph with $\{0,1,\ldots, N\}$ as the set of nodes in which 0 denotes the depot and remaining nodes represent customers. The travel time between nodes i and j is denoted by t_{ij} and assumed to be constant. Since the vehicle capacity is limited and each order comprising of a particular batch has to delivered to associated customer before the lifespan, the vehicle is used for multiple trips.

The problem is to determine which orders will be assigned to which batches in which sequence on each of the M machines and to find the sequence of customer visit along with delivery times to the associated customers for each trip so as to minimize the returning time of the vehicle, before which all customer deliveries has to be made.

3.2 Description of the Sequential Problem

The sequential problem is comprised of two separate problems: (i) production scheduling and (ii) vehicle routing. Production scheduling is to determine order sequences on each machine setup in series along with their production starting times so as to minimize makespan (C_{max}). The reason lying behind the selection of makespan as objective is to use machines as efficiently as possible. Completion time of order i (denoted by r_i) on the last machine found by solving the production scheduling problem (or interpreted as release time) is given as a parameter for the vehicle routing problem. The vehicle can only start the delivery of order i after r_i. Contrary to the integrated problem, in the sequential problem, production and distribution sequences are not necessarily the same, meaning that orders are not grouped into the batches.

In comparing the integrated and sequential models, lifespan constraint is not considered. This is because of the fact that a particular lifespan value, for which a feasible solution is found for the integrated model may be infeasible for the sequential model. A larger lifespan value, under which a feasible solution exists for the sequential model may not be a binding constraint for the integrated model. In order to compare both models under the same experimental conditions, lifespan constraint is not taken into consideration.

3.3 Illustrative Example for Both Models

In this section, we aim to explain both models on an illustrative example. Table 2 shows the dataset generated for an instance involving 3 machines and 8 customer orders. The first column represents the customer nodes. Node 0 indicates the depot, which is located in the middle of the two-dimensional plane. Second and third columns give the coordinates of the associated customer nodes. While d_i represents the demand size of customer i, p_{i1}, p_{i2}, and p_{i3} are the processing times of order i on machines 1, 2, and 3, respectively. The vehicle capacity is set to 204. The transportation time between each pair of nodes i and j is calculated using Euclidian distances. As mentioned before, the lifespan value is not considered.

For the illustrative example, Figure 1 displays a feasible solution involving sequences of customer visitation of the vehicle for integrated decisions. According to the solution, the vehicle completes its delivery service within three tours as follows: 0-4-1-0 (tour 1), 0-8-3-0 (tour2), and 0-7-2-6-5-0 (tour 3). For a particular tour, the load capacity of the vehicle is not exceeded. Figure 2 represents a Gantt chart for production and distribution stages. In the integrated approach, production and distribution sequences are the same. The production of the batch including orders 4 and 1 is completed at time 65, at which vehicle starts the delivery of the batch (tour 1). The vehicle returns to the depot at time 126 after making the delivery of the orders 4 and 1. The production completion time of the second batch including orders 8 and 3 is 161. Although vehicle is ready after time 126, the delivery of the second batch starts only after completion of second production batch (i.e., 161). The vehicle returns to the depot at time 237 after making the delivery of customer orders 8 and 3. Although the completion of the third production batch including orders 7, 2, 6 and 5 completes at time 232, the delivery of this batch can only start at time 237, which is the returning time of vehicle after making the delivery of second batch. Under this production and distribution sequences, the returning time of vehicle after fulfilling all deliveries is found as 338.

Figure 3 is a schematic representation of sequences of customer visitation for each trip of the vehicle when production and distribution decisions are made separately, under which production of orders are not grouped into the batches, implying that sequences of customer visitation and production of orders are not necessarily the same. According to the feasible solution found using the parameters of the instance in Table 2 for this approach, there exist five tours performed by the vehicle. It can be checked that capacity limit of the vehicle is not violated. Figure 4 shows the Gantt chart of the production and delivery time of orders under the sequential approach. The processing of order 5 is completed at time 30, at which its delivery starts. The vehicle returns to the depot following the delivery of order 5 at time 98. Although the production of order 4 completes at time 68, the vehicle can only start its delivery at time 98. The vehicle completes the second trip and returns to the depot at time 146. The third trip including only delivery of order 8 can start 152, which is the completion time of order 8 and returns to the depot at time 202. The fourth trip involving only order 3 can start at time 217, which is the production completion time of order 3 and the vehicle returns to the depot at time 279. The final trip 5 includes the delivery of orders 7, 2, 6, and 1, among which the latest completion time, which is 224, belongs to order 1. The vehicle starts trip 5 at time 279 and returns to the depot after delivering all customer orders at time 403.

Table 2: Random dataset for the illustrative example.

i	*X coordinate*	*Y coordinate*	d_i	p_{i1}	p_{i2}	p_{i3}
0	0	0				
1	– 14	23	31	17	9	7
2	32	22	35	27	26	21
3	–9	–30	48	38	21	9
4	–4	24	31	13	25	20
5	15	30	22	9	14	7
6	31	31	6	16	20	11
7	31	14	28	45	34	18
8	10	–23	48	18	59	45

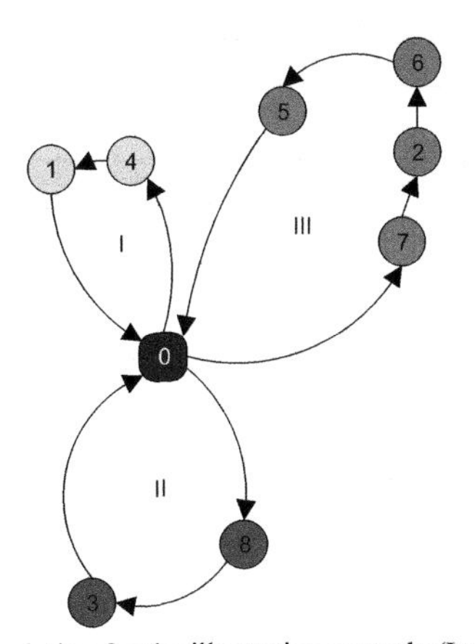

Fig. 1: A feasible solution for the illustrative example (Integrated decisions).

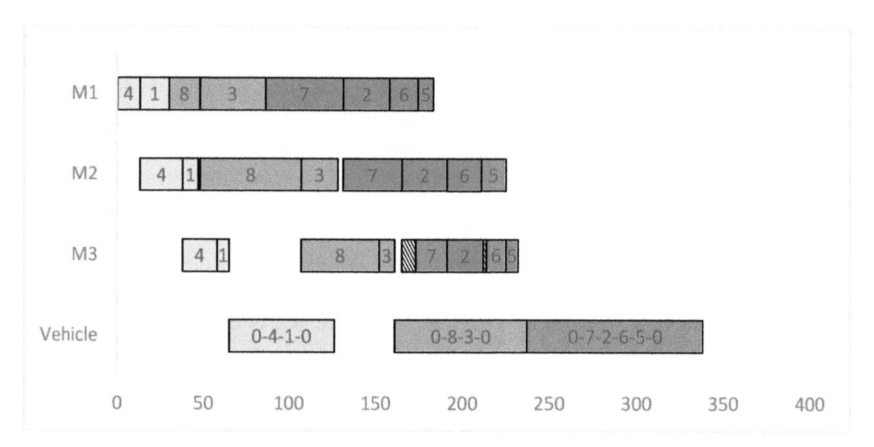

Fig. 2: Gantt chart of the feasible solution for the illustrative example (Integrated decisions).

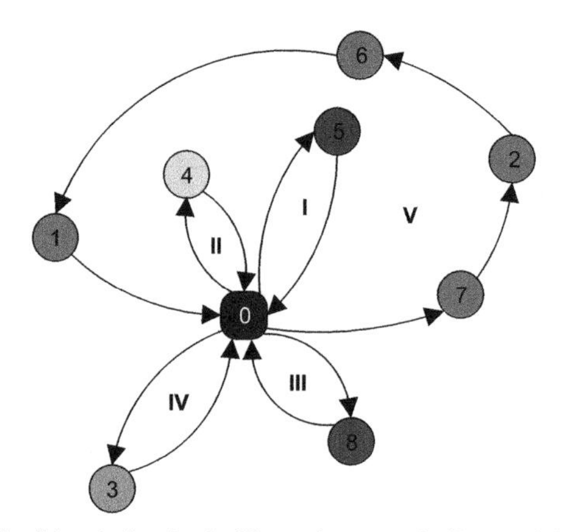

Fig. 3: A feasible solution for the illustrative example (Sequential decisions).

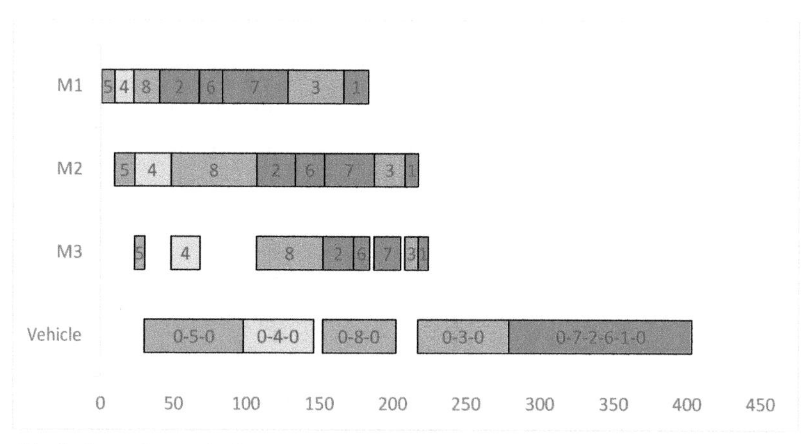

Fig. 4: Gantt chart of the feasible solution for the illustrative example (Sequential decisions).

4. MILP Formulations for Integrated and Sequential Model

In this section, we will present two formulations, one for integrated decisions and one for sequential decisions. We now describe the parameters, decision variables and MILP formulation for integrated model.

4.1 Description of MILP Model for Integrated Decisions

Parameters

d_i : demand of customer $i(i = 1,\dots, N)$

p_{im} : processing time of order i on machine m for all the d_i units ($i = 1,\dots, N$; $m = 1,\dots, M$)

Q : the capacity of vehicle

t_{ij} : travel time between customers and $j(i, j = 0,1,\dots, N;\ i \neq j$; index 0 indicates depot)

B : lifespan value of a any production batch

H : sufficiently large number

Decision variables

X_{ij} = 1 if vehicle goes directly from node i to node $j(i, j = 0,1,\dots,N;\ i \neq j)$, 0 otherwise.

W_{ij} = 1 if vehicle completes preceding tour with node i and starts succeeding tour with node $j(i, j = 0,1,\dots, N;\ i \neq j)$, 0 otherwise.

u_i : the total size of load on vehicle just before visiting node $i(i = 0,1,\dots, N)$

S_{im} : process starting time of order on machine $m(i = 1,\dots, N;\ m = 1,\dots, M)$

C_{im} : process completion time of order i on machine $m(i = 1,\dots, N;\ m = 1,\dots, M)$

Ai : The arrival (or delivery) time of vehicle at customer $i(i = 1,\dots, N)$

Y_i : Production completion time of a batch to which order i belongs ($i = 1,\dots, N$)

D_{max} : The time required for vehicle to return to the depot upon completion of production and delivery of all customer orders.

The formulation for Integrated Model, ***IM*** is given as follows:

$$\textbf{\textit{IM}}\ minimize\ D_{max} \tag{1}$$

subject to

$$\sum_{i=1}^{N} X_{ij} = 1 \quad j = 1,\dots,N \tag{2}$$

$$\sum_{i=0}^{N} X_{ij} = \sum_{j=0}^{N} X_{ji} \quad i = 0,\dots,N \tag{3}$$

$$u_j - u_i + QX_{ij} + (Q - d_i - d_j)X_{ji} \le Q - d_i \quad i, j = 1,\dots,N; i \neq j \tag{4}$$

$$u_i \ge d_i + \sum_{j=1;\, j\neq i}^{N} d_j X_{ij} \quad i = 1,\dots,N \tag{5}$$

$$u_i \le Q - (Q - d_i)X_{i0} \quad i = 1,\dots,N \tag{6}$$

$$\sum_{j=1;\, i\neq j}^{N} W_{ij} \le X_{i0} \quad i = 1,\dots,N \tag{7}$$

$$\sum_{i=1;\, i\neq j}^{N} W_{ij} \le X_{oj} \quad j = 1,\dots,N \tag{8}$$

$$\sum_{j=1}^{N} X_{0j} - \sum_{i=1}^{N}\sum_{j=1}^{N} W_{ij} = 1 \tag{9}$$

$$Y_i \ge C_{iM}\, i = 1,\dots,N \tag{10}$$

$$Y_i \leq C_{iM} + H(1 - X_{i0}) \quad i = 1,\ldots,N \tag{11}$$

$$Y_j - Y_i \leq H(1 - X_{ij} - X_{ji}) \quad i,j = 1,\ldots,N; i \neq j \tag{12}$$

$$A_i - A_j + HX_{ij} + (H - t_{ij} - t_{ji})X_{ji} \leq H - ti_j \quad i,j = 1,\ldots,N; i \neq j \tag{13}$$

$$A_i - A_j + HW_{ij} \leq H - t_{i0} - t_{0j} \quad i,j = 1,\ldots,N; i \neq j \tag{14}$$

$$A_j \geq HW_{ij} - H + Y_j + t_{0j} \quad i,j = 1,\ldots,N; i \neq j \tag{15}$$

$$A_j \geq Y_j + t_{0j} - H\left(1 - X_{0j} + \sum_{i=1}^{N} Wij\right) \quad j = 1,\ldots,N; i \neq j \tag{16}$$

$$A_j \leq Y_j + t_{0j} - H\left(1 - X_{0j} + \sum_{i=1}^{N} Wij\right) \quad j = 1,\ldots,N; i \neq j \tag{17}$$

$$C_{im} - S_{jm} \leq H(1 - X_{ij} - W_{ij}) \quad i,j = 1,\ldots,N; i \neq j; m = 1,\ldots,M \tag{18}$$

$$S_{i(m+1)} \geq C_{im} \quad i = 1,\ldots,N; m = 1,\ldots,M-1 \tag{19}$$

$$C_{im} = S_{im} + P_{im} \quad i = 1,\ldots,N; m = 1,\ldots,M \tag{20}$$

$$D_{max} \geq A_i + t_{i0} \quad i = 1,\ldots,N \tag{21}$$

$$A_i, u_i, S_{im}, C_{im}, Y_i \geq 0 \tag{22}$$

$$X_{ij} \in \{0,1\} \quad i,j = 0,\ldots,N \tag{23}$$

$$W_{ij} \in \{0,1\} \quad i,j = 1,\ldots,N \tag{24}$$

The objective, which is given in Equation (1) is to minimize the returning time of the vehicle to the depot after completing production and distribution of all customer orders. Equation (2) ensures that each node is visited exactly once. Equation (3) indicates that the number of arcs entering and leaving any node must be the same. Equations (4)–(6) are capacity and sub-tour elimination constraints. In particular, Equation (4) states that the total load on vehicle in any tour must not exceed the capacity limit of the vehicle. Equation (5) and Equation (6) indicate the lower and upper limit for auxiliary variables of u_i. Equation (7) and Equation (8) determine the successive tour combinations, using the last customer (let's say customer *i*) of the preceding tour and the first customer (let's say customer *j*) of the succeeding tour. Equation (9) guarantees that difference between the total number of tours and the total number of successive tour combinations is exactly one. Equation (10) and Equation (11) state that production of any particular batch must be completed after operation of the last order in the batch is performed on the last machine *M*. Equation (12) indicates that all orders in a batch must wait until the completion of the last order *i* in a production to get ready for distribution. Equation (13) determines the arrival time of the successive orders in the same batch to the associated customers. Equation (14) along with Equation (15) inter connect the arrival time of the first order *j* in the succeeding batch with the last order *i* in the preceding batch if $W_{ij} = 1$. Equation (14) ensures that the vehicle can only start the delivery of the first customer *i* in the succeeding batch after completing the delivery of last order *j* in the preceding batch and returning to the depot. Equation (15) states that the vehicle can only start the delivery of customer *j* after production of the batch to which order *j* belongs completes. Equation (16) and Equation (17) determines the delivery time of the first customer in the first tour. Equation (18) determines production starting time of the successive orders in the same batch and also determines the production starting time of the first order in the successive batches. Equation (19) guarantees that processing of any order on succeeding machine can only start after its processing completes on the preceding machine. Equation (20) ensures that completion time of any order i on machine m equals to starting time of order i plus its processing time on machine $m(p_{im})$. Equation (21) states that the returning time of the vehicle on the last tour must be greater than or equal to the arrival time of the vehicle to the last customer plus distance between the last customer and the depot. Equation (22)–(24) represent the non-negativity and integrality restrictions on the variables.

4.2 Description of MILP Model for Sequential Decisions

Sequential model composes of describing the formulation for permutation flow shop scheduling, denoted by ***SM*1**, and subsequently presenting the vehicle routing with release dates, denoted by ***SM*2**. In formulating ***SM*1**, we define new decision variables as follows:

Decision variables

X_{ik} : 1 if order i is processed at position k on all machines ($i,k = 1,\ldots, N$), 0 otherwise.
S_{km} : process starting time of the order at position k on machine ($i = 1,\ldots, M$)
C_{km} : process completion time of the order at position k on machine $m(i = 1,\ldots, M)$
C_{max} : completion time of the order at last position on last machine

$$\textbf{\textit{SM}1}\ minimize\ C_{max} \tag{25}$$

subject to

$$\sum_{i=1}^{N} X_{ik} = 1 \quad k = 1,\ldots,N \tag{26}$$

$$\sum_{k=1}^{N} X_{ik} = 1 \quad i = 1,\ldots,N \tag{27}$$

$$S_{(k+1)m} \geq C_{km} \quad k = 1,\ldots,N-1; m = 1,\ldots,M \tag{28}$$

$$S_{k(m+1)} \geq C_{km} \quad k = 1,\ldots,N; m = 1,\ldots,M-1 \tag{29}$$

$$C_{km} = S_{km} + \sum_{i=1}^{N} p_{im} X_{ik} \quad k = 1,\ldots,N; m = 1,\ldots,M \tag{30}$$

$$C_{max} \geq C_{iM} \quad i = 1,\ldots,N \tag{31}$$

$$S_{km}, C_{km} \geq 0 \quad k = 1,\ldots,N; m = 1,\ldots,M \tag{32}$$

$$X_{ik} \in \{0,1\} \quad i,k = 1,\ldots,N \tag{33}$$

The objective defined in Equation (25) is to minimize the makespan value, for which reason we aim to utilize the machine as effectively as possible. Equation (26) and Equation (27) are assignment constraints. Equation (26) ensures that only one order can be assigned to a particular position. Equation (27) guarantees that only one position can be assigned for a particular order. Equation (28) states that processing of the order at position $k + 1$ on any machine can only be started upon the completion of the order at position k on the same machine. Equation (29) enforces that processing of an order at position k on machine $m + 1$ can only be started upon the completion of the same order on machine m. Equation (30) describes that if order i is assigned to position k, (i.e., $X_{ik} = 1$) completion time of order i on machine m is equal to starting time of the order plus processing time on machine m. Equation (31) indicates that C_{max} is bigger than or equal to the completion time of the order at last position on last machine. Equation (32) and Equation(33) are the non-negativity and integrality restrictions on the variables.

In formulating ***SM2***, the completion time of the order at each position k on the last machine M (C_{kM}) is used as release time (r_i) of the associated order i, after which delivery of the order can only be made.

New constraints to model ***SM2*** are given as follows:

$$Y_i \geq r_i \quad i = 1,\ldots,N \tag{34}$$

$$Y_i \leq r_i + H(1 - X_{i0}) \quad i = 1,\ldots,N \tag{35}$$

$$A_i, u_i, Y_i \geq 0 \quad i = 1,\ldots,N \tag{36}$$

Constraints (34) and (35) ensure that the vehicle can only start the deliveries of orders in a particular tour after all orders composing of the tour is ready. Equation (36) is the non-negativity restrictions on the variables. Using the new parameters and constraints, the ***SM2*** formulation can be given as follows:

SM2 *minimize* (1) *subject to* (2)–(9), (12)–(17), (22), (24), (25), (34)–(36).

5. The Impact of Lifespan Restriction on Integrated Model

As we discussed in Section 1, there are many cases where joint decisions of production and distribution activities have to be made involving products with limited lifetime, make-to-order production, assembly-to-order, limited storage area, etc. In this section, we assume that products have a limited lifetime of B, before which they have to be delivered to the destined customer. We further assume that only after the production of orders comprising of a production batch completes, does lifetime start. In order to see the effect of lifetime on integrated model, we add the following constraint:

$$A_i - C_{iM} \leq B + H(1 - X_{i0}) \quad i = 1,\ldots,N \tag{37}$$

The constraint (37) ensures that any order i must be delivered to the destined customer before the lifespan. After defining the constraint (37), integrated model with limited lifespan, denoted as ***IMLL***, can be formulated as follows:

IMLL*minimize* (1) *subject to* (2)–(24), (37).

We study the impact of limited lifespan on integrated model, for which purpose we select three different lifespan values, namely $B = 50$, $B = 60$, and $B = 75$ in addition to the dataset given in Table 2. We solve the illustrative example and report the optimal solution in Table 3, where the first column indicates arbitrary lifespan value chosen, under which second column presents the number of tours with customers visited in third column. Fourth and fifth columns give the starting and ending time of the associated tour, respectively. The last column shows the objective value for a particular lifespan value. It is clear from Table 3 that an increase in the lifespan value results in a decrease in the objective value. Another finding from the table is that change in lifespan value differentiates the number of tours and sequence of customer visitation in each tour. When lifespan value is set to 75, the optimal objective value is found to be the same (i.e., 338) with the integrated model disregarding lifespan, meaning that lifespan constraint is not binding under this value. For more detailed solutions on the illustrative example including Gantt charts, the readers are referred to Appendix A.

Table 3: Optimal solutions of integrated model under different lifespan values.

Lifespan value	# of tour	Customers visited	Starting time of tour	Completion time of tour	Objective value
50	1	0–5-6–0	56	150	368
	2	0–4–1–0	150	211	
	3	0–8–3–0	211	287	
	4	0–7–2–0	287	368	
60	1	0–5–0	30	98	341
	2	0–4–1–0	98	159	
	3	0–8–3–0	170	246	
	4	0–7–2–6–0	246	341	
75	1	0–4-1–0	65	126	338
	2	0–8–3–0	161	237	
	3	0–7–2–6–5–0	237	338	

6. Generation of Test Instances

In this section, we will describe how we generate the test instances in order to evaluate the relative performances of the formulations developed and to examine the effect of lifespan restrictions on the integrated model. Whilst generating our test instances, we use a similar way described by the study of Ullrich (2013).

Test instance generation process in the study starts with a value selection for the parameter ρ, which will be a base in generating all other parameters. We set $\rho = 100$. Specifically, for each order i, p_i value is drawn from a discrete uniform distribution UNIF[$\rho/5$, ρ], which then be used to generate processing time of each order i on machine m. Processing time p_{im} of order i on machine m is drawn from a discrete uniform distribution UNIF[$\rho_i/5$, ρ_i]. The reason why we use $\rho/5$ as a lower bound is to maintain consistency of processing times among machines. If quite lower ρ_i value is chosen (let's say near to 0), then the resulting processing times on all machines will quite be lower, as it is used as an upper bound in generation of processing times. As pointed out by Ullrich (2013), the order with longer processing times on machines are more likely to require larger space in the vehicle, for which reason we associate demand size of orders with their processing times. Demand size d_i of order i is assumed to be uniformly distributed and represented by $UNIF(\rho_i/5, \Sigma_{i=1}^{M} p_{im})$. As regards to vehicle capacity, we reasonably assume that it cannot be lower than maximum demand size and is not allowed unnecessarily to exceed the total demand sizes of all customers. Therefore, the vehicle capacity Q is selected from a uniform distribution $UNIF(max\{d_i\}, \Sigma_{i=1}^{N} d_i)$.

In generation of the travel time matrix, we first randomly select x and y coordinates of each customer on a plane with two-dimensions and then calculate the Euclidian distances between each pair of customers. The distances are assumed to be symmetric (i.e., $t_{ij} = t_{ji}$; $\forall i, j = 0, \ldots, N$). We generate the customer locations in a fashion that the maximum travelling time between each pair of customers does not exceed $\lfloor \xi \rfloor$, where $\lfloor * \rfloor$ denotes the largest integer value less than or equal to $*$. In order effectively to manage both production and distribution operations, balancing between processing and travelling times are obligatory. If processing times are relatively larger than travelling times, machine scheduling problems turns out to be trivial and the integrated problem boils down to the vehicle routing problem. In practice, however, processing and travelling times are well-fitted. It is clear that an increase in the machine and order number results in an increase in the completion time of orders. As we associate processing times with parameter ρ, we use a coefficient $(M + N - 1)$ as multiplier, where M and N denote the number of machines and orders, respectively. As regards to distribution, if all customer orders are delivered in a single tour, the minimum number of arcs that the vehicle traverses is equal to $N + 1$. If each customer order is delivered in a different tour, the vehicle traverses $2N$ arcs. The average number of arcs vehicle traverses is, therefore, calculated as $(3N + 1)/2$. Based on our experimental study, the upper limit used in generating travelling times are suggested as $\xi = \rho(N + M - 1)/([3N + 1]/2)$.

In determining lifespan value B, we follow a simple procedure that it is proportional to upper limit ξ, such as 0.5, 0.6, 0.75.

In generation of test instances, the number of orders is chosen between 3 and 12 with an increment of 1 and number of machines is chosen 1 and 8 with an increment of 1. For each level of order and machine number, five test instances are generated, producing $10 \times 8 \times 5 = 400$ test instances. When observing the effect of lifespan value on integrated model, three levels of lifespan is selected, resulting in $10 \times 8 \times 5 \times 3 = 1200$ test instances.

7. Experimental Results

Our experimental results are based on two different subsections: (i) comparison of integrated and sequential model, and (ii) Effects of varying lifespan levels on integrated model.

7.1 Comparison of Integrated and Sequential Model

This section reports the integrated and sequential model solutions in terms of objective value, CPU times (in sec.) and idle times for orders and vehicle by disregarding the lifespan limitation. We categorize our test instances into two groups as small sized instances and big sized instances with regards to the number of customers. While the small sized instances include customer number varying between 3 and 8, the big sized instances cover customers between 9 and 12. It is found that both *IM* and *SM* are able to find optimal solutions within 2 hours (7200 sec.) limit in 240 test instances, including small sized instances. For big sized instances involving 160 test instances, *IM* is only capable of finding optimal solutions in two instances but *SM* is able to find optimal solution in 50 instances. Although *SM* seems to perform better than *IM* for the number of optimally found solutions, the objective function values found by *IM* are better than *SM* finds. In 21 instances out of 240 ones, *IM* and *SM* find solutions to optimality and these optimal values for both models are found to be the same. While *IM* finds better solutions than that of *SM* in 216 instances for small sized instances, *SM* can only outperform the *IM* in remaining 3 instances. To summarize, *IM* performs better than *SM* in 90 percent of the small sized instances in terms of objective value. As for big sized instances, *IM* outperforms *SM* in all instances. Overall, *IM* is capable of producing better objective value than *SM* in 94 percent of the total test instances. For further details, the readers are referred to Appendix B.

Table 4 is obtained by the individual solutions presented in Appendix B for all combinations of machines and customers. As can be seen in Table 4, an increase in machine and customer number leads to an increase in the objective value. *IM* provides with lower average objective values than *SM* for all machine and customer combinations. If a decision maker prefers using *IM* instead of *SM*, the improvement of $100 \times (586.03 - 534.75)/586.03 = 8.7\%$ in the objective value can be achieved.

Table 5 reports the comparative solutions of *IM* and *SM* in terms of CPU times and is generated using the individual solutions presented in Appendix B. In Table 5, if the results are examined in terms of CPU time, it is seen that an increase in the number of customer leads to a dramatic increase in solution times for both models. It is also seen from Table 5 that *SM* can find optimal solutions much more quickly than does *IM*. This finding is not interesting in that *IM* composes of two NP-Hard problems.

Table 6, which is obtained using Appendix B reports the idle times of both orders and the vehicle. In Table 6, Batch Idle Time (BIT) denotes the total amount of time that batches wait until the vehicle is ready for delivery. Vehicle Idle Time (VIT) represents the total amount of time that the vehicle waits until the batch is ready for being delivered. The reason why we use this metric is to evaluate how balanced the processing and travelling times are generated. From Table 6, it is seen that average idle time produced by *IM* is lower than *SM* produces, excluding when the customer number is three, resulting that *SM* finds better objective values as compared to *IM*. Another remarkable result seen in Table 6 that averaged BIT and VIT values for a particular number of customers in both *IM* and *SM* are closer to each other, meaning that randomly generated processing and travelling time are well-fitted.

7.2 Effects of Varying Lifespan Levels on the Integrated Model

The second part of the experimental study is designed to examine the impact of different lifespan values on IM, for which three different lifespan levels are chosen as a percentage of maximum travelling time (ξ) described in Section 6. Table 7 demonstrates the averaged results in terms of objective value and CPU time found by IM for a particular number of customers. From Table 7, we can observe that increased (or loosed) lifespan value reduces the objective value but at the same time increases the solution time as it enlarges the solution space. The readers are referred to Appendix B for further details.

8. Conclusions

This chapter addressed an operational level problem in a two-stage supply chain, in which production and distribution activities are involved. Orders first undergo a series of operations on machines setup in series (permutation flow shop environment) and subsequently are delivered to the customers located in geographical regions by a single vehicle with limited capacity, which necessitates order consolidation (vehicle routing). The objective was to determine the minimum time required for vehicle to return to the depot after production and deliveries of all orders are made. We investigated the potential economies of joint decisions over sequential decisions, the former of which determines the production sequences of orders on machines and sequences of customer visitation in each tour of the vehicle simultaneously, the latter of which allows for routing decisions to be made based upon completion times of orders on machines. We formulated mixed integer programming formulations to model both cases, for which a wide range of test instances were randomly generated. We elaborated our study by considering effect of orders' lifetime restrictions on the integrated model, which is commonly encountered in practice when production and distribution decisions are jointly made. Computational results showed that the integrated model provided better solutions in terms of objective function value when compared with the sequential model. It, however, required more computational time than did sequential model. Results also indicated that more restrictions on the lifetime worsened the objective function value. In future research, we suggest the bespoke exact and heuristic solution algorithms capable of solving larger scale instances than the ones tackled in this study due to intractability matters of the problem.

Table 4: Comparative results of integrated and sequential models with respect to objective value.

M	1		2		3		4		5		6		7		8		(Avg)	(Avg)
N	*IM*	*SM*	*IM*	*SM*	*IM*	*SM*	*IM*	*SM*	*IM*	*SM*	*IM*	*SM*	*IM*	*SM*	*IM*	*SM*	*IM*	*SM*
3	151.6	171	224.2	239.6	259.4	278.41	371.41	385.21	346.61	388.61	446.81	489.01	611.01	679.79	682.01	746.61	386.63	422.28
4	152.6	174.4	255.81	273.81	355.61	393.61	395.21	470.21	409.01	494.41	566.62	646.42	682.62	738.82	728.02	797.02	443.18	498.58
5	202.8	222.21	273.61	302.21	325.41	361.81	447.21	495.42	483.42	529.02	556.22	618.82	640.62	709.42	695.02	765.63	453.03	500.56
6	263.81	300.41	327.81	365.81	369.21	414.22	458.82	489.02	477.22	541.22	627.82	714.03	676.23	725.63	675.63	747.83	484.56	537.27
7	265.01	305.81	325.61	355.82	371.82	406.62	503.42	552.03	526.03	599.63	661.83	702.83	626.83	657.63	778.64	823.64	507.39	550.50
8	308.81	362.62	387.02	430.82	437.02	474.22	517.43	554.83	632.03	693.24	664.44	697.44	740.24	806.84	776.24	842.05	557.90	607.75
9	352.82	397.22	420.42	459.22	478.83	531.63	564.03	592.43	569.43	631.04	618.24	663.24	737.85	821.45	788.84	906.06	566.30	625.28
10	357.42	395.82	460.22	500.83	536.24	569.04	592.44	621.24	636.64	702.04	700.05	761.65	767.65	861.86	924.25	1000.47	621.86	676.61
11	411.62	459.83	490.03	541.23	601.64	664.24	616.44	665.45	636.85	678.65	762.25	838.86	742.05	823.26	929.07	991.27	648.74	707.84
12	457.03	508.04	490.03	533.43	572.24	629.04	661.65	726.46	749.26	786.66	739.85	807.26	825.47	873.47	928.07	1004.68	677.95	733.63
Avg	292.35	329.73	365.47	400.27	430.74	472.28	512.80	555.23	546.65	604.45	634.41	693.95	705.05	769.81	790.57	862.52	534.75	586.03

Table 5: Comparative results of integrated and sequential models with respect to CPU time (sec.).

M	1		2		3		4		5		6		7		8		(Avg)	(Avg)
N	*IM*	*SM*	*IM*	*SM*	*IM*	*SM*	*IM*	*SM*	*IM*	*SM*	*IM*	*SM*	*IM*	*SM*	*IM*	*SM*	*IM*	*SM*
3	0.21	0.16	0.20	0.06	0.22	0.05	0.19	0.06	0.15	0.05	0.15	0.05	0.24	0.05	0.18	0.05	0.19	0.07
4	0.28	0.06	0.18	0.06	0.32	0.06	0.26	0.06	0.32	0.07	0.26	0.08	0.19	0.07	0.31	0.07	0.27	0.07
5	0.88	0.19	0.84	0.21	0.98	0.25	0.82	0.34	1.11	0.32	0.96	0.29	1.07	0.29	0.97	0.23	0.95	0.27
6	5.82	0.97	5.61	1.60	7.13	1.73	6.91	1.22	7.92	1.82	7.15	1.37	7.26	1.15	8.83	1.74	7.08	1.45
7	87.28	6.83	84.7	7.34	79.14	7.71	65.83	5.53	60.58	8.21	117.99	9.53	101.53	6.71	108.09	8.44	88.14	7.54
8	2294.96	112.12	2145.55	94.24	1781.17	75.10	1058.16	52.44	2730.41	129.63	1279.69	47.33	3137.34	140.74	1819.34	90.94	2030.83	92.82
9	6701.22	883.36	6701.22	1187.44	7200.00	1511.52	7200.00	1303.26	7200.00	1401.27	7136.75	898.43	7200.00	1063.79	7200.00	2090.03	7067.40	1292.39
10	7200.00	6886.74	7200.00	6886.93	7200.00	6747.69	7200.00	4873.39	7200.00	6050.73	7200.00	6991.33	7200.00	6552.23	7200.00	7200.00	7200.00	6523.63
11	7200.00	7200.00	7200.00	7200.00	7200.00	6406.23	7200.00	7200.00	7200.00	7200.00	7200.00	7200.00	7200.00	7200.00	7200.00	7200.00	7200.00	7100.78
12	7200.00	7200.00	7200.00	7200.00	7200.00	7200.00	7200.00	7200.00	7200.00	7200.00	7200.00	7200.00	7200.00	7200.00	7200.00	7200.00	7200.00	7200.00
Avg	3069.06	2229.04	3053.83	2257.78	3066.89	2195.03	2993.21	2063.63	3160.04	2199.21	3014.29	2234.84	3204.76	2216.50	3073.77	2379.15	3079.49	2221.90

Table 6: Comparative results of integrated and sequential models with respect to idle times.

M		1		2		3		4		5		6		7		8		(Avg)	(Avg)
N	Idle Time	*IM*	*SM*	*IM*	*SM*	*IM*	*SM*	*IM*	*SM*	*IM*	*SM*	*IM*	*SM*	*IM*	*SM*	*IM*	*SM*	*IM*	*SM*
3	BIT	7.6	15.6	6.8	31.4	17.4	25.4	25.2	43.2	46.2	68.8	71	67.8	21	55.2	77.6	143	34.1	56.3
	VIT	30.2	19.2	5	5.6	32.8	48.4	50.4	26	16.4	25	55.4	16	94.6	115.4	111	0	49.47	31.95
4	BIT	19	35.2	7.2	13.6	37.4	121	12	41.8	47.2	39.6	29	96.6	29	99.2	76	87.2	32.1	66.77
	VIT	3	8.6	30.4	32.8	12.8	32.6	53	70	28	40	87.2	73.8	89	46.6	76	118.4	47.42	52.85
5	BIT	13	16.6	9.2	31.4	19.4	42.2	30.6	47.2	28.2	55.8	48	51.4	42.6	73.2	89.8	189.8	35.1	63.45
	VIT	9.8	20.2	30.6	47.6	41	22.2	52.6	46.4	33.4	51.2	56.2	52.6	73	96.2	43.2	75	42.47	51.42
6	BIT	22.8	40.8	43	37.8	18.8	25.2	17	31	63.8	82.2	38.8	144.4	107.6	133.6	58.8	50.8	46.32	68.22
	VIT	26.8	17.4	28.2	54.2	26	55.4	73	90.6	58.6	81.4	64	93.6	80.4	101.4	92.6	66.2	56.2	70.02
7	BIT	20.8	56.4	35.4	34	43	57.2	68.2	44.4	40.8	65.2	55.6	57	14.8	50.4	100.4	105.2	47.37	58.72
	VIT	19.4	25.4	24.2	37	45.4	44.6	29.2	63	51.8	77.8	83.4	150	146	84	142.2	102.6	67.7	73.05
8	BIT	21.8	26.2	28	42.4	40.2	41.2	49.8	107.8	56	73	92.4	169.8	92	61.6	57.8	99.2	54.75	77.65
	VIT	22.6	39.2	41	51	35.6	51.2	14.8	22.4	76.4	109.6	66.8	52.8	129	167	138	96.4	65.52	73.7

Table 7: The effect of three different lifespan values on integrated model.

B	**1**		**2**		**3**		**(Avg)**	**(Avg)**
N	Obj. Val.	CPU (sec.)	Obj. Val.	CPU (sec.)	Obj. Val.	CPU (sec.)	Obj.Val.	CPU (sec.)
3	387.41	0.06	386.76	0.06	386.71	0.05	386.96	0.06
4	445.04	0.09	444.31	0.09	443.19	0.10	444.18	0.09
5	456.89	0.58	454.84	0.58	453.14	0.58	454.96	0.58
6	492.22	4.19	487.24	4.06	484.72	4.17	488.06	4.14
7	514.12	53.44	509.97	56.95	507.55	60.35	510.55	56.91
8	564.08	1244.87	560.53	1330.11	558.83	1423.20	561.15	1332.73
9	570.78	7001.78	569.23	7039.91	566.68	7027.34	568.90	7023.01
10	626.91	7200	623.41	7200.02	622.64	7200.02	624.32	7200.14
11	649.52	7209.52	647.42	7204.25	645.27	7201.49	647.40	7205.09
12	682.12	7200	680.12	7200	677.42	7200	679.89	7200.07
Avg	538.91	2991.49	536.38	3003.62	534.61	3011.73	536.64	3002.28

Appendix A. Solutions on the illustrative example including Gantt charts

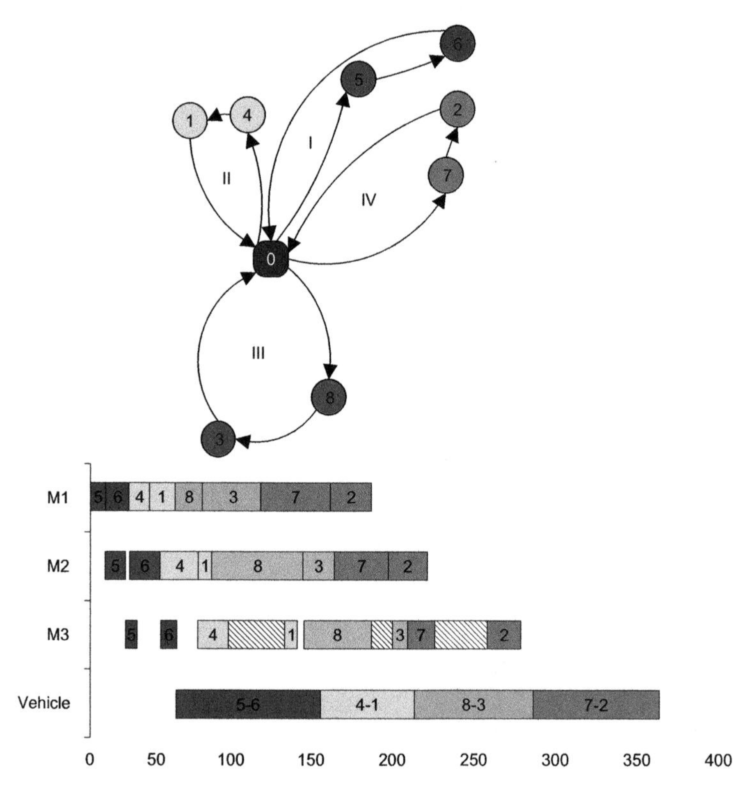

Fig. A1: Optimal solution of illustrative example (Integrated model; $B = 50$).

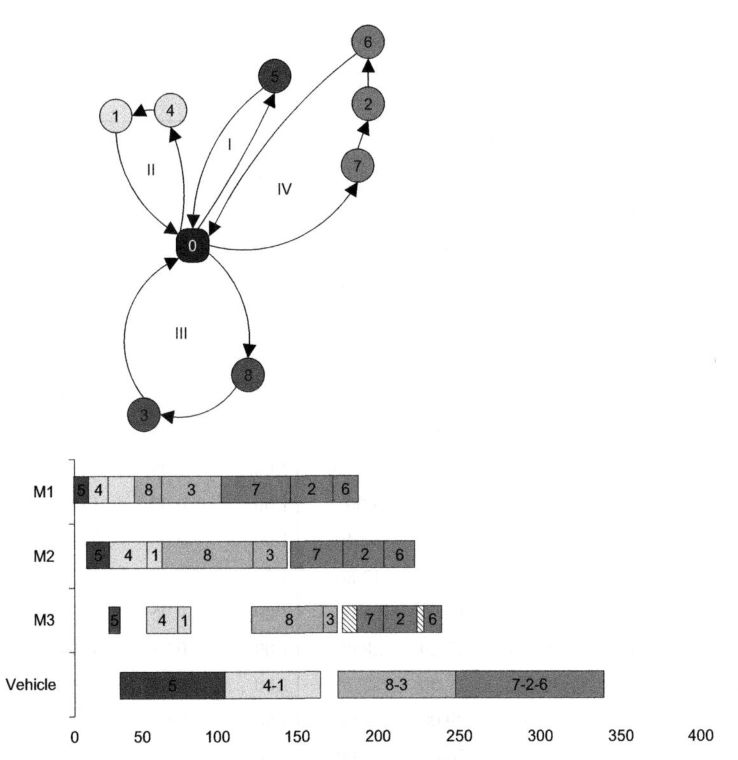

Fig. A2: Optimal solution of illustrative example (Integrated model; $B = 60$).

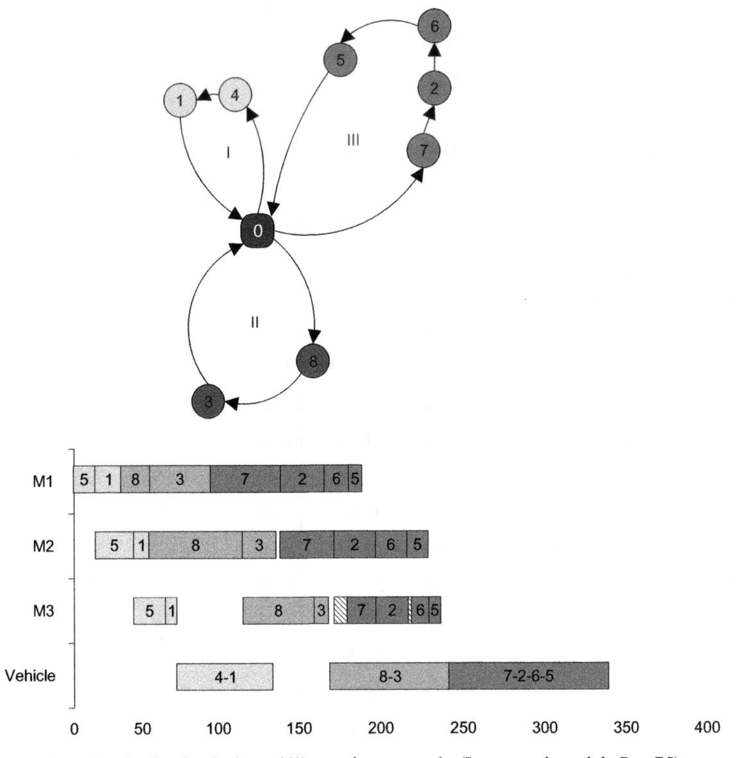

Fig. A3: Optimal solution of illustrative example (Integrated model; $B = 75$).

Appendix B. Experimental results in individual instances bases.

Table B1: Comparative results of integrated and sequential model.

Instance		Integrated Model					Sequential Model				
N	*M*	*Opt. Ratio*	*CPU*	*Obj. Val.*	*BIT*	*VIT*	*Opt. Ratio*	*CPU*	*Obj. Val.*	*BIT*	*VIT*
3	1	1.00	0.21	151.60	7.60	30.20	1.00	0.16	171.00	15.60	19.20
3	2	1.00	0.20	224.20	6.80	5.00	1.00	0.06	239.60	31.40	5.60
3	3	1.00	0.22	259.40	17.40	32.80	1.00	0.05	278.41	25.40	48.40
3	4	1.00	0.19	371.41	25.20	50.40	1.00	0.06	385.21	43.20	26.00
3	5	1.00	0.15	346.61	46.20	16.40	1.00	0.05	388.61	68.80	25.00
3	6	1.00	0.15	446.81	71.00	55.40	1.00	0.05	489.01	67.80	16.00
3	7	1.00	0.24	611.01	21.00	94.60	1.00	0.05	679.79	55.20	115.40
3	8	1.00	0.18	682.01	77.60	111.00	1.00	0.05	746.61	143.00	0.00
4	1	1.00	0.28	152.60	19.00	3.00	1.00	0.06	174.40	35.20	8.60
4	2	1.00	0.18	255.81	7.20	30.40	1.00	0.06	273.81	13.60	32.80
4	3	1.00	0.32	355.61	37.40	12.80	1.00	0.06	393.61	121.00	32.60
4	4	1.00	0.26	395.21	12.00	53.00	1.00	0.06	470.21	41.80	70.00
4	5	1.00	0.32	409.01	47.20	28.00	1.00	0.07	494.41	39.60	40.00
4	6	1.00	0.26	566.62	29.00	87.20	1.00	0.08	646.42	96.60	73.80
4	7	1.00	0.19	682.62	29.00	89.00	1.00	0.07	738.82	99.20	46.60
4	8	1.00	0.31	728.02	76.00	76.00	1.00	0.07	797.02	87.20	118.40
5	1	1.00	0.88	202.80	13.00	9.80	1.00	0.19	222.21	16.60	20.20
5	2	1.00	0.84	273.61	9.20	30.60	1.00	0.21	302.21	31.40	47.60
5	3	1.00	0.98	325.41	19.40	41.00	1.00	0.25	361.81	42.20	22.20
5	4	1.00	0.82	447.21	30.60	52.60	1.00	0.34	495.42	47.20	46.40
5	5	1.00	1.11	483.42	28.20	33.40	1.00	0.32	529.02	55.80	51.20
5	6	1.00	0.96	556.22	48.00	56.20	1.00	0.29	618.82	51.40	52.60
5	7	1.00	1.07	640.62	42.60	73.00	1.00	0.29	709.42	73.20	96.20
5	8	1.00	0.97	695.02	89.80	43.20	1.00	0.23	765.63	189.80	75.00
6	1	1.00	5.82	263.81	22.80	26.80	1.00	0.97	300.41	40.80	17.40
6	2	1.00	5.61	327.81	43.00	28.20	1.00	1.60	365.81	37.80	54.20
6	3	1.00	7.13	369.21	18.80	26.00	1.00	1.73	414.22	25.20	55.40
6	4	1.00	6.91	458.82	17.00	73.00	1.00	1.22	489.02	31.00	90.60
6	5	1.00	7.92	477.22	63.80	58.60	1.00	1.82	541.22	82.20	81.40
6	6	1.00	7.15	627.82	38.80	64.00	1.00	1.37	714.03	144.40	93.60
6	7	1.00	7.26	676.23	107.60	80.40	1.00	1.15	725.63	133.60	101.40
6	8	1.00	8.83	675.63	58.80	92.60	1.00	1.74	747.83	50.80	66.20
7	1	1.00	87.28	265.01	20.80	19.40	1.00	6.83	305.81	56.40	25.40
7	2	1.00	84.70	325.61	35.40	24.20	1.00	7.34	355.82	34.00	37.00
7	3	1.00	79.14	371.82	43.00	45.40	1.00	7.71	406.62	57.20	44.60
7	4	1.00	65.83	503.42	68.20	29.20	1.00	5.53	552.03	44.40	63.00
7	5	1.00	60.58	526.03	40.80	51.80	1.00	8.21	599.63	65.20	77.80
7	6	1.00	117.99	661.83	55.60	83.40	1.00	9.53	702.83	57.00	150.00
7	7	1.00	101.53	626.83	14.80	146.00	1.00	6.71	657.63	50.40	84.00

Table B1: Contd. ...

...Table B1: Contd.

Instance		Integrated Model					Sequential Model				
N	*M*	*Opt. Ratio*	*CPU*	*Obj. Val.*	*BIT*	*VIT*	*Opt. Ratio*	*CPU*	*Obj. Val.*	*BIT*	*VIT*
7	8	1.00	108.09	778.64	100.40	142.20	1.00	8.44	823.64	105.20	102.60
8	1	1.00	2294.96	308.81	21.80	22.60	1.00	112.12	362.62	26.20	39.20
8	2	1.00	2145.55	387.02	28.00	41.00	1.00	94.24	430.82	42.40	51.00
8	3	1.00	1781.17	437.02	40.20	35.60	1.00	75.10	474.22	41.20	51.20
8	4	1.00	1058.16	517.43	49.80	14.80	1.00	52.44	554.83	107.80	22.40
8	5	1.00	2730.41	632.03	56.00	76.40	1.00	129.63	693.24	73.00	109.60
8	6	1.00	1279.69	664.44	92.40	66.80	1.00	47.33	697.44	169.80	52.80
8	7	1.00	3137.34	740.24	92.00	129.00	1.00	140.74	806.84	61.60	167.00
8	8	1.00	1819.34	776.24	57.80	138.00	1.00	90.94	842.05	99.20	96.40
9	1	0.20	6701.22	352.82	34.80	53.80	1.00	883.36	397.22	52.40	57.80
9	2	0.00	7200	420.42	16.60	42.60	1.00	1187.44	459.22	46.40	54.20
9	3	0.00	7200	478.83	40.20	48.60	1.00	1511.52	531.63	74.60	78.60
9	4	0.00	7200	564.03	36.20	56.00	1.00	1303.26	592.43	86.20	53.20
9	5	0.00	7200	569.43	48.40	61.00	1.00	1401.27	631.04	106.40	72.20
9	6	0.20	7136.75	618.24	48.80	58.80	1.00	898.43	663.24	93.00	73.40
9	7	0.00	7200	737.85	42.20	114.40	1.00	1063.79	821.45	62.20	158.40
9	8	0.00	7200	788.84	86.40	92.00	1.00	2090.03	906.06	83.80	157.80
10	1	0.00	7200	357.42	32.00	27.00	0.20	6886.74	395.82	60.80	34.40
10	2	0.00	7200	460.22	27.80	72.40	0.20	6886.93	500.83	84.80	94.20
10	3	0.00	7200	536.24	50.60	53.20	0.20	6748.69	569.04	86.00	51.40
10	4	0.00	7200	592.44	65.00	46.60	0.60	4873.39	621.24	142.80	75.20
10	5	0.00	7200	636.64	79.40	67.80	0.20	6050.73	702.04	103.40	78.00
10	6	0.00	7200	700.05	101.20	55.60	0.20	6991.33	761.65	72.00	162.20
10	7	0.00	7200	767.65	66.20	109.00	0.20	6552.23	861.86	130.80	126.40
10	8	0.00	7200	924.25	64.00	140.40	0.00	7200.00	1000.47	111.60	202.20
11	1	0.00	7200	411.62	24.40	38.40	0.00	7200.00	459.83	40.20	31.00
11	2	0.00	7200	490.03	66.00	44.60	0.00	7200.00	541.23	93.40	78.40
11	3	0.00	7200	601.64	72.40	61.80	0.20	6406.23	664.24	89.80	90.40
11	4	0.00	7200	616.44	76.40	33.60	0.00	7200	665.45	91.60	76.20
11	5	0.00	7200	636.85	47.60	67.60	0.00	7200	678.65	87.20	77.00
11	6	0.00	7200	762.25	87.40	85.40	0.00	7200	838.86	101.60	129.20
11	7	0.00	7200	742.05	109.80	86.00	0.00	7200	823.26	75.40	163.20
11	8	0.00	7200	929.07	96.40	168.00	0.00	7200	991.27	156.20	168.80
12	1	0.00	7200	457.03	47.20	46.60	0.00	7200	508.04	93.40	68.20
12	2	0.00	7200	490.03	34.80	41.80	0.00	7200	533.43	41.60	76.80
12	3	0.00	7200	572.24	55.40	48.40	0.00	7200	629.04	74.40	65.60
12	4	0.00	7200	661.65	65.00	65.20	0.00	7200	726.46	67.60	118.80
12	5	0.00	7200	749.26	75.60	82.80	0.00	7200	786.66	124.20	107.60
12	6	0.00	7200	739.85	87.40	92.00	0.00	7200	807.26	125.20	102.00
12	7	0.00	7200	825.47	128.40	105.60	0.00	7200	873.47	100.80	121.60
12	8	0.00	7200	928.07	78.20	136.00	0.00	7200	1004.68	120.00	192.00
Average		*0.61*	*3085.74*	*534.76*	*49.88*	*61.67*	*0.73*	*2221.92*	*586.03*	*75.11*	*75.26*

Table B2: Comparative results of integrated model with different lifespan value.

Instance			Integrated Model					Instance			Integrated Model				
N	*M*	*B*	*Opt. Ratio*	*CPU*	*Obj. Val.*	*BIT*	*VIT*	*N*	*M*	*B*	*Opt. Ratio*	*CPU*	*Obj. Val.*	*BIT*	*VIT*
3	1	1	1.00	0.11	152.20	8.20	30.20	5	1	1	1.00	0.75	204.21	10.00	19.00
3	1	2	1.00	0.10	152.20	11.40	30.20	5	1	2	1.00	0.87	204.01	12.20	19.60
3	1	3	1.00	0.06	152.20	15.40	30.20	5	1	3	1.00	0.61	202.81	12.20	18.40
3	2	1	1.00	0.05	224.60	10.40	5.00	5	2	1	1.00	0.60	275.81	13.00	9.20
3	2	2	1.00	0.05	224.60	12.40	5.00	5	2	2	1.00	0.45	274.81	16.00	9.40
3	2	3	1.00	0.05	224.20	6.80	5.00	5	2	3	1.00	0.53	274.41	10.20	31.40
3	3	1	1.00	0.05	260.40	18.80	27.80	5	3	1	1.00	0.70	333.01	23.60	36.60
3	3	2	1.00	0.05	259.40	17.40	32.80	5	3	2	1.00	0.65	329.21	21.20	39.80
3	3	3	1.00	0.05	259.40	17.40	32.80	5	3	3	1.00	0.63	325.41	19.40	41.00
3	4	1	1.00	0.05	374.61	15.00	50.40	5	4	1	1.00	0.52	447.41	17.40	52.80
3	4	2	1.00	0.05	371.41	24.40	50.40	5	4	2	1.00	0.57	447.21	24.60	54.00
3	4	3	1.00	0.05	371.41	25.20	50.40	5	4	3	1.00	0.79	447.21	30.60	52.60
3	5	1	1.00	0.05	346.61	32.20	16.40	5	5	1	1.00	0.55	483.42	23.40	17.80
3	5	2	1.00	0.05	346.61	39.00	16.40	5	5	2	1.00	0.59	483.42	20.40	33.40
3	5	3	1.00	0.05	346.61	46.00	16.40	5	5	3	1.00	0.53	483.42	25.00	33.40
3	6	1	1.00	0.05	447.81	43.80	56.00	5	6	1	1.00	0.51	563.42	48.80	51.60
3	6	2	1.00	0.05	446.81	55.40	55.40	5	6	2	1.00	0.52	556.22	42.00	56.20
3	6	3	1.00	0.05	446.81	64.00	55.40	5	6	3	1.00	0.50	556.22	47.20	56.20
3	7	1	1.00	0.05	611.01	4.40	94.60	5	7	1	1.00	0.46	640.62	35.20	66.20
3	7	2	1.00	0.05	611.01	8.80	94.60	5	7	2	1.00	0.49	640.62	44.60	66.20
3	7	3	1.00	0.06	611.01	15.20	94.60	5	7	3	1.00	0.48	640.62	47.80	66.20
3	8	1	1.00	0.05	682.01	66.20	111.00	5	8	1	1.00	0.53	707.22	55.80	36.80
3	8	2	1.00	0.05	682.01	76.40	111.00	5	8	2	1.00	0.53	703.22	50.20	43.60
3	8	3	1.00	0.05	682.01	77.60	111.00	5	8	3	1.00	0.53	695.02	65.60	43.20
4	1	1	1.00	0.08	156.00	12.00	3.00	6	1	1	1.00	3.09	264.61	19.80	31.60
4	1	2	1.00	0.09	156.00	15.60	3.00	6	1	2	1.00	3.44	263.81	19.60	26.80
4	1	3	1.00	0.09	152.60	13.20	3.00	6	1	3	1.00	3.66	263.81	22.80	26.80
4	2	1	1.00	0.09	255.81	9.20	25.80	6	2	1	1.00	3.51	331.01	19.40	36.80
4	2	2	1.00	0.09	255.81	11.20	25.80	6	2	2	1.00	3.24	329.01	24.80	35.40
4	2	3	1.00	0.09	255.81	7.20	30.40	6	2	3	1.00	3.45	327.81	46.40	28.20
4	3	1	1.00	0.09	358.41	25.20	2.60	6	3	1	1.00	5.22	384.01	31.80	24.40
4	3	2	1.00	0.09	356.61	30.60	13.80	6	3	2	1.00	4.50	376.41	21.20	25.00
4	3	3	1.00	0.09	355.61	35.40	12.80	6	3	3	1.00	4.04	369.81	27.20	26.60
4	4	1	1.00	0.08	399.81	21.60	53.00	6	4	1	1.00	3.94	459.42	23.80	66.80
4	4	2	1.00	0.07	399.81	24.40	53.00	6	4	2	1.00	4.09	459.42	16.40	74.80
4	4	3	1.00	0.08	395.21	10.00	53.00	6	4	3	1.00	4.55	459.42	19.00	74.80
4	5	1	1.00	0.09	409.21	35.60	30.40	6	5	1	1.00	4.52	493.42	50.60	53.80
4	5	2	1.00	0.10	409.01	46.60	28.00	6	5	2	1.00	4.40	479.82	58.60	54.20
4	5	3	1.00	0.11	409.01	47.20	28.00	6	5	3	1.00	4.64	477.22	62.20	58.60
4	6	1	1.00	0.10	570.42	13.20	89.00	6	6	1	1.00	4.10	634.62	33.00	98.20
4	6	2	1.00	0.10	566.62	28.60	87.20	6	6	2	1.00	3.87	628.82	34.00	57.00

Table B2: Contd. ...

...Table B2: Contd

Instance			Integrated Model					Instance			Integrated Model				
N	*M*	*B*	*Opt. Ratio*	*CPU*	*Obj. Val.*	*BIT*	*VIT*	*N*	*M*	*B*	*Opt. Ratio*	*CPU*	*Obj. Val.*	*BIT*	*VIT*
4	6	3	1.00	0.10	566.62	29.00	87.20	6	6	3	1.00	4.25	627.82	43.20	58.80
4	7	1	1.00	0.10	682.62	26.00	89.00	6	7	1	1.00	3.95	684.02	56.00	96.80
4	7	2	1.00	0.09	682.62	29.00	89.00	6	7	2	1.00	3.74	677.43	82.80	81.60
4	7	3	1.00	0.10	682.62	29.00	89.00	6	7	3	1.00	3.80	676.23	102.20	80.40
4	8	1	1.00	0.11	728.02	48.00	76.00	6	8	1	1.00	5.18	686.63	65.80	43.00
4	8	2	1.00	0.11	728.02	64.80	76.00	6	8	2	1.00	5.19	683.23	47.80	77.00
4	8	3	1.00	0.11	728.02	72.60	76.00	6	8	3	1.00	4.96	675.63	53.20	92.60
7	1	1	1.00	38.43	267.41	18.20	21.60	9	1	1	0.20	6176.56	353.82	18.00	54.80
7	1	2	1.00	37.24	265.61	18.40	20.00	9	1	2	0.20	6329.97	352.82	32.80	54.00
7	1	3	1.00	45.57	265.01	20.80	19.40	9	1	3	0.20	6381.58	352.82	41.00	51.60
7	2	1	1.00	63.69	329.81	21.00	32.80	9	2	1	0.00	7200	422.02	20.60	39.80
7	2	2	1.00	58.42	326.41	38.00	25.40	9	2	2	0.00	7200	420.42	21.00	40.40
7	2	3	1.00	58.62	326.01	35.00	24.60	9	2	3	0.00	7200	420.42	23.00	40.40
7	3	1	1.00	52.83	379.02	43.20	39.00	9	3	1	0.20	7192.90	479.83	44.60	53.40
7	3	2	1.00	51.23	372.62	45.20	44.00	9	3	2	0.00	7200	479.63	55.40	45.00
7	3	3	1.00	58.56	372.62	52.20	44.80	9	3	3	0.00	7200	478.23	55.40	47.20
7	4	1	1.00	38.98	514.22	35.40	44.20	9	4	1	0.00	7200	564.83	41.80	55.40
7	4	2	1.00	57.79	511.42	47.80	28.00	9	4	2	0.00	7200	563.23	45.20	48.20
7	4	3	1.00	43.27	503.42	60.80	29.20	9	4	3	0.00	7200	563.23	41.20	54.00
7	5	1	1.00	37.57	536.23	34.20	48.00	9	5	1	0.00	7200	575.83	45.00	48.20
7	5	2	1.00	42.65	533.83	47.80	48.00	9	5	2	0.00	7200	575.03	62.20	61.60
7	5	3	1.00	43.74	526.03	41.40	51.80	9	5	3	0.00	7200	570.43	32.20	63.40
7	6	1	1.00	78.92	665.63	56.80	66.00	9	6	1	0.20	6642.86	624.64	46.20	71.00
7	6	2	1.00	79.87	663.03	59.20	67.40	9	6	2	0.20	6789.05	623.84	45.00	53.20
7	6	3	1.00	82.42	661.83	55.60	83.40	9	6	3	0.20	6637.01	618.24	46.20	58.80
7	7	1	1.00	57.37	636.43	19.40	132.00	9	7	1	0.00	7200	744.65	32.20	114.80
7	7	2	1.00	63.60	627.83	25.60	128.40	9	7	2	0.00	7200	739.25	37.80	102.60
7	7	3	1.00	73.91	626.83	16.80	146.00	9	7	3	0.00	7200	739.65	20.80	135.60
7	8	1	1.00	59.71	784.24	68.80	138.40	9	8	1	0.00	7200	800.65	46.20	113.00
7	8	2	1.00	64.81	779.04	69.60	142.20	9	8	2	0.00	7200	799.64	66.40	122.60
7	8	3	1.00	76.72	778.64	89.00	142.20	9	8	3	0.00	7200	790.44	61.00	99.80
8	1	1	1.00	914.71	316.01	13.20	28.00	10	1	1	0.00	7200	362.82	7.00	46.60
8	1	2	1.00	994.32	311.81	15.80	22.00	10	1	2	0.00	7200	357.82	13.60	37.60
8	1	3	1.00	1159.85	311.21	15.80	22.20	10	1	3	0.00	7200	357.42	41.00	38.60
8	2	1	1.00	1110.23	387.42	18.60	38.00	10	2	1	0.00	7200	463.22	22.00	65.20
8	2	2	1.00	1361.35	387.22	22.80	38.00	10	2	2	0.00	7200	460.62	27.20	73.00
8	2	3	1.00	1642.86	387.22	33.40	38.00	10	2	3	0.00	7200	462.83	38.40	68.00
8	3	1	1.00	1225.13	444.62	36.00	31.60	10	3	1	0.00	7200	545.03	40.20	42.80
8	3	2	1.00	1363.01	438.62	51.60	31.00	10	3	2	0.00	7200	538.43	41.60	55.40
8	3	3	1.00	1238.11	437.02	39.80	35.60	10	3	3	0.00	7200	536.24	60.80	52.60
8	4	1	1.00	590.90	526.43	41.20	13.40	10	4	1	0.00	7200	592.64	58.00	36.80

Table B2: Contd. ...

...Table B2: Contd

Instance			Integrated Model					Instance			Integrated Model				
N	*M*	*B*	*Opt. Ratio*	*CPU*	*Obj. Val.*	*BIT*	*VIT*	*N*	*M*	*B*	*Opt. Ratio*	*CPU*	*Obj. Val.*	*BIT*	*VIT*
8	4	2	1.00	612.60	524.23	62.40	12.00	10	4	2	0.00	7200	593.24	76.60	30.40
8	4	3	1.00	791.75	520.03	63.20	14.00	10	4	3	0.00	7200	592.84	69.00	43.20
8	5	1	1.00	1622.06	633.43	44.40	71.00	10	5	1	0.00	7200	639.24	53.20	71.80
8	5	2	1.00	1726.64	632.63	55.00	70.20	10	5	2	0.00	7200	638.44	79.40	75.00
8	5	3	1.00	1965.54	632.03	52.20	76.40	10	5	3	0.00	7200	638.64	65.80	89.60
8	6	1	1.00	749.32	678.04	58.80	50.20	10	6	1	0.00	7200	704.84	83.00	79.20
8	6	2	1.00	859.33	666.64	66.40	26.60	10	6	2	0.00	7200	700.65	76.40	66.80
8	6	3	1.00	945.98	666.64	60.80	30.20	10	6	3	0.00	7200	703.65	84.80	90.00
8	7	1	1.00	1903.58	745.64	73.20	88.60	10	7	1	0.00	7200	772.25	67.80	79.60
8	7	2	1.00	2184.91	745.64	83.80	88.60	10	7	2	0.00	7200	768.65	61.40	85.40
8	7	3	1.00	2274.47	740.24	89.40	129.00	10	7	3	0.00	7200	766.65	75.80	111.00
8	8	1	1.00	1843.06	781.04	65.00	117.40	10	8	1	0.00	7200	935.26	55.00	116.20
8	8	2	1.00	1538.72	777.44	44.80	139.60	10	8	2	0.00	7200	929.45	107.40	113.20
8	8	3	1.00	1367.03	776.24	57.80	138.00	10	8	3	0.00	7200	922.86	110.40	106.40
11	1	1	0.00	7200	417.82	21.40	39.40	12	1	1	0.00	7200	459.03	33.00	58.80
11	1	2	0.00	7200	411.82	31.80	29.80	12	1	2	0.00	7200	457.83	41.60	63.40
11	1	3	0.00	7200	409.42	30.80	26.40	12	1	3	0.00	7200	456.63	53.80	48.20
11	2	1	0.00	7200	497.03	31.80	48.00	12	2	1	0.00	7200	498.03	34.80	35.60
11	2	2	0.00	7200	490.03	62.20	52.00	12	2	2	0.00	7200	488.63	35.40	50.80
11	2	3	0.00	7200	488.83	55.20	40.60	12	2	3	0.00	7200	488.63	39.80	48.80
11	3	1	0.00	7200	603.44	31.40	73.00	12	3	1	0.00	7200	578.24	39.00	51.80
11	3	2	0.00	7200	603.04	45.40	60.80	12	3	2	0.00	7200	573.64	55.40	37.00
11	3	3	0.00	7200	600.84	70.00	54.00	12	3	3	0.00	7200	575.04	57.40	47.40
11	4	1	0.00	7200	623.04	63.20	49.80	12	4	1	0.00	7200	663.25	59.20	59.80
11	4	2	0.00	7200	623.24	43.20	69.40	12	4	2	0.00	7200	657.25	48.00	77.40
11	4	3	0.00	7200	620.84	79.80	25.00	12	4	3	0.00	7200	656.25	52.80	76.80
11	5	1	0.00	7200	643.05	63.00	30.80	12	5	1	0.00	7200	753.66	65.40	50.60
11	5	2	0.00	7200	642.05	80.40	75.80	12	5	2	0.00	7200	750.65	62.00	91.00
11	5	3	0.00	7200	636.05	95.20	40.00	12	5	3	0.00	7200	747.65	64.40	86.60
11	6	1	0.00	7200	743.85	78.80	60.20	12	6	1	0.00	7200	747.05	70.40	96.80
11	6	2	0.00	7200	744.45	75.00	58.80	12	6	2	0.00	7200	746.25	101.20	71.00
11	6	3	0.00	7200	739.05	62.00	97.60	12	6	3	0.00	7200	741.85	71.20	86.60
11	7	1	0.00	7200	743.65	84.00	66.00	12	7	1	0.00	7200	829.06	75.40	79.40
11	7	2	0.00	7200	746.05	77.20	87.80	12	7	2	0.00	7200	832.86	90.40	90.60
11	7	3	0.00	7200	749.25	64.00	118.60	12	7	3	0.00	7200	828.86	86.00	101.20
11	8	1	0.00	7200	924.26	89.80	109.60	12	8	1	0.00	7200	928.67	75.00	109.80
11	8	2	0.00	7200	918.66	96.60	64.00	12	8	2	0.00	7200	933.87	73.00	180.60
11	8	3	0.00	7200	917.86	96.00	90.20	12	8	3	0.00	7200	924.47	67.00	186.60

References

Amorim, P., M. Belo-Filho, F.M.B.D. Toledo, C. Almeder and B. Almada-Lobo. 2013. Lot sizing versus batching in the production and distribution planning of perishable goods. International Journal of Production Economics, 146(1): 208–218.

Armstrong, R., S. Gao and L. Lei. 2008. A zero-inventory production and distribution problem with a fixed customer sequence. Annals of Operations Research, 159(1): 395–414.

Belo-Filho, M., P. Amorim and B. Almada-Lobo. 2015. An adaptive large neighbourhood search for the operational integrated production and distribution problem of perishable products. International Journal of Production Research, 53(20): 6040–6058.

Chen, H.-K., C.F. Hsueh and M.S. Chang. 2009. Production scheduling and vehicle routing with time windows for perishable food products. Computers & Operations Research, 36(7): 2311–2319.

Chen, Z.-L. 2010. Integrated production and outbound distribution scheduling: review and extensions. Operations Research, 58(1): 130–148.

Devapriya, P., W. Ferrell and N. Geismar. 2006. Optimal fleet size of an integrated production and distribution scheduling problem for a perishable product. Paper presented at the IIE Annual Conference. Proceedings.

Devapriya, P., W. Ferrell and N. Geismar. 2017. Integrated production and distribution scheduling with a perishable product. European Journal of Operational Research, 259(3): 906–916.

Farahani, P., M. Grunow and H.O. Günther. 2012. Integrated production and distribution planning for perishable food products. Flexible Services and Manufacturing Journal, 24(1): 28–51.

Geismar, H.N., M. Dawande and C. Sriskandarajah. 2011. Pool-point distribution of zero-inventory products. Production and Operations Management, 20(5): 737–753.

Geismar, H.N., G. Laporte, L. Lei and C. Sriskandarajah. 2008. The integrated production and transportation scheduling problem for a product with a short lifespan. INFORMS Journal on Computing, 20(1): 21–33.

Hurter, A.P. and M.G. Van Buer. 1996. The newspaper production/distribution problem. Journal of Business Logistics, 17(1): 85.

Karaoğlan, İ. and S.E. Kesen. 2017. The coordinated production and transportation scheduling problem with a time-sensitive product: a branch-and-cut algorithm. International Journal of Production Research, 55(2): 536–557.

Kergosien, Y., M. Gendreau and J.C. Billaut 2017. A Benders decomposition-based heuristic for a production and outbound distribution scheduling problem with strict delivery constraints. European Journal of Operational Research, 262(1): 287–298.

Lacomme, P., A. Moukrim, A. Quilliot and M. Vinot. 2018. Supply chain optimisation with both production and transportation integration: multiple vehicles for a single perishable product. International Journal of Production Research, 1–24.

Lee, J., B.I. Kim, A.L. Johnson and K. Lee. 2014. The nuclear medicine production and delivery problem. European Journal of Operational Research, 236(2): 461–472.

Marandi, F. and S. Zegordi. 2017. Integrated production and distribution scheduling for perishable products. Scientia Iranica, 24(4): 2105–2118.

Moons, S., K. Ramaekers, A. Caris and Y. Arda. 2017. Integrating production scheduling and vehicle routing decisions at the operational decision level: a review and discussion. Computers & Industrial Engineering, 104: 224–245.

Park, Y.-B. and S.C. Hong. 2009. Integrated production and distribution planning for single-period inventory products. International Journal of Computer Integrated Manufacturing, 22(5): 443–457.

Reimann, M., R. Tavares Neto and E. Bogendorfer. 2014. Joint optimization of production planning and vehicle routing problems: A review of existing strategies. Pesquisa Operacional, 34(2): 189–214.

Russell, R., W.C. Chiang and D. Zepeda. 2008. Integrating multi-product production and distribution in newspaper logistics. Computers & Operations Research, 35(5): 1576–1588.

Thomas, D.J. and P.M. Griffin. 1996. Coordinated supply chain management. European Journal of Operational Research, 94(1): 1–15.

Ullrich, C.A. 2013. Integrated machine scheduling and vehicle routing with time windows. European Journal of Operational Research, 227(1): 152–165.

Van Buer, M.G., D.L. Woodruff and R.T. Olson. 1999. Solving the medium newspaper production/distribution problem. European Journal of Operational Research, 115(2): 237–253.

Viergutz, C. and S. Knust. 2014. Integrated production and distribution scheduling with lifespan constraints. Annals of Operations Research, 213(1): 293–318.

Chapter 16

Profit-oriented Balancing of Parallel Disassembly Lines with Processing Alternatives in the Age of Industry 4.0

*Seda Hezer** and *Yakup Kara*

1. Introduction

In recent years, environmental issues have been gaining importance due to some important factors such as environmental regulations, enhanced consumer awareness as well as the economic attractiveness of reusing products and parts/subassemblies of the products (McGovern and Gupta 2006, Hezer and Kara 2015). These factors have led the manufacturers to perform product recovery conditions for their post-consumed products (Gungor and Gupta 1999, Agrawal and Tiwari 2008). The aim of product recovery is to recover valuable components and materials from old or outdated products by means of remanufacturing and recycling, thereby reducing the amount of waste sent to landfills (Thierry et al. 1995; Gungor 2006; Hezer and Kara 2015). Disassembly is an important step of the product recovery to achieve this aim. It is essential to disassemble all kinds of goods ranging from electronic goods (computers, printers, etc.) to white goods (dishwashers, refrigerators, etc.), from brown goods (TV sets, sound systems, etc.) to cars, and so on.

In this chapter, Section 1 (Introduction)is divided into six subsections. Section 1.1 briefs the information about disassembly processes and operations. Section 1.2 explains the importance of disassembly lines and the factors affecting the efficiency of the lines. Section 1.3 presents uncertain factors specific to disassembly. Section 1.4 explains industry 4.0 that is applied to solve the problems caused by the uncertainty and key technologies of industry 4.0. Section 1.5 gives the literature of disassembly line balancing problem (DLBP) and emphasizes inadequate issues. Section 1.6 presents the aim and scope of the paper in the light of the information presented in the first five subsections and gaps addressed in the literature.

1.1 Disassembly Processes and Operations

Disassembly is a systematic method for the extraction of valuable components (parts and materials) from discarded products via a series of operations. In disassembly, parts are demanded (as compared to assembly). However, not all parts of the product have to be demanded and disassembled. If a part has a demand, various types of demand sources may emerge and they may affect the performance of the disassembly. Demand sources can be categorized as the first type, second type and third type demand. The first type is that the demand source may accept part "as is". In the second type, the demand source may accept the part without defect. The third type is that the demand source may accept the defective part according to the seriousness of the defect (Gungor and Gupta 2002). According to the demand situations of the parts, all parts of a product may be disassembled (complete disassembly) or disassembly of multiple parts may be required (partial disassembly) (Gungor and Gupta 2002). As disassembly is a costly process, partial disassembly is usually preferred to yield profit (Gungor and Gupta 2001) because even if all parts have positive revenue, the total cost of disassembly may exceed the total revenue of the parts. At each step of the disassembly, disassembly may be performed with one of two processes that are categorized as unfastening (non-destructive) and destructive actions (Das and Naik 2002).

Department of Industrial Engineering, Konya Technical University, Konya, Turkey; Email: ykara@ktun.edu.tr
* Corresponding author: shezer@ktun.edu.tr

Unfastening action (UA) is the exact opposite of the fastening action performed in the assembly process. It separates parts or subassemblies from each other by removing fasteners usually manually with or without the use of equipment (Sonnenberg 2001). For example, removing a screw which links two or more parts using a manual or electric screwdriver is an unfastening action. *Destructive action (DA)*, on the other hand, comprises all other actions other than the unfastening action. In DA, the geometric structure of one or more parts is damaged during disassembly due to the cutting operations (Lee and Gadh 1998). To illustrate, separating two parts welded to each other by using flame cutting is a destructive action. While UA is applied to fasteners, DA is applied to fasteners and/or parts. For this reason, fasteners may be damaged during UA while either fasteners and/or parts may be damaged during DA. The parts which are removed by applying UA can be reused or used for remanufacturing (Das and Naik 2002). DA is usually applied when UA is not possible or only recycling of the components is required (Lee and Gadh 1998). The equipment used for UA and DA are different from each other. For detailed information on the actions and equipment, researchers can utilize the papers of Das et al. (2000) and Das and Naik (2002).

According to the structure of the product and demand type of the parts, disassembly of a part may have one or more than one actions, alternatively. If demanded parts are not defective and the demand type is the second type, these parts are disassembled by UA. Sometimes some or all parts of a product may have different structures and/or operational specifications from their original structures and/or specifications. For instance, a part may get stuck in its place or may be broken (Gungor and Gupta 2001). If demand types of these parts are the first or the third type, these parts may require DA. On the other hand, the non-demanded parts that are preceded by the demanded parts may be disassembled using DA to terminate the disassembly as soon as possible (Koc et al. 2009; Ilgin and Gupta 2010; Bentaha et al. 2018). If a demanded part is not defective and its demand type is the first or third type, disassembly of this part may be performed by UA or DA, alternatively. Moreover, in addition to the alternative actions of a part, each action may also need to have a particular equipment and assistance of another worker to reduce the task time of a task (In the context of this chapter, removing a part is a disassembly task and it is simply referred to as 'a task'). Thus, different processing alternatives, i.e., resource combinations, workers, assistants and equipment, may emerge for the task of a part. Each processing alternative has an actual time, revenue, and cost. While the revenue and time of a processing alternative change depending on the action and equipment assigned, the time is also affected by whether or not an assistant is allocated to perform the task (Kara et al. 2011). Furthermore, a costly processing alternative may usually reduce the task time. If a task can be performed with either UA or DA, performing with UA provides a higher revenue and requires longer time. The reason for the higher revenue is that the part is not damaged, so the revenue remains as determined. However, if the task is performed by using DA, because the part is damaged, the revenue changes according to the degree of the damage which depends on the equipment used, and it is always lower than the determined level. On the other hand, the reason for the longer time is that the worker may act more carefully in order not to damage the part when performing UA (Das et al. 2000; Das and Naik 2002).

According to the information given above, for each task, it is important to choose the best processing alternative that provides the good trade-off between the revenue and cost of the task.

1.2 Disassembly Lines

A disassembly line contains several sequential work stations in which some tasks of an old or outdated product are performed in a cycle time. It provides the best conditions for automated disassembly processes, but some important problems arise about the design and optimization of disassembly lines (Duta et al. 2005). One of these problems is DLBP. DLBP is the act of assigning tasks to sequential workstations by satisfying some constraints and optimizing one or more than one performance measures while meeting the demands for disassembled parts. Basic constraints and assumptions of DLBP are as follows: Each task should be assigned to only one workstation, all precedence relationships among these tasks should be satisfied and the workload of a workstation cannot exceed the cycle time. Single-model product is disassembled and the exact quantity of the parts in product received is known (Hezer and Kara 2015). Supply of each product is infinite. The disassembly times, cycle time, and demand parameters are deterministic and known. The precedence relationships among tasks of a product are known and represented using AND/OR precedence relationships (Gungor and Gupta 2001, 2002; Mcgovern and Gupta 2011).

Disassembly lines can be categorized as single and mixed-model with regards to the number of different products disassembled on the line;straight, U-shaped, two sided, and parallel disassembly lines with regards to the layout of the line. Only one type of product is disassembled on single-model disassembly while different products having different orginal structures or different models of a product type with similar original structures are disassembled on mixed-model disassembly lines (Gungor and Gupta 2002; Lambert and Gupta 2005; Kara et al. 2010; Mcgovern and Gupta 2011). Traditionally, disassembly is usually performed on a straight disassembly line. Ordered sequence of workstations are arranged along the straight line and a worker performs task or tasks on each workstation within the same cycle time (Hezer and Kara 2015). In recycling plants, where there is a large variety or number of products, generally more than one parallel straight disassembly lines must be used. The balancing of the related lines is carried out independently of each other. Because the tasks using the same resources (equipment, assistants) are implemented on different lines, the total cost of the lines is high depending on the number of resources used in these lines. By combining these tasks, cost and time can be saved by providing common resource usage. One way to achieve this is to balance the parallel straight disassembly lines simultaneously (Gokcen et al. 2006; Kara et al. 2010; Ozcan et al. 2010). This problem is referred to as parallel disassembly line balancing problem (PDLBP) (Hezer and Kara 2015).

Basic constraints and assumptions of PDLBP are as follows: Single-model product is disassembled on each disassembly line. Precedence diagrams of each product are known. Task times, cycle times, and demand parameters of each product are deterministic and known. There may be common workstations that contain tasks from two adjacent lines for parallel lines. These common stations provide flexibility in minimizing total idle time and the number of workstations needed in the recovery facility. In addition, the visibility and traceability of the workers increase and the team work environment is created. Workers become multi-skilled. Work can be carried out on each side of any line (Gokcen et al. 2006; Ozcan et al. 2010; Hezer and Kara 2015). Figure 1 presents parallel disassembly lines that contain two parallel disassembly lines.

Figure 1 shows a recovery facility that consists of two parallel lines and four workstations. In workstation 1, a worker performs some tasks of line 1, i.e., product 1, while a worker performs some tasks of the line 2, i.e., product 2 in workstation 2. Workstations 3 and 4 are 'common' workstations. When the workers work in these common workstations, they perform tasks on both disassembly lines. After they perform the tasks on line 1, they move to line 2 and finally return to line 1 at the end of the cycle (Hezer and Kara 2015).

In disassembly line balancing problems, the aim is to optimize one or more performance measures. Minimization of total number of workstations, maximization of line efficiency or profit and revenue, eliminating hazardous parts early and removing highly demanded parts as early as possible are some of these performance measures (Battaia et al. 2018). In recent years, maximizing the profit of the disassembly line is rapidly growing in importance. The problem which deals with maximization of profit is called profit-oriented disassembly line balancing problem, which was firstly proposed by Altekin et al. (2008). In this problem, the single-model product is partially disassembled on a straight line. All or some parts of the product may be demanded and not all demanded parts have to be removed. Quantities of the parts may be the same or different from each other. Revenues of demanded parts are positive, deterministic and known while revenues of non-demanded parts are zero (Altekin et al. 2008; Bentaha et al. 2018).

In real life, many challenges are encountered during the disassembly processes of the products. In order to optimize the total net profit, effective methods are needed for the related challenges. The most important of these challenges is related with the uncertain factors about the discarded products on a disassembly line. The structure and quality of the product to be disassembled are uncertain because these properties may vary depending on customer use, and the product may

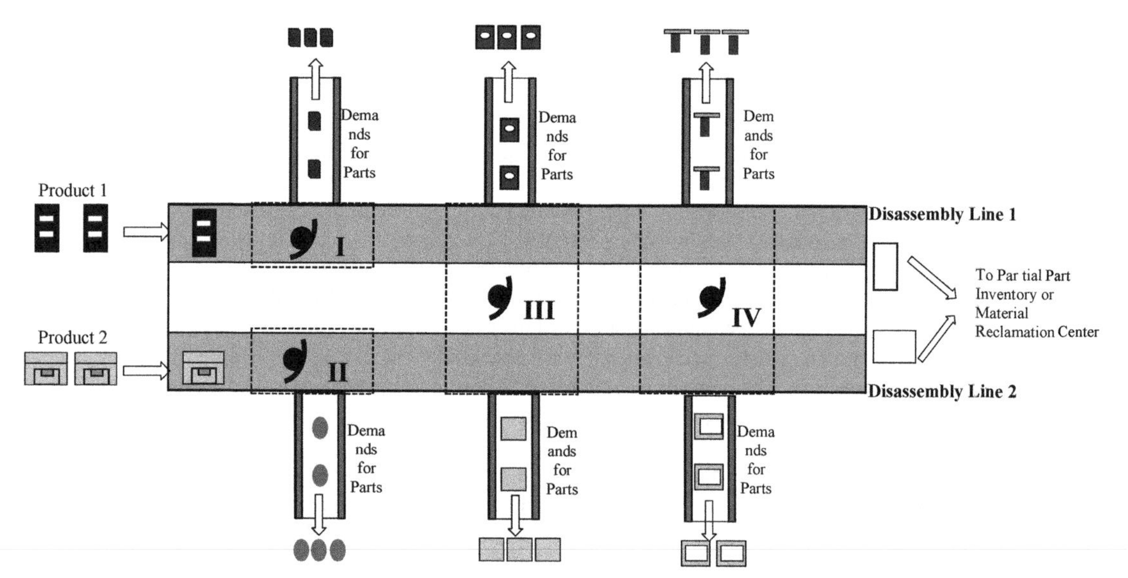

Fig. 1: A sample parallel disassembly lines (Hezer and Kara 2015).

be different from the original structure. Some of these uncertainties may be related to defective parts or fasteners in the product. Depending on this situation, disassembly of such parts may be hazardous or difficult or may cause uneconomical consequences. Some uncertainties may be related to the number of parts of the product. Customers may have removed one or more parts or increased the number of parts. For example, a computer user can upgrade the computer's memory to 64 MB by using an additional 32 MB module. In this case, the computer will have one more 32 MB memory module than the original (Gungor and Gupta 2001). Such uncertain factors about the product are not known in advance. They arise when the product enters into the disassembly processes; therefore, disassembly tasks, which are previously planned,may need to be modified. For example, some tasks may not be completed, so some tasks may not be performed due to precedence relations or may be returned to previous stations or new tasks may be performed. Consequently, unnecessary tasks may be performed or parts that are not needed may be removed (Gungor and Gupta 2001, 2002; Altekin et al. 2008; Duta et al. 2008; Bentaha et al. 2014b; Riggs et al. 2015). These unfavorable situations lead to a waste of time and cost. However, if the properties related to the disassembly tasks, the quality, and structure of the product are known in advance, these negative effects can be minimized or eliminated. Nowadays, the industry 4.0 technologies have been used in predicting necessary information about the products. The relevant components help to make decisions about the level of disassembly and the planning of the disassembly processes by providing the necessary information on the basic characteristics of the products and disassembly tasks.

1.3 Industry 4.0 and its Key Technologies

Three major industrial revolutions have taken place until the present period. In the industrial sense, the first industrial revolution (Industry 1.0), which first started with steam engines in the 18th century and aimed at increasing production, was followed by the second industrial revolution (Industry 2.0), which emerged as a transition to mass production at the beginning of the 20th century and paved the way for the utilization of electrical energy. Then came the third industrial revolution (Industry 3.0), where production systems ceased to be analog and digital systems took place in the industry. Thus, the first three industrial revolutions brought mechanization, electricity, and information technology to human production, respectively (Yıldız 2018). Industry 4.0 means the fourth industrial revolution and the term was first used in 2011 at the Hannover Fair. Industry 4.0 is a collective set of technologies and concepts of value chain organizations.

The basic key technologies of Industry 4.0 can be categorized as cyber-physical systems, big data, cloud computing, and the internet of things (IoT). Cyber-physical systems are defined as the concept of collecting information from physical objects using IoT, computer networks, or accelerated wireless connections(Liao et al. 2017; Zheng et al. 2018). Information from products, machines or production lines generates a significant amount of statistical data that can be changed or analyzed. A large amount of data as a whole is defined as "big data", which is another major idea in industry 4.0. All available information is processed through cloud computing. IoT is a data network that occurs when objects communicate with each other through specific protocols without human intervention. With this network, all objects can be monitored. With IoT, uncertainty can be reduced or virtually eliminated(Lu 2017; Xu et al. 2018). Radio frequency identification (RFID) and sensors are core components of IoT. While passive RFID labels are sufficient for monitoring purposes, active RFID labels with embedded sensors provide more information about the use or status of each object. Such products are named as sensor embedded products (SEPs). Sensors and RFID labels are placed on the objects to track information on the objects. Static information such as sales date, serial number, model, disassembly sequence, and bill of materials are obtained with the RFID label. Dynamic information such as maintenance, repair, insertion or removal of parts are obtained through sensors (Ondemir et al. 2012).

1.4 Evaluation of Industry 4.0 in Terms of Disassembly Processes

Thanks to industry 4.0, uncertain factors, one of the most important problems in the disassembly processes, can be largely eliminated. With the sensor and RFID labels placed on the product, critical components of the product can be monitored and the necessary information about the product is recorded throughout its life cycle. When the products reach the recovery facility, the information can be retrieved (Ondemir and Gupta 2014b). In this way, important information is obtained about the current status of the components, namely, the number of failures, whether they are functional or not, and the operations to be performed accordingly. This eliminates unnecessary tasks (Ondemir and Gupta 2014a). For example, due to the lack of some parts, not performing the tasks related to the removal of those parts, and the waste of resources such as equipment, worker, etc., used in these tasks are prevented. The demands of the parts are updated accordingly and demand sources are notified. In other words, disassembly planning arrangements are made to prevent wasting time and cost. Besides, the position and current status of each product in the disassembly path are monitored instantly, and the information obtained for future analysis is recorded (Gungor and Gupta 2001, 2002; Ondemir et al. 2012).

1.5 Literature Review

The simplest version of DLBP is the single-model and straight DLBP, as described by Gungor and Gupta (2001, 2002). McGovern and Gupta (2007a) provided NP-completeness proof of the decision version of DLBP, and Mcgovern and Gupta (2007b) showed unary NP-completeness. There are a number of studies that aim to optimally solve DLBP by using mathematical programming techniques (Altekin et al. 2008; Koc et al. 2009; Altekin and Akkan 2012; Kalaycilar et al. 2016). Due to combinatorial nature of DLBP, the researchers needed to use metaheuristic approaches, such as genetic algorithms (McGovern and Gupta 2007a; Kalayci et al. 2016; Seidi and Saghari 2016; Pistolesi et al. 2018), ant colony optimization (Agrawal and Tiwari 2008; Ding et al. 2010; Kalayci and Gupta 2013d; Zhu et al. 2014), simulated annealing (SA) (Kalayci and Gupta 2013a; Kalayci and Gupta 2013c; Fang, Ming et al. 2019), artificial bee colony (Kalayci and Gupta 2013b; Kalayci et al. 2015; Liu and Wang 2017), and tabu search (Kalayci and Gupta 2014). Some researchers focused on the uncertainty of product quality and task times and they proposed solution approaches, such as stochastic methods (Bentaha et al. 2014a,b; Riggs et al. 2015; Altekin 2016, 2017; He et al. 2019) and fuzzy methods (Paksoy et al. 2013; Ozceylan and Paksoy 2014; Kalayci et al. 2015; Seidi and Saghari 2016; Zhang et al. 2017). In recent years, robotic disassembly line balancing has attracted attention due to high disassembly productivity and researchers have proposed various solution models (Cil et al. 2017; Liu et al. 2018; Fang, Liu et al. 2019; Fang, Ming et al. 2019). A more comprehensive and detailed review of DLBP papers can be found in Ozceylan et al. (2019) and Deniz and Ozcelik (2019).

When DLBP literature is reviewed, it is seen that much research has focused on balancing individual straight, complete and non-destructive disassembly lines. However, there are numerous versions of DLBP with very specific considerations in industrial practice. Several studies focused on these considerations, such as U-type layout (Agrawal and Tiwari 2008; Avikal and Mishra 2012; Avikal et al. 2013), station paralleling (Aydemir-Karadag and Turkbey 2013), balancing parallel lines simultaneously (Hezer and Kara 2015), two sided disassembly lines (Wang et al. 2019b), destructive process (Duta et al. 2008, Igarashi et al. 2014), and partial disassembly (Altekin et al. 2008; Altekin and Akkan 2012; Ren et al. 2017; Bentaha et al. 2018; Wang et al. 2019a,b).

It is observed that the literature on DLBP has to be improved by considering industrial facts, such as processing alternatives and uncertainty. In real systems, it is very important to know necessary information about the condition of the parts of the product before the disassembly and choose the most appropriate processing alternative which provides the best return with respect to revenue, time, and cost parameters for each task to optimise the disassembly line. For instance, by using the information of the parts obtained from key technologies of the industry 4.0, the right decision may be to choose the alternative with high revenue but longer task time in some cases, or to choose the alternative with low revenue but shorter task time, which decreases the station cost. Choosing the most appropriate processing alternative also allows for obtaining a certain amount of data regarding the disassembly plan, which is an output of the disassembly planning process.

For these reasons, this chapter focuses on the topics about parallel lines, disassembly actions, processing alternatives and industry 4.0. In addition to DLBP literature, the papers of Das and Naik (2002), Kara et al. (2011) and the literature of parallel line balancing were utilized for more information about these topics.

Problem of balancing parallel lines simultaneously and its solution techniques were firstly presented by Gokcen et al. (2006) for assembly lines. The problem is called the parallel assembly line balancing problem (PALBP). The problem was further developed by many papers, such as Benzer et al. (2007); Baykasoglu et al. (2009); Cercioglu et al. (2009); Guo and Tang (2009); Kara et al. (2010); Ozcan et al. (2010); Ozbakir et al. (2011); Baykasoglu et al. (2012); Araujo et al. (2015); Ozcan (2018). Readers are referred to the paper of Lusa (2008) for detailed information about the problem. In DLBP literature, only one paper (Hezer and Kara 2015) about the problem of balancing parallel disassembly lines, PDLBP, has been presented and introduced.

Kara et al. (2011) focused on balancing straight and U-shaped assembly lines with resource dependent task times. Das and Naik (2002) aimed to commentate disassembly actions. They introduced a descriptive model for solving the disassembly process planning problem which is about identifying unfastening and destructive actions, and the required equipment.

1.6 Aim and Scope of the Paper

This chapter deals with the profit-oriented parallel disassembly line balancing problem with processing alternatives (PDLBP_PA) which utilizes industry 4.0 technology. PDLBP_PA is defined as partial and single-model parallel disassembly lines with processing alternatives for disassembly actions including unfastening and destructive actions. The aim of the problem is to simultaneously address the assignment of tasks to workstations on parallel disassembly lines and selection of the most appropriate processing alternative for each task to maximize the total net recovery profit. It is also aimed to avoid unnecessary operations, waste of time, and cost by using information collected by embedded sensors and devices.

A 0-1 integer programming formulation for PDLBP_PA has been proposed and a family of valid inequalities adapted from the literature have been presented to strengthen the formulation. Due to the combinatorial nature of the problem, development of heuristic approaches is required to solve medium- and large-size test problems. Therefore, an effective heuristic approach based on adaptive simulated annealing (ASA) meta-heuristic, called PASA, has been proposed to solve medium- and large-size PDLBP_PA in a reasonable amount of time. The contribution of this chapter mainly includes the following:

1) Processing alternatives are thought with the parallel lines that are balanced simultaneously.
2) Disassembly actions are considered with equipment and assistants. Thus, in addition to deciding on which action to perform, which equipment will be used and whether the assistant will be assigned or not for a task are also decided on.
3) Solving the problem allows for obtaining a certain amount of useful data regarding the disassembly plan, which is an output of the disassembly planning process.

The remainder of this chapter is organized as follows: Section 2 defines the problem. In Section 3, a 0-1 integer linear programming formulation is developed and an illustrative example and a family of valid inequalities for the PDLBP_PA are presented. Section 4 details PASAfor PDLBP_PA. Computational results are reported in Section 5. Some concluding remarks and future perspectives are given in Section 6.

2. Problem Definition

In our problem, a task results in the removal of one part. For this reason, the number of tasks equals to the number of parts. All or some parts of the product are demanded. A part is may be demanded by only one demand source. Any part may be undamaged or damaged or may be damaged during the disassembly processes. As explained in detail in Section 1.1, UA and DA occur for a task in three different ways. The first is the removal of some parts by applying only UA. The second one relates only to the application of DA to parts. The third one is that both applications may be applicable, alternatively. Each action is carried out with or without equipment, with assistant or without assistant, so one or more processing alternatives occur for a task. According to the Section 1.1., and explanations given above, all possible processing alternatives for a task are given in Figure 2. It is seen that there are six possible processing alternatives for a task.

It is assumed that important information on disassembly actions, products, and parts are obtained by using Industry 4.0 applications before the product enters into disassembly processes. Disassembled products are sensor embedded products. That is, each product is monitored throughout the life cycle via active RFID labels and sensors placed thereon. In this way, besides the static information about the products and parts such as sales date, serial number, bill of material, etc., dynamic information such as maintenance, repair or renewal of the product, and adding or removing parts from the product are continuously updated. All static and dynamic information are retrieved when the product,which has completed its life cycle, arrives at the recovery facility. Thus, which action or actions can be applied to which part is known before the disassembly starts. Parameters such as time, revenue, and cost are determined or predicted by using the information previously saved for other products. Obtaining the relevant information from any product is as follows: After the product is loaded onto the conveyor belt, it starts to move towards the stations. Each workstation is equipped with a PC, which receives work instructions and transmits the results of disassembly. Each PC is connected to a server. The product is identified on each workstation by the RFID reader of the station. The required disassembly instructions and information on the disassembled part(s) are displayed on the PC monitor of the station. The tasks of each workstation and the status of the currently disassembled parts are displayed in real time. Information about the disassembled parts and tasks are stored on a computer server for future utilization and statistical analysis.

In addition to the assumptions explained in PDLBP and given above, the following assumptions of the developed PDLBP_PA are given below (Kara et al. 2011; Hezer and Kara 2015):

- A workstation can perform both unfastening and destructive actions.
- Products are disassembled with the same cycle time.
- The equipment used in UA and the equipment used in DA are generally different from each other. Therefore, it can be said that the equipment used for the task represents the type of action.
- The task time is deterministic, but depends on the resources (equipment type and assistant) allocated to perform the task.
- The time of some tasks can be reduced by performing them with the help of an assistant.

- Some tasks should be performed using specific equipment. Alternative types of equipment may be used for a task. Some tasks can be performed without any equipment or with a type of equipment, each of which has a cost value.
- The number of workers required for operating the workstations is sufficient. However, the number of other resources (equipment types and assistants) is limited.
- The idle time of the workers are ignored.

3. Mathematical Formulations

In this section, firstly a 0-1 integer linear programming formulation for PDLBP_PA (referred to as PLF in the sequel) is proposed by referring to the papers of Das and Naik (2002); Gokcen et al. (2006); Kara et al. (2011). Then PLF is validated on an illustrative problem and valid inequalities are introduced to strengthen PLF.

The notation used to describe the proposed 0-1 integer linear programming formulation, PLF, and valid inequalities are given as follows:

Indices

h	disassembly line
i,l	task (part)
j,v	workstation
e	equipment

Parameters and sets

H	set of disassembly lines
J	set of workstations
I_h	set of tasks on line h
E	set of equipment
E_{hi}	set of equipment which can be used to process task i on line h
PA_{hi}	set of immediate "AND" predecessors of task i on line h
PO_{hi}	set of immediate "OR" predecessors of task $i$on line h
PAL_{hi}	set of all "AND" predecessors of task $i$on line h
POL_{hi}	set of all "OR" predecessors of task i on line h
SAL_{hi}	set of all successors of task $i$on line h
NH	available number of disassembly lines
n_h	available number of tasks on line h
$\|\|S_{hj}\|\|$	total number of tasks (that can be) assigned to station j on line h
N_{he}	available number of equipment e which can be used on line h
NA	available number of assistants of disassembly lines
NWS_{max}	maximum number of workstations
t_{hie0}	task time of task i if it is processed with equipment e without assistant on line h
t_{hie1}	task time of task i if it is processed with equipment e with assistant on line h
r_{hie}	revenue that meets per unit demand of part $i$on line h with equipment e
d_{hi}	demand for part i on line h
C	cycle time
cw	utilization cost of a workstation (worker + fixed costs)

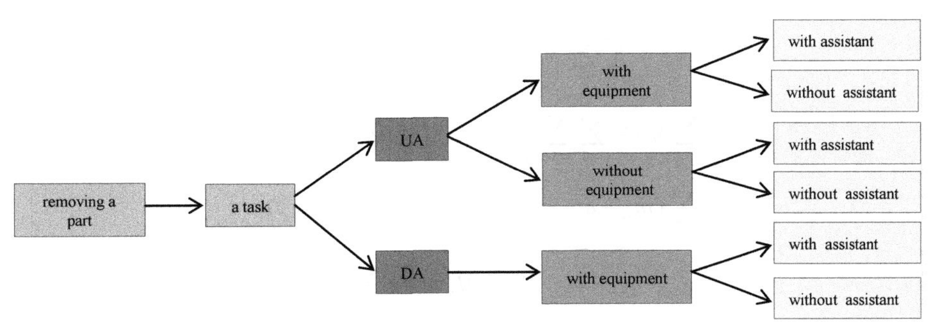

Fig. 2: All possible processing alternatives for a task.

ca	employment cost of an assistant
c_e	operating cost of equipment e
EL_{hi}	the earliest station for task i on line h
LT_{hi}	the latest station for task $i$on line h
tS_{hi}	the shortest time for task i on line h
tm_{hi}	the longest time for task i on line h
top_{hi}	the sum of task times of all predecessors of task i on line h and the task time of i
$topm_{hi}$	the sum of task times of all successors of task i on line h and the task time of i
M	a big number

Variables

x_{hiej}	1, if task $i$on line h is assigned to workstation j with equipment e; 0, otherwise
p_{hiej}	1, if task i on line h is assigned to workstation j with equipment e without assistant; 0, otherwise
q_{hiej}	1, if task i on line h is assigned to workstation j with equipment e with assistant; 0, otherwise
z_{hej}	1, if equipment e is assigned to workstation j on line h; 0, otherwise
a_j	1, if an assistant is assigned to workstation j; 0, otherwise
f_j	1, if workstation j is utilized; 0, otherwise
U_{hj}	1, if workstation j is utilized on line h; 0, otherwise

3.1 Proposed 0-1 Integer Linear Programming Formulation

Following the assumptions mentioned in Section 2, the resources to be assigned to the tasks and the stations to which the tasks will be assigned should be decided depending on these assumptions. These decisions should be made in a way that satisfies the constraints on precedence relationships, cycle time, and resources. Solving PLF allows determining stations to be opened, tasks to be performed, parts to be removed, disassembly action to be processed for each task and the most appropriate processing alternative to be selected for each task performed. It also allows the tasks that require the assistants and/or the same equipment type may be assigned to the same workstation, so the total cost of the line, which is associated with cw, ca, c_e, may be reduced. Furthermore, PLF determines the number of lines to be opened because in some cases disassembling the product(s) may not be profitable at any disassembly level. Therefore, the right decision in this case is not to disassemble such product/products. The proposed integer programming formulation, PLF, for PDLBP_PA is as follows:

$$Maximize \sum_h \sum_{i \in I_h} \sum_{e \in E_{hi}} \sum_j d_{hi} r_{hie} x_{hiej} - \sum_j (cwf_j + caa_j) - \sum_h \sum_{e \in E} \sum_j c_e z_{hej} \quad (1)$$

$$\sum_j \sum_{e \in E_{hi}} x_{hiej} \leq 1 \quad \forall h \in H; \forall i \in I_h \quad (2)$$

$$\sum_{e\in E_{hi}} x_{hiej} \le \sum_{v=1}^{j}\sum_{e\in E_{hl}} x_{hlev} \quad \forall h\in H;\forall i\in I_h;\forall l\in PA_{hi};\forall j\in J \tag{3}$$

$$\sum_{e\in E_{hi}} x_{hiej} \le \sum_{v=1}^{j}\sum_{l\in PO_{hi}}\sum_{e\in E_{hl}} x_{hlev} \quad \forall h\in H;\forall i\in I_h;\forall j\in J \tag{4}$$

$$\sum_{e\in E_{hi}} (p_{hiej}+q_{hiej}) = \sum_{e\in E_{hi}} x_{hiej} \quad \forall h\in H;\forall i\in I_h;\forall j\in J \tag{5}$$

$$p_{hiej}+q_{hiej} = x_{hiej} \quad \forall h\in H;\forall e\in E_{hi};\forall i\in I_h;\forall j\in J \tag{6}$$

$$\sum_{i=1}^{nh}\sum_{e\in E_{hi}} (t_{hie0}p_{hiej}+t_{hie1}q_{hiej}) + \sum_{i=1}^{n(h+1)}\sum_{e\in E(h+1)i} (t_{(h+1)ie0}p_{(h+1)iej}+t_{(h+1)ie1}q_{(h+1)iej}) \le Cf_j \quad \forall j\in J; j=1,\ldots,H-1 \tag{7}$$

$$\sum_{i\in I_h}(p_{hiej}+q_{hiej}) - Mz_{hej} \le 0 \quad \forall h\in H;\forall e\in E_{hi};\ \forall j\in J \tag{8}$$

$$\sum_{h}\sum_{i\in I_h}\sum_{e\in E_{hi}} q_{hiej} - Ma_j \le 0 \quad \forall j\in J \tag{9}$$

$$\sum_{j\in J} z_{hej} \le N_{he} \quad \forall h\in H;\forall e\in E_{hi} \tag{10}$$

$$\sum_{j\in J} a_j \le NA \tag{11}$$

$$\sum_{i\in I_h}\sum_{e\in E_{hi}} x_{hiej} - \| S_{hj} \| U_{hj} \le 0 \quad \forall h\in H;\forall j\in J \tag{12}$$

$$\sum_{i\in I_h}\sum_{e\in E_{hi}} x_{hiej} \ge U_{hj} \quad \forall h\in H;\forall j\in J \tag{13}$$

$$U_{hj} + U_{(h+b)j} \le 1 \quad H\ge 3; h=1.....H-2; b=2,\ldots,H-h;\forall j \tag{14}$$

$$x_{hiej}, p_{hiej}, q_{hiej}, z_{hej}, f_j, a_j, U_{hj} \in \{0,1\} \qquad \forall h,i,j,k \tag{15}$$

The objective function (1) maximizes the total net recovery profit associated with the total revenue earned from released parts, workstation utilization, assistant assignment, and equipment allocation. Equation (2) indicates that a task can be assigned to at most one work station. Precedence relations among tasks of each line are satisfied by the sets of constraints given in Equations (3) and (4). Equation (3) ensures that task i cannot be assigned until its AND predecessors are assigned to station 1 through j on line h. Equation (4) ensures that task i cannot be assigned to station j until at least one of its OR predecessors is assigned to workstation 1 through j on line h. Equations (5) and (6) determine the resources (equipment type and assistant) allocated to a workstation. Equation (7) ensures that the workload of a workstation does not exceed the predetermined cycle time. Equation(8) determines whether equipment e is allocated for line h to workstation j. Equation (9) determines whether an assistant is assigned to workstation j. Equation (10) restricts the allocated number of equipment type e on line h by the available number of this equipment type. Equation (11) ensures that the number of assistants assigned to workstations does not exceed the available number of assistants. Equations (12) and (13) determine which stations should be opened on the lines. Equation (14) ensures that a common workstation is utilized for only two adjacent lines. Finally, Equation (15) denotes that all variables in the formulation are binary variables.

3.2 Illustrative Example

In this section, an illustrative example, which consists of two disassembly lines (two products)each with 10 tasks, is used to describe PDLBP_PA and also to show the efficiency of parallel balancing of the lines simultaneously according to the independent balancing of the lines. The number of workstations opened, types and numbers of equipment used, numbers of assistants assigned for each line and the most appropriate processing alternative of each task performed are obtained by solving the problems. Also, the numbers of lines opened are determined. Figure 3 illustrates the precedence diagrams for line 1 and line 2.

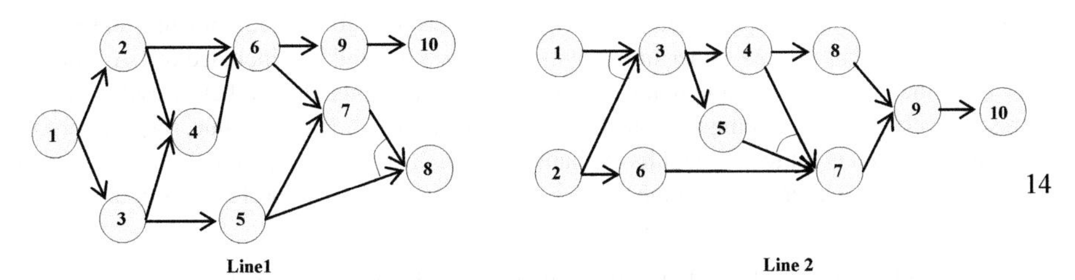

Fig. 3: The first and second products disassembled in line 1 and line 2, respectively.

Table 1 presents the problem data and describes processing alternatives of tasks. The task times, costs of equipment, revenues and available number of resources are generated randomly. The lines work for 8 h in a day for 20 workdays. Cycle time is calculated according the paper of Gungor and Gupta (2002) ($C = (8 \times 60 \times 20)/300 = 32$ min per unit).

Table 1: Problem data.

Line 1

i	d_{hi}	Task times and revenues		Equipment 0	1	2	3
1	160	Assistant	Yes	15	13		10
			No				
		r_{hie}		15	15		11
2	220	Assistant	Yes				
			No				10
		r_{hie}					12
3	0	Assistant	Yes	13	10		
			No		12		
		r_{hie}		0	0		
4	0	Assistant	Yes			12	
			No			14	
		r_{hie}				0	
5	200	Assistant	Yes		13		
			No				
		r_{hie}			19		
6	300	Assistant	Yes				
			No		14		
		r_{hie}			16		
7	0	Assistant	Yes				
			No			12	
		r_{hie}				0	
8	230	Assistant	Yes				
			No	13	11		
		r_{hie}		15	15		
9	0	Assistant	Yes				
			No		12		11
		r_{hie}			0	0	0
10	300	Assistant	Yes	15		10	
			No		13	12	12
		r_{hie}		20	20	17	13

Line 2

i	d_{hi}	Task times and revenues		Equipment 0	1	2	3
1	280	Assistant	Yes				
			No		15		10
		r_{hie}			20		15
2	0	Assistant	Yes		12		
			No				
		r_{hie}			0		
3	140	Assistant	Yes	15	12		
			No				
		r_{hie}		12	12		
4	230	Assistant	Yes				
			No				10
		r_{hie}					17
5	0	Assistant	Yes				11
			No				13
		r_{hie}					0
6	300	Assistant	Yes				
			No		14	10	
		r_{hie}			18	14	
7	0	Assistant	Yes	10			
			No	12			
		r_{hie}		0			
8	300	Assistant	Yes				
			No		15	10	
		r_{hie}			18	12	
9	100	Assistant	Yes				
			No	14	12		10
		r_{hie}		15	15		12
10	0	Assistant	Yes				
			No	12			
		r_{hie}		0			

$cw = 2000$, $ca = 1200$, $c_1 = 950$, $c_2 = 1050$ and $c_3 = 850$ cost units,
$NA = 2$, $N_{11} = 2$, $N_{12} = 1$, $N_{13} = 1$, $N_{21} = 1$, $N_{22} = 2$, $N_{23} = 1$

There are three different types of equipment. While equipment 1 is used for UA,equipment 2 and 3 are used for DA. The equipment type 0 means that the task is performed without equipment and it is included in UA. The last row shows the costs (*cw, ca, e*) defined in PLF, available number of assistants (*NA*), available number of each equipment (N_{he}) and cycle time (*C*), respectively. *i* and d_{hi} columns denote the task (part) number and part revenue, respectively. As a matter of course, the cost of equipment 0 is set to zero and its available amount is not restricted in the solution methods. The other cells of Table 1 show processing alternatives of the tasks. For example, task 10 of line 1 results in the removal of part 10 of line 1, and it has five processing alternatives given below:

- The first processing alternative includes UA with equipment 0 and with an assistant.
- The second one includes UA with equipment 1 and without assistant.
- The third one includes DA with equipment 2 and with assistant.
- The fourth one includes DA with equipment 2 and with assistant.
- The fifth one includes DA with equipment 3 and without assistant.

The task time of task 10 changes depending on these processing alternatives and each takes time of five processing alternatives is 15, 13, 10, 12, 12 min, respectively. It should be noted here that the task time without an assistant is greater than that of the one without an assistant for equipment 2 (i.e., the third and fourth processing alternatives). The revenues of the processing alternatives depending on the equipment and actions of the part 10 are 20, 20, 17, 17, 13 units, respectively. While the revenue remains the same for UA (equipment 0 and 1), it changes for DA (equipment 2 and 3) and revenues obtained by using equipment 2 and equipment 3 are different from each other. When equipment 2 is used, part 10 is less damaged, so the revenue obtained with equipment 2 is more than the one obtained with equipment 3.

Firstly, two lines are balanced independently by using PDLBP_PA (i.e., H is set at 1 and some constraints are modified according to the single line) and Figure 4 shows the optimal solutions for line 1 and line 2, respectively.

As it is seen in Figure 4, independent solutions of line 1 and line 2 consist of 3 and 3 workstations, respectively. While the tasks 3, 4, 5, 7, 8 of line 1 and 5, 7, 9, 10 of line 2 are not performed, utilized number of assistants, equipment 1 and equipment 3 are 2, 1 and 1, respectively for line 1 and 1, 1, and 1, respectively for line 2. Equipment 2 is not allocated for both line 1 and line 2. The total net recovery profits of line 1 and line 2 are 5640 and 10640 cost units, respectively. If a facility, in which line 1 and line 2 are balanced independently, is considered, the utilized number of workstations, assistants, equipment 1 and equipment 3 will be 6, 3, 3 and 2, respectively with a total net recovery profit of 16280 cost units (sum of line 1 and line 2).

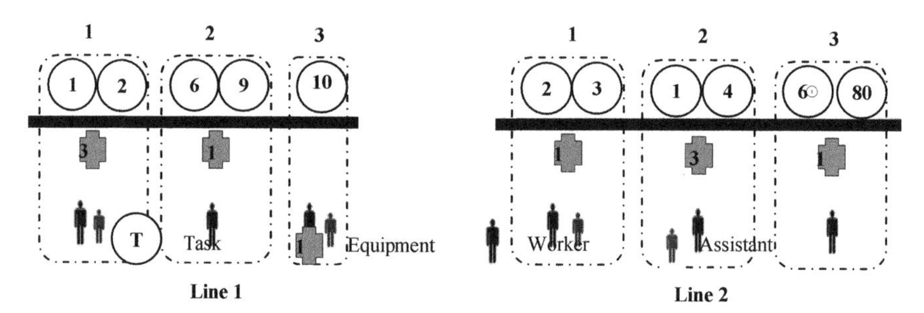

Fig. 4: Independent balancing of line 1 and line 2.

Then, the example is solved considering balancing the parallel lines simultaneously. The optimal PDLBP_PA solution is given in Figure 5. It consists of five work stations with a total net recovery profit of 18840 units. When the results of the independent solutions shown in Figure 4 are examined, while utilized number of equipment 1 and equipment 3 are the same, number of workstations and assistants are reduced to 5 and 2, respectively.

These results show that PDLBP_PA provides 13.5% upper total net recovery profit compared to independent balancing of the lines for the cycle time value of 32 min. As mentioned before, information about the processing alternative of the each task performed on each line are obtained. For example, task 10 of line 1 and task 3 of line 2 are performed with the cooperation of worker and assistant with equipment 0 (no equipment). Thus, both tasks can be performed with UA.

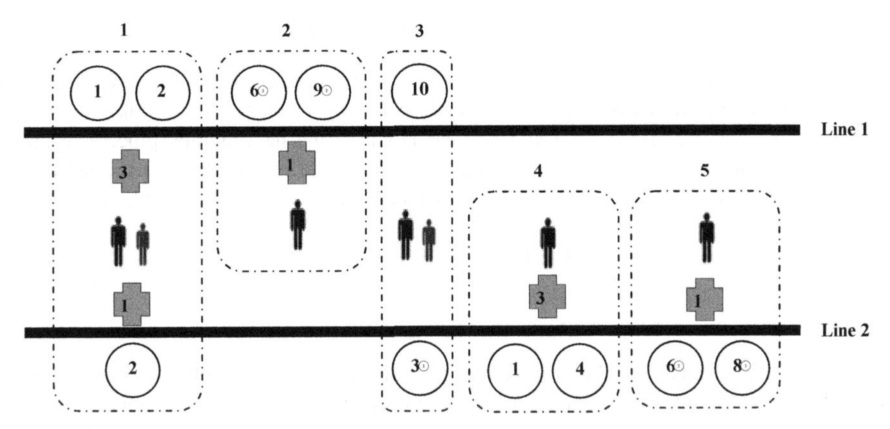

Fig. 5: Parallel balancing of line 1 and line 2.

3.3 Valid Inequalities

In this section, valid inequalities that can be added to PLF are proposed to strengthen the weak upper limits obtained by linear programming relaxation. These inequalities eliminate some fractional solutions from the solution space, providing stronger upper limits (Karaoglan et al. 2012). The following five polynomial-size valid inequalities are utilized. The first and second valid inequalities are given as follows:

$$\sum_{j\in J}\sum_{e\in E_{hi}} x_{hiej} \le \sum_{j\in J}\sum_{e\in E_{hl}} x_{hlej} \quad \forall h\in H; \forall i\in I_h; \forall l\in PA_{hi} \tag{16}$$

$$\sum_{j\in J}\sum_{e\in E_{hi}} x_{hiej} \le \sum_{j\in J}\sum_{l\in PO_{hi}}\sum_{e\in E_{hl}} x_{hlej} \quad \forall h\in H; \forall i\in I_h \tag{17}$$

As a result of relaxing constraints of PLF, tasks are assigned to stations as fractional numbers; therefore, precedence constraints presented in Equations (3) and (4) are violated. Equations (16) and (17), based on the paper of Altekin et al. (2008), are used to prevent the violation of "AND" and "OR" precedence relationships. Thus, Equation (16) ensures that the total fractional task assignment of *i* cannot exceed the total fractional assignment of each "AND predecessors' of *i*. Equation (17) ensures that the total fractional assignment of task *i* is equal to or less than the total fractional assignment of all OR predecessors of *i*.

Another three valid inequalities are related to the earliest and the latest stations to which each task can be assigned are given as follows:

$$\sum_{j=EL_{hi}}^{LT_{hi}}\sum_{e\in E_{hi}} x_{hiej} \le 1 \quad \forall h\in H; \forall i\in I_h \tag{18}$$

$$\sum_{j>EL_{hi}}\sum_{e\in E_{hi}} x_{hiej} = 0 \quad \forall h\in H; \forall i\in I_h \tag{19}$$

$$\sum_{j>LT_{hi}}\sum_{e\in E_{hi}} x_{hiej} = 0 \quad \forall h\in H; \forall i\in I_h \tag{20}$$

Since the "OR" precedence relationship type is used in the proposed PDLBP_PA problem, the calculation of the earliest and latest stations to which task *i* can be assigned on line *h* has certain differences compared to general calculation methods. First of all, one task time must be determined for each task, since the task time may vary depending on processing alternatives of a given task. While the task time is taken directly for tasks with one processing alternative, the shortest task time (ts_{hi}) is selected for tasks that have multiple processing alternatives. After the sum of task times of all predecessors of each task and the task time of the related task (top_{hi}) is calculated according to Equation (21), the earliest station for each task is calculated according to Equation (22), based on the paper of Kalaycilar (2016).

$$top_{hi} = \left(\sum_{l\in PAL_{hi}\cup POL_{hi}} ts_{hl}\right) + ts_{hi} \quad \forall h, \quad \forall i\in I_h \tag{21}$$

$$EL_{hi} = \lceil top_{hi} / C \rceil \quad \forall h, \quad \forall i \in I_h \tag{22}$$

The calculation method for the latest station for a task is developed in this chapter as follows: As in the procedure for determining the earliest station, one task time must be set for each task. While the task time is taken directly for tasks with one processing alternative, the longest task time (tm_{hi}) is selected for tasks that have multiple processing alternatives. After determining the task time for each task, the sum of task times of its successors of each task and the task time of the related task ($topm_{hi}$) and the latest station for each task (LT_{hi}) are calculated according to Equations (23) and (24), respectively.

$$topm_{hi} = tm_{hi} + \sum_{l \in SAL_{hi}} tm_{hl} \forall h, \quad \forall i \in I_h \tag{23}$$

$$LT_{hi} = NWS_{max} + (1 - \lceil topm_{hi} / C \rceil \quad \forall h, \quad \forall i \in I_h \tag{24}$$

For the illustrative example, task 3 of line 1 has three processing alternatives as follows: "manual and with assistant (t_{1301})", "using equipment 1 without an assistant (t_{1310})", and "using equipment 1 with an assistant (t_{1311})". The task time for each processing alternative is $t_{1301} = 13$, $t_{1310} = 12$, $t_{1311} = 10$. According to the methods described above, ts_{13} and tm_{13} are determined as 10 and 13, respectively. Then, top_{hi}, EL_{hi}, $topm_{hi}$, LT_{hi} values of task 3 on line 1 are calculated as follows: $top_{13} = 20$, $EL_{13} = 1$, $topm_{13} = 106$, $LT_{13} = 17$.

4. Introduction to Simulated Annealing and Adaptive Simulated Annealing

SA is an efficient metaheuristic approach that starts from a solution formed with a constructive algorithm and produces a neighbour solution with small changes in the current solution in each iteration. If the neighbour solution is better than the current solution, the algorithm accepts the better neighbour solution as the current solution; otherwise it accepts the neighbour solution with a certain probability of $exp(\Delta–/T)$ as the current solution. Δ is calculated by $(f(neighbour\ solution) - f(current\ solution)$ and T is the control parameter called temperature.

Allowing the transition to neighbourhood solutions worse than the current solution with certain probabilities enables the algorithm to escape from the local optimum points. At the beginning of the search, the value of the T parameter is kept high and decreased during the search according to a function known as the cooling schedule. That is, poor neighbourhood solutions are more likely to be accepted initially, but the possibility decreases over time. Thus, from the start the aim is, a mechanism which searches for the solution space more generally and focuses on the good solution regions is aimed at. The initial value of the parameter T and the rate of cooling schedule affect the dependence and performance of the algorithm on the initial solution. The search continues until a certain termination criterion is met (Guden and Meral 2016).

In ASA, the algorithm corrects the value of the control parameter T using the information obtained during the search. This creates a smarter and more flexible structure that adjusts to the knowledge based on the search history. In the classical SA, the value of T parameter is continuously decreased over time while it can be increased in ASA, if needed. It allows the exploration of the different regions of solution space, so the dependence of the performance of the algorithm on the parameter T is reduced (Guden and Meral 2016).

5. A Heuristic Approach for PDLBP_PA

Since DLBP belongs to the class of NP-hard problems (McGovern and Gupta 2007a,b), PDLBP_PA is also an NP-hard problem. Therefore, PLF is not directly applicable in finding optimal solutions to medium- and large-size problems. Consequently, meta-heuristic approaches are needed in finding quick solutions to such problems (Kesen et al. 2010). In DLBP literature, Kalayci and Gupta (2013a), Kalayci and Gupta (2013c), Fang, Ming et al. (2019) proposed an SA based approach for solving straight DLBP, and they demonstrated the superior functionality of SA over DLBP. Thus, in this chapter, a heuristic approach based ASA, called PASA is developed to solve medium- and large-size problems.

The PASA starts with an initial solution. It consists of two inner loop levels and an outer loop level. The temperature is updated according to certain rules in each step of the outer loop and the neighborhood solution procedure is repeated throughout each outer loop according to special rules. When the rules for completion of the outer loop are met, PASA is terminated. The notation used in PASA is as follows:

SAT set of assignable tasks

SCT set of tasks obtained from the solution of linear relaxation of the results of the formulation after adding valid inequalities to PLF

SOS	set of objective functions
SIT	set of insertable tasks
SST	set of selected tasks
k	iteration counter of outer loop
T_b	initial temperature
T_k	current temperature of current iteration k
T_s	final temperature
$iter_{out}$	maximum iteration number of outer loop
$iter_{in1}$	maximum iteration number of the first inner loop
$iter_{in2}$	maximum iteration number of the second inner loop
t_{in1}	iteration counter of the first inner loop
t_{in2}	iteration counter of the second inner loop
gar	general acceptance rate
S_0	the initial solution
S_b	the best solution
S_c	the current solution
S_n	the neighbour solution
A_0	objective function of the initial solution
A_b	objective function of the best solution
A_c	objective function of the current solution
A_n	objective function of the neighbour solution
α_1	the first cooling rate
α_2	the second cooling rate

The details of PASA are given including the generation of an initial solution and neighbourhood solutions in the following subsections.

5.1 Initial Solution

As previously mentioned in Section 1.1, in practical applications, removing all parts of the product is not usually profitable. This is because there may not be a demand for some parts or they may not be valuable. For this reason, valueless parts should not be removed as far as possible. For this purpose, instead of starting the initial solution with all tasks, a method, which involves some data related to the optimal solution of the proposed formulation, is utilized. It starts with the total fractional task assignments from the solution of semi-linear programming relaxation of the strong formulations (referred to as SLF in the sequel) in which decision variables related with workstation and assistant are kept as binary while remaining decision variables are relaxed (detailed information about this relaxation type is given in Section 5.2) (Altekin et al. 2008).

First of all, it is assumed that all tasks and resource(s) of the tasks with positive fractions obtained as a result of the solution of SLF are 'selected'. Then it continues with the initial solution consisting of two stages considering these tasks and resources. The tasks are sequenced randomly considering the precedence relationships in the first stage. In the second stage, tasks are assigned to stations using "station oriented assignment procedure". The general steps of the initial solution are given below iteratively:

Step 1: Create *SCT* using solution of SLF

Step 2: Add all tasks, which have not any predecessors, from *SCT* to *SAT*.

Step 3: Update *SCT*.

Step 4: Select a task randomly from *SAT* and add the task to *SST*.

Step 5: Update *SAT*.

Step 6: Update *SCT* according to *SST* in terms of precedence relationships.

Step 7: If $SCT = \phi$ and $SAT = \phi$ then go to Step 9. Otherwise, go to Step 8.

Step 8: If there are tasks which have not any predecessors in *SCT*, go to Step 2. Otherwise go to Step 4.

Step 9: Apply station-oriented procedure using the tasks of *SST*, create and go to Step 10.

Step 10: Calculate *A*0. Set $S_b = S_c = S_0$, $A_b = A_c = A_0$ and go to Step 11.

Step 11: Stop.

The data obtained from the solution of SLF for the illustrative example is presented as "line.task.equipment.assistant" in *SCT* as follows (Step 1):

$$\text{SCT} = \left\{ \begin{matrix} 1.1.0.1, 1.2.3.0, 1.3.1.1, 1.5.1.1, 1.6.1.0, 1.8.0.0, 1.9.3.0, 1.10.1.0, \\ 2.1.1.0, 2.2.1.1, 2.3.1.1, 2.4.3.0, 2.6.1.0, 2.7.0.1, 2.8.1.0, 2.9.0.0 \end{matrix} \right\}$$

According to *SCT*, tasks 4 and 7 of line 1, and tasks 5 and 10 of line 2 are not included in the solution. It should be noted here that in PASA, if an assistant helps the worker, it is represented as "1", otherwise it is represented as "0". When the steps from 2 to 11 are applied to SCT, the initial solution (S_0), which is given in Figure 6, is obtained. As seen in Figure 6, workstations 1, 3, 5 and 6 are common workstations. In workstation 1, task 1 of line 1 is done without any equipment and task 2 of line 2 is done with equipment 1. Both tasks are performed with the help of an assistant. Work station 3 consists of task 3 of line 1 and task 3 of line 2 that are processed with the cooperation of worker and assistant with equipment 1. In workstation 5, task 5 of line 1 is performed using equipment 1 with the help of an assistant while task 4 of line 2 is performed with equipment 3 without any assistance. In workstation 6, task 8 of line 1 is done without any equipment and

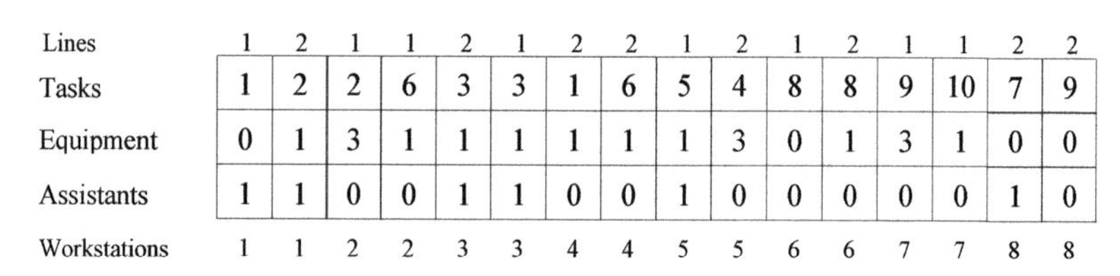

Lines	1	2	1	1	2	1	2	2	1	2	1	2	1	1	2	2
Tasks	1	2	2	6	3	3	1	6	5	4	8	8	9	10	7	9
Equipment	0	1	3	1	1	1	1	1	1	3	0	1	3	1	0	0
Assistants	1	1	0	0	1	1	0	0	1	0	0	0	0	0	1	0
Workstations	1	1	2	2	3	3	4	4	5	5	6	6	7	7	8	8

Fig. 6: The initial solution of the illustrative example with cycle time 32 min.

task 8 of line 2 is done with equipment 1. Both tasks are performed without any assistance. The objective function of the initial solution (A_0) is calculated as 15630 TL per mon. It is set as $A_c = A_b = A_0 = 15630$ TL per mon and $S_c = S_b = S_0$.

5.2 Accepting the Neighborhood Solution

In this chapter, the equation of the metropolitian criterion, which is usually presented for minimization problems in SA literature, needs to be updated referring to the papers about maximization problems (Etgar et al. 1997; Attiya and Hamam 2006; Gharan and Vondrak 2011). Neighbour solution is compared with the current solution according to $\Delta = (A_c - A_n)$. If Δ is greater than zero ($\Delta > 0$), neighbour solution is accepted, otherwise, it may be accepted as a new current solution with the probability of exp (Δ/T_k) in which case, a random number (*rnd*) uniformly between zero and one is generated. If min{1, $exp(\Delta/T_k)$} > *rnd*[0,1] neighbour solution is accepted, it is rejected.

5.3 Neighbourhood Generation

A neighbour solution (A_n) was obtained from a current solution (A_c) by using a specific move. In this chapter, in addition to *insert* and *swap* moving mechanisms, which are usually used in SA, three additional moving mechanisms were developed specific to the structure of the problem. The first two mechanisms, which are called *task removing* and *task adding*, were developed based on the paper of Altekin et al. (2008). The last mechanism is new and called resource changing.

In *insert* mechanism, a randomly selected task is inserted into a station with the minimum station time. In *swap* mechanism, two randomly selected tasks from different stations, which are selected randomly, are exchanged. In each case, precedence relationships, cycle time, and special limitations are checked and new feasible neighborhood solutions are obtained in both cases (Erel et al. 2001).

Task *removing* involves removing tasks from the solution sequence according to a certain procedure. This procedure is repeated in each iteration according to the number of tasks in the solution sequence. Each task is assessed based on the current solution and the current objective function value. This moving mechanism facilitates the elimination of tasks that the feasible or optimal solution does not contain (Altekin et al. 2008). The steps of the task removing moving mechanism are as follows:

Step 1: Select the first task of S_c.

Step 2: Remove the selected task and its successor/successors from S_c, temporarily.

Step 3: Calculate A_c and save A_c to *SOS.*

Step 4: Add again the removed task/tasks to S_c.

Step 5: Pass onto the next task of S_c.

Step 6: Repeat the Steps 2 to 5 for all tasks of S_c. Then go to Step 7.

Step 7: Select the solution which had the highest total net recovery profit. Set it as A_n and its solution as S_n, respectively.

Step 8: Let $\Delta = A_n - A_c$.

Step 9: If $\Delta > 0$, then accept the neighbour solution and set $S_c = S_n$, $A_c = A_n$ and go to Step 11.

Step 10: If $\Delta \leq 0$, then accept the neighbour solution as a new current solution with the probability of $\exp(\Delta/T)$ and set $S_c = S_n$, $A_c = A_n$ and go to Step 12. Otherwise S_c and A_c remain unchanged and go to Step 13.

Step 11: If $A_c > A_b$, then $A_b = A_c$; $S_b = S_c$. Otherwise A_b and S_b are not changed and go to Step 12.

Step 12: Delete the task/tasks of selected solution from S_c, permanently. Update S_c. Set $SOS = \{\phi\}$ and go to Step 13.

Step 13: Stop.

When *task removing* moving mechanism is applied to the initial solution, which is given in Figure 6, it must begin with task 1 of line 1. Task 1 is "AND" precedecessor of all tasks of line 1, so task 1 and all remaining tasks (successors of task 1 of line 1) are deleted from the solution temporarily (Step 1 and Step 2). Then the objective function of this temporary solution is calculated and saved in *SOS* (Step 3). The temporary solution applied after the first three steps of task removing moving mechanism is shown in Figure 7.

The deleted tasks are re-added to the solution and the next task, i.e., task 2 of line 2 is selected and deleted temporarily. When task 2 of line 2 is deleted, tasks to be deleted from line 2 are 3, 6, 7, and 9. Task 2 is one of the "or" precedecessors of task 3. The other "or" predecessor of task 3 is task 1. The point to consider when deleting tasks is the criterion for evaluating tasks with an "or" precedence relationship. In case of deletion of a task with a predecessor, the deletion of the task(s), to which it has a predecessor, is determined by the status of other "or" predecessors. In this example, if task number 1 was assigned to the station to which task 3 was assigned, or to stations that preceded it, task 3 would not have been deleted. However, in this example, since task 1 is assigned after task 3, task 2 becomes the predecessor "and" of task 3. Therefore, when task 2 is deleted, task 3 must also be deleted. Steps 1 to 4 are repeated for all tasks of S_c. All objective function values of all temporary solutions (Step 1 to Step 6) are given in Table 2.

According to Step 7, the highest total net recovery profit is obtained when task 7 and its successor task 9 are deleted from line 2. The relevant section is shown in Table 2 in bold. When steps 8–13 are applied for this total net recovery profit, task 7 and task 9 ofline 2 are deleted from the solution permanently because $\Delta = A_n - A_c = 17330 - 15630 > 0$. Then the objective functions of the best solution and current solution are updated as $A_b = A_c = A_n = 17330$ and the best and the current solutions are set as $S_b = S_c = S_n$. Figure 8 shows the updated current solution.

The *task adding* moving mechanism involves adding new tasks to the solution sequence according to a certain procedure. The tasks to be added are selected from tasks that have never been included in the initial solution and have

Lines	2	2	2	2	2	2	2	2
Tasks	2	3	1	6	4	8	7	9
Equipment	1	0	1	1	3	1	0	0
Assistants	1	1	0	0	0	0	1	0
Workstations	1	2	3	3	4	5	6	6

Fig. 7: The temporary solution after the first three steps.

Table 2: The results of the first six steps.

Line	Deleted task	Deleted successor tasks	Total net recovery profit
1	1	2,3,5,6,8,9,10	3280
2	2	3,6, 7, 9	5590
1	2	6,9,10	9790
1	6	9,10	9580
2	3	4, 7, 8, 9	10290
1	3	5,8	9090
2	1	-	10030
2	6	7, 9	11930
1	5	8	10530
2	4	7, 8, 9	9820
1	8	-	12180
2	8	9	9680
1	9	10	13430
1	10	-	10580
2	7	9	17330
2	9	-	14130

Lines	1	2	1	1	2	1	2	2	1	2	1	2	1	1
Tasks	1	2	2	6	3	3	1	6	5	4	8	8	9	10
Equipment	0	1	3	1	1	1	1	1	1	3	0	1	3	1
Assistants	1	1	0	0	1	0	0	0	1	0	0	0	0	0
Workstations	1	1	2	2	3	3	4	4	5	5	6	6	7	7

Fig. 8: Updated current solution.

been permanently deleted in the task removing moving mechanism. This mechanism may provide an increase in the profit by decreasing idle times of workstations. The steps of the task adding mechanism are as follows:

Step 1: Create *SIT* considering the tasks that are not in S_c.

Step 2: Add the assignable tasks from *SIT* to *SAT* according to precedence relationships.

Step 3: Apply following procedure for each task of *SAT*.

Step 3.1: Select the first task of *SAT* and add it to S_c temporarily.

Step 3.2: Calculate the total net recovery profit of temporary solution and save it to *SOS*.

Step 3.3: Remove the task from S_c and add it to the end of the sequence of *SAT*.

Step 3.4: Repeat the Step 3.1 to Step 3.3 for each task of *SAT* and go to Step 4.

Step 4: Select the solution which had the highest total net recovery profit from *SOS*. Set it as A_n and its solution as S_n, respectively.

Step 5: Let $\Delta = A_n - A_c$.

Step 6: If $\Delta > 0$, then accept the neighbour solution as a new current solution, set $S_c = S_n$, $A_c = A_n$. Then go to Step 8.

Step 7: If $\Delta \leq 0$, then accept the neighbour solution as a new current solution with the probability of $\exp(\Delta/T)$ and set $S_c = S_n$, $A_c = A_n$ and go to Step 8. Otherwise S_c and A_c remain unchanged and go to Step 10.

Step 8: Add the task/tasks of selected solution to S_c permanently and update S_c and go to Step 9.

Step 9: If $A_c > A_b$, then $A_b = A_c$; $S_b = S_c$. Otherwise A_b and S_b are not changed. Set SOS = $\{\phi\}$ and go to Step 10.

Step 10: Stop.

Resource changing was developed to make use of alternative resources for a given task. In every iteration, the resource is changed randomly for all tasks that have more than one processing alternatives and the objective function is evaluated. The steps of the resource changing mechanism are as follows:

Step 1: Start with the first task of S_c.

Step 2: Randomly select a resource for the task and calculate A_n.

Step 3: Let $\Delta = A_n - A_c$.

Step 4: If $\Delta > 0$, then accept the neighbour solution as a new current solution and set $S_c = S_n$, $A_c = A_n$ and go to Step 6.

Step 5: If $\Delta \leq 0$, then accept the neighbour solution as a new current solution with the probability of exp (Δ/T) and set $S_c = S_n$, $A_c = A_n$. Otherwise S_c and A_c remain unchanged and go to Step 7.

Step 6: If $A_c > A_b$, then $A_b = A_c$; $S_b = S_c$. Otherwise A_b and S_b are not changed and go to Step 7.

Step 7: Repeat Step 2 to Step 7 for each task of S_c and go to Step 8.

Step 8: Stop.

After all moving mechanisms are applied, a repair procedure is used for the current solution. According to this procedure, tasks, which are assigned with an excess number of resources, are controlled and reassessed depending on other alternative resources that they have.

5.4 Cooling Schedule

The cooling schedule to be applied for temperature update was developed based on the paper of Guden and Meral (2016). In PASA, the temperature value is updated by observing results obtained at the end of each outer iteration and assessing the number of current solutions (*NCS*), and also the number of neighborhood solutions which are worse than the current solution (*NSWS*), and the number of worse neighborhood solutions which are accepted (*NWSA*). The calculation method given in Equation (25) is used to update the temperature value. According to the first of three conditions in Equation (25), it means that the algorithm considers the region as a local optimal region and the possibility of escaping from this region is too small. Thus, an updating method, which aims to keep the temperature value fixed rather than reducing, was developed in order to increase the possibility of escaping from this region. The method ensures that NWSA increases, so the larger solution space can be searched for good solutions. In the other two conditions, the possibility of acceptance of inferior solutions decreases and the temperature is reduced more rapidly from the second condition to the third condition.

$$T_{k+1} = \begin{cases} T_k/1, & NSWS/NCS \geq 0{,}8 \text{ and } NWSA/NSWS \leq 0{,}1 \\ \alpha_1 T_k, & 0{,}5 \leq NSWS/NCS < 0{,}8 \text{ and } NWSA/NSWS \leq 0{,}1 \\ \alpha_2 T_k, & NSWS/NCS < 0{,}5 \end{cases} \tag{25}$$

5.5 Stop Condition

PASA terminates by any one of the following criteria:

- The current temperature drops below a predetermined final temperature value,
- The maximum number of iterations is reached (for outer loop),
- The general acceptance rate (*gar* = the number of accepted neighborhood solutions/the number of tried neighborhood solutions) drops below a predetermined minimum value.

General steps of PASA are as follows:

Step 0: Start by determining T_b, T_s, $iter_{out}$, $iter_{in1}$, $iter_{in2}$, gar, a_1, a_2 values and assigning $k = 1$, $t_{in1} = 1$, $t_{in2} = 1$.

Step 1: Create the initial solution, S_c and calculate A_c.

Step 2: Apply the following neighbourhood moving mechanisms.

Step 2.1: Apply *task removing* mechanism.

Step 2.2: Apply *insert* mechanism.

Step 2.3: Apply *swap* mechanism.

Step 2.4: Apply *resource changing* mechanism.

Step 2.5: If $t_{in1} = iter_{in1}$ or $s(S_c) \leq 1$, go to Step 3, otherwise go to Step 2.6.

Step 2.6: Set $t_{in1} = t_{in1} + 1$ and go back to Step 2.

Step 3: Set $S_c = S_b$, $A_c = A_b$.

Step 4: Apply the following neighbourhood moving mechanisms.

Step 4.1: Apply *task adding* mechanism.

Step 4.2: Apply *insert* mechanism.

Step 4.3: Apply *swap* mechanism.

Step 4.4: Apply *resource changing* mechanism.

Step 4.5: If $t_{in2} = iter_{in2}$ go to Step 5, otherwise go to Step 4.6.

Step 4.6: Set $t_{in2} = t_{in2} + 1$ and go back to Step 4.

Step 5: Apply the repair procedure.

Step 6: Set $S_c = S_b$ and $A_c = A_b$.

Step 7: If $k = iter_{out}$ or $T = T_s$ or $gar \leq 0.2$, go to Step 10, otherwise go to Step 8.

Step 8: Set $k = k + 1$ and update T_k according to Equation (25).

Step 9: Set $t_{in1} = 1$, $t_{in2} = 1$ and go back to Step 2.

Step 10: Stop.

6. Computational Analysis

In this section,firstly, the test problems created by the authors are presented for evaluating the performance of the solution approaches. Then, experimental studies which are organized into three stages are presented to evaluate the performance of PLF, valid inequalities, PASA, and the effects of simultaneous balancing of the parallel lines. In the first stage, after the integrated effects of the valid inequalities in PLF are investigated, PLF and the strong formulation, i.e., the formulation after adding inequalities (16)–(20) to PLF, are examined on small-size problems in terms of their ability to optimally solve PDLBP_PA. The second stage investigates the performance of PASA. The last stage investigates the effect of balancing parallel lines simultaneously, instead of independently, on the total net recovery profit.

In this chapter, the state-of-the-art LP/MIP solver CPLEX (version 10.2) was used to solve the formulation and the relaxations. PASA was coded using Visual Basic 6.0 programming language. All experiments were performed on Intel Xeon 8 Duo, 3, 30 GHz equipped (with 8 GB RAM). The CPU time required to obtain optimal solutions was limited to 2 h.

6.1 Test Problems

Currently, there is not a general benchmark used for PDLB_PA; thus, the test problems were created to analyze the performance of the proposed solution approaches. The following assumptions were used to develop test problems:

- Two lines are balanced simultaneously.
- The number of tasks for each line is determined as 10 levels (10, 20,.., 100).
- Three precedence diagrams are generated randomly for each level.
- For each precedence diagram, the demand for parts to be disassembled is determined as three levels showing discrete uniform distribution with U(10–150), U(10–200), U(10–300).
- In connection with the highest demand (i.e., 150, 200, 300), three different cycle time levels (32, 48, 64) are determined for all test instances in minutes. It should be noted here that the cycle times are calculated according to the paper of Gungor and Gupta (2002).
- The numbers of different equipment types are selected as 3, 6, 9, 12, 15, 18, 21, 24, 27 and 30 for 10, 20, 30, 40, 50, 60, 70, 80, 90, 100-task problems, respectively. Available numbers of equipment for each line (N_{he}) in all problems are randomly restricted by 1, 2 or 3. Available numbers of assistants (NA) are restricted by 2, 4, 6, 8, 10, 12, 14, 16, 18 and 20 for 10, 20, 30, 40, 50, 60, 70, 80, 90, 100-task problems, respectively.
- The deterministic task times are randomly generated by following a discrete uniform distribution with U(1–15). It should be noted here that, time of a task without an assistant should be greater than that of with an assistant.

- It is assumed that some parts have no demand and the demands of these parts are determined as zero. A random number in the [0–1] range is produced for each part to determine such parts and the demand for a given part is accepted to be zero if this value is equal to or smaller than 0.4.
- Revenues of parts with no demand are accepted as zero. Revenues for demanded parts are randomly generated by following a discrete uniform distribution with *U*(1–15). As stated in Section 2, if a part can be disassembled by performing either UA or DA, removing by UA offers higher revenue. Equipment costs are randomly generated by following a discrete uniform distribution with *U*(100–1000).
- Cycle times of the lines are equal to each other.
- Available number for each equipment type is used without any changes for each line.
- Available number of assistants of both lines is equal to the number of assistants of the line with a large number of assistants available.
- All combinations of precedence diagrams are considered when combining problems for parallel lines and only problems with the same cycle time are combined. Therefore, PDLB_PA 243 test instances were created.

The test problems are classified according to the total number of tasks of two lines: small-size (20–70 test instances) and medium-size (80–120 test instances).Total number of test instances for small- and medium-size test problems are 135 and 108, respectively. It should be noted here that since PDLBP_PA is an NP-hard problem, the lower and upper bounds obtained by formulations in large-size problems are far from the best solutions, making it difficult to make an unbiased comparison. Therefore, small- and medium-size test problems are taken into consideration for use in experimental studies.

6.2 Computational Results of PLF and Relaxations

In this section, the effects of the valid inequalities are investigated on PLF and for this purpose, small-size test problems are considered. To analyze the computational results, the following performance measures are used: relaxed percent deviation (*RPD*) and CPU time in seconds averaged over all instances for each test problem. *RPD* is the deviation between the upper bound obtained by the linear programming (LP) relaxation bound (Z^{RB}) produced by a particular formulation and lower bound (Z^{LB}) obtained as a result of running PLF, i.e., $RDP = 100.\ [(Z^{RB} - Z^{LB})/Z^{RB}]$. The lower bound is the optimal or best known solution obtained by solving PLF with CPLEX for a maximum of 2h.

Table 3 summarizes the integrated effect of the valid inequalities on PLF. The first column represents the problem sets. A problem set is shown with the number of tasks ($n_1\_n_2$) of line 1 and line 2. The numbers in parentheses indicate the total number of test instances contained in each test problem set. In the following two columns, named PLF, *RPD* and CPU time for PLF are reported. In a similar manner, successive two columns report the *RPD* and CPU time for PLF with extra valid inequalities. For example, columns 6 and 7 with caption (16)–(20) in Table 3 presents *RPD* and CPU time obtained by solving the relaxations of PLF with the valid inequalities (16)–(20).

When Table 3 is analyzed, it is seen that valid inequalities are effective in improving upper bounds obtained from PLF. While *RPD* values of PLF are 67.71% on average for the small-size test problems, successive inclusion of valid inequalities in PLF improves the upper bounds and *RPD* values reduce to 40.64% on average,so the reduction rate is 27.07% on average for the small-size problems, respectively. These results show that valid inequalities added to PLF improve the upper bounds and the best upper bound values are obtained with valid inequalities (16)–(20). When the results

Table 3: Effects of the valid inequalities on PLF on small-size test problems.

$n_1\_n_2$	PLF		(16), (17)		(16), (17), (18), (19), (20)	
	RPD	**CPU**	***RPD***	**CPU**	***RPD***	**CPU**
10_10 (18)	60.95	0.02	55.62	0.02	49.9	0.02
10_20 (27)	66.69	0.02	54.82	0.02	40.79	0.03
20_20 (18)	70.85	0.02	57.76	0.04	41.95	0.03
20_30 (27)	68.07	0.03	55.74	0.07	42.59	0.07
30_30 (18)	65.54	0.05	49.58	0.21	32.26	0.12
30_40 (27)	72.22	0.06	54.42	0.28	37.11	0.24
Average (small)	67.71	0.04	54.72	0.11	40.64	0.09

Table 4: Computational results of the formulations on small-size test problems.

$n_1\_n_2$	*PD*		CPU		*NOS*	
	PLF	VLF	PLF	VLF	PLF	VLF
10_10 (18)	0	0	0.47	0.23	18(18)	18(18)
10_20 (27)	0	0	2.38	2.51	27(27)	27(27)
20_20 (18)	0	0	64.52	181.18	18(18)	18(18)
20_30 (27)	0	0	275.5	188.12	27(27)	27(27)
30_30 (18)	3.48	0	1901.96	487.31	15(18)	18(18)
30_40 (27)	16.88	2.15	3461.21	1797.47	16(27)	24(27)
Average	3.84	0.43	1010.08	486.78	121(135)	132(135)

are evaluated in terms of CPU times, it is observed that the addition of all valid inequalities to PLF does not affect CPU times much in solving LP relaxation.

In the last computational study in this section, as a consequence of the results that are obtained from Table 3, strong formulation is considered, i.e., the formulation after adding inequalities (16)–(20) to PLF (referred to as VLF in the sequel) within 2 h of CPU time, and the results of PLF and VLF are compared in terms of optimality using small-size test problems. To analyze the computational results, the following performance measures are used: the optimality gap (*PD*) and CPU time in seconds averaged over all instances for each test problem set and the number of optimal solutions (*NOS*) obtained within 2 h of CPU time. *PD* is the gap between the best lower bound (Z^{LB}) and the best upper bound (Z^{UB}) when the corresponding formulation (i.e., PLF or VLF) is solved by CPLEX within 2 h of CPU time, i.e., $PD = 100 \times ((Z^{UB} - Z^{LB})/Z^{UB})$. The lower bound is the optimal or best known solution obtained by solving the corresponding formulation with CPLEX for a maximum of 2 h.

Table 4 presents the results of PLF and VLF. The first column of the table is the same as in Table 3. Every successive two columns of the table report *PD*, CPU time, and *NOS* for the corresponding formulation. For example, the 30_40 test problem set consisting of 27 test instances are solved with PLF, optimal solutions are obtained in 16 test instances while the same test problem set is solved with VLF, 24 test instances reach the optimal solutions.

As seen in Table 4, 132 out of 135 test instances (97.78%) reach optimal solutions in 486.78 is on average with VLF, while PLF produces 121 (89.63%) optimal solutions in 1010.08 s. The number of optimal solutions increased by 9.09% and deviation improved by 3.41% with VLF on average. In addition, VLF consumes an average of 52% less CPU time to solve all small-size test problems than PLF. These results show once again that valid inequalities are successful in limiting the solution space and improving the solutions in less computation time.

Depending on these results, all small- and medium-size test problems are solved with a time limit of 2 h using VLF in order to evaluate the solutions of PASA.

6.3 Computational Results for PASA

PASA is run five times for each test instance. The average of five runs and the best of five runs for each test instance are considered to analyze the solutions of PASA. The parameters used in the approach are determined by preliminary analyses to reach a satisfactory solution quality in a reasonable timespan. Functions of initial temperature and final temperature are determined with reference to the paper of Jayaswal and Agarwal (2014). The most appropriate parameter levels used for test problems are illustrated in Table 5.

In addition to *PD*, CPU and *NOS*, four deviation values given in Equations (27)–(29) are used to analyze the effectiveness of the approach to investigate the performance of PASA. In these equations, Z^{AV} stands for the average solution of five runs and Z^{BT} is the best solution over five runs. Z^{LB} and Z^{UB} are obtained as a result of running VLF. Z^{LB} is the optimal or best known solution obtained by solving VLF with CPLEX for a maximum of 2 h. Table 6 summarizes the computational results of PASA and VLF on small-size test problems.

$$PD^{AV} = 100 \times ((Z^{LB} - Z^{AV}) / Z^{LB}) \tag{26}$$

$$PD^{BT} = 100 \times ((Z^{LB} - Z^{BT}) / Z^{LB}) \tag{27}$$

$$PDU^{AV} = 100 \times ((Z^{UB} - Z^{AV}) / Z^{UB}) \tag{28}$$

Table 5: Levels of the parameters of PASA approach.

Parameter	**Parameter value**
Initial temperature (T_b)	$(ace + cw + ca) \times tnt$
Final temperature (T_s)	$(T_b \times 2)/(tnt)$
Number of outer loop ($iter_{out}$)	$2 \times tnt$
Number of first inner loop ($iter_{in1}$)	$tnt/2$
Number of second inner loop ($iter_{in2}$)	$tnt/2$
The first rate of cooling (a_1)	0.9
The second rate of cooling (a_2)	0.8
The general acceptance rate(gar)	0.2

ace: the average of the total cost of the equipment of the related test instance
Itnt: total number of tasks of the related test instance

Table 6: Comparison table of PASA solutions with VLF solutions.

	NOS			***PD***	***PD^AV***	***PDU^AV***	***PD^BT***	***PDU^BT***	**CPU**	
	VLF	**PASA**								
		Z^{AV}	Z^{BT}						**VLF**	**PASA**
10_10 (18)	18(18)	15(18)	18(18)	0	0.17	0.52	0	0	0.23	0.07
10_20 (27)	27(27)	17(27)	24(27)	0	0.44	0.44	0.08	0.08	2.51	0.27
20_20 (18)	18(18)	9(18)	11(18)	0	0.8	0.8	0.61	0.61	181.18	0.33
20_30 (27)	27(27)	15(27)	23(27)	0	0.41	0.41	0.16	0.16	188.12	0.46
30_30 (18)	18(18)	7(18)	12(18)	0	0.77	0.77	0.43	0.43	487.31	0.83
30_40 (27)	24(27)	10(27)	17(27)	2.15	–0.14	2.23	–0.67	1.71	1797.47	1.93
Average (small)	132(135)	73 (135)	105 (135)	0.43	0.37	0.85	0.05	0.53	486.78	0.7
40_40 (18)	12(18)	10(18)	10(18)	15.34	–3.79	14.06	–4.69	13.42	2894.1	3.84
40_50 (27)	11(27)	7(27)	9 (27)	20.96	–7.39	17.35	–9.22	16.19	5010.24	6.62
50_50 (18)	3(18)	0(18)	2(18)	27.17	–4.81	24.63	–11.91	19.92	6717.28	14.04
50_60 (27)	3(27)	1 (27)	2(27)	32.29	–7.09	28.45	–15.08	24.25	7041.31	19.41
60_60 (18)	4(18)	2(27)	4(18)	37.22	–15.59	29.63	–17.83	28.43	6664.73	28.88
Average (medium)	32(108)	22(108)	27(108)	26.6	–7.65	22.84	–11.81	20.41	5725.57	14.3

$$PDU^{BT} = 100 \times ((Z^{UB} - Z^{BT}) / Z^{UB}) \tag{29}$$

The first column of Table 6 is the same as in the previous tables. Successive three columns, named *NOS*, show the number of optimal solutions obtained with overall instances for each test problem. The next five columns report the results of *PD*, PD^{AV}, PDU^{AV}, PD^{BT}, PDU^{BT}, respectively. The last two columns report the CPU time in seconds averaged over all instances for each test problem.

Before evaluating all the results presented in Table 6, the evaluation of the deviation values used is explained through a selected test problem set. For example, 18 of 18 small-size test instances of 10_10 test problem set reach the optimal solutions for VLF, so *PD* is calculated as zero. Each instance of these setsis solved with PASA for five times. 18 of 18 test instances reached the best solution as optimal solution (Z^{BT}), and average values of PD^{BT} and PDU^{BT} are calculated as zero. On the other hand, 15 of 18 test instances reach the optimal solutions for each solution of five solutions (Z^{AV}). That is, each instance of 15 test instances is obtained as optimal solution five times and near optimal solutions are obtained for three test instances for each run. Therefore, according to these three test instances, average values of PD^{AV} and PDU^{AV} are calculated as 0.17% and 0.52% for 18 test instances.

When Table 6 is examined, the results are summarized as follows:

- When the results are evaluated in terms of average values (Z^{AV}) and the best solutions (Z^{BT}) for PASA;

- o According to Z^{AV} vaues, 73 (54.07%) of 135 (small-size) and 22 (20,37%) of 108 (medium-size) test instances reach the optimal solutions,
- o According to Z^{BT} values, 105 (77.78%) of 135 (small-size) and 27 (25%) of 108 (medium-size) reach the optimal solutions.

• When the results of PASA is compared with lower bound values of VLF;

- o According to PD^{AV} values, the deviations from optimal solutions are 0.37% and –7.65% on average for small and medium-size test problems, respectively,
- o According to PD^{BT} values, the deviations from optimal solutions are 0.05% and –11.82% on average for small and medium-size test problems, respectively.

It should be noted here that negative deviation values mean that the proposed PASA achieves better solutions (lower bounds) than VLF.

• When the comparison is applied according to the upper bound values of VLF;

- o According to PDU^{AV}, the deviations from optimal solutions are 0.85% and 22.84% on average for small- and medium-size test problems, respectively,
- o According to PDU^{BT}, the deviations from optimal solutions are 0.53% and 20.41% on average small- and medium-size problems, respectively.

All results indicate that PASA improves the lower bounds. It produces optimal solutions at a significant rateand feasible solutions that are quite close to the optimal solutions.

6.4 Investigation of the Effect of Parallel Lines on Total Net Recovery Profit

In this section, the effect on the total net recovery profit of simultaneous balancing instead of independent balancing of parallel lines with processing alternatives is investigated using 132 small-size problems where the optimal solutions are obtained. The results are given in Figure 9. The rows show the test problem sets, and numbers in parenthesis show the number of optimal solutions obtained by solving VLF. The columns show the deviation values (%) of the optimal solutions obtained by balancing parallel lines simultaneously from the solutions obtained by balancing the same lines independently.

According to Figure 9, the average improvement for 132 small-size problems is calculated as 42.87% on the total net recovery profit. It is clear that the simultaneous balancing of parallel lines increases the efficiency of the system and is quite effective on the total net recovery profit according to the calculated rates.

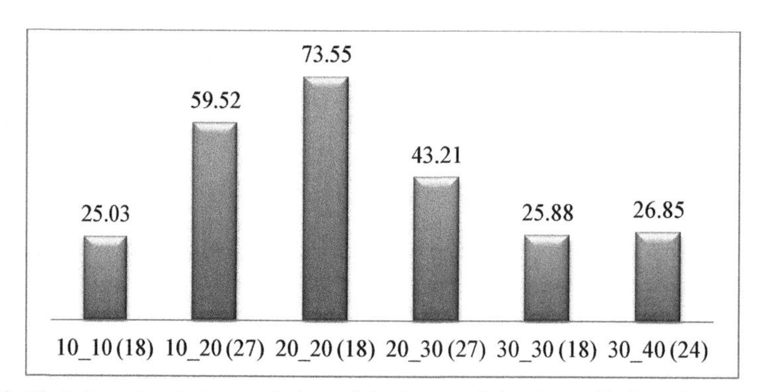

Fig. 9: Deviation values between solutions of simultaneous balancing and independent balancing.

7. Conclusion and Future Directions

In this chapter, the parallel disassembly line balancing problem with processing alternatives, PDLBP_PA, which allows single-model and partial disassembly, has been considered. In PDLBP_PA,industry 4.0 technologies were utilized in terms of acquiring necessary information about the parts of the products. After a 0-1 integer linear programming formulation, PLF was proposed for solving PDLBP_PA, valid inequalities were defined in order to improve the upper bounds. A heuristic approach based ASA, named PASA, was proposed to solve medium- and large-size problems.

Three experimental studies were carried out to evaluate the performance of PLF, valid inequalities, PASA, and effects of simultaneous balancing of the parallel lines using 243 test instances created by the authors. Computational results show that valid inequalities are quite effective in improving upper bounds and they also speed up the solver. It is also observed that the strong formulation, named VLF, increases number of optimal solutions and provides good upper bounds for small- and medium-size test problems. Computational results of all test instances also reveal that PASA yields good quality solutions in efficient solution times. Thus, it is a preferable approach for solving PDLBP_PA instances successfully. Lastly, it is clearly observed that the recovery lines where more than one disassembly lines are installed and used at the same time must be balanced simultaneously in order to save the use of resources and benefit from system efficiency at the highest rates.

This chapter is limited to PDLBP_PA with a single-model and a deterministic task time. There are several interesting directions for further research on the problem as follows: Since one of the most characteristic features of the disassembly processes is uncertain, problem parameters such as task time, demand and revenue can be considered stochastic or fuzzy in order to reflect this situation. Different valid inequalities or algorithms can be developed to obtain better upper bounds. PASA can be used with different meta-heuristics to improve lower bounds. In this chapter, the cycle time is determined according to the highest demand and it does not change during the disassembly processes. According to this situation, total idle times are ignored. In the future, the cycle time as well as the number of stations can be determined as a decision variable and the idle times can be minimized. There can be more than one conflicting objective for such problems in real life. In this case, the multi-objective PDLBP_PA can be considered. The problem can also be updated taking into account the hazardous parts specific to disassembly.In addition to the SEPs, other industry 4.0 technologies such as big data, augmented reality, cloud computing, robots, etc., can be used to obtain information about disassembly processes and lines. PDLBP_PA can be expanded according to these technologies and information obtained. So, PLF can be adapted to specific applications in real-world industries, and some additional constraints can be added or available ones can be removed.

References

Agrawal, S. and M.K. Tiwari. 2008. A collaborative ant colony algorithm to stochastic mixed-model u-shaped disassembly line balancing and sequencing problem. International Journal of Production Research, 46(6): 1405–1429. doi: 10.1080/00207540600943985.

Altekin, F.T., L. Kandiller and N.E. Ozdemirel. 2008. Profit-oriented disassembly-line balancing. International Journal of Production Research, 46(10): 2675–2693. doi: 10.1080/00207540601137207.

Altekin, F.T. and C. Akkan. 2012. Task-failure-driven rebalancing of disassembly lines. International Journal of Production Research, 50(18): 4955–4976. doi: 10.1080/00207543.2011.616915.

Altekin, F.T. 2016. A piecewise linear model for stochastic disassembly line balancing. Ifac Papersonline, 49(12): 932–937. doi: 10.1016/j.ifacol.2016.07.895.

Altekin, F.T. 2017. A comparison of piecewise linear programming formulations for stochastic disassembly line balancing. International Journal of Production Research, 55(24): 7412–7434. doi: 10.1080/00207543.2017.1351639.

Araujo, F.F.B., A.M. Costa and C. Miralles. 2015. Balancing parallel assembly lines with disabled workers. European Journal of Industrial Engineering, 9(3): 344–365. doi: 10.1504/Ejie.2015.069343.

Attiya, G. and Y. Hamam. 2006. Task allocation for maximizing reliability of distributed systems: A simulated annealing approach. Journal of Parallel and Distributed Computing, 66(10): 1259–1266. doi: 10.1016/j.jpdc.2006.06.006.

Avikal, S. and P.K. Mishra. 2012. A new u-shaped heuristic for disassembly line balancing problems Pratibha: International Journal of Science. Spirituality, Business and Technology, 1(1): 21–27.

Avikal, S., R. Jain and P. Mishra. 2013. A heuristic for u-shaped disassembly line balancing problems. International Journal of Mechanical Engineering, 3(1): 51–56.

Aydemir-Karadag, A. and O. Turkbey. 2013. Multi-objective optimization of stochastic disassembly line balancing with station paralleling. Computers & Industrial Engineering, 65(3): 413–425. doi: 10.1016/j.cie.2013.03.014.

Battaia, O., A. Dolgui, S.S. Heragu, S.M. Meerkov and M.K. Tiwari. 2018. Design for manufacturing and assembly/disassembly: Joint design of products and production systems. International Journal of Production Research. 56(24): 7181–7189. doi: 10.1080/00207543.2018.1549795.

Baykasoglu, A., L. Ozbakir, L. Gorkemli and B. Gorkemli. 2009. Balancing parallel assembly lines via ant colony optimization. Cie: 2009 International Conference on Computers and Industrial Engineering, 506–511. doi: Doi 10.1109/ Iccie.2009.5223867.

Baykasoglu, A., L. Ozbakir, L. Gorkemli and B. Gorkemli. 2012. Multi-colony ant algorithm for parallel assembly line balancing with fuzzy parameters. Journal of Intelligent & Fuzzy Systems, 23(6): 283–295. doi: 10.3233/Ifs-2012-0520.

Bentaha, M.L., O. Battaia and A. Dolgui. 2014a. Lagrangian relaxation for stochastic disassembly line balancing problem. Variety Management in Manufacturing: Proceedings of the 47th Cirp Conference on Manufacturing Systems, 17: 56–60. doi: 10.1016/j.procir.2014.02.049.

Bentaha, M.L., O. Battaia and A. Dolgui. 2014b. A sample average approximation method for disassembly line balancing problem under uncertainty. Computers & Operations Research, 51: 111–122. doi: 10.1016/j.cor.2014.05.006.

Bentaha, M.L., A. Dolgui, O. Battaia, R.J. Riggs and J. Hu. 2018. Profit-oriented partial disassembly line design: Dealing with hazardous parts and task processing times uncertainty. International Journal of Production Research, 56(24): 7220–7242. doi: 10.1080/00207543.2017.1418987.

Benzer, R., H. Gokcen, T. Cetinyokus and H. Cercioglu. 2007. A network model for parallel line balancing problem. Mathematical Problems in Engineering, doi: Artn 1010610.1155/2007/10106.

Cercioglu, H., U. Ozcan, H. Gokcen and B. Toklu. 2009. A simulated annealing approach for parallel assembly line balancing problem. Journal of the Faculty of Engineering and Architecture of Gazi University, 24(2): 331–341.

Cil, Z.A., S. Mete, E. Ozceylan and K. Agpak. 2017. A beam search approach for solving type ii robotic parallel assembly line balancing problem. Applied Soft Computing, 61: 129–138. doi: 10.1016/j.asoc.2017.07.062.

Das, S.K., P. Yedlarajiah and R. Narendra. 2000. An approach for estimating the end-of-life product disassembly effor and cost. international Journal of Production Research, 38(3): 657–673.

Das, S.K. and S. Naik. 2002. Process planning for product disassembly. International Journal of Production Research 40(6): 1335–1355.

Deniz, N. and F. Ozcelik. 2019. An extended review on disassembly line balancing with bibliometric & social network and future study realization analysis. Journal of Cleaner Production, 225: 697–715. doi: 10.1016/j.jclepro.2019.03.188.

Ding, L.P., Y.X. Feng, J.R. Tan and Y.C. Gao. 2010. A new multi-objective ant colony algorithm for solving the disassembly line balancing problem. International Journal of Advanced Manufacturing Technology, 48(5-8): 761–771. doi: 10.1007/s00170-009-2303-5.

Duta, L., F. Gh. Filip and J.M. Henrioud. 2005. Applying equal piles approach to disassembly line balancing problem 16th Triennial World Congress Prague, Czech Republic.

Duta, L., F.G. Filip and I. Caciula. 2008. Real time balancing of complex disassembly lines. In Proceedings of the 17th World Congress the International Federation of Automatic Control, eoul.

Erel, E., I. Sabuncuoglu and B.A. Aksu. 2001. Balancing of u-type assembly systems using simulated annealing. International Journal of Production Research, 39(13): 3003–3015. doi: Doi 10.1080/00207540110051905.

Etgar, R., A. Shtub and L.J. LeBlanc. 1997. Scheduling projects to maximize net present value—the case of time-dependent, contingent cash flows. European Journal of Operational Research, 96(1): 90–96. doi: Doi 10.1016/0377-2217(95)00382-7.

Fang, Y.L., Q. Liu, M.Q. Li, Y.J. Laili and D.T. Pham. 2019. Evolutionary many-objective optimization for mixed-model disassembly line balancing with multi-robotic workstations. European Journal of Operational Research, 276(1): 160–174. doi: 10.1016/j.ejor.2018.12.035.

Fang, Y.L., H. Ming, M.Q. Li, Q. Liu and D.T. Pham. 2019. Multi-objective evolutionary simulated annealing optimisation for mixed-model multi-robotic disassembly line balancing with interval processing time. International Journal of Production Research, doi: 10.1080/00207543.2019.1602290.

Gharan, S.O. and J. Vondrak. 2011. Submodular maximization by simulated annealing. Proceedings of the Twenty-Second Annual Acm-Siam Symposium on Discrete Algorithms: 1098–1116.

Gokcen, H., K. Agpak and R. Benzer. 2006. Balancing of parallel assembly lines. International Journal of Production Economics, 103(2): 600–609. doi: 10.1016/j.ijpe.2005.12.001.

Guden, H. and S. Meral. 2016. An adaptive simulated annealing algorithm-based approach for assembly line balancing and a real-life case study. International Journal of Advanced Manufacturing Technology, 84(5-8): 1539–1559. doi: 10.1007/s00170-015-7802-y.

Gungor, A. and S.M. Gupta. 1999. Issues in environmentally conscious manufacturing and product recovery: A survey. Computers & Industrial Engineering, 36(4): 811–853. doi: Doi 10.1016/S0360-8352(99)00167-9.

Gungor, A. and S.M. Gupta. 2001. A solution approach to the disassembly line balancing problem in the presence of task failures. International Journal of Production Research, 39(7): 1427–1467. doi: Doi 10.1080/00207540110052157.

Gungor, A. and S.M. Gupta. 2002. Disassembly line in product recovery. International Journal of Production Research, 40(11): 2569–2589. doi: 10.1080/00207540210135622.

Gungor, A. 2006. Evaluation of connection types in design for disassembly (dfd) using analytic network process. Computers & Industrial Engineering, 50(1-2): 35–54. doi: 10.1016/j.cie.2005.12.002.

Guo, Q.X. and L.X. Tang. 2009. A scatter search based heuristic for the balancing of parallel assembly lines. Proceedings of the 48th Ieee Conference on Decision and Control, 2009 Held Jointly with the 2009 28th Chinese Control Conference (Cdc/Ccc 2009): 6256–6261. doi: Doi 10.1109/Cdc.2009.5399991.

He, J.K., F. Chu, F.F. Zheng, M. Liu and C.B. Chu. 2019. A multi-objective distribution-free model and method for stochastic disassembly line balancing problem. International Journal of Production Research, doi: 10.1080/00207543.2019.1656841.

Hezer, S. and Y. Kara. 2015. A network-based shortest route model for parallel disassembly line balancing problem. International Journal of Production Research, 53(6): 1849–1865. doi: 10.1080/00207543.2014.965348.

Igarashi, K., T. Yamada and M. Inoue. 2014. 2-stage optimal design and analysis for disassembly system with environmental and economic parts selection using the recyclability evaluation method. Industrial Engineering and Management Systems, 13(1): 52–66

Ilgin, M.A. and S.M. Gupta. 2010. Environmentally conscious manufacturing and product recovery (ecmpro): A review of the state of the art. Journal of Environmental Management, 91(3): 563–591. doi: 10.1016/j.jenvman.2009.09.037.

Jayaswal, S. and P. Agarwal. 2014. Balancing u-shaped assembly lines with resource dependent task times: A simulated annealing approach. Journal of Manufacturing Systems,33(4): 522–534.

Kalayci, C.B. and S.M. Gupta. 2013a. Simulated annealing algorithm for solving sequence-dependent disassembly line balancing problem. 7th IFAC Conference on Manufacturing Modelling, Management, and Control, Saint Petersburg.

Kalayci, C.B. and S.M. Gupta. 2013b. Artificial bee colony algorithm for solving sequence-dependent disassembly line balancing problem. Expert Systems with Applications, 40(18): 7231–7241. doi: 10.1016/j.eswa.2013.06.067.

Kalayci, C.B. and S.M. Gupta. 2013c. Balancing a sequence-dependent disassembly line using simulated annealing algorithm. Applications of Management Science, 16: 81–103. doi: 10.1108/S0276-8976(2013)0000016008.

Kalayci, C.B. and S.M. Gupta. 2013d. Ant colony optimization for sequence-dependent disassembly line balancing problem. Journal of Manufacturing Technology Management, 24(3): 413–427.

Kalayci, C.B. and S.M. Gupta. 2014. A tabu search algorithm for balancing a sequence-dependent disassembly line. Production Planning & Control, 25(2): 149–160. doi: 10.1080/09537287.2013.782949.

Kalayci, C.B., A. Hancilar, A. Gungor and S.M. Gupta. 2015. Multi-objective fuzzy disassembly line balancing using a hybrid discrete artificial bee colony algorithm. Journal of Manufacturing Systems, 37: 672–682. doi: 10.1016/j.jmsy.2014.11.015.

Kalayci, C.B., O. Polat and S.M. Gupta. 2016. A hybrid genetic algorithm for sequence-dependent disassembly line balancing problem. Annals of Operations Research, 242(2): 321–354. doi: 10.1007/s10479-014-1641-3.

Kalaycilar, E.G., M. Azizoglu and S. Yeralan. 2016. A disassembly line balancing problem with fixed number of workstations. European Journal of Operational Research, 249(2): 592–604. doi: 10.1016/j.ejor.2015.09.004.

Kara, Y., H. Gokcen and Y. Atasagun. 2010. Balancing parallel assembly lines with precise and fuzzy goals. International Journal of Production Research, 48(6): 1685–1703. doi: 10.1080/00207540802534715.

Kara, Y., C. Ozguven, N. Yalcin and Y. Atasagun. 2011. Balancing straight and u-shaped assembly lines with resource dependent task times. International Journal of Production Research, 49(21): 6387–6405. doi: 10.1080/00207543.2010.535039.

Karaoglan, I., F. Altiparmak, I. Kara and B. Dengiz. 2012. The location-routing problem with simultaneous pickup and delivery: Formulations and a heuristic approach. Omega-International Journal of Management Science, 40(4): 465–477. doi: 10.1016/j.omega.2011.09.002.

Kesen, S.E., S.K. Das and Z. Gungor. 2010. A genetic algorithm based heuristic for scheduling of virtual manufacturing cells (vmcs). Computers & Operations Research, 37(6): 1148–1156. doi: 10.1016/j.cor.2009.10.006.

Koc, A., I. Sabuncuoglu and E. Erel. 2009. Two exact formulations for disassembly line balancing problems with task precedence diagram construction using an and/or graph. Iie Transactions, 41(10): 866–881. doi: 10.1080/07408170802510390.

Lambert, A.J.D. and S.M. Gupta. 2005. Disassembly modelling for assembly, maintenance, reuse and recycling, Boca Raton, FL: CRC Press.

Lee, K. and R. Gadh. 1998. Destructive disassembly to support virtual prototyping. IIE Transactions, 30(10): 959–972.

Liao, Y.X., F. Deschamps, E.D.R. Loures and L.F.P. Ramos. 2017. Past, present and future of industry 4.0-a systematic literature review and research agenda proposal. International Journal of Production Research, 55(12): 3609–3629. doi: 10.1080/00207543.2017.1308576.

Liu, J. and S.W. Wang. 2017. Balancing disassembly line in product recovery to promote the coordinated development of economy and environment. Sustainability, 9(2). doi: ARTN 30910.3390/su9020309.

Liu, J.Y., Z.D. Zhou, D.T. Pham, W.J. Xu, J.W. Yan, A.M. Liu, C.Q. Ji and Q. Liu. 2018. An improved multi-objective discrete bees algorithm for robotic disassembly line balancing problem in remanufacturing. International Journal of Advanced Manufacturing Technology, 97(9-12): 3937–3962. doi: 10.1007/s00170-018-2183-7.

Lu, Y. 2017. Industry 4.0: A survey on technologies, applications and open research issues. Journal of Industrial Information Integration, 6: 1–10. doi: 10.1016/j.jii.2017.04.005.

Lusa, A. 2008. A survey of the literature on the multiple or parallel assembly line balancing problem. European Journal of Industrial Engineering, 2(1): 50–72. doi: Doi 10.1504/Ejie.2008.016329.

McGovern, S.M. and S.M. Gupta. 2006. Ant colony optimization for disassembly sequencing with multiple objectives. International Journal of Advanced Manufacturing Technology, 30(5-6): 481–496. doi: 10.1007/s00170-005-0037-6.

McGovern, S.M. and S.M. Gupta. 2007a. A balancing method and genetic algorithm for disassembly line balancing. European Journal of Operational Research, 179(3): 692–708. doi: 10.1016/j.ejor.2005.03.055.

Mcgovern, S.M. and S.M. Gupta. 2007b. Combinatorial optimization analysis of the unary np-complete disassembly line balancing problem. International Journal of Production Research, 45(18-19): 4485–4511. doi: 10.1080/00207540701476281.

Mcgovern, S.M. and S.M. Gupta. 2011. The disassembly line: Balancing and modeling, New York, McGraw Hill.

Ondemir, O., M.A. Ilgin and S.M. Gupta. 2012. Optimal end-of-life management in closed-loop supply chains using rfid and sensors. Ieee Transactions on Industrial Informatics, 8(3): 719–728. doi: 10.1109/Tii.2011.2166767.

Ondemir, O. and S.M. Gupta. 2014a. A multi-criteria decision making model for advanced repair-to-order and disassembly-to-order system. European Journal of Operational Research, 233(2): 408–419. doi: 10.1016/j.ejor.2013.09.003.

Ondemir, O. and S.M. Gupta. 2014b. Quality management in product recovery using the internet of things: An optimization approach. Computers in Industry, 65(3): 491–504. doi: 10.1016/j.compind.2013.11.006.

Ozbakir, L., A. Baykasoglu, B. Gorkemli and L. Gorkemli. 2011. Multiple-colony ant algorithm for parallel assembly line balancing problem. Applied Soft Computing, 11(3): 3186–3198. doi: 10.1016/j.asoc.2010.12.021.

Ozcan, U., H. Cercioglu, H. Gokcen and B. Toklu. 2010. Balancing and sequencing of parallel mixed-model assembly lines. International Journal of Production Research, 48(17): 5089–5113. doi: Pii 91398950610.1080/00207540903055735.

Ozcan, U. 2018. Balancing stochastic parallel assembly lines. Computers & Operations Research, 99: 109–122. doi: 10.1016/j.cor.2018.05.006.

Ozceylan, E. and T. Paksoy. 2014. Fuzzy mathematical programming approaches for reverse supply chain optimization with disassembly line balancing problem. Journal of Intelligent & Fuzzy Systems, 26(4): 1969–1985. doi: 10.3233/Ifs-130875.

Ozceylan, E., C.B. Kalayci, A. Gungor and S.M. Gupta. 2019. Disassembly line balancing problem: A review of the state of the art and future directions. International Journal of Production Research, 57(15-16): 4805–4827. doi: 10.1080/00207543.2018.1428775.

Paksoy, T., A. Gungor, E. Ozceylan and A. Hancilar. 2013. Mixed model disassembly line balancing problem with fuzzy goals. International Journal of Production Research, 51(20): 6082–6096. doi: 10.1080/00207543.2013.795251.

Pistolesi, F., B. Lazzerini, M.D. Mura and G. Dini. 2018. Emoga: A hybrid genetic algorithm with extremal optimization core for multiobjective disassembly line balancing. Ieee Transactions on Industrial Informatics, 14(3): 1089–1098. doi: 10.1109/Tii.2017.2778223.

Ren, Y.P., D.Y. Yu, C.Y. Zhang, G.D. Tian, L.L. Meng and X.Q. Zhou. 2017. An improved gravitational search algorithm for profit-oriented partial disassembly line balancing problem. International Journal of Production Research, 55(24): 7302–7316. doi: 10.1080/00207543.2017.1341066.

Riggs, R.J., O. Battaia and S.J. Hu. 2015. Disassembly line balancing under high variety of end of life states using a joint precedence graph approach. Journal of Manufacturing Systems, 37: 638–648. doi: 10.1016/j.jmsy.2014.11.002.

Seidi, M. and S. Saghari. 2016. The balancing of disassembly line of automobile engine using genetic algorithm (ga) in fuzzy environment. Industrial Engineering and Management Systems, 15(4): 364–373. doi: 10.7232/iems.2016.15.4.364.

Sonnenberg, Manuela. 2001. Force and effort analysis of unfastening actions in disassembly processes. Mechanical Engineering, New Jersey Institute of Technology.

Thierry, M., M. Salomon, J. Vannunen and L. Vanwassenhove. 1995. Strategic issues in product recovery management. California Management Review, 37(2): 114–135. doi: Doi 10.2307/41165792.

Wang, K.P., X.Y. Li and L. Gao. 2019a. Modeling and optimization of multi-objective partial disassembly line balancing problem considering hazard and profit. Journal of Cleaner Production, 211: 115–133. doi: 10.1016/j.jclepro.2018.11.114.

Wang, K.P., X.Y. Li and L. Gao. 2019b. A multi-objective discrete flower pollination algorithm for stochastic two-sided partial disassembly line balancing problem. Computers & Industrial Engineering, 130: 634–649. doi: 10.1016/j.cie.2019.03.017.

Xu, L.D., E.L. Xu and L. Li. 2018. Industry 4.0: State of the art and future trends. International Journal of Production Research, 56(8): 2941–2962. doi: 10.1080/00207543.2018.1444806.

Yıldız, A. 2018. Endüstri 4.0 ve akıllı fabrikalar Sakarya Üniversitesi Fen Bilimleri Enstitüsü Dergisi. 22 (2):546–556.

Zhang, Z.G., K.P. Wang, L.X. Zhu and Y. Wang. 2017. A pareto improved artificial fish swarm algorithm for solving a multi-objective fuzzy disassembly line balancing problem. Expert Systems with Applications, 86: 165–176. doi: 10.1016/j.eswa.2017.05.053.

Zheng, P., H.H. Wang, Z.Q. Sang, R.Y. Zhong, Y.K. Liu, C. Liu, K. Mubarok, S.Q. Yu and X. Xu. 2018. Smart manufacturing systems for industry 4.0: Conceptual framework, scenarios, and future perspectives. Frontiers of Mechanical Engineering, 13(2): 137–150. doi: 10.1007/s11465-018-0499-5.

Zhu, X., Z. Zhang and J. Hu. 2014. An ant colony optimization algorithm for multi-objective disassembly line balancing problem. China Mechanical Engineering, 25(8): 1075–1079.

SECTION 10

Maturity Models and Analysis for Industry 4.0 and Logistics 4.0

CHAPTER 17

A Study of Maturity Model for Assessing the Logistics 4.0 Transformation Level of Industrial Enterprises

Literature Review and a Draft Model Proposal

Kerem Elibal,[1] *Eren Özceylan*[2,*] and *Cihan Çetinkaya*[3]

1. Logistics

1.1 Introduction

The term logistics is a common expression in the World's social and industrial life, and generally used in many areas. Despite its usage popularity, reviewed literature about logistics showed us that there is still an absence of consent in real meaning and definition of the term.

Reviewed literature (Coyle et al. 1992; Bowersox and Closs 1996; Razaqque 1997; Ghiani et al. 2004; Ballou et al. 2007; Keskin 2011; Küçük 2016) showed us that there are many definitions about logistics which can describe it in a business perspective. The common agreement on logistics definition belongs to the Council of Logistics Management (CLM), a leading organization for logistics professionals with a current membership of over 15.000. The CLM defines logistics as "*the process of planning, implementing and controlling the efficient, effective flow and storage of goods, services and related information from point of origin to point of consumption for the purpose of conforming to customer requirements*".

[1] BCS Metal Industry, Third Industry Zone, Gaziantep 27500, Turkey, Email: keremelibal@gmail.com
[2] Department of Industrial Engineering, Gaziantep University, Gaziantep 27310, Turkey.
[3] Department of Management Information Systems, Adana Alparslan Türkeş Science and Technology University, Adana 01250, Turkey, Email: ccetinkaya@atu.edu.tr
* Corresponding author: eozceylan@gantep.edu.tr

The infrastructure of a logistic management system is defined in Figure 1, which has been inspired from logistical integration of Bowersocks and Closs (1996).

Like the variety of logistic definitions in the literature, there are also a variety of logistic activities for a production system. Findings from the literature helped us to construct key activities of logistics as shown in Table 1 which will be investigated in the manner of Industry 4.0 in this study.

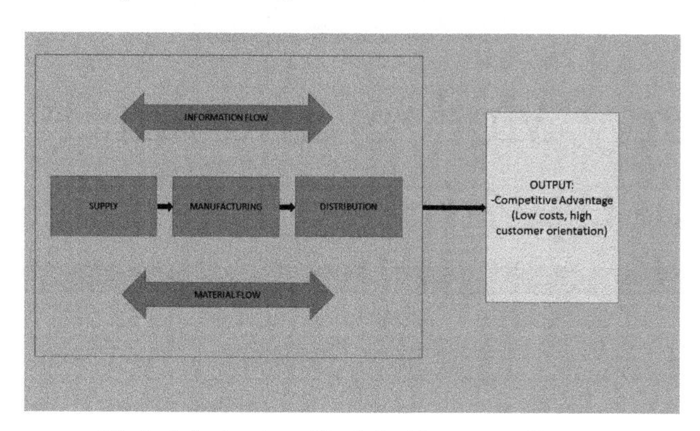

Fig. 1: Infrastructure of Logistics Management System.

Table 1: Reviewed Literature for Logistics Key Activities and Sub-activities.

Supply	**Manufacturing**	**Distribution**
Procurement: *Stock and Lambert 2001, Blanchard 2004*	Warehousing: *Bowersocks and Closs 1996, Stock and Lambert 2001, Baki 2004, Keskin 2011, Stroh 2011, Gümüş 2012, Küçük 2016*	Transportation: *Bowersocks and Closs 1996, Stock and Lambert 2001, Baki 2004, Blanchard 2004, Keskin 2011, Stroh 2011, Gümüş 2012*
Purchasing: *Baki 2004, Blanchard 2004, Keskin 2011*	Packaging: *Bowersocks and Closs 1996, Stock and Lambert 2001, Baki 2004, Blanchard 2004, Keskin 2011, Gümüş 2012*	Customer Service: *Stock and Lambert 2001, Baki 2004, Blanchard 2004, Keskin 2011, Stroh 2011*
Inventory Management: *Bowersocks and Closs 1996, Stock and Lambert 2001, Baki 2004, Blanchard 2004, Keskin 2011, Stroh 2011, Gümüş 2012, Küçük 2016*	Inner distribution: *Gümüş 2012*	Inventory Management: *Bowersocks and Closs 1996, Stock and Lambert 2001, Baki 2004, Blanchard 2004, Keskin 2011, Stroh 2011, Gümüş 2012, Küçük 2016*
Transportation: *Bowersocks and Closs 1996, Stock and Lambert 2001, Baki 2004, Blanchard 2004, Keskin 2011, Stroh 2011, Gümüş 2012*	Maintenance: *Keskin 2011*	
Demand Forecasting: *Stock and Lambert 2001, Baki 2004, Blanchard 2004, Keskin 2011*	Quality Control: *Keskin 2011*	
	Handling: *Bowersocks and Closs 1996, Stock and Lambert 2001, Baki 2004, Blanchard 2004, Keskin 2011, Küçük 2016*	
	Production Planning: *Baki 2004, Blanchard 2004*	
	Inventory Management: *Bowersocks and Closs 1996, Stock and Lambert 2001, Baki 2004, Blanchard 2004, Keskin 2011, Stroh 2011, Gümüş 2012, Küçük 2016*	

2. Logistics 4.0

The term "Industry 4.0" is a new phenomenon, presented first at the Hannover Fair, 2011. Despite the fact that the term showed up in Germany, because of its promising benefits, it spread out to many countries with advanced economies

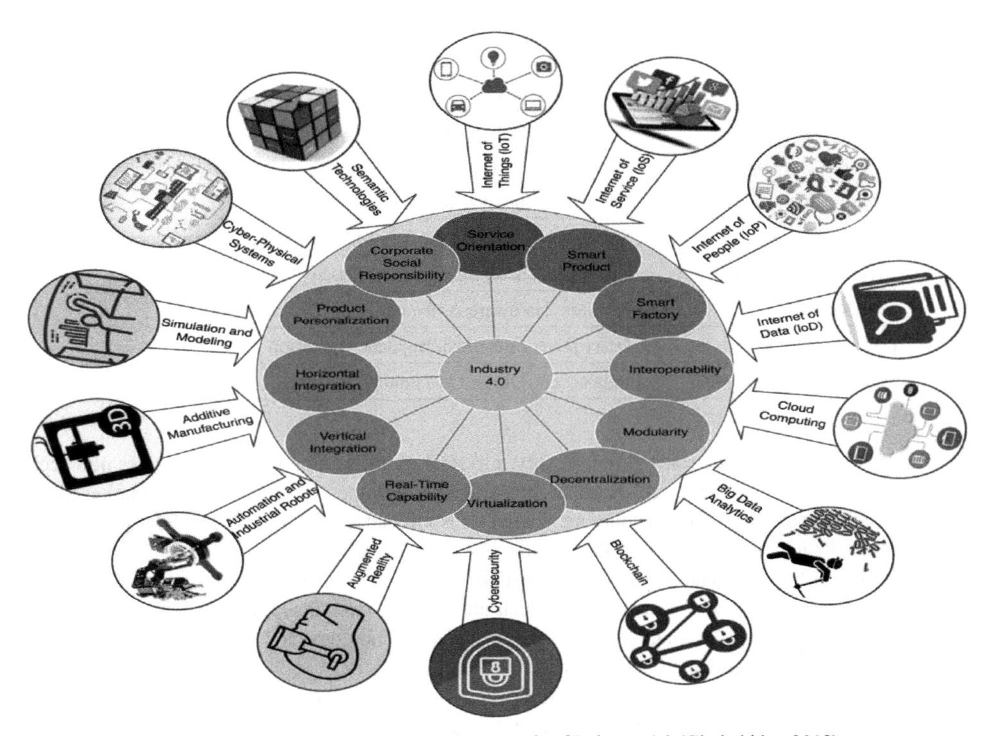

Fig. 2: Design principles and technology trends of Industry 4.0 (Ghobakhloo 2018).

quickly. Generally, all industrial revolutions aim at operational and productional efficiency for a profitable and sustainable economy, so does Industry 4.0. But Industry 4.0 also provides gains to satisfy developed economies' sensitivities about environment, work-life balance and demographic changes (Elibal et al. 2018).

Today, emerging technologies especially about IT systems and electronics allow digital integration of all the components in an organization, starting from the product development till reaching the finished product to the customer. Internet of things (IoT), big data analysis, additive manufacturing (3D printing), manufacturing execution systems (MES) applied with artificial intelligence (AI), cloud computing, autonomous manufacturing assets, augmented reality systems, smart sensors, smart products are some of the key concepts of Industry 4.0 and provide not only the integration and communication of the organization in a cyber physical system (CPS), but also delivers the customized production in low batches in affordable and profitable levels (Elibal et al. 2018).

It should also be considered that the term industry 4.0 is not only about connecting machines and/or products with each other via internet or it is not only about fully automated systems, but it is a technological philosophy in which the final goal is making fast and right decisions by using digitalized data. Digitization of manufacturing and/or other departments of a company, and giving fast and agile decisions by using these data is the main concept of Industry 4.0. Companies should consider automated decision systems with automated physical systems, the peak point of a "Dark Factory" should not be fully automated physical resources, and also there should be autonomous and learning decision systems, in a decentralized manner (Elibal et al. 2018).

2.1 Logistics 4.0 Literature Review

There are many researches which defines key technologies of Industry 4.0. Lu (2017) provides a significant amount of literature survey about Industry 4.0 technologies. Also Bibby and Dehe (2018) give information about the attributes of "Factory of the Future".

In the concept of Industry 4.0, logistics 4.0 can be summarized as implementing industry 4.0 concepts into logistical functions. Logistics 4.0 is an emerging topic in both academia and practitioners. There are a large amount of studies which describes logistics 4.0 and logistics 4.0 technologies.

Bamberger et al. (2017) claims that, for the management, it is hard to decide which technological trends should be selected for logistics 4.0. In addition, it presents logistics 4.0 categories. These categories are; data, new methods of physical transportation, digital platforms marketplace and new production methods. Data category includes data collection

Table 2: Important technologies of logistics 4.0 (Glistau et al. 2018).

Technology	Characteristics
Identification	Smart card, barcode, RFID, Sensor technologies, biometrics
Mobile communication	5G network, UMTS/LTE, GSM/GPRS, Wlan, satellite based
Electronic data interchange	EDI, XML, Internet, Telematics
Terminals	Smartphones, tablets, Special hand-held units, On-board computer
Architecture paradigm	Centralized, decentralized, agent based, blockchain
Architecture	Network, Hardware, Software, Database, Virtualization
Data analysis methods	Descriptive, Inferential, Big data, regression
Data analytics processing	Data access, OLAP, Data mining

and treatment, logistics control tower, augmented reality. Physical transportation category includes new technological physical methods like driverless trucks/vehicles, handling robots and drones. Digital platforms market place category includes shared warehouse capacity, crowd sourcing, shared transport capacity and big cross-order platform. Finally the last category new production methods include 3D printing.

Hofman and Rüsch (2017) indicated that there are two dimensions that explains how logistics is affected by Industry 4.0; the first is the physical supply chain dimension, which includes autonomous logistic sub-systems like autonomous picking robots or order processing via smart contracts on block chain technology, and the second one is digital data value chain dimension which includes collected data via IoT enabled technologies.

Glistau and Machado (2018) presents some important logistics 4.0 technologies as; identification, mobile communication, localization, electronic data interchange, terminals, architecture paradigm, architecture, data analysis method, and data analytics processing that are reflected in Table 2.

Glistau and Machado (2018) also indicated some typical solutions of logistics 4.0 as; smart logistical objects, autonomous driving, CPS solutions like augmented reality for picking, big data and automatic video control of logistics objects, traffic control and new business processes which includes e Procurement platform. They also investigated smart material solutions under a specific topic.

Gocmen and Erol (2018) presented some technologies which are conducted in a Turkish logistics company within transition of logistics 4.0., which are;

- Proof of delivery which includes sending delivery approval via electronic signature sent with PDA
- Electronic data interchange used for data interchange in Transport Management System
- Information of delivery which includes real time monitoring of loads and unloads
- The trailer monitoring system which provides the tracking inside of the trailer by taking video or photo automatically
- Sensor trailer locking system which controls the door to not unlock without approval
- Geo-fencing which defines the route for the vehicle
- Warehouse management system
- Material flow control system
- Portal applications
- Automated storage and retrieval system
- Automation systems for garments on hanger
- Vertical lift systems
- Pick to light systems
- Automated sorter systems
- Product pick by voice
- RF handheld terminal

Besides these technologies, image processing projects, satellite tracking systems and storage technologies via face recognition, voice or light guideness have been indicated as emerging technologies in the company. In the study of Gocmen and Erol (2018) also, technologies as autonomous transportation, autonomous inventory management, 3D warehouse management, global resource planning, and real time routing are selected as logistics 4.0 technologies and a priority research has been done by using the fuzzy logic approach to determine which technologies have more importance in transition to logistics 4.0. Results showed that autonomous transportation, inventory management and real time routing must be prioritized by the company for successful logistics 4.0 transition. Sekkeli and Bakan (2018) presented some visions about logistics 4.0. These can be summarized as;

- Transportation will be made by autonomous robots. Via the sensors or RFIDs on the products and/or the transportation equipment, material flow will be conducted in an autonomous way with a real time information flow.
- Suppliers will get and/or send real-time information about the orders automatically and supply procedures will be executed in a more efficient way.
- Warehousing, receiving, storage operations will be done by the communication of wireless equipment automatically.
- Picking and packaging operations will be done by augmented reality systems.
- Driverless transportation carriers or semi-autonomous trucks will be used.
- To prevent accidents, technological equipment will be used to maximize driving security.
- Data flow for traffic management will improve the transportation efficiency.

Radivojević and Milosavljević (2019) identified components and technologies of logistics 4.0 as seen in Figure 3. They indicated that there are 6 components of logistics 4.0; automatic identification, automatic data collection, business service, connectivity and integration, data processing and analysis, and real-time locating. They investigated several technologies which can be implemented in these components.

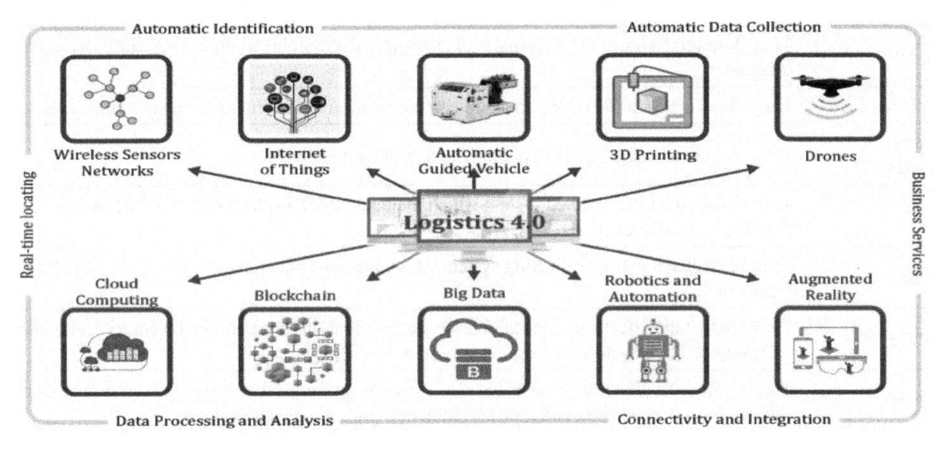

Fig. 3: Components and Technologies of Logistics 4.0. Radivojević and Milosavljević (2019) Other reviewed literature about logistics 4.0 can be summarized as in Table 3.

2.2 Findings

From the reviewed literature about industry 4.0 and logistics 4.0, findings can be summarized as;

- Logistics 4.0 is the concept of industry 4.0 technologies implementation to logistics functions to ensure the aim of logistics.
- Because of the ongoing and emerging technologies, it is hard to make a strict definition of logistics 4.0.
- For a mature logistics 4.0 system, not only autonomized physical assests, but also cognitive and decisive systems should be improved. Autonomozation is not the only aim for physical resources; also autonomozation of decision making should be constructed.
- Among many defined technologies for industry 4.0, IoT, big data applications and analytics, and CPS are the main ones that have been studied and implemented into logistics 4.0.
- A mature logistics environment;
    - o Must be digitalized
    - o Must enable to collect data and communicate with each other via IoT technologies

Table 3: Literature summary for logistics 4.0.

Paper	Findings
Reif and Walch 2008	They evaluated picking systems with augmented reality and virtual reality.
Windt et al. 2008	They described the autonomous logistics as the logistics objects which have the ability of processing and rendering information to make decisions.
Alves and Roßman 2009	They described a discrete-event simulation system which simulates transportation models using real time resource routing and collision avoidance.
Reif and Günthner 2009	They investigated an augmented reality supported picking system, described hardware properties and showed some efficiency results of the analyzed picking system.
Lang et al. 2010	They presented an intelligent container which can measure parameters such as temperature and humidity via sensors, make cognitive analysis about the transported good losses or quality and communicate with logistics headquarter.
Liu and Sun 2011	They claimed that information flow is large and complex in a vendor-managed inventory system in automobile parts inbound logistics, and RFID based IoT applications can improve the effectiveness and convenience of the information.
Schwerdtfeger et al. 2011	They described the augmented reality for order picking and identified which facts should be considered for a robust and effective system.
Circulis and Ginters 2013	They investigated the augmented reality in logistics activities such as packaging, handling, storage and transportation.
Ginters and Martin-Gutierrez 2013	They presented an augmented reality RFID solution in a warehouse environment.
Chen et al. 2014	They described a logistical information management system based on IoT (RFIDs) and software integration.
Gupta and Jones 2014	They investigated an RFID based cloud computing system for order picking in a warehouse.
Krajcovic and Gabajova 2014	They studied an order picking system with augmented reality.
Gan 2015	He introduced an IoT enabled system which includes RFID, GPS, 3G and GIS technologies to provide real time surveillance of logistics.
Kim et al. 2015	They presented a multiple tracking system which includes materials and personnel, with RFID and GPS technologies.
Pan et al. 2015	They studied the effects of physical internet and simulations showed that inventory and transportation costs reduced by physical internet's flexibilities
Qiu et al. 2015	They proposed a physical IoT enabled model which can be used in a Supply Hub in Industrial Park that can integrate warehousing and transportation activities via the customers.
Twist 2015	He investigated the effects and challenges of RFID usage on supply chain facilities, and especially concluded that labour, accuracy and transportation cost savings in warehouse management is expected.
Zhong et al. 2015	They introduced a big data approach for RFID enabled logistics data.
Addo-Teknorang and Helo 2016	They investigated the literature for big data applications in supply chain management systems. They subdivided the concepts as; big data acquisition sources, big data storage, big data analytics, big data applications and big data value adding in supply chain management systems.
Bertsimas et al. 2016	They proposed a model for inventory management in the era of big data. They described producers and formulas to predict demand with the collected data from internet for managing the inventory.
Chukwuekwe et al. 2016	They proposed predictive maintenance within an industry 4.0 environment and defined key drivers as, internet of things, big data, cloud computing, cyber-physical systems, and 3D printing.
Culler and Long 2016	They presented a smart material warehouse with autonomous robots and camera tracking systems which the robots are used for shipping, receiving and storage.
Galindo 2016	Identification, real time locating, sensing, networking, data collection and analysis, and business services are indicated as technical components of logistics 4.0 and design principles of a logistics 4.0 system has been presented.

Table 3 contd. ...

...Table 3 contd.

Paper	Findings
Geng and He 2016	They indicated that main IoT technologies are RFID, GPS/GIS, video and image sensing and sensor technology. Contributions on transportation, storage, packaging, distribution and information management have been highlighted.
Kwon et al. 2016	They introduced the IoT based asset health system which includes 4 dimensions; sensing, diagnosis, prognosis and management. Study highlighted that not only data collection, but also big data analysis is needed for an effective asset maintenance system.
Pujo et al. 2016	They proposed a wireless intelligent control of a production system.
Qu et al. 2016	They investigated an IoT based cloud manufacturing system for production logistics which include customer service (order making), warehousing and delivery activities and conducted a case study.
Wang 2016	He defined logistics 4.0 as the use of IoT devices through information processing and network communication technology platform which are used for logistics activities like transportation, warehousing, distribution, packaging and handling.
Wang et al. 2016	They emphasized big data analytics in logistics and supply chain management and indicated big data analytics techniques as statistical analysis, simulation and optimization.
Yu 2016	He proposed a design and development of RFID based logistics tracking management system which realizes the tracking and monitoring goods in transportation.
Zhong et al. 2016a	They introduced an approach for visualization of RFID enabled logistics big data in a shop floor for advanced managerial decisions.
Zhong et al. 2016b	They highlighted the big data applications in supply chain management systems in the perspective of data collection methods, data transmission, data storage, processing technologies, big data enabled decision making models and applications.
Alwadi et al. 2017	They described RFID technology in detail, both in aspects of hardware and software, and analyzed RFID architecture for inventory management.
Baretto et al. 2017	They defined logistics 4.0 as combination of activities as resource planning, warehouse management systems, transportation management systems, intelligent transportation systems and information security with the CPS applications.
Ben-Daya et al. 2017	They examined the usage IoT technologies in supply chain functions such as warehousing, order management, inventory management, and transportation and impacts of IoT on these functions by a literature review.
Kache and Seuring 2017	They investigated the impact of big data concept in the context of supply chain management by Delphi research technique. Results showed that the logistics has the highest rank for the opportunities of big data analytics in supply chain management level. Also, maintenance, inventory management, demand management and production planning have been highlighted as the big data analytics opportunities.
Lee et al. 2017	They designed an IoT enabled warehouse system with not only physical IoT assets but also with data clustering, data mining and machine learning procedures for an effective warehouse.
Mahmud 2017	He declared that IoT assets such as RFID sensors, which provide the tracking of position and condition of materials and transportation assets integrated with SAP software can improve the effectiveness of the warehouse and production management system. Big data and cloud computing have been stressed in this paper, also the benefits for predictive maintenance by IoT technologies have been indicated.
Mathaba et al. 2017	They described an inventory control architecture via RFID technology and Web 2.0 tools which is able to detect misplaced products and low stock levels.
Mladineo et al. 2017	They presented a metaheuristic algorithm for an NP-hard partner selection problem that can be used in a cyber physical production network.
Papert and Pflaum 2017	They made suggestions for logistics companies to design and realize the IoT ecosystem. They identified some major guideline points on how to build the IoT enabled logistics environment.
Qu et al. 2017	They compared classical production logistics systems with IoT based systems and made quantitative analysis of transport costs and inventory costs of different IoT based scenarios.
Schuhmacher et al. 2017	They investigated intelligent intralogistics equipment.

Table 3 contd. ...

...Table 3 contd.

Paper	Findings
Szymańska et al. 2017	They defined logistics 4.0 as a narrower term of industry 4.0 which includes autonomous and self organizing systems with other systems.
Trappey et al. 2017	They investigated the patents for technological concepts like RFID, wireless sensor network, cloud computing which may be assumed as tools of logistics in the context of industry 4.0. Results showed that major projects are about inventory status, shipping notification, merchandise tracking, transportation optimization, optimization of resource allocation, warehousing, quality control, sales information, accounting information and purchasing information.
Wan et al. 2017	They investigated context-aware cloud robotics, which collect information via IoT technologies and make decisive actions according to the predetermined constraints by cloud computing for handling operations.
Witkowski 2017	He highlighted IoT and big data as innovative solutions in logistics and supply chain management.
Yerpude and Singhal 2017	They aimed to study the effects of IoT data on vendor-managed inventory systems, and showed the improvement of the system by the IoT data usage.
Zhang et al. 2017	They presented a cyber-physical system based smart control model for shop floor handling system.
Zheng and Wu 2017	They proposed a smart inventory management system for spare parts with IoT and big data analytics.
Bukova et al. 2018	They defined logistics 4.0 as; connecting production with consumption with artificial intelligence and digitization of supply chain.
Chen and Zao 2018	They described a logistical automation management based on IoT. They constructed a prototype hardware platform and analyzed different anti-collision algorithms for RFID signals.
Edirisuriya et al. 2018	They made a systematic literature review about logistics 4.0 and indicated that IoT and automation, cloud computing, big data analytics, simulation, and augmented/virtual reality are the industry 4.0 technologies which can be applied in logistics.
Kousi et al. 2018	They presented a design and prototype implementation of autonomous material supply process which is responsible for the movement of the consumables from the warehouse to the production line.
Lin and Wang 2018	They claimed that intelligent logistics centers which have cyber physical systems that are equipped with IoT based technologies and cloud computing systems have heavy computing loads, and they proposed a mathematical model to construct an effective fog computing system.
Skapinyecz et al. 2018	They claimed that cyber physical systems equipped with IoT technologies will ensure the real time data collection and using the right optimization methods will improve the logistics processes.
Tu et al. 2018a and Tu et al. 2018b	They proposed a detailed IoT system modeling approach for production logistics and supply chain system, designed an experimental system and conducted a prototype IoT based cyber physical system architecture.
Wang et al. 2018	They presented a smart packaging with RFID system to detect the abnormal internal changes in the package and to test the internal status of the package. They explained the system design within the aspects of mathematical modeling.
Zhu 2018	They discussed the IoT and big data based logistics scheduling methods. Results of this study shows that customer experience, utilization of logistics resources and the logistics costs have been improved.
Kozma et al. 2019	They declared that resource and material management, inventory management, intelligent transport, and material handling/distribution should be improved in the context of Industry 4.0 technologies like IoT and cyber physical systems.
Tang and Veelenturf 2019	They claimed that the factors as faster speed by drones or delivery robots, higher reliability by robotic storage and retrieval systems, lower operation costs by inventory monitoring and replenishment systems via sensors and efficiency improving by blockchain enabled container shipping may be assumed as economical benefits of industry 4.0 applications to logistics.
Wahrmann et al. 2019	They presented an intelligent framework for pick-up-and place tasks, which includes modules as object recognition, environment modeling, motion planning, and collision avoidance.

- o Must use CPS, big data, cloud computing concepts for decision and action autonomozation
- o Must visualize its environment
- o Must coordinate physical assests as autonomusly as possible in a prescriptive but preferrebly cognitive way.
- o Must behave flexible for both pyhsical and information aspects to meet internal and external customer demands.

According to the reviewed literature, Table 4 has been constructed to show what kind of technologies of industry 4.0 have been studied in logistics operations.

A number of papers have been investigated for the logistics 4.0 as shown in Table 4 and it is seen that the most studied logistics activities are warehousing, transportation, handling and inventory management. Also IoT, big data analytics and Cyber Physical Systems are the most studied concepts of logistics 4.0.

As it can be seen in Table 5, IoT concept is the most investigated one for the warehousing and transportation activities. This table may give us an idea about what kind of logistics activities and technologies are important for logistics 4.0 implementation.

3. Logistics 4.0 Maturity

Since the concept of maturity measuring was introduced by Paulk et al. (1993) which was introduced for software development process assessment, many domains have been inspired from this study. De Bruin et al. (2005) indicates that more than 150 maturity models have been developed for domains such as IT service capability, strategic alignment, innovation management, programme management, enterprise architecture and knowledge management maturity.

Kohlegger et al. (2009) defines maturity models as rating capabilities of maturing elements and selecting appropriate actions to take the elements to a higher level of maturity. Becker et al. (2009)states that a maturity model consists of a sequence of maturity levels for a class of object and these objects are organizations or processes.

Besides developing maturity models for different domains, phases of developing a maturity assessment procedure has also been investigated. Among many researches; Knowledge Management Capability Assessment by Freeze and Kulkani (2005), and Business Process Maturity Model by Fisher (2004) may be assumed as the advanced ones which describes maturity modeling phases. In this study, methods of De Bruin et al. (2005) are used to construct the proposed maturity model.

Maturity modeling of Industry 4.0 is being studied both by academicians and practitioners. VDMA (2015), Acatech (2017), Capgemini Group Consulting (2017), Forrester's Digital Maturity Model 4.0 by Gill and VanBoskirk (2016) are some important studies done by practitioners. Schumacher et al. (2016), Gokalp et al. (2017), Colli et al. (2018), Canetta et al. (2018), Akdil et al. (2018), Bittighofer et al. (2018), Berghaus et al. (2016), Stefan et al. (2018), Horvat et al. (2018), and Elibal et al. (2018) have studied maturity models for industry 4.0 assessment. These studies are in the general context and they try to evaluate an organization regarding their functions.

3.1 Logistics 4.0 Maturity Literature

Sternad et al. (2018) proposed a maturity model for logistics 4.0 based on NRW's[1] maturity model for industry 4.0. Among all the questions in the model, they selected 6 logistics related questions and assessed the logistics system according to 4 functional characteristics which are; purchase logistics, internal logistics, distribution logistics and after sales logistics. Besides the quickness and easy implementation of the model, it lacked identifying the maturity levels of all logistics functions and only limited concepts of industry 4.0 are evaluated like data collection or analyzing.

Szłapka and Stachowiak (2018) constructed a maturity model with five maturity levels in 3 logistics 4.0 dimensions. These dimensions and evaluation areas are as in Table 6.

The study of Szłapka and Stachowiak (2018) investigates a wide range of the industry 4.0 concepts, relatively, but there are still missing logistics dimensions.

Felch et al. (2018) proposed a maturity model for logistics 4.0 in a narrower area. They only evaluated outbound logistics and determined three activity dimensions which are; order processing, warehousing and shipping. The study interviews the determined processes via detailed key techs of Industry 4.0 and detailed assessment criteria, but only for three functions of logistics.

Behrendt et al. (2018) constructed a maturity model based on Fraunhofer IFF's Industry 4.0 stage model by Häberer et al. (2017). They transformed this model into logistics 4.0 maturity assessment for the logistics functions transport, handle, store, pick, distribute, pack and administrate. Evaluation is made according to the 4 stages of Fraunhofer IFF's Industry 4.0 model. Although this model investigates many areas of logistics, there are still some missing functions like procurement, order management, maintenance, etc., and also logistics 4.0 has been evaluated only in the area of data collection and analyzing aspects. Study does not include physical autonomy.

---

[1] KompetenzzentrumMittelstand NRW. Quick Check Industries 4.0 Reifegrad [available at: https://indivsurvey.de/umfrage/53106/uHW7XM-7ca92c0323f74cac34a7af8bfbb654a2, access 05-09-2019]

Table 4: Literature summary of industry 4.0 technologies in logistics functions.

Logistics Activities	**IoT**	**Big data analytics**	**CPS**	**Cloud computing**	**Augmented/ Virtual Reality**	**Autonomus vehicles/ Automozation-Robotization**	**Image/ vision processing**	**3D printing**
Warehousing	Qiu et al. 2015; Twist 2015; Geng and He 2016; Qu et al. 2016; Wang 2016; Baretto et al. 2017; Ben-Daya et al. 2017; Lee et al. 2017; Mahmud 2017; Trappey et al. 2017; Sekkeli and Bakan 2018; Skapinyecz et al. 2018	Wang 2016; Lee et al. 2017; Mahmud 2017	Culler and Long 2016; Skapinyecz et al. 2018; Wang 2016	Mahmud 2017; Trappey et al. 2017	Circulis and Ginters 2013; Ginters and Martin-Gutierrez 2013	Culler and Long 2016; Skapinyecz et al. 2018	Culler and Long 2016; Geng and He 2016	
Transportation	Lang et al. 2010; Gan 2015; Kim et al. 2015; Pan et al.2015; Qiu et al. 2015; Geng and He 2016; Qu et al. 2016; Yu 2016; Wang 2016; Baretto et al. 2017; Ben-Daya et al. 2017; Qu et al. 2017; Trappey et al. 2017; Skapinyecz et al. 2018; Sekkeli and Bakan 2018; Kozma et al. 2019	Wang 2016	Wang 2016; Baretto et al. 2017; Skapinyecz et al. 2018; Kozma et al. 2019	Trappey et al. 2017	Alves and Roßman 2009; Circulis and Ginters 2013	Skapinyecz et al. 2018; Sekkeli and Bakan 2018	Geng and He 2016	
Inventory management	Liu and Sun 2011; Pan et al. 2015; Alwadi et al. 2017; Ben-Daya et al. 2017; Mathaba et al. 2017; Qu et al. 2017; Trappey et al. 2017; Yerpude and Singhal 2017; Kozma et al. 2019	Bertsimas et al. 2016; Kache and Seuring 2017		Alwadi et al. 2017; Trappey et al. 2017				
Maintenance	Chukwuekwe et al. 2016; Kwon et al. 2016; Zheng and Wu 2017	Chukwuekwe et al. 2016; Kwon et al. 2016; Kache and Seuring 2017; Zheng and Wu 2017	Chukwuekwe et al. 2016	Chukwuekwe et al. 2016				Chukwuekwe et al. 2016
Logistics general concepts	Chen et al. 2014; Zhong et al. 2015; Zhong et al. 2016a; Witkowski 2017; Chen and Zao 2018; Edirisuriya et al. 2018; Glistau and Machado 2018; Lin and Wang 2018; Skapinyecz et al. 2018; Tu et al. 2018a; Tu et al. 2018b; Zhu 2018	Zhong et al. 2015; Addo-Teknorang and Helo 2016; Wang et al. 2016; Zhong et al. 2016a; Zhong et al. 2016b; Bamberger et al. 2017; Witkowski 2017; Edirisuriya et al. 2018; Glistau and Machado 2018; Zhu 2018	Chen et al. 2014; Mladineo et al. 2017; Lin and Wang 2018; Skapinyecz et al. 2018; Tu et al. 2018a; Tu et al. 2018b	Edirisuriya et al. 2018; Lin and Wang 2018	Bamberger et al. 2017; Edirisuriya et al. 2018	Windt et al. 2008; Bamberger et al. 2017; Edirisuriya et al. 2018; Glistau and Machado 2018; Skapinyecz et al. 2018		Bamberger et al. 2017

Intralogistics	Kim et al. 2015; Geng and He 2016; Wang 2016; Kozma et al. 2019	Wang 2016	Pujo et al. 2016; Wang 2016; Schuhmacher et al. 2017; Kozma et al. 2019			Kousi et al. 2018	Geng and He 2016	
Handling	Gupta and Jones 2014; Wang 2016; Wan et al. 2017; Kozma et al. 2019	Wang 2016; Zhang et al. 2017	Wang 2016; Witkowski 2017; Kozma et al. 2019	Gupta and Jones 2014; Wan et al. 2017	Reif and Günthner 2009; Schwerdtfeger et al. 2011; Circulis and Ginters 2013; Krajcovic and Gabajova 2014; Glistau and Machado 2018; Sekkeli and Bakan 2018			
Order management	Ben-Daya et al. 2017							
Packaging	Geng and He 2016; Wang 2016; Wang et al. 2018	Wang 2016	Wang 2016		Circulis and Ginters 2013; Sekkeli and Bakan 2018		Geng and He 2016	
Production planning		Kache and Seuring 2017	Baretto et al. 2017					
Demand management		Kache and Seuring 2017						
Customer Service	Qu et al. 2016; Trappey et al. 2017			Trappey et al. 2017				
Quality Control	Trappey et al. 2017			Trappey et al. 2017				
Procurement	Skapinyecz et al. 2018	Glistau and Machado 2018	Skapinyecz et al. 2018					

Table 5: Amount of related papers according to Table 4.

	IoT	Big data analytics	CPS	Cloud computing	Augmented/ Virtual Reality	Autonomous vehicles/ Autonomozation-Robotization	Image/ vision processing	3D printing	Total
Warehousing	11	3	3	2	2	2	2		25
Transportation	15	1	4	1	2	2	1		26
Inventory management	8	2		2					12
Maintenance	3	4	1	1				1	9
Production logistics	1	1		1					3
Logistics general concepts	10	9	5	2	2	4		1	33
Intralogistics	3	1	3			1	1		9
Handling	4	2	3	2	5				16
Order management	1								1
Packaging	3	1	1		2		1		8
Demand management		1							1
Customer Service	2			1					3
Procurement	1	1	1						3
Total	63	27	21	13	13	9	5	2	

Table 6: Logistics 4.0 dimensions and areas of evaluation. Szłapka and Stachowiak 2018.

Logistics 4.0 dimensions	Areas of evaluation
Management	Investments, innovations management, integration of value chains.
Flow of material	Degree of automation and robotization in warehouse and transportation, Internet of things, 3D printing, 3D scanning, advanced materials, augmented reality, smart products.
Flow of information	Data driven services, big data (data capturing and usage), RFID, RTLS (real-time locating systems), IT systems.

Table 7: Overview of the three dimensions and their respective elements. Felch et al. (2018).

Order Processing	Warehousing	Shipping
Process inquiry and quote	Receive product from source or make	Build loads
Receive, enter and validate order	Pick product	Route shipments
Reserve inventory and determine delivery date	Pack product	Select carriers and rate shipments
Consolidate orders		Load product/vehicle and generate shipping docs
Invoice		Ship product
		Receive and verify product by customer
		Install product

As a result, according to the reviewed literature about logistics 4.0, studies are still in infancy stage and in the reviewed studies there are still unanalyzed logistics functions and industry 4.0 technologies in logistics. The aim of this paper is to fill this gap by assessing a wide range of logistics functions to enable accurate road map for transition to logistics 4.0

4. Methodology

Study of De Bruin et al. (2005) inspired this study in terms of methodology. According to De Bruin et al. (2005), the maturity model development phase includes 6 stages.

Fig. 4: Model Development Phases (De Bruin et al. 2005).

The stages of proposed assessment procedure can be briefly described by Figure 5.

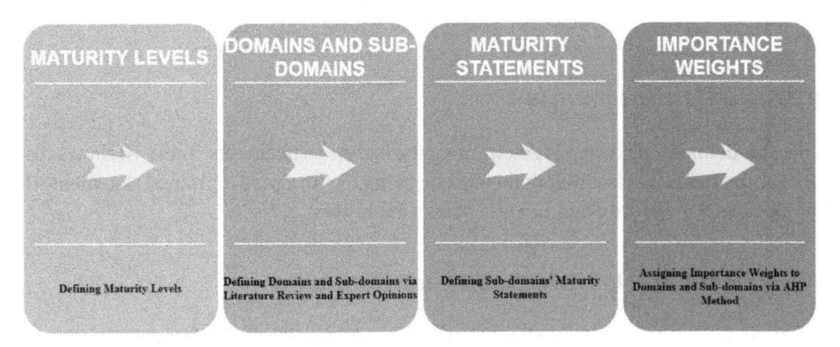

Fig. 5: Steps of Proposed Logistics 4.0 Maturity Assessment.

4.1 Maturity Levels Definition

Maturity level will be defined with a five- point Likert scale in which 5 represents the highest level of maturity scale, as shown at Table 8.

Table 8 shows the lingual representation of the maturity. Each logistics sub-domain will be assigned to a maturity level from 1 to 5, and so also the main domain. Maturity of the sub-domains and domains can be identified and reported linguistically with the help of Table 8.

Level-1 defines that there is no awareness and application of digitalization. Procedures are done with very classical methods; generally depending on the current employee's experience. Level-2 defines that there are some digitalized procedures but there is no awareness or willingles to improve them. For example, the inventory is traced on an Excel sheet which is created manually but possible gains of more digitalization like barcode implementations are not applied. Level-3 defines that digitalization is in the agenda of the management; some applications are done but still insufficient. Level-4 defines that there is a high awareness of logistics 4.0 and there are many applications but only in physical functions. For

Table 8: Logistics 4.0 levels.

Level	Level Definition
Level-1	Functions are not suitable for Logistics 4.0 concepts, there is no awareness of digitalization.
Level-2	Insufficient digitalized applications are available, there is a limited awareness of digitalization.
Level-3	There is an awareness and there are some applications of digitalization, management strategies are in their infancy.
Level-4	There is a high management support on digitalization; functions are preceding with highly structured logistics 4.0 concepts, but mostly in physical functions.
Level-5	There is a high degree of automation for both physical and information aspects. Besides autonomated physical activities, cognitive decisions can be done with applied technologies.

example, automated robotic movements or RFID tracking methods are implemented but learning and cognitive methods for decision making are not used. Finally, level-5 defines that there is a high degree of automation for both physical and information aspects. Not only automated physical assests, but also cognitive methods like data analytics and artificial intelligence are used for decision making.

With these descriptions, management can identify the current status and can create an infrastructure for an accurate road map.

4.2 Defining Domains and Sub-domains

Among the mentioned logistics functions in Table 1, 11 of them are determined as domains for each logistics functions, also sub-domains are composed as in Table 9. De Bruin et al. (2005) indicates that determining the domains, sub-domains and questionnaire statements need an extensive literature review and also some other techniques like the Delphi technique, Nominal group technique, case study interviews and brain storming studies. In this study, an extensive literature review has been done for logistics and logistics 4.0 and also corporate and academic experience are taken into account to construct domains, sub-domains and questionnaire statements.

Domains and sub-domains are generally determined according to the logistics literature of Bowersocks and Closs (1996), Baki (2004), Blanchard (2004), Stock and Lambert (2001), Keskin (2011), Stroh (2011), Gümüş (2012), and Küçük (2016).

4.3 Defining Maturity Statements

After identifying the sub-domains, next step is determining the maturity statements. Table 10 shows the proposed maturity level statements for sub-domains. Each sub-domain must be defined with a predetermined statement which Level-1 shows the lowest level of logistics 4.0 maturity and Level-5 shows the highest.

4.4 Assigning Weights to Domains and Sub-domains

Schumacher et al. (2016) indicate that all items should have different importance while measuring the industry 4.0 maturity. For example, the procurement function may have more importance than the customer service function for logistics. Also for the procurement function, the supplier selection procedure may be more important than the supplier evaluation function. To clarify this issue, AHP method is used to determine the weights of domains and sub-domains.

Table 11 shows the pairwise relations between the domains which are constructed according to the expert opinions. By using these relations weights are determined by AHP and the results can be seen in Table 12. According to these results, D11 (Customer Service) domain is determined as the most important activity of logistics and D7 (Packaging) is determined as the least important. It should be remembered that AHP method includes subjectivity. Many other methods can be used to determine these weights. Table 13 shows the important weights of sub-domains which are calculated with AHP method.

5. Conclusion

The aim of this study was the, construction of a maturity model for logistics 4.0. First, the activities of logistics was defined, then by reviewing current maturity models for logistics 4.0, a model was proposed. This model included maturity statements for each sub-domain of logistics. As it mentioned, e-logistics has a wide range of definitions and these definitions are changing according to the dynamic economical and technological trends. Discussing some other activities as logistics' activities may be a future topic and may be implemented into this study or into other similar studies. This discussion is also valid for the mentioned sub-activites of logistics in this study.

Defining maturity statements has also expert subjectivity and may be changed according to the future technological developments and applications. Industry 4.0 and also Logistics 4.0 are emerging phenomenons and the maturity statements should be revised according to the future technologies and trends.

Table 9: Proposed Domains and Subdomains.

Domains	**Sub-domains**
D1-Procurement	D1.1-Supplier Selection D1.2-Supplier Evaluation D1.3-Inventory Investment Management (raw material or semi-products) D1.4-Standardization D1.5-Make or Buy Analysis D1.6-Purchasing Decisions (JIT- Milk-Run-etc.) D1.7-Quote Management
D2-Purchasing	D2.1-Order processing D2.2-Order Confirmation and Tracking D2.3-Order Completion Approval D2.4-Documentary Management D2.5-Purchasing Efficiency Management
D3-Inventory Management	D3.1-Inventory Management(Risk management, order decision, costs management)
D4-Transportation	D4.1-Mode, Vehicle, Route Selection D4.2-Vehicle Loading D4.3-Vehicle Tracking, Vehicle Condition Monitoring D4.4-Regularity and Documentary Management D4.5-Transportation Costs Management
D5-Demand Forecasting	D5.1-Forecasting Data Management and Integration Management D5.2-Forecasting Techniques D5.3-Forecasting Efficiency
D6-Warehousing	D6.1-Inventory Quantity/Quality Tracking D6.2-Movement Management/Technology D6.3-Warehouse Capacity/Condition Tracking D6.4-Warehouse Efficiency Management
D7-Packaging	D7.1-Labeling management D7.2-Handling management D7.3-Apportionment Management and Packaging Efficiency Management
D8-Inner Distrubition	D8.1-Movement Tracking D8.2-Movement Technology Management D8.3-Movement Costs/Efficiency Management
D9-Handling	D9.1-Handling Standardization and Handling Information Management D9.2-Handling Technology Management
D10-Maintenance	D10.1-Maintenance Planning Management and Asset Condition Tracking D10.2-Predictive Maintenance Management D10.3-Maintenance Efficiency Management D10.4-Spare Part Inventory Management D10.5-Maintenance Technology Management
D11-Customer Service	D11.1-Written Statement of Policy, Customer Receipt of Policy Statement D11.2-Organization Structure D11.3-System Flexibility D11.4-Management Services D11.5-Stockout Level D11.6-Order Information and Elements of Order Cycle D11.7-Expedited Shipments and Transshipments D11.8-System Accuracy and Order Convenience D11.9-Installation, Warranty, Alterations, Repairs and Parts D11.10-Product Tracing D11.11-Customer Claims, Complaints and Returns

Table 10: Proposed Maturity Statements of Logistics Sub-Domains.

Sub-domains	Level-1	Level-2	Level-3	Level-4	Level-5
D1.1-Supplier Selection	We directly contact with new suppliers or suppliers contact us without any preliminary information about them. Information about suppliers is obtained by face to face meetings or phone calls. There is not a structured procedure or method to evaluate the supplier. Decision of the supplier selection is made according to the price and convenience of the first batch	Before the first contact, we send suppliers a form by e-mail or fax to get preliminary information. The terms and agreements are given to the customer manually. The first convenience decision is made by analyzing this information manually (by reviewing the form).	Before the first contact, we send suppliers a form by e-mail or fax to get preliminary information. Data are put into a database (via solutions like MS Office) manually and the first decision is made according to this information.	Before the first contact, we send suppliers a form by e-mail or fax to get preliminary information. We are using some mathematical models and methods for ranking and evaluation the suppliers. These models are used by manipulating the collected data in different softwares.	Candidate suppliers can fill a detailed form by our website, including a variety of information such as financial issues, production capacity, shareholder information, human resource information, etc. We have a decision support system which evaluates the supplier autonomously according to this information and give a preliminary rank to the supplier. Also suppliers can see our terms of agreement and procurement policy from our website.
D1.2-Supplier Evaluation	We don't have a structured evaluation system. We are considering any evaluation in case of any negative feedback about the supplied material.	We have key performance indicators about our suppliers which are collected from different departments manually and evaluations are done in a predetermined period.	Each department inputs related information about the supplier into a database and evaluations are done according to these information by manipulating the data in softwares like MS Office solutions	We have an ERP system and we can reach any related data about our suppliers, evaluations are done according to this information by manipulating the data in softwares like MS Office solutions.	We have an ERP system, all related data about our suppliers are stored in it and evaluations are done autonomously by our supplier management system.
D1.3-Inventory Investment Management(raw material or semi-products)	We don't have an inventory management system.	Inventory investment strategies/decisions are done according to the annual budget and reviewed at predetermined periods via manually collected data from sales, finance and other related departments.	Inventory investment strategies/ decisions are done according to the annual budget and reviewed at predetermined periods via data which is put into databases by related departments.	We have an ERP system and can reach any related data about sales, finance and also external economic parameters. Via manipulating these data in softwares like MS Office solutions inventory investment decisions are made.	We have an ERP system and an integrated decision support systems which simultaneously evaluate all related data about inventory investment.

Table 10 contd. ...

...Table 10 contd.

Sub-domains	Level-1	Level-2	Level-3	Level-4	Level-5
D1.4-Standardization	We don't have standard specifications and technical documents of supplied materials.	Technical specifications of supplied materials are declared to the supplier during purchasing process for every order batch by phone, e-mail or fax.	Technical specifications of supplied materials are defined and declared to the supplier at the contract phase.	Technical specifications of supplied materials are defined and declared to the supplier at the contract phase. In case of revisions, information are sent to the suppliers by e-mail or fax.	We have an internet based supplier portal which our suppliers can reach all own related technical documents and specifications at any time. When revisions are made an autonomous alert system informs our suppliers.
D1.5-Make or Buy Analysis	We don't have a make or buy analysis procedure.	We are doing make or buy analysis sometimes via collecting and/or calculating production costs and purchasing costs manually.	We can trace our production and purchasing costs via manually data entering by related departments and doing make or buy analysis at predetermined financial periods.	We have an ERP system which makes tracing and comparing our costs available at predetermined financial periods.	We can simultaneously trace and predict our costs via our ERP system which can produce instantaneous signals for make or buy analysis.
D1.6-Purchasing Decisions (JIT-Milk-Run- etc.)	We can not collect any relevant data for supporting the purchasing decisions.	At predetermined periods we are analyzing our suppliers' performance and supplied amounts to make specific purchasing decisions via easy collectable key performance indicators.	Supplier evaluation and analysis like ABC analyzes are done via manually data input by related departments and analysis are done by softwares.	We have an ERP system which makes tracing and collecting all relevant data for purchasing decisions, we can reach any relevant data at any time and manipulate data outside for analysis by software solutions like MS Office.	We have an ERP system which makes collecting all relevant data and an autonomous decision support system which can highlight suppliers and supplied parts for specific purchasing decision like JIT, Milk-Run, Kanban, etc.
D1.7-Quote Management	We don't take price requests. We review the prices at predetermined periods or after invoicing.	We are taking quotations by phone for each order batch.	We are sending quote documents to our suppliers and they send price requests by fax.	We send requests and take quotes via e-mail.	We have an internet based supplier portal. Suppliers are alerted for a new quote. They can see all supply information and technical documents online and send price requests online with detailed price break down.

Table 10 contd. ...

...Table 10 contd.

Sub-domains	Level-1	Level-2	Level-3	Level-4	Level-5
D2.1-Order processing	Orders are given by phone according to the predetermined order plans.	Orders are given by fax or e-mail according to the predetermined order plans.	Orders are given by e-mail after checking the related data (like inventory or production plan) manually.	Orders are given by e-mail after checking the related data (like inventory or production plan) from the ERP system.	Our ERP system can generate automatic orders according to the inventory level and production plan and send the orders to the suppliers via supplier portal.
D2.2-Order Confirmation and Tracking	Order confirmation and tracking are done by phone.	Suppliers confirm and give information about the order via fax or e-mail when we request.	Suppliers confirm and give information about the order without our request via e-mail.	Suppliers confirm and give information about the order via entering contemporary data into supplier portal.	We can physically trace our order via IoT technologies like barcode, GPS, etc.
D2.3-Order Completion Approval	Order completion information is taken by phone from warehouse and from the quality control department.	Warehouse and quality control department give information about the received order via e-mail.	We have a database like an Excel sheet or Access table which trace our order approval. Warehouse and quality control enters data into this database about the order.	We can trace order approvals from our ERP system. Every department enters related data into the system.	We can physically trace our order approvals via IoT technologies like barcode, Qrcode, etc. Warehouse and quality control department use reader devices to approve the received orders.
D2.4-Documentary Management	We don't send any documents to our suppliers about the orders.	We send related documents (bill of order, etc.) by fax or e-mail which is prepared manually by the purchasing department.	We use MS Office solutions to prepare the related documents.	Our ERP system can prepare the related documents and send to the supplier by e-mail automatically.	Suppliers can reach all related document from our internet based supplier portal which is generated automatically by our ERP system.
D2.5-Purchasing Efficiency Management	We don't measure the purchasing efficiency, we don't have any KPIs.	We manually collect and store information about purchasing like liability of delivery times or quality, and anaylze at predetermined periods.	We use MS Office solutions to store purchasing KPIs and analyze at predetermined periods.	We can trace purchasing KPIs and generate reports at anytime via our ERP system.	KPIs like lead time and quality liabilities are defined and our ERP system can autonomously alert for an inconvenience.
D3.1-Inventory Management (Risk management, order decision, costs management)	We are replenishing our inventory according to the predetermined periods specified in annual budget.	We are using historical data to replenish our raw material inventory. Historical data is collected manually.	We are using historical data collected on Excel (or similar solutions) and using some ordering techniques like Economic Order Quantity formula.	We can reach many data on our ERP system like historical sales, orders and we can predict future sales and/ or future raw material prices by manipulating ERP data, also we are using ordering techniques.	Our ERP system can use big data relevant with inventory replenishment, not only data about our corporation but also environmental data like price of material or economical trends, risks. We generate appropriate choices for ordering inventory via integrated software with our ERP.

Table 10 contd. ...

...Table 10 contd.

Sub-domains	Level-1	Level-2	Level-3	Level-4	Level-5
D4.1-Mode,Vehicle, Route Selection	We don't decide mode and/or vehicle. Goods are carried to the customers by our contracted carriers.	We search feasible vehicles and modes manually for every customer and keep records manually.	We search feasible vehicles and modes manually for every customer and document information on a computer database.	Our ERP system can generate decision supportive reports for mode, vehicle and route selection when we get order from our customers.	Our ERP system can give appropriate set of solutions for mode, vehicle and route selection; can combine loads of different customers and locations in an efficient way by using all relevant data autonomously.
D4.2-Vehicle Loading	We are making loading decisions manually.	Loading instructions of every product is documented on paper and shipment department can reach them at their station.	Loading instructions of every product is documented on computer database and shipment department can reach them at their station.	Our ERP system has the information of loading instructions and can generate reports for shipment department.	We have a loading software integrated with our ERP system and can generate feasible loading for trucks when the bill of shipment is generated.
D4.3-Vehicle Tracking,Vehicle Condition Monitoring	We don't trace the location or condition of the vehicle. We get information when it delivers to the customer.	We are taking daily reports by phone or e-mail and we keep records manually.	We are taking daily reports by phone or e-mail and input information on a computer database or our ERP system.	We are tracing the locations by GPS simultaneously.	We are tracing the locations by GPS and condition of the trucks/ containers by GPS communicated sensors simultaneously.
D4.4-Regularity and Documentary Management	We generally don't have information about transportation regularities; we get external information if needed. We are preparing transportation documents manually.	We have documented papers about regularities of each location that we ship. We are preparing transportation documents manually.	Regularity information for each location is on a computer database and we can prepare transportation documents via Excel, Word (or similar solutions)	We have regularity information in our ERP system which can be updated manually when it is needed. Our ERP system can generate transportation documents according to the input regularity data.	We have regularity information in our ERP system which can be updated autonomously by third party data centers. Our ERP system can generate transportation documents according to these data.
D4.5-Transportation Costs Management	We are calculating an overall transportation cost and calculate it for all products for a determined period.	We are collecting transportation cost (only freight) for each shipment and calculate it for the related products via Excel (or similar) based solutions. Other costs like insurance, customs, etc., are calculated by overall method.	We are collecting transportation cost (not only freight but also other transportation costs) for each shipment and calculate it to the related products via Excel (or similar) based solutions.	We can trace exact transportation costs via our ERP system for each shipment which includes freight, customs, insurance, and by manipulating data via external software we can calculate transportation cost for each product.	Our ERP system can generate exact reports for each product for all transportation costs and we can predict future transportation costs by using historic data.

Table 10 contd. ...

...Table 10 contd.

Sub-domains	Level-1	Level-2	Level-3	Level-4	Level-5
D5.1-Forecasting Data Management and Integration Management	To make forecasts, we use manually collected data from related departments.	Each department sends related data via a Excel (or similar solution) by e-mail. This data is regulated for analysis.	Each department fills a database for related data where sales department can reach and use for analysis.	Sales department can reach all data from our ERP system. Manipulation of data is needed if it is not contemporary.	Sales department can reach all data from our ERP system, even the data is not contemporary, and system can generate predictive data autonomously.
D5.2-Forecasting Techniques	We are making forecasts by only reviewing past years sales manually.	We analyze the trend of sales on Excel (or similar solutions) and try to understand the trend by graphical methods.	We are trying to make sales forecast by time series techniques.	We are trying to construct mathematical models for regression which includes more parameters	We have machine learning and/or artificial intelligence models for prediction.
D5.3-Forecasting Efficiency	We don't review forecasting efficiency.	We review forecasting accuracy at predetermined periods manually.	We review forecasting accuracy via collected data on a computer database at predetermined periods.	We can generate efficiency reports about forecasting on our ERP system when it is needed.	We can simultaneously trace the forecasting efficiency via machine learning or other techniques and if any unexpected efficiency is predicted, system alerts us.
D6.1-Inventory Quantity/Quality Tracking	We record and trace inventories manually.	We record and trace inventories in software. Data are recorded manually.	We record and trace inventories in software. Data are collected by devices like barcode scanners and put in software.	The inventory is recorded and traced automatically by several technologies like RFID. Data are put on our ERP system.	The inventory is recorded and traced automatically by several technologies like RFID. Data are put on our ERP system. Condition of the inventories are also collected and analyzed by sensors (humidity, heat, etc.). Data are put on our ERP system and ERP system can make alerts.
D6.2-Movement Management/ Technology	Inventories are distributed to production via verbal orders by phone or face to face communication.	Distribution lists are prepared manually and given to the warehouse manually.	Distributions lists are prepared via MS Excel (or similar solutions) and sent to the warehouse by e-mail	Distribution lists are generated via ERP system by an operator at needed periods and the warehouse can reach these lists.	Automatic distribution orders are generated according to the production flow simultaneously via our ERP system.
D6.3-Warehouse Capacity/Condition Tracking	We don't trace warehouse capacity and/or condition.	We are reporting warehouse capacity and/or condition manually at predetermined periods.	We are reporting warehouse capacity and/ or condition on Excel (or similar solutions) at predetermined periods.	We can trace warehouse capacity on our ERP system. Conditional data is put manually into our ERP system.	Via IoT technology, capacity and/or condition (heat, humidity, etc.) can be traced simultaneously via sensors.

Table 10 contd. ...

...Table 10 contd.

Sub-domains	Level-1	Level-2	Level-3	Level-4	Level-5
D6.4-Warehouse Efficiency Management	We don't trace warehouse efficiency.	We are tracing warehouse efficiency at predetermined periods on paper. (We are collecting limited data like how much material being handled, how much labour cost has been occurred, etc. Most of other costs are calculated by overall method.)	We are tracing warehouse efficiency at predetermined periods on Excel (or similar solutions). (We are collecting limited data like how much material being handled, how much labour cost has been occurred, etc. Most of other costs are calculated by overall method.)	We can trace warehouse efficiency on our ERP system. We can collect and analyze wider data.	We have a system which traces warehouse efficiency simultaneously; handled amount, labour cost, label or package consumption, energy consumption, etc. This system also can predict future data for the warehouse.
D7.1-Labeling management	Packages do not include special labels.	Packages include labels which are written manually.	Packages include labels which are written via computer (via softwares like Excel, Word, or similar solutions)	Our ERP system creates labels automatically when the shipment or warehousing orders are triggered.	Our ERP system creates labels automatically when the shipment or warehousing orders are triggered. Packages include labels with RFID, barcode or other technologies which can allow tracing the package and monitoring the information about the content of package.
D7.2-Handling management	There are no handling instructions on packages.	Packages include handling instructions which are written manually if needed.	Packages include standard handling instructions on it.	Standard handling instructions are printed on labels which are created by ERP system.	Specialized handling instructions are written according to the content of packages by our ERP system.
D7.3-Apportionment management and Packaging Efficiency Management	We are making package apportionment manually.	Apportionment instructions and dimensions of every product is documented manually and the instructions.	Apportionment instructions and dimensions of every product are documented on computer database and packaging is done according to the instructions.	Our ERP system has the instructions of apportionment instructions and dimensions of the products, and the packaging orders are created on ERP system.	We have a apportionment software integrated with our ERP system and can generate feasible nesting for packaging according to the order location.

Table 10 contd. ...

...Table 10 contd.

Sub-domains	Level-1	Level-2	Level-3	Level-4	Level-5
D8.1-Movement Tracking	We can not trace the inner movements.	Inner movements can be traced by daily reports manually which are generated manually.	Inner movements can be traced by daily reports on MS Excel (or similar solutions) which generated manually.	Inner movements can be traced on our ERP system.	Inner movements can be traced via IoT technologies, RFID or other solutions simultaneously.
D8.2-Movement Technology Management	We are not using autonomously guided vehicles for inner movement.	We are not using autonomously guided vehicles but researches are going on.	We have pilot implementations for autonomously guided inner movement.	We are using autonomously guided vehicles in selected areas.	We are using autonomously guided vehicles in a wide range.
D8.3-Movement Costs/Efficiency Management	We do not calculate inner movement costs.	We calculate inner movement costs with a simple overall method.	We calculate inner movement costs via manipulating ERP data on Excel (or similar solutions) at any time for past periods.	Our ERP system can generate reports about inner movement costs and efficiency at any time for past periods.	Via IoT based tracing integrated with our ERP system. We calculate inner movement costs and efficiency simultaneously.
D9.1-Handling Standardization and Handling Information Management	We don't have handling instructions.	Handling instructions are given verbal.	Handling instructions are documented manually which are located in determined areas.	Operators can reach handling instructions as Excel (or similar) files.	Handling instructions can be reached in our ERP system.
D9.2-Handling Technology Management	We are using classical handling technologies.	We are aware of new handling technologies and studying for implementations.	There are pilot implementations for technological handling.	Industry 4.0 technologies are used in selected areas.	Industry 4.0 technologies are widely used for handling and improvement strategies are going on.
D10.1-Maintenance Planning Management and Asset Condition Tracking	We make when a machine failure occurs.	We have maintenance plans, prepared manually.	We can set machine alert times. Machines alerts for maintenance according to these settings.	Machines have self diagnosis systems.	Machines have self diagnosis systems. Maintenance schedule adjusts itself based on real time data like failure importance, maintenance team workload, production planning, etc.
D10.2-Predictive Maintenance Management	We don't make any predictive maintenance.	We collect data manually and try to predict maintenance manually.	We collect data on computer database and via manipulating these data we predict maintenance.	We collect data into our ERP system and we can reach data for predictive maintenance. Prediction is done out of ERP system via manipulating the data.	Machines and/ or equipments send information automatically via IoT technologies and prediction for maintenance is done automatically. Also product quality data is used for predictions. Alerts are sent to the operators.

Table 10 contd. ...

...Table 10 contd.

Sub-domains	Level-1	Level-2	Level-3	Level-4	Level-5
D10.3-Maintenance Efficiency Management	We don't trace maintenance efficiency.	We generate manual work orders for maintenance actions but we don't have standard maintenance times.	We generate work orders for maintenance actions and trace maintenance efficiency by using standard maintenance times via Excel (or similar solutions)	We generate work orders for maintenance actions via ERP system and trace efficiency by predetermined standard times.	Our maintenance system can compare defined standard maintenance times with historical data and calculate new times if necessary via machine learning, artificial intelligence, etc.
D10.4-Spare Part Inventory Management	We trace all our spare parts manually.	We trace only some spare parts manually which are decided by ABC analysis	We trace all our spare parts on a computer database.	We trace all our spare parts in our ERP system, information like lead times, minimum stock values are also defined and inventory management is done via manipulating these data.	Our system can give autonomous decisions for spare parts ordering by using simultaneous data like asset condition, maintenance prediction, etc.
D10.5-Maintenance Technology Management	We use classical equipments for maintenance. There are no written instructions.	We have maintenance instructions on paper and/or on computer.	We are aware of new maintenance technologies in the concepts of Industry 4.0, researches are going for it.	We have pilot implementations for new maintenance technologies.	Technologies like virtual reality, augmented reality, and robotic maintenance are used in our corporation.
D11.1-Written Statement of Policy,Customer Receipt of Policy Statement	We don't have any written customer policy.	We have a general written statement of policy which is declared to our customers.	Customers can reach our general customer policy on our web-site.	Besides a general customer policy, we have different policies for our customers according to their needs which are kept in CRM system.	Customer can reach all specified statements about themselves on our customer portal online, any change in statements are automatically alerted to our customers via e-mail.
D11.2-Organization Structure	There is no specified contact information for our customers. In any case, they contact by phone with who is available in our corporation at that time.	Customers first contact with an operator by phone and they are directed to the related department by the operator.	Customers have the written information about whom they should contact on paper or on our website.	Customers are directed to the related department via automatic reply when they call.	We have an autonomous answering system which can be filler from our web site and/or live chat module and/or voice recognition telephone help system.

Table 10 contd. ...

...Table 10 contd.

Sub-domains	**Level-1**	**Level-2**	**Level-3**	**Level-4**	**Level-5**
D11.3-System Flexibility	We don't have an emergency plan for force majeure or unexpected events.	We have a determined team which is established to predict force majeure or unexpected events.	Our emergency plan is written on paper for unexpected events.	All our emergency plans and actions are kept in ERP system (for example alternative suppliers and/ or contractors). We can decide for alternative solutions in a fast way.	We have a system which can predict and alert us via historic data for unexpected situations and create alternative set of solutions automatically.
D11.4-Management Services	We don't have any actions to inform our customers about efficient relations with us.	Our customers contact us in case of any information needed.	Our sales department organizes some seminars and/ or visits about current/new practices of collaboration.	We have written statements about our practices like how to give new order, how to feedback us, or how to use products or training manuals and send them to the customers via e-mail.	All our customers can reach many information about collaboration practices, can make feedbacks to us. Online training webinars are organized for any topic if it is needed.
D11.5-Stockout Level	We don't keep any stockout data.	Stockout data is kept manually on paper when it occurs but analyses are done at predetermined future periods.	Stockout data is kept computer database like Excel (or similar solutions) when it occurs but analysis are done at predetermined future periods.	We can reach stockout data on our ERP system and analysis are done via manipulating data out of the ERP system.	We have an ERP integrated intelligent system which can predict and alert us for stockouts and also can generate set of solutions to compensate it.
D11.6-Order Information and Elements of Order Cycle	We can not monitor order status and don't give any information about the orders to our customers.	Customers are informed about their order status by phone if they request. Orders are monitored manually.	Customers are informed about their order status by mail if they request. Order status is entered to a computer database manually.	Customers are informed about their order status via automatic mails for every step via our CRM system which data is put manually.	Customers can reach any information about their order via our web based customer portal online. Order status is monitored simultaneously via barcode, RFID, or GPS technology.

Table 10 contd. ...

...Table 10 contd.

Sub-domains	Level-1	Level-2	Level-3	Level-4	Level-5
D11.7-Expedited Shipments and Transshipments	We don't make any expedited shipments or transshipments.	We do expedited shipments or transshipments in case of preventing any stockouts with calculating extra costs manually.	We do expedited shipments or transshipments in case of preventing any stockouts with calculating extra costs via Excel based (or similar solutions) softwares.	We decide expedited shipments or transshipments according to the data in our ERP/CRM. We can calculate extra transport costs and give approximate decision if the customer is worth for these extra costs. These calculations are done out of ERP/ CRM system via manipulating data.	Our ERP/CRM system can make exact decisions for expedited shipment or transshipments via using the big data in it by machine learning and/or artificial intelligence techniques.
D11.8-System Accuracy and Order Convenience	We don't collect any information about the order accuracy or convenience	We collect information about order accuracy and/ or convenience manually and analyze them at predetermined periods.	We collect information about order accuracy and/ or convenience on a computer database and analyze them at predetermined periods.	Order accuracy and convenience data can be monitored on our ERP/CRM system and measurement reports can be generated whenever needed.	Order accuracy and convenience can be monitored simultaneously and we have an intelligent system can generate alerts for errors and or inconveniences. Our system can measure and analyze the order giving time and can give information that which customers are facing troubles during order giving process.
D11.9-Installation, Warranty,Alteration s,Repairs and Parts	Instructions of installation, warranty information, repair information is given to the customer when the product is sold, on paper.	Customers can take any information about installation, warranty, and repair by phone.	Customers can take any information about installation, warranty, and repair by e-mail.	Customers can take any information about installation, warranty, repair via our web site or internet based customer portal.	Customers can take any information about installation, warranty, repair via our web site or internet based customer portal. Also our system keeps the information like sold date, repair request and can give autonomous information about warranty expiration or repair.

Table 10 contd. ...

...Table 10 contd.

Sub-domains	Level-1	Level-2	Level-3	Level-4	Level-5
D11.10-Product Tracing	We don't have a traceability system for raw material to product.	We can trace from product to raw material via the manually filled forms or work orders.	We can trace from product to raw material via a computer database like Excel which is filled manually.	We can trace from product to raw material via an ERP system is filled manually.	All raw materials are being readed by barcodes or RFID readers when it is used and we can make an exact trace from product to raw material via product ids.
D11.11-Customer Claims, Complaints and Returns	We take customer complaints and claims via phone but don't record them.	We record the customer complaints and claims which is given via phone or e-mail manually and analyze at predetermined periods.	We record the customer complaints and claims manually on a computer database like Excel and analyze at predetermined periods.	We record the customer complaints and claims manually on our ERP/ CRM system which is given via phone or e-mail and analyze them whenever it is needed.	Our customers can fill a complaint and claim form on our customer portal, we have an intelligent system which can make inferences about the claims and complaints.

Table 11: Pairwise relation table of logistics domains.

	D1	D2	D3	D4	D5	D6	D7	D8	D9	D10	D11
D1	1	5	2	1/3	6	1/2	7	2	5	2	1
D2		1	1/5	1/6	4	1/6	3	0.25	1	1/3	1/7
D3			1	1	6	1	7	1	7	2	1/5
D4				1	5	2	7	2	8	2	1/2
D5					1	1/5	1	1/7	1/3	1/2	1/7
D6						1	7	4	7	3	1/4
D7							1	1/8	1	1/7	1/9
D8								1	8	3	1/6
D9									1	1/7	1/9
D10										1	1/6
D11											1

Table 12: Weights of logistics domains with CR value of 0.079.

	Domains										
	D1	D2	D3	D4	D5	D6	D7	D8	D9	D10	D11
WEIGHT	0.119	0.028	0.098	0.15	0.018	0.136	0.015	0.09	0.02	0.06	0.26

Table 13: Weights of logistics sub-domains with CR values.

Sub-Domain Weights	CR Value
D1.1-0.33, D1.2-0.14, D1.3-0.34, D1.4-0.04, D1.5-0.02, D1.6-0.05, D1.7-0.08	0.050
D2.1-0.24, D2.2-0.34, D2.3-0.28, D2.4-0.07, D2.5-0.04	0.056
D3.1-1	0.000
D4.1-0.34, D4.2-0.08, D4.3-0.43, D4.4-0.08, D4.5-0.04	0.016
D5.1-0.62, D5.2-0.27, D5.3-0.09	0.089
D6.1-0.62, D6.2-0.09, D6.3-0.24, D6.4-0.03	0.069
D7.1-0.19, D7.2-0.72, D7.3-0.07	0.010
D8.1-0.72, D8.2-0.18, D8.3-0.09	0.000
D9.1-0.2, D9.2-0.8	0.000
D10.1-0.56, D10.2-0.17, D10.3-0.03, D10.4-0.11, D10.5-0.11	0.099
D11.1-0.08, D11.2-0.02, D11.3-0.02, D11.4-0.01, D11.5-0.05, D11.6-0.04, D11.7-0.05, D11.8-0.05, D11.9-0.10, D11.10-0.29, D11.11-0.23	0.083

References

Abdur Razzaque, M. 1997. Challenges to logistics development: the case of a third world country-Bangladesh. International Journal of Physical Distribution & Logistics Management, 27(1): 18–38.

Acatech (National Academy of Science and Engineering). 2017. Industry 4.0 maturity index. Managing the digital transformation of companies.

Addo-Tenkorang, R. and P.T. Helo. 2016. Big data applications in operations/supply-chain management: A literature review. Computers & Industrial Engineering, 101: 528–543.

Akdil, K.Y., A. Ustundag and E. Cevikcan. 2018. Maturity and readiness model for industry 4.0 strategy. In Industry 4.0: Managing The Digital Transformation (pp. 61–94). Springer, Cham.

Alves, G., J. Roßmann and R. Wischnewski. 2009. A discrete-event-simulation approach for logistic systems with real time resource routing and VR integration. World Academy of Science, Eng., and Tech., 58: 821–826.

Alwadi, A., A. Gawanmeh, S. Parvin and J.N. Al-Karaki. 2017. Smart solutions for RFID based inventory management systems: A survey. Scalable Computing: Practice and Experience, 18(4): 347–360.

Baki, B. 2004. Lojistik Yönetimive Lojistik Sektör Analizi. Volkanmatbaacılık.

Ballou, R.H. 2007. Business logistics/supply chain management: planning, organizing, and controlling the supply chain. Pearson Education India.

Bamberger, V., F. Nanse, B. Schreiber and M. Zintel. 2017. Logistics 4.0–Facing digitalization-driven disruption. Prism, 38: 39.

Barreto, L., A. Amaral and T. Pereira. 2017. Industry 4.0 implications in logistics: an overview. Procedia Manufacturing, 13: 1245–1252.

Becker, J., R. Knackstedt and J. Pöppelbuss. 2009. Developing maturity models for IT management—a procedure model and its application. Business and Information Systems Engineering, 1: 213–222.

Behrendt, F., L.K. Lau, M. Müller, T. Assmann and N. Schmidkte. 2018. Development of a concept for a smart logistics maturity index. In PROLOG 2018: International Conference on Project Logistics (pp. 1–13).

Ben-Daya, M., E. Hassini and Z. Bahroun. 2017. Internet of things and supply chain management: a literature review. International Journal of Production Research, 1–24.

Berghaus, S. and A. Back. 2016 (September). Stages in Digital Business Transformation: Results of an Empirical Maturity Study. In MCIS (p. 22).

Bertsimas, D., N. Kallus and A. Hussain. 2016. Inventory management in the era of big data. Production and Operations Management, 25(12): 2006–2009.

Bibby, L. and B. Dehe. 2018. Defining and assessing industry 4.0 maturity levels-case of the defence sector. Production Planning and Control. doi: 10.1080/09537287.2018.1503355.

Bittighofer, D., M. Dust, A. Irslinger, M. Liebich and L. Martin. 2018 (June). State of Industry 4.0 Across German Companies. In 2018 IEEE International Conference on Engineering, Technology and Innovation (ICE/ITMC) (pp. 1–8). IEEE.

Blanchard, B.S. 2004. Logistics Engineering and Management-Sixth Edition, Pearson Education Inc.

Bowersox, D.J. and D.J. Closs. 1996. Logistical management: the integrated supply chain process. McGraw-Hill College.

Bukova, B., E. Brumercikova, L. Cerna and P. Drozdziel. 2018. The Position of Industry 4.0 in the Worldwide Logistics Chains. LOGI-Scientific Journal on Transport and Logistics, 9(1): 18–23.
Canetta, L., A. Barni and E. Montini. 2018. Development of a digitalization maturity model for the manufacturing sector. IEEE International Conference of Engineering, Technology and Innovation.
Capgemini Group Consulting, Technology, Outsourcing. 2017. Asset performance management maturity model. Retrieved from https://www.capgemini.com/wpcontent/uploads/2017/08/asset_performance_management_maturity_model_paper_web_version.pdf.
Chen, J. and W. Zhao. 2018. Logistics automation management based on the Internet of things. Cluster Computing, 1–8.
Chen, S.L., Y.Y. Chen and C. Hsu. 2014. A new approach to integrate int-ernet-of-things and software-as-a-service model for logistic systems: A case study. Sensors, 14(4): 6144–6164.
Chukwuekwe, D.O., P. Schjoelberg, H. Roedseth and A. Stuber. 2016. Reliable, robust and resilient systems: towards development of a predictive maintenance concept within the industry 4.0 environment. In EFNMS Euro Maintenance Conference.
Cirulis, A. and E. Ginters. 2013. Augmented reality in logistics. Procedia Computer Science, 26: 14–20.
Colli, M., O. Madsen, U. Berger, C. Møller, B.V. Wæhrens and M. Bockholt. 2018. Contextualizing the outcome of a maturity assessment for Industry 4.0. Ifac-papersonline, 51(11): 1347–1352.
Coyle, J.J., E.J. Bardi and C.J. Langley. 1992. The management of business logistics, St. Paul, Minnesota, USA.
Crowell, J.F. 1901. Report of the industrial commission on the distribution of farm products (Vol. 6). US Government Printing Office.
Culler, D. and J. Long. 2016. A prototype smart materials warehouse application implemented using custom mobile robots and open source vision technology developed using emgucv. Procedia Manufacturing, 5: 1092–1106.
De Bruin, T., R. Freeze, U. Kaulkarni and M. Rosemann. 2005. Understanding the Main Phases of Developing a Maturity Assessment Model.
Domingo Galindo, L. 2016. The Challenges of logistics 4.0 for the Supply chain management and the Information Technology (Master's thesis, NTNU).
Edirisuriya, A., S. Weerabahu and R. Wickramarachchi. 2018 (December). Applicability of lean and green concepts in logistics 4.0: a systematic review of literature. In 2018 International Conference on Production and Operations Management Society (POMS) (pp. 1–8). IEEE.
Elibal, K., E. Özceylan, M. Kabak and M. Dağdeviren. 2018. Current situation of gaziantep industry in the perspective of industry 4.0:development of a maturity index. pp. 77–88. *In*: Kutlu, Y., S. Alkan, Y. Daşdemir, I.P. Turgut, A. Gökçen and G. Altan [eds.]. International Conference on Artificial Intelligence towards Industry 4.0 (ICAII4.0), İskenderun, Hatay, Turkey.
Felch, V., B. Asdecker and E. Sucky. 2018. Digitization in outbound logistics–Application of an Industry 4.0 Maturity Model for the Delivery Process.
Fisher, D.M. 2004. The business process maturity model: a practical approach for identifying opportunities for optimization. Business Process Trends, 9(4): 11–15.
Frazelle, E. 2002. Supply Chain Strategy: The Logistics of Supply Chain Management. McGrraw Hill.
Freeze, R. and U. Kulkarni. 2005 (January). Knowledge management capability assessment: Validating a knowledge assets measurement instrument. In Proceedings of the 38th Annual Hawaii International Conference on System Sciences (pp. 251a–251a). IEEE.
Gan, Q. 2015. Research on multi-dimensional logistics based on the Internet of Things. Open Automation and Control Systems Journal, 7: 2051–2056.
Geng, J. and Z. He. 2016. Innovation and development strategy of logistics service based on internet of things and rfid automatic technology. International Journal of Future Generation Communication and Networking, 9(12): 251–262.
Ghiani, G., G. Laporte and R. Musmanno. 2004. Introduction to logistics systems planning and control. John Wiley & Sons.
Ghobakhloo, M. 2018. The future of manufacturing industry: a strategic roadmap toward Industry 4.0. Journal of Manufacturing Technology Management, 29(6): 910–936.
Gill, M. and S. VanBoskirk. 2016. The Digital Maturity Model 4.0. Benchmarks: Digital Transformation Playbook.
Ginters, E. and J. Martin-Gutierrez. 2013. Low cost augmented reality and RFID application for logistics items visualization. Procedia Computer Science, 26: 3–13.
Glistau, E. and N.I. CoelloMachado. 2018. Industry 4.0, logistics 4.0 and materials-Chances and solutions. In Materials Science Forum (Vol. 919, pp. 307–314). Trans Tech Publications.
Göçmen, E. and R. Erol. 2018. The transition to industry 4.0 in one of The Turkish Logistics Company. International Journal of 3d Printing Technologies and Digital Industry, 2(1): 76–85.
Gökalp, E., U. Sener and E. Eren. 2017. Development of an assessment model for industry 4.0: Industry 4.0-MM. International Conference on Software Process Improvement and Capability Determination, 128–142.
Gümüş, Y. 2012. Lojistikfaaliyetlervemaliyetler. GaziKitabevi.
Gupta, S. and E.C. Jones. 2014. Optimizing supply chain distribution using cloud based autonomous information. International Journal of Supply Chain Management, 3(4): 79–90.

Häberer, S., L. Lau and F. Behrendt. 2017. Development of an Industrie 4.0 Maturity Index for Small and Medium-Sized Enterprises. In 7th IESM Conference, Saarbrücken, Germany.
Hofmann, E. and M. Rüsch. 2017. Industry 4.0 and the current status as well as future prospects on logistics. Computers in Industry, 89: 23–34.
Horvat, D., T. Stahlecker, A. Zenker, C. Lerch and M. Mladineo. 2018. A conceptual approach to analysing manufacturing companies' profiles concerning Industry 4.0 in emerging economies: Paper ID: 1213. Procedia Manufacturing, 17: 419–426.
Kache, F. and S. Seuring. 2017. Challenges and opportunities of digital information at the intersection of Big Data Analytics and supply chain management. International Journal of Operations & Production Management, 37(1): 10–36.
Keskin, M.H. 2011. Kavramlar, Prensipler, Uygulamalar Lojistik El KitabıKüreselTedarikZinciriPratikleri. Ankara: GaziKitabevi.
Kim, J.S., H.J. Lee and R.D. Oh. 2015. Smart integrated multiple tracking system development for iot based target-oriented logistics location and resources services. International Journal of Smart Home, 9(5): 195–204.
Kohlegger, M., R. Maier and S. Thalmann. 2009. Understanding maturity models results of a structured content analysis. Proceedings of I-KNOW '09 and I-SEMANTICS '09.
Kousi, N., S. Koukas, G. Michalos and S. Makris. 2019. Scheduling of smart intra–factory material supply operations using mobile robots. International Journal of Production Research, 57(3): 801–814.
Kozma, D., P. Varga and C. Hegedüs. 2019 (March). Supply chain management and logistics 4.0-a study on arrowhead framework integration. In 2019 8th International Conference on Industrial Technology and Management (ICITM) (pp. 12–16). IEEE.
Krajcovic, M., G. Gabajova and B. Micieta. 2014. Order picking using augmented reality. Communications-Scientific letters of the University of ˇZilina, 16(3A): 106–111.
Küçük, O. 2016. Lojistikilkeleriveyönetimi :lojistiğintemelleri, lojistikkarması (7L), lojistikaraçları. Ankara, Türkiye.
Kwon, D., M.R. Hodkiewicz, J. Fan, T. Shibutani and M.G. Pecht. 2016. IoT-based prognostics and systems health management for industrial applications. IEEE Access, 4: 3659–3670.
LaLonde, B.J. and P.H. Zinzer. 1976. Customer Service: Meaningand Measurements, Chicago: National Council of Physical Distribution Management.
La Londe, B.J. and M.C. Cooper. 1988. Customer service: a management perspective. The Council.
Lang, W., R. Jedermann, D. Mrugala, A. Jabbari, B. Krieg-Brückner and K. Schill. 2010. The "intelligent container"—a cognitive sensor network for transport management. IEEE Sensors Journal, 11(3): 688–698.
Lee, C.K.M., Y. Lv, K.K.H. Ng, W. Ho and K.L. Choy. 2018. Design and application of Internet of things-based warehouse management system for smart logistics. International Journal of Production Research, 56(8): 2753–2768.
Lin, C.C. and J.W. Yang. 2018. Cost-efficient deployment of fog computing systems at logistics centers in industry 4.0. IEEE Transactions on Industrial Informatics, 14(10): 4603–4611.
Liu, X. and Y. Sun. 2011. Information flow management of vendor-managed inventory system in automobile parts inbound logistics based on Internet of Things. JSW, 6(7): 1374–1380.
Lu, Y. 2017. Industry 4.0: A survey on Technologies, applications and open research issues. Journal of Industrial Information Integration, 6: 1–10.
Mahmud, B. 2017. Internet of Things (IoT) for Manufacturing Logistics on SAP ERP Applications. Journal of Telecommunication, Electronic and Computer Engineering (JTEC), 9(2-6): 43–47.
Mathaba, S., M. Adigun, J. Oladosu and O. Oki. 2017. On the use of the Internet of Things and Web 2.0 in inventory management. Journal of Intelligent & Fuzzy Systems, 32(4): 3091–3101.
Mladineo, M., I. Veza and N. Gjeldum. 2017. Solving partner selection problem in cyber-physical production networks using the HUMANT algorithm. International Journal of Production Research, 55(9): 2506–2521.
Oleśków-Szłapka, J. and A. Stachowiak. 2018 (September). The framework of Logistics 4.0 maturity model. In International Conference on Intelligent Systems in Production Engineering and Maintenance (pp. 771–781). Springer, Cham.
Pan, S., M. Nigrelli, E. Ballot, R. Sarraj and Y. Yang. 2015. Perspectives of inventory control models in the Physical Internet: A simulation study. Computers & Industrial Engineering, 84: 122–132.
Papert, M. and A. Pflaum. 2017. Development of an ecosystem model for the realization of Internet of Things (IoT) services in supply chain management. Electronic Markets, 27(2): 175–189.
Paulk, M.C., B. Curtis, M.B. Chrissis and C.V. Weber. 1993. Capability maturity model, version 1.1. IEEE software, 10(4): 18–27.
Pujo, P., F. Ounnar, D. Power and S. Khader. 2016. Wireless Holon Network for job shop isoarchic control. Computers in Industry, 83: 12–27.
Qiu, X. and G.Q. Huang. 2013. Supply Hub in Industrial Park (SHIP): The value of freight consolidation. Computers & Industrial Engineering, 65(1): 16–27.
Qiu, X., H. Luo, G. Xu, R. Zhong and G.Q. Huang. 2015. Physical assets and service sharing for IoT-enabled Supply Hub in Industrial Park (SHIP). International Journal of Production Economics, 159: 4–15.

Qu, T., S.P. Lei, Z.Z. Wang, D.X. Nie, X. Chen and G.Q. Huang. 2016. IoT-based real-time production logistics synchronization system under smart cloud manufacturing. The International Journal of Advanced Manufacturing Technology, 84(1-4): 147–164.

Qu, T., M. Thürer, J. Wang, Z. Wang, H. Fu, C. Li and G.Q. Huang. 2017. System dynamics analysis for an Internet-of-Things-enabled production logistics system. International Journal of Production Research, 55(9): 2622–2649.

Radivojević, G. and L. Milosavljević. 2019. The Concept of Logistics 4.0. 4th Logistics International Conference, Belgrad, Serbia.

Reif, R. and D. Walch. 2008. Augmented & Virtual Reality applications in the field of logistics. The Visual Computer, 24(11): 987–994.

Reif, R. and W.A. Günthner. 2009. Pick-by-vision: augmented reality supported order picking. The Visual Computer, 25(5-7): 461–467.

Schuhmacher, J., W. Baumung and V. Hummel. 2017. An intelligent bin system for decentrally controlled intralogistic systems in context of Industrie 4.0. Procedia Manufacturing, 9: 135–142.

Schumacher A., S. Erol and W. Sihn. 2016. A maturity model for assessing industry 4.0 readiness and maturity of manufacturing enterprises. Procedia CIRP, 52: 161–166.

Schwerdtfeger, B., R. Reif, W.A. Günthner and G. Klinker. 2011. Pick-by-vision: there is something to pick at the end of the augmented tunnel. Virtual reality, 15(2-3): 213–223.

Şekkeli, Z.H. and I. Bakan. 2018. Endüstri 4.0'in etkisiylelojistik 4.0. Journal of Life Economics, 5(2): 17–36.

Skapinyecz, R., B. Illés and A. Bányai. 2018 (November). Logistic aspects of Industry 4.0. In IOP Conference Series: Materials Science and Engineering (Vol. 448, No. 1, p. 012014). IOP Publishing.

Stefan, L., W. Thom, L. Dominik, K. Dieter and K. Bernd. 2018. Concept for an evolutionary maturity based Industrie 4.0 migration model. Procedia CIRP, 72: 404–409.

Sternad, M., T. Lerher and B. Gajšek. 2018. Maturity levels for logistics 4.0 based onnrw's industry 4.0 maturity model. Business Logistics in Modern Management, 695–708.

Stock, J.R. and D.M. Lambert. 2001. Strategic Logistics Management-Fourth Edition, McGraw-Hill Irwin.

Stroh, M.B. 2001. A practical guide to transportation and logistics. Logistics Network.

Szymańska, O., M. Adamczak and P. Cyplik. 2017. Logistics 4.0-a new paradigm or set of known solutions? Research in Logistics & Production, 7.

Tang, C.S. and L.P. Veelenturf. 2019. The strategic role of logistics in the industry 4.0 Era. Forthcoming in: Transportation Research Part E: Logistics and Transportation Review.

Trappey, A.J., C.V. Trappey, C.Y. Fan, A.P. Hsu, X.K. Li and I.J. Lee. 2017. IoT patent roadmap for smart logistic service provision in the context of Industry 4.0. Journal of the Chinese Institute of Engineers, 40(7): 593–602.

Tu, M., M. Lim and M.F. Yang. 2018a. IoT-based production logistics and supply chain system–Part 2: IoT-based cyber-physical system: a framework and evaluation. Industrial Management & Data Systems, 118(1): 96–125.

Tu, M., M.K. Lim and M.F. Yang. 2018b. IoT-based production logistics and supply chain system–Part 1: Modeling IoT-based manufacturing supply chain. Industrial Management & Data Systems, 118(1): 65–95.

Twist, D.C. 2005. The impact of radio frequency identification on supply chain facilities. Journal of Facilities Management, 3(3): 226–239.

VDMA. 2015. Impuls-Industrie 4.0 readiness. Retrieved from https://industrie40.vdma.org/documents/4214230/26342484/Industrie_40_Readiness_Study_1529498007918 .pdf/0b5fd521-9ee2-2de0-f377-93bdd01ed1c8.

Wahrmann, D., A.C. Hildebrandt, C. Schuetz, R. Wittmann and D. Rixen. 2019. An Autonomous and flexible robotic framework for logistics applications. Journal of Intelligent & Robotic Systems, 93(3-4): 419–431.

Wan, J., S. Tang, Q. Hua, D. Li, C. Liu and J. Lloret. 2017. Context-aware cloud robotics for material handling in cognitive industrial Internet of Things. IEEE Internet of Things Journal, 5(4): 2272–2281.

Wang, G., A.W. Gunasekaran, E.W. Ngai and T. Papadopoulos. 2016. Big data analytics in logistics and supply chain management: Certain investigations for research and applications. International Journal of Production Economics, 176: 98–110.

Wang, G., J. Han, C. Qian, W. Xi, H. Ding, Z. Jiang and J. Zhao. 2018. Verifiable smart packaging with passive RFID. IEEE Transactions on Mobile Computing, 18(5): 1217–1230.

Wang, K. 2016 (November). Logistics 4.0 solution-new challenges and opportunities. In 6th International Workshop of Advanced Manufacturing and Automation. Atlantis Press.

Windt, K., F. Böse and T. Philipp. 2008. Autonomy in production logistics: Identification, characterisation and application. Robotics and Computer-Integrated Manufacturing, 24(4): 572–578.

Witkowski, K. 2017. Internet of things, big data, industry 4.0–innovative solutions in logistics and supply chains management. Procedia Engineering, 182: 763–769.

Yerpude, S. and T.K. Singhal. 2017. Augmentation of effectiveness of Vendor Managed Inventory (VMI) operations with IoT data–a research perspective. International Journal of Applied Business and Economic Research, 15(16): 469–482.

Yu, Q. 2016. Design of logistics tracking and monitoring system based on Internet of Things. Journal of Residuals Science & Technology, 13(5): 43–1.

Zhang, Y., Z. Zhu and J. Lv. 2017. CPS-based smart control model for shopfloor material handling. IEEE Transactions on Industrial Informatics, 14(4): 1764–1775.

Zheng, M. and K. Wu. 2017. Smart spare parts management systems in semiconductor manufacturing. Industrial Management & Data Systems, 117(4): 754–763.

Zhong, R.Y., G.Q. Huang, S. Lan, Q.Y. Dai, X. Chen and T. Zhang. 2015. A big data approach for logistics trajectory discovery from RFID-enabled production data. International Journal of Production Economics, 165: 260–272.

Zhong, R.Y., S. Lan, C. Xu, Q. Dai and G.Q. Huang. 2016a. Visualization of RFID-enabled shopfloor logistics big data in cloud manufacturing. The International Journal of Advanced Manufacturing Technology, 84(1-4): 5–16.

Zhong, R.Y., S.T. Newman, G.Q. Huang and S. Lan. 2016b. Big data for supply chain management in the service and manufacturing sectors: Challenges, opportunities, and future perspectives. Computers & Industrial Engineering, 101: 572–591.

Zhu, D. 2018. IoT and big data based cooperative logistical delivery scheduling method and cloud robot system. Future Generation Computer Systems, 86: 709–715.

SECTION 11

Smart and Sustainable/Green SCM

Chapter 18

Smart and Sustainable Supply Chain Management in Industry 4.0

Gökhan Akandere[1,*] and *Turan Paksoy*[2]

1. Introduction

In today's accepted scenario, while businesses have economic opportunities such as global and local supply chains, they also face population growth, climate change, water and energy, pollution, health and safety, human rights and poverty, increasing environmental degradation caused by economic growth and the depletion of resources. On the one hand industry is increasingly confronted with environmental and social problems. On the other hand, industries aim to meet the preferences of changing customers, which are created by the continuous growth of global demand in capital and consumer goods, and to ensure sustainable development in the business world. One of the biggest challenges facing businesses in globally distributed supply chains is the integration of social and environmental protection into trade.

Due to Digitalization and intelligent technologies brought about by Industry 4.0 industries are experiencing a digital and intelligence transformation. Businesses must adopt intelligent technologies to optimize product life cycles and adapt to environmental changes. Because in general, smart technologies can be said to be directly related to the sustainability of a country's economic, environmental and social activities. In addition, economic, environmental and social based technical progress can improve the quality of sustainability. Since the emergence of intelligent technological solutions, businesses have transformed themselves to take advantage of these innovations and improve their sustainability performance. Research has also shown that the adoption of smart technologies in sustainable supply chain management practices is a prerequisite. Recent studies show that the development of innovations can improve sustainability.

Pressures from different stakeholders such as different regulatory agencies, regulations, media, customers, suppliers, competitors, investors, stakeholders, shareholders, local and global communities and non-governmental organizations, by integrating processes and planetary-oriented economic, environmental and social concerns into supply chains, forms the basis for the adoption and use of smart, flexible, automated and sustainable technologies. The actors that influence the smart and sustainable supply chain are shown in Figure 1.

[1] Selcuk University, Vocational School of Social Sciences.
[2] Konya Technical University, Faculty of Engineering, Konya Technical University, Email: tpaksoy@yahoo.com
* Corresponding author: gakandere@selcuk.edu.tr

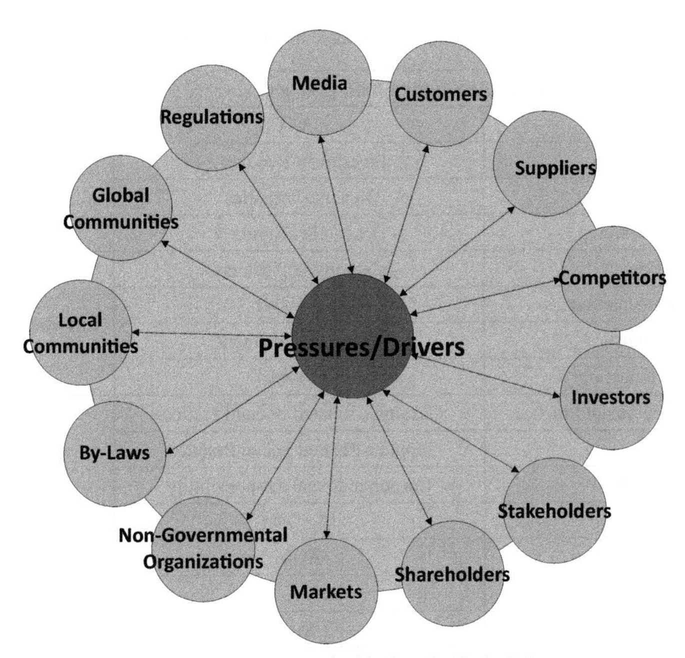

Fig. 1: Smart And Sustainable Supply Chain Drivers.

Parallel to the development of Industry 4.0, corporate sustainability, social values and ethics, social responsibility, economic stability, reducing global carbon footprint, logistics optimization, environmental protection, economic value sharing among supply chain actors, increasing environmental and social responsibility awareness of customers, company transparency, employee benefits and security concerns, employee rights, intensification of competition, such as economic, environmental and social issues in the operations of business operations management decision-making process have been integrated.

Organizations are also motivated to make environmentally friendly products using renewable raw materials, build close-loop supply chains and achieve end-of-life products to reduce carbon footprints and improve economic performance.

The industrial revolution 1.0 corresponds to the production of steam-powered machines. Industry 2.0 corresponds to the period in which the division of labor, assembly lines, and machines operating with electrical energy were used after the 1970s, the main philosophy of Taylorism, the scientific philosophy. With the development of electronics, computers and robots with the millennium, the industry 3.0 era began, focusing on quality and cost performance. Automation provides opportunities to optimize production processes and increase productivity by designing more flexible, ergonomic and safer machines. With the emergence of intelligent production systems following all these changes in industry history, industry 4.0 will be able to respond to existing issues related to the principles of sustainable development. In this sense, industrial production systems need to be balance environmental, social and economics in the use of modern technologies. Industry 4.0 can provide opportunities for the implementation of sustainability.

In order to achieve the Industry 4.0 paradigm, businesses should adopt smart and sustainable supply chain practices as shown in Table 1.

Industry 4.0 based smart and sustainable oriented paradigm can provide businesses with the benefits shown in Table 2.

Traditional supply chains consist of geographically dispersed physical facilities to help establish and maintain transport links between them. Supply chains can be defined as a set of interconnected activities involving coordination, planning and control of products and services between suppliers and customers.

From the 1970s onwards, human resources and industry have increased worldwide through the economic, scientific and technological development of our society, which has an increasing impact on the environment. Being aware of these ecological problems has led to new ideas for environmental sustainable development.

The Brundtland report identified the beginning of sustainable development; Gelişme a development that meets current needs without compromising the ability of future generations to meet their own needs. Elkington proposed the 3P formulation "human, planet and profit". Babu and Mohan expanded the concept of sustainability by adding cultural and governance dimensions.

In this context, businesses should evaluate their business and strategic decisions that are equally and simultaneously optimized for economic efficiency, environmental sustainability and social justice. In the sustainable supply chain, the

Table 1: Smart and Sustainable Supply Chain Practices.

Smart And Sustainable Supply Chain Practices		
5S	LCA	**Economic Practices**
JIT	Investment Recovery	
Risk Management	Reverse Logistics	
Flexible Sourcing	Lean Six Sigma	
Eco Design	Carbon Management	**Environmental Practices**
Green Procurement	ISO 50001	
Green Logistics	Waste Management	
Green Production	ISO 14001	
Employee-Related Social Practices	Customer-Related Social Practices	**Social Practices**
TQM	Supplier-Related Social Practices	
OHSAS 18001	Corporate Social Responsibility	
Smart Factory	Big Data	**Smart Practices**
Cloud Computing	AGV	
3DP	AI	
CPS	IoT	

Table 2: Smart and Sustainable Supply Chain Benefits.

Benefits	
Increase productivity	Reducing supply chain complexity through design consolidation
Decrease unnecessary transactions and waste of time	Redundancy cost reduction for parts and transport through local production
Decrease costs	Optimization of value chains
Ensure zero error in processes	Improved supply chain agility
Optimize inventory	Long-term management of renewable resources
Decrease in labor cost	Reduction of waste generation and pollution
Lower energy consumption	Achieving sustainable development
Decrease emissions	New automation processes
Decrease risk	Product customization
Increase market share	Increasing capacity for product quality
Increase profitability	Global accessibility
Positive providing brand image	Employee and community well-being
Increasing competitiveness	Merging processes to ensure procurement security through smarter and flexible processes
Providing financial gain	Ensuring interaction and synchronization between businesses
Creating customer loyalty and satisfaction	Adding products and services important economic, environmental and social benefits such as making it valuable, accessible and affordable
Efficiency in energy consumption	Develop a sustainable economy
Increasing business sustainability performance	Bringing together supply chain innovation
Ensuring product life cycle traceability	Offer opportunities for smart and sustainable development
Minimizing scrap formation and material waste	

material, in which enterprises integrate and evaluate social, economic and environmental issues into supply chains, can be defined as the management philosophy of economic and information flows. Sustainable Supply Chain Management extends the supply chain scope and balances the three dimensions of sustainability in a supply chain. The STZY focus is on the exchange or potential win-win opportunities between economic performance, environmental footprint and social impact.

Smart supply chain management can be defined as the philosophy of integrating capabilities such as digitalization, collaboration, governance, transparency, flexibility, responsiveness, integrity and automation into the production system and the supply chain is based on cyber-physical interaction.

The linking of people, parts and systems creates self-driven, dynamic, real-time value-added links along the value chain. Industry 4.0 is a mix of digital technology that lifts industrial production to the next level. Under the influence of Industry 4.0, industrial production systems are expected to perform 30% faster and 25% more efficient than before. In the supply chain context, Industry 4.0 presents several challenges, such as data quality and reliability, unemployment, complexity issues, fewer people control, and higher adverse environmental impacts.

It consists of five dimensions and includes various aspects of organization, management, employees and systemic interaction. While technology, research, development and design are evaluated in the areas of design, production, procurement and logistics, the size of management and strategy includes elements such as IT management, embedding Industry 4.0 in company strategies, or product lifecycle management. The organization of production and logistics is a dimension to consider in the context of Industry 4.0, and production means monitoring of quality control as well as inventory and delivery management. The dimension of employees and communication focuses on work order management, skills and competence management and training measures. Finally, the fifth dimension of the model concerns the horizontal integration of companies into international or national value chains and innovation systems, which can either be supported by Industry 4.0 or can be a prerequisite.

The green business adopts the principles, procedures, practices and policies of all elements of the business to improve business continuity, environmental protection, social welfare and social responsibility. There are two main reasons why producers are involved in green businesses. These are internal (social responsibility, competitiveness) and external motivation (customers and government pressures or stakeholder pressures). However, most companies have incentives to implement green practices, not only because of their social responsibility, but because of their competitiveness and cost concerns. Green businesses are associated with green economy, green society and environmental planning, which will come from an environmental chain system. In addition to social business and profit maximization, green business practices have a significant impact on the use of natural resources at different stages of activities and productions.

2. Industries 4.0

Supply chain applications are very important for businesses. They have a significant impact on the costs and profits of enterprises. SCM applications are processes, infrastructure and systems that manage the flow of information, materials and services from the supplier to the end consumer.

One of the key factors of economic, environmental and social growth in the world is the industrial sector. The industrial sector operates as an important locomotive, accounting for 75% of all exports and 80% of all innovations. In Europe, especially Eastern Europe and Germany have a constantly growing industrial sector, while many Western European countries, such as Great Britain or France, have been losing market shares over the past 20 years. While Europe has lost nearly 10% of its sectoral market share in the last 20 years, developing countries have managed to double their share by making up 40% of global manufacturing.

With the effect of this sectoral decline, the term Industry 4.0 was introduced to the public at the Hannover Trade Fair in 2011 as part of Germany's smart and sustainability strategy to prepare and strengthen the industrial sector in relation to future production requirements to maintain and strengthen its leading role in the industrial sector in Germany. Due to the differences between countries and industries, it is very difficult to make a clear definition of Industry 4.0. Industry 4.0 is the ecosystem after industry integration and merger.

Industry 4.0 is a modular structure for the advancement of the sustainable development of businesses, organizations and society, through the virtual copy of a physical world with the software used, and the communication of systems, machines, devices and people in real time through a digital network, using big data. It is defined as the integration of intelligent technologies that enables planning, optimizing, controlling and making decentralized decisions within the value chain among the supply chain stakeholders.

The effects of Industry 4.0 applications in the context of supply chain, havent yet been understood by the enterprises and this is reason why they have difficulty in adopting it. The lack of regulations, lack of vision and strategies related to smart applications of enterprises, lack of top management support, lack of data quality, complexity problems, lack of digital culture, has led to a reluctance to adopt these applications. Enerproses also hesitate because it requires high investment, uncertain economic benefits, lack of infrastructure and internet-based networks, lack of expertise in implementing new business models, lack of integration of technology platforms, security issues, coordination and cooperation issues, lack of government support and financial constraints and such similar obstacles.

3. Literature

In this section, literature on concepts such as Industry 4.0, smart supply chain management, sustainable supply chain management, digital supply chain management, IoT, CPS, Smart Factory, Big Data, 3DP, CC, AGV and AI are examined.

The aim here is to help both academicians and practitioners to determine the main characteristics, success factors and barriers required for the application and development of SSSCM by evaluating the relevant literature. Based on the results of the studies conducted in the literature on smart, digital and sustainable supply chain management, it is revealed that there is the need to determine the applicability of the framework to the supply chains of enterprises. Smart and sustainable supply chain targets may also differ as a result of the different objectives of varied businesses. The subject, purpose, method and outputs of the researches related to digital, green and sustainable supply chain management and Industry 4.0 in the literature are shown in Table 3.

With the framework in Figure 2, executives will be able to help create a vision and roadmap for how to induct smart and sustainable practices into the existing processes of supply chains.

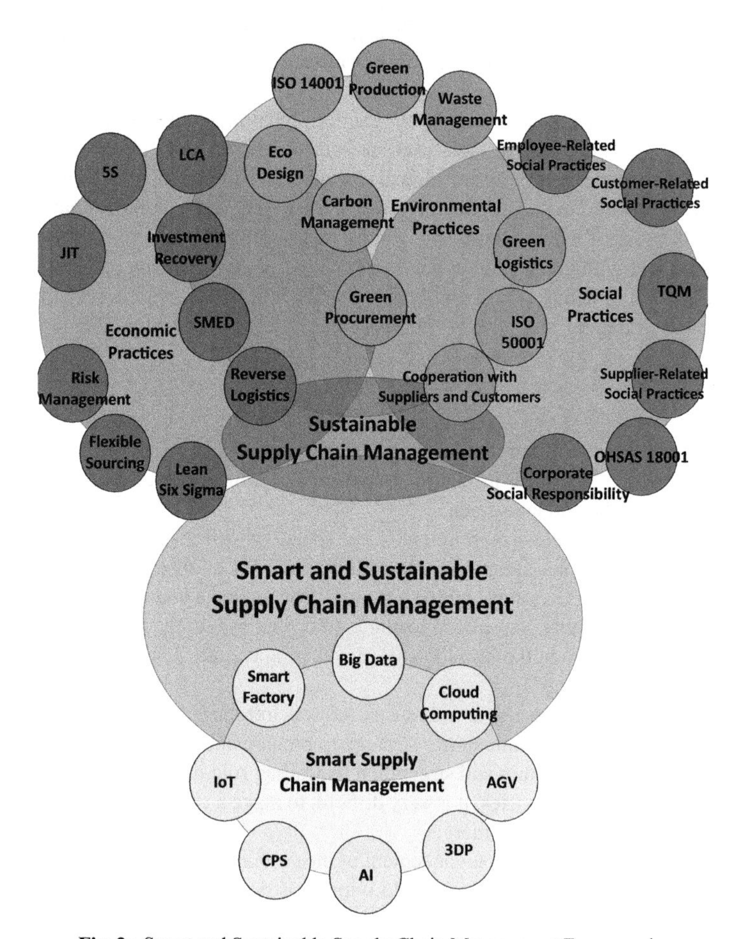

Fig. 2: Smart and Sustainable Supply Chain Management Framework.

4. Framework

Businesses will be able to reach the Industry 4.0 paradigm framework, supply chain strategies, mission and vision by successfully adopting the smart and sustainable practices proposed in the research.

This section provides a framework for intelligent and sustainable supply chain management. A proposed model for SSSCM is shown in the figure above by reviewing the studies in the literature. The features, components, driving forces, obstacles and benefits in the literature have been taken into consideration while forming the framework.

5. Smart and Sustainable Supply Chain Management Applications

5.1 Cyber Physical Systems (CPS)

Businesses have started to use smart applications to collaborate with devices that can reduce costs, real-time business processes, and collaboration.

Industry 4.0 is the idea of systems that bring the physical and virtual world together. In this context, the machines and devices of the production lines and cells are connected to each other via a network of intelligent controllers integrated with information systems and interact with each other.

RFID technology, sensor, actuator, control unit and communication device are used in cyber physical system processes. The data obtained with these devices are stored in the cloud system and transmitted to the network system. With the help of smart applications and mobile devices, seamless integration in the supply chain has been achieved. With this integration, human-machine interaction (HMI) with the Internet communicates with users in real time and transmits data from anywhere. Interconnected ecosystems can automatically adapt to and respond to changing demands. CPS processes the enterprises, real-time information assessment, self-monitoring and control, anticipating users' actions or needs, and consequently performs self-organizing production steps.

In CPS, the goal is to create digital reality that is as similar to physical reality as possible. With CPS, businesses can make real-time decisions such as production prioritization of orders, optimization of tasks and maintenance requirements. Businesses can achieve positive outcomes such as control, oversight, transparency and efficiency in their processes. Again, it can guarantee just-in-time delivery, thus reducing waste caused by safety stocks. In addition, with CPS, businesses can extend the equipment life by examining the environmentally friendly features of sustainable production.

5.2 Internet of Things (IoT)

The Internet of Things (IoT), which develops on the philosophy that supply chain stakeholders are virtually connected to each other through processes, has a major impact on the global economic platform.

Businesses will be able to sustainably develop their supply chains as a result of real-time integration of basic technologies such as Internet of Things (IoT), robotics, artificial intelligence, interoperable systems, cloud computing, big data, 3D printing and digital payment with intelligent and sensing devices.

The prevalence and access of smart and easy-to-interact devices will enable businesses to make more sustainable decisions. The success of IoT technology depends on consistency and scalability. IoT applications offer reliable, intuitive and effective technological processes that can operate internationally.

In addition, supporting government policies to strengthen sustainable practices and promote digital technologies encourage organizations to move forward with a circular economy that meets industry 4.0 requirements.

IoT is a self-supporting tool that enables all members to communicate, collect, identify, localize, monitor, and connect data, using cloud and internet technologies in real-time between machines, components, devices and users owned by actor enterprises in the supply chain. It is a smart information technology network infrastructure that envisages a society which can access the internet environment through intelligent interfaces that integrate perfectly into the information system, filled with configuration, self-management, and smart technology.

IoT evolved from the powerful formation of wireless technologies, sensors and the Internet. IoT connects networked systems and other devices over the Internet. The systems are aware of the environment and the devices are intuitive with the help of sensors that transmit enormous amounts of data every day. These systems are easier to use and understand with the help of the Internet.

Businesses use technologies such as barcodes, smart and wireless sensors and radio frequency identification (RFID), mobile terminals, intelligent embedded devices, network of smart computer technologies, software, hardware, databases, virtual and physical objects in the IoT process of objects.

IoT offers businesses the ability to connect basically all physical objects as a smart thing over a global network of smart things. In this context, objects are not just machines, but all devices and people who interact with detection, identification, processing, communication and networking capabilities. It is an active network infrastructure capable of making the best decisions and self-structuring with information centralization technology processes.

Table 3: Literature matrix of smart and sustainable supply chain management studies.

References	Subject	Objective	Method	Findings
Manavalan and Jayakrishna (2019)	Internet of Things (IoT) practices in sustainable supply chain for industry 4.0	Opportunities for transformation of Industry 4.0 through ıot implementations in the sustainable supply chain of enterprises were explored.	Modeling	In this research, it is concluded that enterprises can improve their sustainability performance by extending their product lifecycles and reducing their carbon footprints with green products produced using edible raw materials, and that Industry 4.0 processes can be integrated into supply chain stakeholders such as suppliers, manufacturers, retailers and customers in real time.
Ivanov et al. (2018)	Impact of Control Theory applications on operational systems, supply chain management and Industry 4.0	Applicability of control theory to engineering and management problems in supply chain operations	Modelling	In this study, the control theory can be adapted to Industry 4.0, where planning and control processes are adaptable, operational and SC performance can be created and optimized which are difficult to express in static and discrete time models, dynamics and changes in supply chain processes can be handled at different levels of detail in the process design functions.this can be created and the entire system can be brought about.The timing problems can be integrated and stability, durability, flexibility in application can be achieved as important performance outputs.
Cornelis de Man and Strandhagen (2017)	In the Industry 4.0 world that is digitalizing and automating sustainable business models	How the Industry 4.0 paradigm can be used to build sustainability business models	Case Study	In the study, the potential of Industry 4.0 applications on sustainable business models was determined.
Rajput and Singh (2019)	Circular Economy (CE) and Industry 4.0	Identify the hidden link between Circular Economy (CE) and Industry 4.0 in the context of the supply chain	PCA and DEMATEL	In the study, it was concluded that the application of Circular Economy and Industry 4.0 can be achieved not only by technical factors but also by environmental, social and economic factors such as sustainability. In addition, improvement in the use of resources and energy, repair, reuse and integration of processes, machines and products, and optimization of logistic optimization covering time, cost and quality factors were achieved.
Liboni et al. (2018)	How industries can meet the environmental protection and process safety objectives of sustainable operations within the framework of Industry 4.0	Identify the challenges posed by the industry 4.0 paradigm in the Brazilian electricity industry with the Soft System Methodology of the process of environmental protection and process safety	Case Study and Modelling	The study concluded that the implementation of new policies for innovation, reduction of bureaucracy and investments in education in Brazil could contribute to the modernization of electrical facilities and make industries more sustainable.
Luthra and Mangla (2018)	Evaluating challenges to Industry 4.0 initiatives for supply chain sustainability in emerging economies	Identify key challenges facing Industry 4.0 initiatives and examine key challenges identified for effective Industry 4.0 concepts for supply chain sustainability in emerging economies for India	Survey and AHP	In the study, it was concluded that the adoption of Industry 4.0 initiatives for supply chain sustainability could be achieved by providing the necessary support to the practitioners, policy makers, regulatory agencies and managers on organizational difficulty, technological challenge, strategic challenge and legal and ethical issues.
Abdel-Baset et al. (2019)	Evaluation of the green supply chain management practices	Identify GSCM practices leading to better economic and environmental performances	Case Study and Neutrosophic Approach	In this study, it was concluded that managers and decision makers can improve their environmental performance through GSCM applications such as reverse logistics, supplier environment cooperation, carbon management.

Govindan et al. (2014)	Supply chain sustainability	The impact of lean, resilient and green supply chain management practices on supply chain sustainability.	Conceptual and Modelling	The study concluded that waste disposal, supply chain risk management and cleaner production practices have a significant impact on supply chain sustainability and that flexible transport, flexible sourcing, ISO 14001 certification and reverse logistics practices have no impact.
Hofmann and Rüsch (2017)	Industry 4.0 on logistics	Identify opportunities for Industry 4.0 in the context of logistics management	Conceptual	In the study, it is concluded that the concept of Industry 4.0 may create opportunities in terms of decentralization, self-regulation and efficiency, but there is still a lack clear understanding of the concept and its applicability is insufficient.
Dallasega et al. (2018)	A systematic literature review on Industry 4.0	Explain the concepts that increase or decrease the proximity of Industry 4.0	Literature Review	In the research, it was determined that technological, organizational, geographical and cognitive proximity dimensions affect Industry 4.0 technologies.
Pranab et al. (2019)	A bibliometric analysis on Industry 4.0	To summarize the growth structure of Industry 4.0 in the last 5 years	Bibliometric Analysis	In the research, the most common keywords in Industry 4.0 were identified as cyber physical system, internet of objects, intelligent production and simulation.
Witkowski (2017)	IoT and Big Data	Providing innovative solutions for Internet of Things and Big Data applications	Survey	In the research, it was concluded that Internet of Things, Big Data and Industry 4.0 applications contribute to the competitive advantage of logistics and supply chain management development in meeting the needs of customers.
Mayr et al. (2018)	lean management and Industry 4.0	Whether lean management and Industry 4.0 complement each other and how Industry 4.0 can support specific lean methods	Case Study	The study concluded that lean management and Industry 4.0 complement each other at a conceptual level and that Industry 4.0 tools support lean methods.
Tjahjono et al. (2017)	Industry 4.0	Analyze the impact of Industry 4.0 on the supply chain	Literature Review	In the research, it was concluded that the applications that will be most affected by Industry 4.0 are order fulfillment and logistics.
Prinz et al (2018)	Lean magament on Industry 4.0	To communicate the benefits and advantages of lean management and digitization solutions to employees at various levels	Conceptual	In the study, it is concluded that digitization helps to reduce productivity losses and increase overall productivity in a simple optimized production.
Stock and Seliger (2016)	Industry 4.0 development	To investigate the micro and macro perspective of Industry 4.0	Conceptual	The study concluded that Industry 4.0 is an opportunity to strengthen sustainable production.
Monizza et al. (2018)	Industry 4.0 approaches and technologies	To examine the potentials and criticisms of parametric and productive design techniques in the construction sector	Case Study	It is concluded that Industry 4.0 approaches and technologies (such as Cyber-Physical Systems) have an effect on increasing the effects of parametric and productive design techniques on the value chain system.
Ding (2018)	Industry 4.0 in Pharma Sector	Identify potential sustainability barriers of PSC and explore how Industry 4.0 can be applied to sustainable PSC paradigms	Literature Review	High costs and time consumption, inadequate expertise and training, implementation of regulations, inadequate business incentives, ineffective cooperation and coordination in the supply chain, lack of objective criteria and lack of customer awareness were the most important obstacles preventing the inclusion of sustainability in PSCs. It has been concluded that coordination and cross-cooperation, sustainability of product lifecycle, proactive product recall management, measurement of sustainable performance, design of new regulatory system and incentives for sustainable activities can help overcome these obstacles.

Table 3 contd. ...

...Table 3 contd.

References	Subject	Objective	Method	Findings
Beltagui et al. (2019)	Social Sustainability and 3DP	Examine how the design of 3D printed mobile phone accessories helps to improve resource utilization, market share growth, and social sustainable supply chain practices driven by consumer demand	Case Study	The study concluded that 3D printing can help overcome resource constraints to support the spread of socially sustainable supply chain innovation, and that there is a relationship between consumer attitudes towards social sustainability and market entry.
Sachin et al. (2018)	Sustainable Industry 4.0 framework	To examine conceptual literature on Industry 4.0, human-machine interactions, machine-machine interactions, Industry 4.0 technologies and sustainability	Literature Review	In the research, a sustainable Industry 4.0 framework based on process integration and sustainable results in technical components of this framework has been determined.
Bechtsis et al. (2017)	AGV and SSCM	Identify what needs to be done to facilitate the incorporation of AGV systems at strategic, tactical and operational levels into supply chain design and planning	Literature Review	In the research, it is concluded that the use of AGVs is widespread in order to increase the corporate productivity and sustainability performance of enterprises in the supply chain.
Hong et al. (2018)	SSCM	Determine the mediation effect of supply chain capabilities on the link between SSCM applications and enterprise performance, and determine the impact of SSCM applications on economic, environmental and social performance	Survey and SEM	In the study, it was concluded that SSCM applications had a positive effect on the three dimensions of performance and SC dynamic features partially mediated the relationship between SSCM applications and corporate performance.
Müller and Voigt (2018)	Industry 4.0	Supply chain management research in the context of Industry 4.0	Case Study	The research contributes to identifying the various potentials that can be achieved by applying Industry 4.0.
Wu et al. (2015)	SSCM	Develop an integrated sustainable practices model that includes lean, green and social practices	Case Study and Survey	In the study, it is concluded that the implementation of integrated applications will have a stronger effect on the performance of 3BL than the implementation of individual applications.
Das (2018)	SSCM	To investigate the adoption of Sustainable Supply Chain Management practices and their impact on firm performance	Modelling (SEM)	In the study, it is concluded that environmental management practices lead to competitiveness, social practices are negatively related to competitiveness and affect operational practices and competitiveness.
Günther et al. (2018)	SSCM	Sustainable supply chain design from raw material sales to recycling	Modelling	In the study, it is concluded that the total emissions and costs in the supply chain can be reduced by a sustainable supply chain optimization covering economic, environmental and social objectives.
Mrugalska and Wyrwicka (2017)	Lean Production and Industry 4.0	Identify the benefits that businesses with an intelligent factory can provide in the context of Industry 4.0	Conceptual	In the research, it is concluded that the benefits of integration of CPS applications with IT by sharing production systems with customers and suppliers.

Lin (2018)	Industry 4.0	Evaluation of human factor in product design based on consumer experience	Case Study	In the research, it was concluded that the user acceptance and user satisfaction of a new product should be re-evaluated in order to increase company resources and service success probability.
Jabbour et al. (2018)	Industry 4.0 and Sustainability	Integrating Industry 4.0 and environmentally sustainable production concepts	Conceptual and Framework	In the study, it is concluded that the technologies related to Industry 4.0 have a high potential for fulfilling environmentally sustainable production and that success factors in the integration of Industry 4.0 and environmentally sustainable production must be fulfilled.
Horvat et al. (2018)	Industry 4.0	Systematically analyze and monitor the readiness of manufacturing companies for Industry 4.0 in developing economies	Conceptual	Where manufacturing companies have different degrees of preparation for Industry 4.0.
Gómez et al. (2019)	SSCM	Proposing a scalable, adaptable and integrable holonic framework for Sustainable Supply Chain Management	Conceptual and Framework	With the model proposed in the research, it is concluded that a sustainable supply chain management can be established which supports a cyclical, integrated natural environment, smart connected social metabolism.
Dileep (2019)	Smart Grid	Usage and measurement of Smart Grid technologies	Survey	In the study, it was concluded that Smart Grid applications can help environmental protection and energy sustainability and provide time series estimation, power quality and reliability, power flow optimization and renewable energy integration.
Novais et al. (2019)	Cloud Computing	Analyze the current research situation in the field of Cloud Computing and Supply Chain Integration	Literature Review	In the research, it has been concluded that technology / process integration and supply chain stakeholder integration can be improved and information, physical and financial flows of the supply chain can be improved by adopting Cloud Computing applications.
Mital et al. (2018)	Internet of Things	Suggest a Structural Equation Model (SEM) approach to test three models in the context of the Internet of Things in India	SEM Modelling	In the study, it was concluded that the use of other smart devices such as smart wearable devices, smart environments, elderly support and welfare and smart health became widespread and perceived behavior control was not a significant predictor of intention to use IoT.
Büyüközkan and Göçer (2018)	Digital Supply Chain	Evaluate the multi-criteria nature of the supplier selection process	ARAS Methodology	In the research, it was concluded that the methodology functionality of GDM and ARAS based on IVIF values for DSC problems is an effective decision making tool for evaluating supplier selection in IVIF environments.
Tönnissen and Teuteberg (2019)	Blockchain-Technology	Develop a descriptive model for the interaction of players in an operational supply chain, including block chain technology	Multiple Case Analysis	The research concluded that blockchain technology shows a high dynamic and may lead to permanent changes in the business models of companies, but it is not more widespread in practice.
Tiwari et al. (2018)	Big Data	To investigate big data analytics and applications in supply chain management	Literature Review	In the study, it is concluded that big data analytics can provide more useful solutions in decision making and provide opportunities for supply chain management.
Nguyen et al. (2018)	Big Data	To investigate the current literature on where and how the BDA is applied in the context of SCM with a new classification framework	Literature Review	Review of current research and development efforts to develop a new agenda

Table 3 contd. ...

...Table 3 contd.

References	Subject	Objective	Method	Findings
Addo and Helo (2016)	Big Data	To propose a framework that provides added value and applications and analysis of big data in operations and supply chain management	Literature Review	In the study, the literature has been expanded by adding the four main features of big data, i.e., Value-Adding alongside Diversity, Speed, Volume and Veracity.
Li and Liu (2019)	Big Data	Provide a data-driven supply chain management framework	Conceptual	In the study, it was concluded that the frameworks where data processes are slow, predictive, stable and fast may affect supply chain processes.
Belaud et al. (2019)	Big Data	Evaluate various processes and technology panels using environmental impacts from a sustainability and agri-food perspective	Conceptual	In the study, it is concluded that big data can improve sustainability management inthe supply chain design by evaluating agricultural wastes.
Nowicka (2016)	Cloud Computing	To research cloud computing applications in sustainable transport	Conceptual	In the research, it has been concluded that using cloud computing in sustainable mobility can provide real-time data access, react quickly and flexibly to events, and provide efficiency in socio-economic issues related to energy, environment and security.
Wong and Ngai (2019)	Supply Chain Innovation	Systematically review supply chain innovation literature over the last 18 years	Literature Review	In the research, it was concluded that supply chain innovations can be studied at more than one analytical level and there is a need to develop measurement scales.
Tseng et al. (2019)	SSCM	Evaluate data-driven sustainable supply chain management performance in a hierarchical structure under uncertainties	Modelling	In the study, it was concluded that social development has the most important effect on long-term relationships;sustainable information and technology, reverse logistics, product recovery techniques, logistics integration and joint development are the most effective criteria in increasing sustainable supply chain management performance.
Srai andLorentz (2019)	Digitalisation of purchasing and supply management	Evaluate designs for digitalization interventions in procurement and supply management	Conceptual	It was concluded that the use of the proposed framework in the research, the design of a future-oriented strategy can be developed with the digitalization element for communicating current and future PSM digitalization situations to stakeholders and for PSM function
Büyüközkan and Göçer (2018)	Digital Supply Chain	Examine in detail the academic studies and the industrial perspective related to DSC in the literature	Literature Review	The main limitations and potential customers of DSC, the advantages, weaknesses and limitations of the methods were identified in the research.
Sarc et al. (2019)	Digitalization And Robot Technologies in Waste Management	To examine Industry 4.0 approaches in the field of waste management	Conceptual	In the study, it was concluded that the use of big data in waste management, smart boxes with sensors for material detection or level measurement, digital image analysis and methods for new business models can contribute to increase efficiency.
Müller (2019)	Industry 4.0 and Digitalization	To investigate the general trends of Industry 4.0 application examples in the research and development phase in the smart wood supply chain and construction sectors.	Literature Review	It was found that Industry 4.0 applications contribute to the literature on digitization in wood supply and provide a comprehensive overview of the future technological advances, applications and challenges of wood supply.

Patra(2018)	GSCM	Examine a two-member supply chain consisting of a manufacturer and a telecom service provider	Modelling	In the study, it is concluded that the supply chain stakeholder, which has the power of greening investment, customer sensitivity towards greening improvement level and customer sensitivity to prices, can make more profit.
Bicocchi et al.(2019)	Digital/Smart Factories	To analyze the basic requirements for a scalable factory architecture characterized by access to services, data collection and production processes for digital factories	Case Study	In the research, it was concluded that innovative technological approaches aiming at the dynamic discovery of the service and the most appropriate data flow to the requirements expressed in the business process specifications and development can be supported by agile supply chains.
Taghikhah et al.(2019)	SSCM	Analyze in the context of sustainability to expand the supply chain concept	Literature Review	In this study, it was determined how the concept of SC developed to include additional processes and actors in order to take into account the requirements of sustainable development.
Basset et al. (2018)	Internet of Things	Implement the Internet of Things (ıot) to consolidate data, information, products, physical objects and all supply chain processes	Dematel and AHP	In the study, the model determined by DEMATEL and AHP technique in the neutrophilic environment was found to be able to overcome all the difficulties of traditional SCM and provide a safe environment for SCM processes.
Kamble et al.(2019)	Internet of Things	Identify the barriers that affect the adoption of ıot by businesses in the Indian retail sector	ISM and Dematel	The research concluded that the most important factors in the adoption of IoT were the lack of government regulations and poor internet infrastructure.
Nižetić et al.(2019)	Smart Technologies	To examine research topics related to Smart City/Environment, Energy, Engineering Modeling and ehealth	Literature Review	In the research, it has been concluded that the concepts of improving efficiency, smart and sustainable resource management and cleaner production concepts can create a sustainable future.
Le et al.(2019)	Internet Of Things	In this paper, the authors use the Business Model Canvas to evaluate ıot applications in Vietnam and create a Business Model Canvas for entrepreneurs.	Conceptual	The research concluded that local enterprises have high competitive advantages and that vertical integration in joint ventures is proposed as an entry strategy for foreign investors.
Schniederjans et al. (2019)	Supply Chain Digitization And Industry 4.0	Develop a research paradigm for researchers to broaden their perspectives and supply chain digitization	Modelling	In the research, it has been determined that issues related to optimizing supply chain digital performance, supply chain digitization and how to use human dimension can guide future research.
Bechtsis et al. (2017)	AGV And SSCM	Facilitate the inclusion of AGV systems in strategic design, tactical and operational levels in SC design and planning	Sustainable Supply Chain Cube	The study concluded that market opportunities and the potential to integrate AGVs into an SC context can be more easily understood using the proposed model.
Oh and Jeong (2019)	Smart Manufacturing	To determine the functional and structural characteristics of intelligent production supply chain	Modelling	In the study, it was concluded that the integration of SMSC (Smart Manufacturing Supply Chain) and ICPT (Information, Communication, and Production Technologies), and with a high degree of flexibility, could shorten the production lead-time and the point of exchange between the appropriate profit and lead-time in order to respond faster to customer orders.

Table 3 contd. ...

...Table 3 contd.

References	Subject	Objective	Method	Findings
Arunachalam et al. (2018)	Big Data	Systematic literature review and model development of Big Data capabilities	Literature Review	The study concluded that the BDA can be useful if businesses develop the right skills to use big data effectively, this will guide practitioners in understanding their current BDA maturity, and can be a guideline to improve BDA capabilities by taking into account the potential difficulties associated with the adoption process.
Muñuzuri et al. (2019)	IoT	Design an ıot system to optimize, manage and monitor container transport operations along an intermodal corridor	Case Study	It was concluded that the design determined in the research can be used to optimize the use of available resources, to make minimum investment requirements and to be cost effective.
Verdouw et al.(2016)	IoT	Analyze the concept of virtual food supply chains from the perspective of the Internet of Things and propose an architecture for the implementation of information systems	Case Study	In the research, it was concluded that the Internet of Things applications in food supply chains can be self-adaptive systems where smart objects work, decide and learn autonomously.
Papetti et al.(2019)	SSCM	Propose a web-based platform that can monitor suppliers and related processes throughout the supply chain	Case Study	With the platform proposed in the research, it was concluded that businesses can improve their portfolio management, decision-making processes and SC performance.

In the supply chain processes, cyber security applications, management of intelligent production networks, intelligent design and production control, virtual factory, technology, sustainable development, cooperation, resource and systems management strategy will transform the environmental risks into Industry 4.0 with an intuitive perspective. After the completion of this transformation, businesses can become a completely intelligent organization.

Traditionally, actuators in production plants are connected directly to stations via a supercomputer that sends data independently. In the new digitizing environment, these actuators can directly communicate with each other by passing the supercomputer through sensor capabilities. In addition, data via sensors can be collected and shared across the supply chain network through information technologies and systems such as cloud computing. This trend is not only for fast and real-time information sharing, but also for automation can be controlled without human intervention.

IoT takes decentralized decisions and real-time responses in industry 4.0, leading towards new functionality, greater reliability, much higher product utilization and intelligent product capabilities, maximizing operational efficiency, carbon footprint reduction. It also results in competitive advantage, sustainability to respond to customer demand, increase productivity, increase profit margins, improve production process, accelerate market entry process, protect the environment, adopt new business models, reduce costs, increase dynamic capacity, facilitate information integration, and produce innovative solutions and new business opportunities. Inventory management with 100% accuracy provides benefits such as maximizing logistics transparency and real-time supply chain management.

IoT-based applications as a developing market, have opened a growing market for business models in the fields of smart agriculture, smart logistics, smart transportation, smart network, smart environmental protection, smart building, smart security, smart medical care and smart home. Enterprises can provide real-time advantages such as efficient energy consumption, monitoring and safety control, temperature, humidity, light, biological properties, irrigation, air quality, noise, vehicle and device lighting control.

These initiatives will enrich the organization's sustainability objectives. Smart applications,

- Use of Internet of Things (IoT) in various sectors such as pharmaceuticals, petrochemicals, aviation, construction, electrical and electronics, manufacturing, retail, e-commerce, logistics,
- Real-time data traceability on the identity of the product, date and time of the event, location and reason for the event in the supply chain processes of foodstuffs,
- The use of self-driving trucks in the logistics and transport sector, the use of the vehicle to track the place and speed of shipment,
- Use for warnings for drivers and stakeholders on traffic changes, travel interruptions due to road conditions, risk of delays and breakdown/maintenance,
- The use of wearable devices such as eyeglasses to inform operators that indicate where to go, how much to collect, and to record this information during storage processes,
- Use of wearable and traceable systems and devices in health management processes to monitor physical health data of patients, workers and employees,
- Efficient and comfortable use of devices for the management of home security and water/electricity/energy consumption by using intelligent control system in household/home and office management processes,
- Intelligent city management processes, car park system, traffic monitoring and control systems, road monitoring, noise, lighting, weather forecasting and resource management,
- Use for intelligent library management,
- The effects of water and energy resources on the environment, their use and protection gaps, the regulation of rivers and waterways, the use of hydroelectric power plants for safeguards against floods in the Energy and Water management processes,

Potential barriers to the use of IoT applications for businesses

Lack of government regulations: The government, institutions and organizations should work together to promote and support technological initiatives and solutions to promote regulatory information systems, jurisdictional laws, efficient use of energy, network capacity development and guidance on network use that clearly identifies restrictions on sensitive frequency bands.

Lack of standardization: In the literature, the lack of standardization in the adoption and application processes of new technologies is seen as an important obstacle. Setting global standards for intelligent technologies (objects and systems) across the entire IoT communication platform, enabling identification and communication security, integration in the implementation process, will enable enterprises to adopt, invest and implement IoT applications.

High energy consumption: The main competence of IoT is that businesses proactively evaluate and make energy consumption processes sustainable. Expansion of smart IoT objects and systems will increase the need for energy consumption and increase costs. In this context, it is important to use smart devices, RFID tags and power units that can use self-charging, renewable energy.

Security and privacy: In the literature, it is emphasized that risks may occur in access, security and confidentiality processes such as data encryption, internet connection, software protection and authorization, flexibility against attacks, data validation, access control and customer privacy on the IoT platform.

High operating and return costs: The high cost of initial investment, use and maintenance of the technology superstructure and infrastructure required for the implementation of IoT technologies, and the lack of information about the return on investment processes are obstacles for enterprises.

Lack of Internet infrastructure: It is likely that the internet infrastructure does not exist at all stakeholder facilities for real-time data monitoring, evaluation and decision-making on the supply chain. Investments by enterprises for internet infrastructure can eliminate these disabilities.

Lack of intellectual capital with educated technological competence: The IoT system requires highly trained professionals who require advanced technical and functional skills in network interface, installation, management and applicability. In this context, enterprises may have difficulty in employing intellectual capital with technology-based skills.

Integration and compatibility issues: Due to the heterogeneity between existing industrial automation systems, including businesses, software, hardware, tools, machinery and equipment, integration and compatibility issues can pose an obstacle to the adoption of intelligent technologies.

Scalability issues: As businesses expand the applicability of IoT technologies, problems with the complexity of scalability of data from smart devices can be hindered.

Lack of verification and detection: The uncertainty of the benefits that enterprises can gain with the adoption of IoT may be an obstacle.

Internet of services (IoS): Internet of services can be understood as an opportunity to provide services as well as production technologies over the internet. Similar to IoT, it is defined as a "service society for businesses and users to integrate, create, and deliver value-added smart services, based on the idea that services are easily accessible via web technologies".

5.3 Smart Factory

Smart Factory is the basic prerequisite for dealing with the challenges of the increasing complexity of the present and the future. Industry 4.0 philosophy will inevitably transform the existing production philosophy through the connection and communication between products, machinery, transportation systems and people. In particular, the roles of employees are expected to change dramatically.

In this context, smart factories,with their potential, will be the center of this change. Smart factories are seen as the point of value where they can analyze and understand the automation of production lines, a certain level of production, and solve them with minimal human participation and communicate with each other.

Smart factories involve a variety of sustainability practices, such as the optimal use of resources and technology. These new systems focus on intelligent products and industrial processes where customers can react quickly to changing habits.

Intelligent factory, software systems in which raw materials, semi-finished products and requested materials are ordered automatically from suppliers, intelligent production systems based on pre-defined instructions, intelligent production systems for real-time communication, self-controlled machines and autonomous robots, and logistics. An advanced factory is defined as using technologies for automatic processing.

Enterprises can create cost-effective, flexible and efficient improvements in smart factory processes with applications such as flexible and individualized serial production, production of smart products, renewable energy usage obtained from smart grids, and water supply from fresh water reservoirs.

Smart Factory philosophy aims to be smart, flexible and dynamic and benefit enterprises by making the increasing complexity of production processes manageable for employees, while being sustainable and profitable, and increasing resource efficiency. It is also thought that the automated and self-regulating nature of smart factories can lead to heavy work destruction and socially affect the lives of future workers.

Robots and machine learning for waste management will revolutionize waste sorting and product (dis) assembly systems. The use of applications and sensors, supports and improves waste separation and collection applications

Smart products are defined as materials that can be connected to the machines, devices and systems with the help of the internet and sensor of the production processes and can communicate and be controlled in a decentralized manner.

5.4 Big Data

Businesses' desire to analyze economic, environmental and social (sustainable) data to provide smart products and services by monitoring and digitizing has led to the emergence of big data applications.

The concept of big data is defined as an extremely large data set that can be analyzed by digitizing with innovative computing methods to reveal patterns, trends and relationships related to human behavior and interactions, and to make predictions and decisions.

Forrester stated that it is a reference point for large data control and usage, i.e., volume, speed, diversity and value.

- Volume (amount of data) refers to the size of the data produced,
- Diversity (data diversity) refers to structural heterogeneity in a data set,
- Speed (speed of new data creation and analysis) is expressed as the speed of production, analysis and application of real-time data,
- Value (value of data) refers to the identification of important data from the whole body of information.

The Big Data for Business application is defined as the ability to process high volumes of integrated data from various data sources at high speed. Businesses should build data infrastructures to develop data generation capabilities, integrate these data into supply chains and develop analytical capabilities in a data-driven culture perspective, and develop methods and tools to analyze and integrate structurally heterogeneous data types.

The benefits that can be obtained by enterprises performing big data applications are shown in Table 4.

The construction of the big data architecture in the supply chain, data collection and extraction, data enrichment and storage, data improvement, data analytics and data visualization processes have to be performed.

Data generation (DG) is the ability of businesses to search, identify, create and access data from heterogeneous data sources. Data integration and management (DIM) capability is the ability of enterprises to use tools and techniques to collect, integrate, transform, and store data from heterogeneous data sources. Advanced analysis capability is the ability of enterprises to use tools and techniques to analyze supply chain data on a batch basis in real time. Data visualization is the ability to use tools and techniques to provide business information visuals and to provide decision-makers with timely,

Table 4: Big Data Practices Benefits.

Benefits	
Adjust their resource strategies according to the organization's strategy	Optimize resource utilization decision
Improve inventory decisions by analyzing unknown customer demand and monitor consumer behavior	Support supplier management by providing accurate information
Data accuracy	Forecasting
Clarity and insight with proactive planning capability provide an intelligent intelligence network	Increase the speed and flexibility of supply chain decisions and facilitate real-time monitoring
Provide accurate information on any return on investment (roi)	Provide competitive advantage
Provide sustainable products and services	Reduce risk taking costs
Improve process monitoring performance	Reduce waste, overproduction and energy consumption
Help them plan future events better	Maximize productivity
Reduce downtime	Re-use environmentally
Provide next-generation control and computing power	Increase product compatibility
Give designer greater confidence, identify market trends, identify best prices, and track consumer loyalty data	Improve supply chain performance with quality data usage

intuitive data-driven information. A data-driven culture is an intangible resource that represents people's beliefs, attitudes and opinions about data-based decision-making.

Organizational challenges

Long implementation time: It consists of several long-lasting stages, such as developing, testing and adapting the BD process to different contexts. The results of BDA implementation in the supply chain can occur between 12–18 months under senior management and stakeholder support.

Insufficient resources: Lack of IT infrastructure among supply chain stakeholders may make a difference in BD capabilities. Businesses may encounter challenges such as competition, representative conflicts, incentive arrangements, and data sharing policies when establishing cross-functional collaboration.

Confidentiality and security concerns: the public is prejudiced due to the unethical use of Big Data by leading businesses. The supply chains of multinational enterprises, which have to comply with the laws of different countries, may have difficulties in privacy and security when sharing data. However, these challenges can be overcome by using effective data governance initiative in the process of data integration and management.

Causality: Since Big Data primarily moves through correlation, the risk of identifying significant but irrelevant correlations poses a risk. This risk can be solved with an intuitive approach.

Return on Investment (ROI) Issues: The lack of sufficient output regarding the return benefits and uncertainties of investments made for Big Data applications can be an obstacle.

Lack of skills: Lack of sufficient quality and quantity of data required for application with Big Data, and lack of necessary expertise and experiential knowledge of managers and employees may prevent adoption and implementation.

Technical Challenges

Data scalability: Scalability of data is a major technical problem in a large data application. To overcome this problem, businesses may need to revise or modify their database management and optimize their data collection methods.

Data Quality: poor data quality can hamper data analysis processes and make management decision-making difficult. In this context, enterprises can make data management applications effective by using statistical techniques.

Lack of technical and procedural: insufficient infrastructure of smart applications may pose an obstacle to large applications for businesses wishing to use the flow of data across the supply chain properly.

Today, the emergence of big data has enabled cloud computing to progress and expand operationally. Cloud technology, also called cloud computing, is defined as a digital integration platform where data is stored, analyzed and made available for collaborative use within the enterprise and between its supply chain partners. Cloud computing can enable big data to be processed in a cheaper, more efficient and more advanced way, virtual intelligent storage and the ability to analyze and manage effectively.

The differences between cloud computing and Big Data are shown below;

- Enables transformation of cloud computing technology architecture while supporting big data operational decision making processes,
- Cloud computing can be open source, while large data is usually composed of aggregated data,
- While large data acts as a database management system for efficient data processing capacity, cloud computing provides resources at the analytical and operating systems level,
- Integrated big data and cloud computing technologies complement each other.

Sustainability can provide specific data sources, methods and visualizations for environmental, economic and social areas to improve data stocks and assessment methods.

5.5 Cloud Technologies

Nowadays, during the production and service processes, the acquisition of big data by the interaction of machines with machines, devices with devices and people has enabled the advancement and operational expansion of cloud technology. Cloud technology, also called cloud computing, is a scalable digital integration platform (networks, networks, data storage, analysis, and availability) where fast, minimal management of data is stored and analyzed for centralized and virtualized collaboration within the enterprise and between supply chain partners.

The main features of the cloud computing model are;

Optional usage: Users can access and use data and information that are useful for them through virtual network storage from anywhere on the servers when they need it, without human interaction,

Wide network access: CC services can be heterogeneously accessed on the integrated network with mobile phones, tablets, laptops and workstations.

Pool Resource pooling: CC service server can be smartly assigned digitally and physically to multiple users on demand without location constraint by gathering intelligent computing resources in a single center. Resources include storage, processing, memory, and network bandwidth.

Elastic Quick flexibility: The CC provider can automatically deliver scalable capabilities to the user quickly and efficiently when requested.

Measured service: Cloud systems can automatically optimize the use of service-appropriate resources, transparency, monitoring, control and reporting processes for both the provider and the consumer of the service used.

The benefits that can be achieved by enterprises performing cloud computing applications are shown in Table 5.

It is used as a common platform for the transformation of stocks by exchanging data between stakeholders in different parts of the world. Globally used to reduce transport activities. Accenture's 2010 study confirms that cloud computing can reduce carbon emissions by 30% to 90% for today's manufacturing and service sectors, and that future energy savings are likely. With the Industry 4.0 framework, cloud computing applications can be implemented in a sustainable manner in hospital health management, logistics and the automotive industry. It is used to overcome realizing aggravating pains on patients' health and to perform real-time health control on a fundamental and significant basis. It is used to monitor production and stock levels in the agricultural industry.

Businesses and other organizations have the ability to choose among the types of services in the cloud computing model. Types of CC;

Private cloud: Infrastructure is designed for private enterprise,

Public Cloud: The infrastructure belongs to a single provider and is intended for a range of businesses or the general public,

Hybrid Cloud: Infrastructure is a combination of private and public cloud applications.

5.6 AGV

The term AGV, "Intelligent Autonomous Vehicle" (IAV), "Autonomous Vehicle", "Automatic Guided Vehicle" and "Automatic Guided Vehicle" are used in the literature, radio waves, cameras, sensors, software systems, magnets and lasers.

Table 5: Cloud Computing Benefits.

Benefits	
Process large data in a cheaper	Effectively reduce its capital expenditure
More efficient and more advanced way, analyze and manage intelligent storage	Faster access to faster computing capabilities
Unified resources serving multiple users increased capacity through IT	The rapid launch of products and services
Faster return on investment	Payment for the capabilities needed with the optional use model
Reduction of barriers to entry through reduced fixed investment costs	Reduction of costs
Mobile to establish new business opportunities	Low-energy servers and data centers to use green energy
Reduce energy and water consumption and greenhouse gas emissions	Improvement in production
Economies of scale	Contribution to a more sustainable world and economic growth

Businesses adopt intelligent robotic technologies in line with the dynamic driving forces of cost, health, safety and environmental factors, as well as for reasons such as government sustainability regulations, tax incentives and corporate social responsibility.

Numerous types of AGVs include forklifts, unit loads, tractors, clamps, hybrid vehicles and specialized vehicles. AGVs, frame, steering controls, engines and transmission systems, equipment such as special purpose robotic parts, main processor, microcontroller, sensors and electrical system, such as electronic and electrical components and electricity, diesel, liquefied petroleum gas, biofuels and hybrid methods using energy power supply.

AGVs have intelligent software that allows them to intuitively learn the logic of business processes such as planning, routing, scheduling, and transferring. In navigation systems used in routing processes, wire, band, laser markers tracking techniques and triangulation, inertial, visualization, laser-guided techniques such as GPS are used. AGVs can be centrally, hierarchically or completely decentralized to provide maximum flexibility in processes.

The adoption of the concept of sustainability by enterprises has increased the use of AGVs in intelligent production and intelligent logistics systems. In this context, businesses can reduce energy consumption, increase sustainability, increase productivity, increase cost and energy efficiency and effectively tackle health, safety and environmental issues, improve environmental performance, reduce work accident compensation, minimize human error, reduction, training and insurance costs, reducing the number of accidents at work, reducing risks, reducing waste and scrap generation, and the ability to work 24/7.

Although AGVs operate autonomously, everything depends on people, as technology itself cannot guarantee production outputs, as it requires the use of information and data sharing for communication, cooperation and coordination between people and machines.

It applies a state-of-the-art energy information system in the context of the digital factory and uses energy cards to provide energy consumption details about all parts in the production system, thereby promoting energy consumption visibility and optimization. In use scenarios, electrical AGVs are recognized as low-energy vehicles that minimize energy consumption.

AGVs are widely used in container terminals, flexible production systems, warehouse management, material handling, transportation processes, automotive production, high technology products, agriculture, mines, and in health management. It is stated that the transportation and logistics cost of the goods in the container terminals is 50% of the total terminal operating cost. In this context, better maneuverability and improved pick-up and loading performance of containers will result in cost reduction.

The German port of Hamburg has become a sustainable port using electric AGV to reduce greenhouse gas emissions, noise levels and costs. Furthermore, the port of Hamburg has increased its global competitiveness, environmental and economic performance with increasing energy efficiency. It was determined that the use of electric AGV at the Altenwerder Container Terminal could reduce the generated CO_2 emissions by 70%. It has been found that the use of AGV in the toothpaste industry in Brazil increases productivity and minimizes hazards and accidents by reducing human errors.

5.7 3DP

Because of the distance between consumption and production and new entrants to the market, enterprises with little market share and SMEs have insufficient resources, it will be very difficult for them to maintain sustainability in their supply chains.

It is known that the impact of consumers and stakeholders on the adoption of sustainability activities by enterprises is quite prominent today. The complex structure of supply chains, the lack of transparency and the fact that manufacturers act more in line with the philosophy of selling products and services can be reduced by the impact of these effects for businesses. In this regard, 3DP can be an intelligent application to overcome these obstacles.

The concept of 3DP, also known as additive manufacturing (AM), is defined as an intelligent application that creates physical objects in layers of plastic, metal or other material by a wide process and technology that uses light or heat directly and is digitally under computer control.

With 3DP applications, businesses can create more personalized products, incorporate customers into processes, create the desired product design and prototype without requiring too much resources, allow SMEs to share resources and innovate more efficiently, enabling them to appreciate and respond to sustainability issues better.

Currently, 3DP applications are widely used to produce motor sports parts, medical implants such as hip prostheses, manufacturing, and custom-designed military aircraft parts.

5.8 Social Practices

Social sustainability is the fulfillment of legal, economic, ethical, and philanthropic and behavior values such as the rights of social capital and people, labor practices, establishment and supervision of occupational health and safety standards, product responsibility, interaction with the training of employees, suppliers and customers.

The support of the senior management is the most important factor for the success of the implementations and aims at taking the necessary steps to ensure social sustainability and allocating resources.

With the increasing awareness in senior management about social sustainability in the supply chain, enterprises are encouraged by health, safety and training practices, and problems like inadequate wages, poor working conditions, long working hours, absence of disaster and emergency management scenarios, lack of hazard prevention plans, gender inequality, can overcome. In addition, all stakeholders in businesses must adopt and implement these practices in to ensure integrated social sustainability throughout the supply chain.

Businesses determine the processes of employment, health and safety factors, and internal and external social sustainability processes through agreement between stakeholders and communities. Businesses and supply chain stakeholders can anticipate, evaluate and mitigate risks that may affect the safety, health and well-being of their employees, customers and suppliers by focusing on OHSAS 18001 standards and the proper implementation of laws. Businesses can provide equal social benefits to their employees and provide sustainability for employment by paying appropriate wages. In this way, businesses can reduce employee absenteeism and increase total productivity.

With risk management practices, enterprises can reduce accidents that may occur in production processes at work as a result of excessive working hours and inadequate rest periods that may increase accident risk. Appropriate training of employees in risk management and OHSAS 18001 practices can help them become aware of and adopt the importance of social sustainability.

It emphasizes the need for equal treatment of human rights, workplace, race, language, gender, religion, national and social origin, age, disability, marital and family status without any discrimination. Within the scope of these requirements, enterprises provide the highest welfare conditions such as food, shelter, heating, clean and proper working environment.

6. Conclusion

Smart and sustainable supply chain management is a new and up-to-date concept for both literature and practitioners. Businesses will be able to integrate smart and sustainable applications into their supply chain and business processes to integrate more flexible, economic and environmentally friendly systems. Today, the estimated investments made by enterprises in integrated technologies are around USD 907 billion.

Administrators are considering or using 10% of intelligent and sustainable technologies, and think that these applications can create a competitive advantage of 72% when supported by artificial intelligence. Leading businesses that incorporate the smart and sustainable change brought by the era into their visions and missions will be able to provide more competitive advantages and sustainable performance in the market than their competitors. Unfortunately, the uncertainties created by the lack of application and research results enable enterprises to resist this transformation.

This research provides a roadmap to address the need to understand the importance and necessity of intelligent and sustainable practices within the supply chain.

In order to ensure sustainable and intelligent transformation of previously unknown and unused technology and adopt it in the whole system structure, businesses, need to be convinced of that its function and form will optimize the use of business resources. It must identify and forecast risks in processes, protect information and people, integrate customer needs into all business processes, enrich business and life experiences, measure productivity, security and perfection, measure progress towards predetermined goals, determine the basic behaviors and change the structure of the system and thus its application would strengthen positive behaviors.

Businesses will be able to achieve smart and sustainable transformation effectively with the support of senior management and evaluating employees, customers, suppliers, partners and stakeholders as a whole. 75% of the supply chain stakeholders think that efficient instant data sharing is the most valuable element. However, many still share information by fax and telephone. Even simple operations such as virtualization and digitization can improve sustainability performance.

The key drivers and success factors for smart and sustainable practices in the research will make the structure of SSSCM more flexible, effective, intelligent, economic, environmental friendly and durable.

By adopting smart and sustainable practices within the framework of the research proposed framework, it can provide technological, economic, environmental and social performance superiority to its competitors in the sector. In addition, the proposed framework can be a guide for academics to work on intelligent and sustainable supply chain management.

Businesses that can make their supply chains smart and sustainable can effectively meet their customer needs while minimizing their environmental impacts, creating outputs by following the most appropriate process for the society and people and maximizing their economic benefits.

This research is concerned with the smoothing and sustainability of the supply chain. The obstacles in the implementation of the proposed framework, drivers and the most efficient and effective practices were identified.

The increase in performance achieved through the coordination of supply chain activities through digital collaborations with all stakeholders will provide added value for the profitability of the enterprises and also for the planet and people. With the proposed framework, businesses can integrate technologies, enable data exchange, automated control mechanisms and integrated planning, including intelligent processes and products, and ensure the sustainability of supply chains.

References

Abdel-Baset, M., V. Chang and A. Gamal. 2019. Evaluation of the green supply chain management practices: A novel neutrosophic approach. Computers in Industry, 108: 210–220.

Abdel-Basset, M., G. Manogaran and M. Mohamed. 2018. Internet of Things (IoT) and its impact on supply chain: A framework for building smart, secure and efficient systems. Future Generation Computer Systems, 86: 614–628.

Addo-Tenkorang, R. and P.T. Helo. 2016. Big data applications in operations/supply-chain management: A literature review. Computers & Industrial Engineering, 101: 528–543.

Arunachalam, D., N. Kumar and J.P. Kawalek. 2018. Understanding big data analytics capabilities in supply chain management: Unravelling the issues, challenges and implications for practice. Transportation Research Part E: Logistics and Transportation Review, 114: 416–436.

Babu, S. and U. Mohan. 2018. An integrated approach to evaluating sustainability in supply chains using evolutionary game theory. Computers & Operations Research, 89: 269–283.

Bechtsis, D., N. Tsolakis, D. Vlachos and E. Iakovou. 2017. Sustainable supply chain management in the digitalisation era: The impact of Automated Guided Vehicles. Journal of Cleaner Production, 142: 3970–3984.

Belaud, J.P., N. Prioux, C. Vialle and C. Sablayrolles. 2019. Big data for agri-food 4.0: Application to sustainability management for by-products supply chain. Computers in Industry, 111: 41–50.

Beltagui, A., N. Kunz and S. Gold. 2019. The role of 3D printing and open design on adoption of socially sustainable supply chain innovation. International Journal of Production Economics.

Bicocchi, N., G. Cabri, F. Mandreoli and M. Mecella. 2019. Dynamic digital factories for agile supply chains: An architectural approach. Journal of Industrial Information Integration.

Büyüközkan, G. and F. Göçer. 2018a. An extension of ARAS methodology under interval valued intuitionistic fuzzy environment for digital supply chain. Applied Soft Computing, 69: 634–654.

Büyüközkan, G. and F. Göçer. 2018b. Digital supply chain: literature review and a proposed framework for future research. Computers in Industry, 97: 157–177.

Dallasega, P., E. Rauch and C. Linder. 2018. Industry 4.0 as an enabler of proximity for construction supply chains: A systematic literature review. Computers in Industry, 99: 205–225.

Das, D. 2018. The impact of sustainable supply chain management practices on firm performance: Lessons from Indian organizations. Journal of Cleaner Production, 203: 179–196.

De Man, J.C. and J.O. Strandhagen. 2017. An Industry 4.0 research agenda for sustainable business models. Procedia Cirp, 63: 721–726.

De Sousa Jabbour, A.B.L., C.J.C. Jabbour, C. Foropon and M. Godinho Filho. 2018. When titans meet–Can industry 4.0 revolutionise the environmentally-sustainable manufacturing wave? The role of critical success factors. Technological Forecasting and Social Change, 132: 18–25.

Dileep, G. 2020. A survey on smart grid technologies and applications. Renewable Energy, 146: 2589–2625.

Ding, B. 2018. Pharma industry 4.0: Literature review and research opportunities in sustainable pharmaceutical supply chains. Process Safety and Environmental Protection, 119: 115–130.

Elkington, J. 1998. Partnerships from cannibals with forks: the triple bottom line of 21st-century business. Environ. Qual. Manag., 8(1): 37–51.

Govindan, K., S.G. Azevedo, H. Carvalho and V. Cruz-Machado. 2014. Impact of supply chain management practices on sustainability. Journal of Cleaner Production, 85: 212–225.

Günther, H.O., M. Kannegiesser and N. Autenrieb. 2015. The role of electric vehicles for supply chain sustainability in the automotive industry. Journal of Cleaner Production, 90: 220–233.

Hofmann, E. and M. Rüsch. 2017. Industry 4.0 and the current status as well as future prospects on logistics. Computers in Industry, 89: 23–34.

Hong, J., Y. Zhang and M. Ding. 2018. Sustainable supply chain management practices, supply chain dynamic capabilities, and enterprise performance. Journal of Cleaner Production, 172: 3508–3519.

Horvat, D., T. Stahlecker, A. Zenker, C. Lerch and M. Mladineo. 2018. A conceptual approach to analysing manufacturing companies' profiles concerning Industry 4.0 in emerging economies: Paper ID: 1213. Procedia Manufacturing, 17: 419–426.

Ivanov, D., S. Sethi, A. Dolgui and B. Sokolov. 2018. A survey on control theory applications to operational systems, supply chain management, and Industry 4.0. Annual Reviews in Control, 46: 134–147.

Kamble, S.S., A. Gunasekaran and S.A. Gawankar. 2018. Sustainable Industry 4.0 framework: A systematic literature review identifying the current trends and future perspectives. Process Safety and Environmental Protection, 117: 408–425.

Kamble, S.S., A. Gunasekaran, H. Parekh and S. Joshi. 2019. Modeling the internet of things adoption barriers in food retail supply chains. Journal of Retailing and Consumer Services, 48: 154–168.

Le, D.N., L. Le Tuan and M.N.D. Tuan. 2019. Smart-building management system: An Internet-of-Things (IoT) application business model in Vietnam. Technological Forecasting and Social Change, 141: 22–35.

Li, Q. and A. Liu. 2019. Big data driven supply chain management. Procedia CIRP, 81: 1089–1094.

Liboni, L.B., L.H. Liboni and L.O. Cezarino. 2018. Electric utility 4.0: Trends and challenges towards process safety and environmental protection. Process Safety and Environmental Protection, 117: 593–605.

Lin, K.Y. 2018. User experience-based product design for smart production to empower industry 4.0 in the glass recycling circular economy. Computers & Industrial Engineering, 125: 729–738.

Luthra, S. and S.K. Mangla. 2018. Evaluating challenges to Industry 4.0 initiatives for supply chain sustainability in emerging economies. Process Safety and Environmental Protection, 117: 168–179.

Manavalan, E. and K. Jayakrishna. 2019. A review of Internet of Things (IoT) embedded sustainable supply chain for industry 4.0 requirements. Computers & Industrial Engineering, 127: 925–953.

Martín-Gómez, A., F. Aguayo-González and A. Luque. 2019. A holonic framework for managing the sustainable supply chain in emerging economies with smart connected metabolism. Resources, Conservation and Recycling, 141: 219–232.

Mayr, A., M. Weigelt, A. Kühl, S. Grimm, A. Erll, M. Potzel and J. Franke. 2018. Lean 4.0-A conceptual conjunction of lean management and Industry 4.0. Procedia Cirp, 72: 622–628.

Mital, M., V. Chang, P. Choudhary, A. Papa and A.K. Pani. 2018. Adoption of Internet of Things in India: A test of competing models using a structured equation modeling approach. Technological Forecasting and Social Change, 136: 339–346.

Monizza, G.P., C. Bendetti and D.T. Matt. 2018. Parametric and Generative Design techniques in mass-production environments as effective enablers of Industry 4.0 approaches in the Building Industry. Automation in Construction, 92: 270–285.

Mrugalska, B. and M.K. Wyrwicka. 2017. Towards lean production in industry 4.0. Procedia Engineering, 182: 466–473.

Muhuri, P.K., A.K. Shukla and A. Abraham. 2019. Industry 4.0: A bibliometric analysis and detailed overview. Engineering Applications of Artificial Intelligence, 78: 218–235.

Müller, F., D. Jaeger and M. Hanewinkel. 2019. Digitization in wood supply–A review on how Industry 4.0 will change the forest value chain. Computers and Electronics in Agriculture, 162: 206–218.

Müller, J.M. and K.L.Voigt. 2018. The impact of industry 4.0 on supply chains in engineer-to-order industries-an exploratory case study. IFAC-PapersOnLine, 51(11): 122–127.

Muñuzuri, J., L. Onieva, P. Cortés and J. Guadix. 2019. Using IoT data and applications to improve port-based intermodal supply chains. Computers & Industrial Engineering.

Nguyen, T., Z.H.O.U. Li, V. Spiegler, P. Ieromonachou and Y. Lin. 2018. Big data analytics in supply chain management: A state-of-the-art literature review. Computers & Operations Research, 98: 254–264.

Nižetić, S., N. Djilali, A. Papadopoulos and J.J. Rodrigues. 2019. Smart technologies for promotion of energy efficiency, utilization of sustainable resources and waste management. Journal of Cleaner Production.

Novais, L., J.M. Maqueira and A. Ortiz-Bas. 2019. A systematic literature review of cloud computing use in supply chain integration. Computers & Industrial Engineering, 129: 296–314.

Nowicka, K. 2016. Cloud computing in sustainable mobility. Transportation Research Procedia, 14: 4070–4079.

Oh, J. and B. Jeong. 2019. Tactical supply planning in smart manufacturing supply chain. Robotics and Computer-Integrated Manufacturing, 55: 217–233.

Papetti, A., M. Marconi, M. Rossi and M. Germani. 2019. Web-based platform for eco-sustainable supply chain management. Sustainable Production and Consumption, 17: 215–228.

Patra, P. 2018. Distribution of profit in a smart phone supply chain under green sensitive consumer demand. Journal of Cleaner Production, 192: 608–620.

Prinz, C., N. Kreggenfeld and B. Kuhlenkötter. 2018. Lean meets Industrie 4.0–a practical approach to interlink the method world and cyber-physical world. Procedia Manufacturing, 23: 21–26.

Rajput, S. and S.P. Singh. 2019. Connecting circular economy and Industry 4.0. International Journal of Information Management, 49: 98–113.

Sarc, R., A. Curtis, L. Kandlbauer, K. Khodier, K.E. Lorber and R. Pomberger. 2019. Digitalisation and intelligent robotics in value chain of circular economy oriented waste management–A review. Waste Management, 95: 476–492.

Schniederjans, D.G., C. Curado and M. Khalajhedayati. 2019. Supply chain digitisation trends: An integration of knowledge management. International Journal of Production Economics.

Srai, J.S. and H. Lorentz. 2019. Developing design principles for the digitalisation of purchasing and supply management. Journal of Purchasing and Supply Management, 25(1): 78–98.

Stock, T. and G. Seliger. 2016. Opportunities of sustainable manufacturing in industry 4.0. Procedia Cirp, 40: 536–541.

Taghikhah, F., A. Voinov and N. Shukla. 2019. Extending the supply chain to address sustainability. Journal of Cleaner Production, 229(2019): 652–666.

Tiwari, S., H.M. Wee and Y. Daryanto. 2018. Big data analytics in supply chain management between 2010 and 2016: Insights to industries. Computers & Industrial Engineering, 115: 319–330.

Tjahjono, B., C. Esplugues, E. Ares and G. Pelaez. 2017. What does industry 4.0 mean to supply chain? Procedia Manufacturing, 13: 1175–1182.

Tönnissen, S. and F. Teuteberg. 2019. Analysing the impact of blockchain-technology for operations and supply chain management: An explanatory model drawn from multiple case studies. International Journal of Information Management.

Tseng, M. L., K.J. Wu, M.K. Lim and W.P. Wong. 2019. Data-driven sustainable supply chain management performance: A hierarchical structure assessment under uncertainties. Journal of Cleaner Production, 227: 760–771.

Verdouw, C.N., J. Wolfert, A.J.M. Beulens and A. Rialland. 2016. Virtualization of food supply chains with the internet of things. Journal of Food Engineering, 176: 128–136.

WCED (World Commission on Environment and Development), 1987. Our Common Future.

Witkowski, K. 2017. Internet of things, big data, industry 4.0–innovative solutions in logistics and supply chains management. Procedia Engineering, 182: 763–769.

Wong, D. T. and E.W. Ngai. 2019. Critical review of supply chain innovation research (1999–2016). Industrial Marketing Management.

Wu, L., N. Subramanian, M. Abdulrahman, C. Liu, K.H. Lai and K. Pawar. 2015. The impact of integrated practices of lean, green, and social management systems on firm sustainability performance—evidence from Chinese fashion auto-parts suppliers. Sustainability, 7(4): 3838–3858.

CHAPTER 19

A Content Analysis for Sustainable Supply Chain Management Based on Industry 4.0

Yesim Deniz Ozkan-Ozen and *Yucel Ozturkoglu**

1. Introduction

The new industrial revolution, namely Industry 4.0 (I4.0) can be seen as the new industrial stage, where an integration between traditional manufacturing processes and information and communication technologies occur (Dalenogare et al. 2018). The term "Industry 4.0" was initially used in 2011 in the Hannover Fair, and announced officially in 2013 as the beginning of Fourth Industrial Revolution (Xu et al. 2018). As it has happened in all industrial revolutions so far, i.e., steam power in the first industrial revolution (Industry 1.0), electricity in the second industrial revolution (Industry 2.0), automation in the third industrial revolution (Industry 3.0); a new technology or concept triggers the new industrial revolution, and for the fourth industrial revolution, internet based technologies have the greatest impact (Lasi et al. 2014). Therefore, I4.0 can be defined as the integration and connection of virtual and real world through cyber physical systems (CPS) and Internet of Things (IoT), where smart objects communicate with each other constantly (Fatorachian and Kazemi 2018).

I4.0 affects all the processes in manufacturing and thus it changes structure of the industry, and customer demands, which results in the need of new business models and competition rules (Dalenogare et al. 2018). Moreover,it encourages factory environments with minimum human power and promotes globally integrated supply chains through digital technologies (Oztemel and Gursev 2018). As a significant part of these tremendous changes, supply chain management (SCM) practices also faced with technological innovations caused by I4.0, became increasingly data dependent (Brinch 2018). Introducing digitalization and automation of operations in SCM increase the importance of collaboration between stakeholders including suppliers, producers and customer, and transparency in all stages from the receiving order to end of life of the product (Tjahjono et al. 2017). Therefore, SCM in I4.0 is major topic in the new industrial era.

Although the economic benefits of I4.0 are highly accepted in terms of increasing efficiency, productivity, quality, and speed; effects on environment and society are still questioning and there are different views. With this view, sustainability of organizations, with its all dimensions, i.e., environment, social, economic, and SCM in I4.0 can be seen as two major topics that should be considered together for having economic gains for protecting environment and increasing the wellbeing of the society. In academia, different researches have been conducted to investigate the relationship between I4.0 and sustainability (Stock and Seliger 2016; Song and Moon 2017; de Sousa Jabbour et al. 2018; Bonilla et al. 2018; Braccini and Margherita 2019) and I4.0 and SCM (Barata et al. 2018; Ben-Daya et al. 2019; Büyüközkan and Göçer 2018; Tu 2018; Frederico et al. 2019). However, intersection of all these concepts has not got the attention that it deserves so far.

The main aim of this study is to propose a theoretically and practically grounded conceptual framework, integrating and filtering the frequently mentioned concepts and new concepts. From this point of view, the aim of this chapter is assessing the impacts of fourth industrial revolution on sustainable supply chain management by conducting a content analysis, based on a literature review, and proposing a framework for sustainable supply chain management (SSCM) in I4.0. This chapter clearly gives comprehensive insight about how to apply industrial 4.0 conceptual models to promote sustainable SCM to the reader. The proposed framework aims to achieve a principles based framework, which harmonizes the frequently mentioned concepts with the new concepts. In light of discussions in the previous paragraphs, we wished

Yasar University, Department of International Logistics Management, Izmir/Turkey, Email: yesim.ozen@yasar.edu.tr
* Corresponding author: yucel.ozturkoglu@yasar.edu.tr

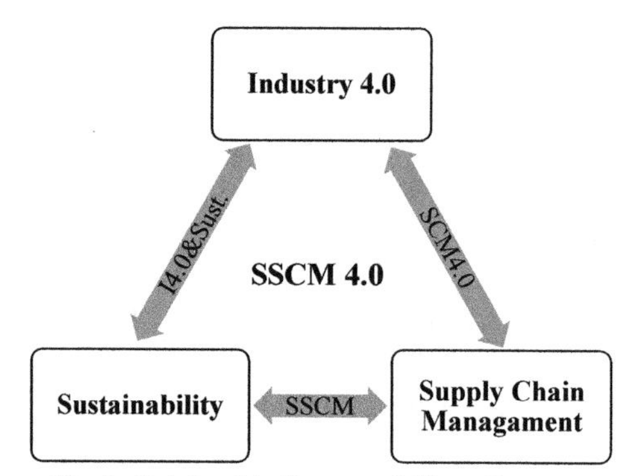

Fig. 1: Relationship Between all Three Concepts.

to identify and understand which cognitive approaches had been used in merging the content of SSCM with industrial 4.0 in the literature. This study therefore investigated the main determinants of SSCM 4.0 documents in practice. Firstly, a vocabulary of a set of words regarding to the issues is built. We then develop a Java code using a library provided by Apache Tika[1] to count the number of occurrences of each predetermined word in the vocabulary within the papers related with *sustainability, supply chain management,* and *industry 4.0*. Next, the collected data that shows how many times each issue was mentioned in each paper was analyzed. In Figure 1, the relationship between concepts are schematized.

This chapter is structured as follows: this introduction section gives the brief descriptions for main concepts, after that in Section 2, literature review related to SCM and I4.0, and sustainable SCM in I4.0 are given, in order to present the gap in the literature. Section 3 explains the methodology. Section 4 includes the details of empirical study. Section 5 and 6 presents the discussions and conclusions respectively.

2. Literature Review

Literature review in this chapter is divided into two sections. Initially, studies related to SCM and I4.0 are presented, secondly SSCM and I4.0 is covered.

2.1 Supply Chain Management in Industry 4.0

Literature related to SCM and I4.0 is extending rapidly, and different approaches have been followed so far. Based on the literature review, it has been noticed that the current literature mostly covers theoretical studies and can be categorized under some headings. For instance, IoT, Big Data and Blockchain are some of the technologies that are highly investigated in terms of applications in SCM. Moreover, some studies followed a more general point of view, and focused on digital supply chains, and I4.0 and SCM in a contextual way. There are also studies that especially focused on supply chain functions including procurement, marketing, and demand chain.

To start with theoretical studies with broader perspective; Tjahjono et al. (2017) conducted one of the initial studies that is directly related to I4.0 and SCM, where the impact of main I4.0 technologies such as VR, AR, 3D printing CPS, RFID, Big Data, and IoT, simulation, etc., on four main supply chain levers, namely; procurement, warehouse, fulfillment, and transport logistics had been investigated and opportunities and threats were presented. Hahn (2019) approached I4.0 from the supply chain innovation perspective, and investigated implications of I4.0 on SCM. Key findings of this study showed that, firstly, productivity improvements is the key goal of I4.0 implementations in supply chain processes, secondly, analytics and smart things come forward as I4.0 solutions and human-centric approaches are neglected, finally, startup companies radically change their processes according to I4.0 while established companies prefer to sustain their existing business architectures during the I4.0 adaptation (Hahn 2019). Furthermore, Frederico et al. (2019) conducted a systematic literature review and proposed the term "Supply Chain 4.0" in order to conceptualize SCM in I4.0. A framework for supply chain 4.0 is proposed with a headings of managerial and capability supporters, technology levers, processes performance requirements and strategic outcomes, and several research questions were presented for future studies at the end of the study (Frederico et al. 2019).

[1] The Apache Tika™ toolkit detects and extracts metadata and text from over a thousand different file types (such as PPT, XLS, and PDF). All of these file types can be parsed through a single interface, making Tika useful for search engine indexing, content analysis, translation, and much more (source: http://tika.apache.org/index.html, accessed October, 25, 2019).

Digital supply chain is a directly related term while explaining SCM and I4.0 relationship, and at some point, it can be said that digital supply chain is a broader term and covers I4.0 technologies and their implementations in SCM. Büyüközkan and Göçer (2018) defined digital supply chain as; *"a smart, value-driven, efficient process to generate new forms of revenue and business value for organizations, ... and, it is about the way how supply chain processes are managed with a wide variety of innovative technologies"*. Büyüközkan and Göçer (2018) conducted an extend literature review, and identified key challenges, opportunities and research directions related to digital supply chains. Similarly, Iddris (2018) also conducted a literature review related to digital supply chain, and presented the main drivers of digital supply chains as technologies, digitization, integration, collaboration and coordination. Related to digital supply chains, Korpela et al. (2017) integrated Blockchain technology with digital supply chain for cost effective and interoperable supply chain structure.

Smart SCM is another concept that is related to I4.0, and used to define more intelligent and highly automated supply chains. As one of earliest attempt, Wu et al. (2016) conducted a literature survey related to smart SCM and proposed key research topics including IT, IT infrastructure, process automation, and supply chain integration. Similarly, Lin et al. (2016) investigated the impacts of smart manufacturing on SCM and proposed that smart manufacturing should highlight not only the manufacturing factory but the entire supply chain. In line with the smart supply chains, Barata et al. (2018) worked on mobile SCM in I4.0, and proposed a bibliography for future studies.

Application of IoT to the supply chain is the most common research area in the field so far. Ben-Daya et al. (2019) conducted a literature review to explore the role and impact of IoT in SCM, they used the SCOR framework, which divides supply chain processes as plan, source, make, deliver, and enable and investigated the impact of IoT on these processes separately. Abdel-Basset et al. (2018) also investigated impacts of IoT on SCM, however they specifically focused on supply chain security by proposing a framework. On the other hand, Haddud et al. (2017) examined benefits and challenges of IoT integration in SCM, where they conducted a survey. Results showed that, IoT integration may have potential benefits such as reduction of date distortion and improvement of business intelligence, reducing delays in data collecting, and better integration in processes. Contrarily some potential challenges were also presented as network security risks and vulnerabilities, lack of understanding about IoT benefits, and risk related to implementation of new business models (Haddud et al. 2017). Similarly, Tu (2018) conducted an exploratory to understand firm's intention to adopt or reject IoT implementation in SCM by presenting benefit-cost aspects, uncertainties, and external motivating factors. Tu et al. (2018a and 2018b) conducted a two-staged research with a focus of IoT based production logistics and supply chain system, and proposed an IoT architecture and its implementations in SCM. On the other hand, Majeed and Rupasinghe (2017) approached to the IoT implementation in SCM form a different perspective, and they integrated IoT with ERP in order to improve inbound and outbound operations and to optimize supply chain processes.

Big Data, another important technology that shapes I4.0, is also a popular topic, which is integrated with SCM. For instance, Queiroz and Telles (2018) conducted a survey to understand the awareness of implementation of big data analytics in SCM. They proposed a Big Data Analytics-SCM triangle to guide implementation projects, which integrates supply chain partnerships, human knowledge, and innovative culture and intersect at the Big Data Analytics-SCM triangle. Furthermore, Wang et al. (2016) presented a literature review of big data analytics in logistics and SCM, and they proposed a maturity framework for supply chain analytics. Similarly, Brinch (2018) also presented a framework for understanding the value of big data in SCM, and proposed that the adaptation of big data improves the data utilization, changes business process configurations, have positive effect on business processes and quality, provides better decision making, and strengthen the supply chain practices. With a broader perspective, Witkowski (2017) integrated IoT and Big Data under the roof of I4.0 and presented innovative solutions for SCM.

Beside these, Ivanov et al. (2016) focused on a more specific area under SCM and I4.0, and they proposed a dynamic model and algorithm for supply chain scheduling in smart factories, in order to overcome challenges caused by altering job arrivals and processing speeds, and temporal machine structures in short term supply chain scheduling activities in I4.0. Similarly, Dolgui et al. (2019) also conducted a study related to scheduling in production supply chain and I4.0. They especially focused on control engineering, presented optimal control models for job scheduling in supply chain, production and I4.0, and analyzed the application areas.

There are also studies that focus on different functions of the supply chain. Procurement is one of these areas that has been studied by Glasand Kleemann (2016) and Bienhaus and Haddud (2018). Glasand Kleemann (2016) presented different observations related to altering procurement activities in I4.0 environment, while Bienhaus and Haddud (2018) focused on the impacts and barriers of I4.0 on procurement. These studies used a common term "Procurement 4.0" and stated that digitalizing procurement activities by I4.0 technologies would be beneficial in terms of improving supply chain efficiency. On the other hand, Ardito et al. (2019) focused on SCM—marketing integration and presented the joint benefits of Industrial IoT, cloud computing, data analytics and customer profiling, and cyber security. Furthermore, under SCM, Ganji et al. (2018) focused on demand chain management and investigated impacts of I4.0 on demand driven supply chains, they also introduced the term "Demand Chain 4.0".

In addition to these studies, Jayaram (2016) integrated the lean six-sigma approach and I4.0 and implemented it to global supply chain management. They came up with the idea of optimizing global supply chain management with Industrial IoT's ability of autonomous and connected processes and providing a zero defect and free from wastes environment with a philosophy of lean six sigma (Jayaram 2016). Chhetri et al. (2018) focused on product life cycle security in I4.0 environment, and presented I4.0 trends for security requirements including confidentiality, integrity, and availability. In addition, Dallasega et al. (2018) conducted a sectoral-based study, which focused on the construction supply chain, and proposed a framework for explaining I4.0 technologies and impacts on supplier relations in the construction sites.

2.2 Sustainable Supply Chain Management in Industry 4.0

Nowadays, I4.0 and sustainability are two major topics in the field. In line with this view, de Sousa Jabbour et al. (2018) suggested that these two terms cannot be considered as industrial revolutions individually, however, synergy between them is going to change the worldwide production system in a revolutionary way. Similarly, Stock et al. (2018) stated that industrial organizations have to shift their processes according to the new industrial revolution while this transformation must be built on economic, social and ecological development, in other words they should be in line with sustainability goals.

Although topics related to I4.0 are trending now, where, studies related to I4.0 & SCM, and I4.0 & sustainability are increasing day by day; there are only few researches focuses on I4.0 and SSCM at the same time. According to the literature review, it is revealed that majority of the current studies are conference papers and there are only few journal articles, which solely focus on I4.0 and SSCM so far.

To start with,in one of the rare studies, Manavalan and Jayakrishna (2019) made a review of IoT embedded SSC for I4.0 requirements, and proposed a framework for assessing SSCM for I4.0. In their study, they made a detailed examination for I4.0 design principles, technologies, and influences of these on supply chain processes. Finally, they proposed a framework for SSCM in I4.0 that contains enablers including business based smart operations; technology based smart products, sustainable development and collaboration (Manavalan and Jayakrishna 2019).

Luthra and Mangla (2018) conducted a study with macro perspective from an emerging economy point of view. They focused on challenges to I4.0 for supply chain sustainability. In total, 18 challenges were defined by the authors and by using explanatory factor analysis those challenges were grouped under four main headings namely, organizations, legal and ethical issues, strategic and technological. Moreover, they used AHP method to rank the challenges for managerial implications (Luthra and Mangla 2018). Similarly, Bhagawati et al. (2019), focused on identifying key success factors of SSCM for I4.0. They defined 13 key success factors, and used DEMATEL method to investigate the relationship between them and the importance order for managerial decisions (Bhagawati et al. 2019). Takhar and Liyanage (2018), focused on identifying key concepts and issues related to impact of I4.0 on SSCM by following a detailed literature review related to I4.0, supply chain, and sustainability, made another research with macro perspective.

Saberi et al. (2019) focused on blockchain technology and its relationships to SSCM. They focused on challenges and opportunities of blockchain adaptation in application of SSCs. Challenges were categorized under intra-organizational barriers, system related barriers, and external barriers. Moreover, some of the advantages of blockchain on SSCM practices were presented as, economic gains through recusing costs, increasing business reliability and competitive advantage; social sustainability gains through preventing corruption of stakeholders from society to employees; environmental gains through reducing rework, recall, greenhouse gas emissions caused by transportation, and waste of supply chains (Saberi et al. 2019).

In a more sectoral point of view, related to I4.0 and its technologies, Ma et al. (2017) presented a collaborative cloud service platform for a more sustainable make to order supply chain in the apparel industry, and a heuristic for service provider selection was designed, where sustainability assessment can be made. They used multi-agent simulation technology for building the proposed platform, and evaluating new SSC model. Results of the simulation showed that, proposed technology contributes the sustainability of the supply chain (Ma et al. 2018). Similarly, Gamage and Rupasinghe (2017) also focused on the apparel industry; used the simulation based modeling approach for SSCs. They proposed a model for assessing smart collaboration activities and their effects on the sustainability of the apparel supply chain. At the end of the study, they concluded that I4.0 applications in supply chain collaboration increases the supply chain sustainability (Gamage and Rupasinghe 2017). On the other hand, Ojo et al. (2018) focused on sustainable food supply chains, they used the TBL approach to discuss impact of I4.0 technologies including automation and robotics, IoT, cloud computing, CPS and big data on typical food supply chain stages namely, production, processing and manufacturing, distribution, and sales. This study contributed to the literature by presenting a deeper understanding of I4.0 in sustainable food supply chains. Finally, Daú et al. (2019) focused on healthcare sector, and they proposed a circular model for a sustainable healthcare supply chain in I4.0. Results of their study showed that integration of TBL, I4.0, and corporate social responsibility allows a transition to the circular model and helps to improve sustainable healthcare supply chain in I4.0.

In Table 2, current literature related to SSCM and I4.0 is summarized by including the type, and the aim of the study.

Table 1: Literature Review of the Supply Chain Management in Industry 4.0.

Author(s)	Title	Name of the Publishing Journal/ Conference	Aim of the Study
Glas and Kleemann (2016)	The impact of industry 4.0 on procurement and supply management: A conceptual and qualitative analysis	Journal of Business and Management Invention	Exploring impacts of I4.0 for the procurement, supply chain and distribution functions
Jayaram (2016)	Lean six sigma approach for global supply chain management using industry 4.0 and IIoT	International Conference on Contemporary Computing and Informatics, IEEE.	Proposing a global supply chain model by integrating lean six sigma and IIoT
Ivanov et al. (2016)	A dynamic model and an algorithm for short-term supply chain scheduling in the smart factory industry 4.0	International Journal of Production Research, Taylor & Francis	Presenting a dynamic model and algorithm for short-term supply chain scheduling in smart factories I4.0
Lin et al. (2016)	Smart Manufacturing and Supply Chain Management	Proceedings of the International conference on Logistics, Informatics and Service Sciences, IEEE	Exploring the impacts of smart manufacturing to SCM
Wang et al. (2016)	Big data analytics in logistics and supply chain management: Certain investigations for research and applications	International Journal of Production Economics, Elsevier	Reviewing the literature on the application of big data business analytics on logistics and supply chain management
Wu et al. (2016)	Smart supply chain management: a review and implications for future research	International Journal of Logistics Management, Emerald	Exploring the currents status and remaining issues of smart SCM
Barata et al. (2018)	Mobile supply chain management in the Industry 4.0 era An annotated bibliography and guide for future research	Journal of Enterprise Information Management, Emerald	Identifying future research areas in mobile supply chain in I4.0
Ben-Daya et al. (2019)	Internet of things and supply chain management: a literature review	International Journal of Production Research, Taylor & Francis	Exploring the role of IoT and its impact on SCM
Haddud et al. (2017)	Examining potential benefits and challenges associated with the Internet of Things integration in supply chains	Journal of Manufacturing Technology Management, Emerald	Examining impact of IoT in organizational supply chain
Korpela et al. (2017)	Digital Supply Chain Transformation toward Blockchain Integration	Proceedings of the 50th Hawaii International Conference on System Sciences	Investigating requirements and functionalities of supply chain integration through the Blockchain
Majeed and Rupasinghe (2017)	Internet of Things (IoT) Embedded Future Supply Chains for Industry 4.0: An Assessment from an ERP-based Fashion Apparel and Footwear Industry	Journal of Supply Chain Management	Proposing a conceptual framework for IoT embedded supply chains
Tjahjono et al. (2017)	What does Industry 4.0 mean to Supply Chain?	Manufacturing Engineering Society International Conference, Elsevier	Analyzing impacts of I4.0 on SCM
Witkowski (2017)	Internet of Things, Big Data, Industry 4.0 – Innovative Solutions in Logistics and Supply Chains Management	7th International Conference on Engineering, Project, and Production Management, Elsevier	Presenting innovative I4.0 solutions for Logistics and SCM

Table 1 contd. ...

...Table 1 contd.

Author(s)	Title	Name of the Publishing Journal/ Conference	Aim of the Study
Abdel Basset et al. (2018)	Internet of Things (IoT) and its impact on supply chain: A framework for building smart, secure and efficient systems	Future Generation Computer Systems, Elsevier	Application of IoT in SCM for smart and secure system
Bienhaus and Haddud (2018)	Procurement 4.0: factors influencing the digitisation of procurement and supply chains.	Business Process Management Journal, Emerald	Identifying the impact of digitisation on procurement and its role within the area of SCM
Brinch (2018)	Understanding the value of big data in supply chain management and its business processes: Towards a conceptual framework.	International Journal of Operations & Production Management, Emerald	Presenting the value of Big Data in SCM
Büyüközkan and Göçer (2018)	Digital Supply Chain: Literature review and a proposed framework for future research	Computers in Industry, Elsevier	Identifying key limitations and prospect of digital supply chains and presenting future research directions
Chhetri et al. (2018)	Manufacturing Supply Chain and Product Lifecycle Security in the Era of Industry 4.0	Journal of Hardware and Systems Security, Springer	Presenting security challenges and trends related to supply chain in I4.0
Dallasega et al. (2018)	Industry 4.0 as an enabler of proximity for construction supply chains: A systematic literature review	Computers in industry, Elsevier	Proposing a framework for I4.0 concepts in construction supply chains
Ganji et al. (2018)	DCM 4.0: integration of Industry 4.0 and demand chain in global manufacturing.	IEEE International Conference on Engineering, Technology and Innovation (ICE/ITMC	Exploring the opportunities of I4.0 in demand-driven supply chains
Iddris (2018)	Digital Supply Chain: Survey of the Literature	International Journal of Business Research and Management	Presenting the main drivers of digital supply chains
Tu (2018)	An exploratory study of Internet of Things (IoT) adoption intention in logistics and supply chain management: A mixed research approach	The International Journal of Logistics Management	Exploring the determinant factors affecting IoT adoption in logistics and SCM
Tu et al. (2018a)	IoT-based production logistics and supply chain system–Part 1: Modeling IoT-based manufacturing supply chain	Industrial Management & Data Systems, Emerald	Proposing a framework for IoT-based production logistics and supply chain system
Tu et al. (2018b)	IoT-based production logistics and supply chain system – Part 2: IoT-based cyberphysical system: a framework and evaluation	Industrial Management & Data Systems, Emerald	Implementation of IoT based CPS architecture framework for production logistics and supply chain system
Queiroz and Telles (2018)	Big data analytics in supply chain and logistics: an empirical approach.	The International Journal of Logistics Management, Emerald	Presenting current state of Big Data Analytics in SCM
Artido et al. (2019)	Towards Industry 4.0: Mapping digital technologies for supply chain management-marketing integration	Business Process Management Journal, Emerald	Presenting digital technologies for managing the interface between SCM and marketing processes

Table 1 contd. ...

...Table 1 contd.

Author(s)	**Title**	**Name of the Publishing Journal/ Conference**	**Aim of the Study**
Dolgui et al. (2019)	Scheduling in production, supply chain and Industry 4.0 systems by optimal control: fundamentals, state-of-the-art and applications	International Journal of Production Research, Taylor & Francis	Presenting a survey on the applications of optimal control to scheduling in production, supply chain and I4.0.
Frederico et al. (2019)	Supply Chain 4.0: concepts, maturity and research agenda	Supply Chain Management: An International Journal, Emerald	Proposing a framework for "Supply Chain 4.0".
Hahn (2019)	Industry 4.0: a supply chain innovation perspective	International Journal of Production Research, Taylor & Francis	Investigating the implications of I4.0 on SCM with a focus on supply chain innovations

Table 2: Literature Review of the SSCM in Industry 4.0.

Author(s)	Type	Aim
Gamage and Rupasinghe (2017)	Conference Paper	Simulating the applications of Industry 4.0 technologies in the supply chain by focusing on sustainability in the apparel industry
Dossou (2018)	Conference Paper	Measuring impact of sustainability on the supply chain 4.0 performance in SMEs
Luthra and Mangla (2018)	Research Article	Identifying and analyzing key challenges of Industry 4.0 for supply chain sustainability in emerging economies
Ma et al. (2018)	Research Article	Proposing a collaborative cloud service platform for developing sustainable make-to-order apparel supply chain
Ojo et al. (2018)	Conference Paper	Examining challenges and opportunities of Industry 4.0 implications towards sustainable food supply chain
Takhar and Liyanage (2018)	Conference Paper	Establishing key concepts and issues related to Industry 4.0, and the potential impacts on sustainability and supply chains
Bhagawati et al. (2019)	Conference Paper	Identifying key success factors of sustainability in supply chain management for Industry 4.0
Daú et al. (2019)	Research Article	Analyzing SSCM 4.0 in healthcare by proposing a circular economy transition framework
Manavalan and Jayakrishna (2019)	Research Article	Reviewing different aspects of SCM, ERP, IoT and Industry 4.0, and exploring potential opportunities of Industry 4.0 transformation on SSCM
Saberiet al. (2019)	Research Article	Examination blockchain technology and smart contracts and their potential application on SSCM

As it can be understood from the literature review, although there are studies that cover SCM in I4.0, there are only few studies related to SSCM and I4.0. Therefore, there is a big gap in the literature about related topics. In order to fulfill the gap, this study identifies the most important keywords and propose a road map about future research about SSCM in I4.0 literature.

In the next section, the method and the details of the keywords that should be used for the future studies to fill the literature gap for the researchers will be explained.

3. Methodology

The aim of this study is to propose a theoretical and practical conceptual framework that integrates and filters frequently mentioned concepts and new concepts related to the subject of SSCM in I4.0. Despite all mentioned studies in the previous section, a comprehensive review of peer-reviewed journal articles was conducted to determine future research topics for the SSCM in I4.0 issue, which still has a large gap in the literature.

The following procedure was applied to identify the keywords and study topics related to these concepts.

- A list of the words identified as a result of a detailed literature review was made on the three related concepts.
- A focus group, consisting of five people, was formed in order to categorize the defined keywords under three concepts. Experts can be made with minimum three and maximum twelve participants (Krueger 1994). Two academic and three private sector employees came together to maintain impartiality in the discussions and to give opinions about the word list created. Experts were asked to classify each words under three concepts.
- Finally, they were asked to propose words related to these three concepts that were not in the current list.

Fifty-seven terms were determined in accordance with the information given by expert opinions. Eight of them are classified under *sustainability*, ten of them are classified under *supply chain* and thirty-nine of them are classified under *industry 4.0* term (Table 3). In order to include the peer-reviewed journal article in the analysis, each of the three concepts should take the title as single, double or triple.

In the next section, content analysis will be carried out using the most cited articles with reference to these fifty-seven words.

Table 3: List of Words Based on Classification of Concepts.

Supply Chain Management	**Sustainability**	**Industry 4.0**		
Chain	Circularity	3-Dimension	Digitization	Modular
Demand Chain	Circular Economy	Additive	Factory of Future	Real-time
Procurement	Economic	Artificial Intelligence	Human-Machine	RFID
Supply	Environment	Augmented Reality	Human-Robot	Robot
Supplier	Green	Automation	Infor. Commun. Techn.	Security
Supply Chain	Social	Big Data	Industrial Internet of Things	Sensor
Supply Chain Man.	Sustainability	Blockchain	Industry 4.0	Service-orientation
Supply Network	Triple Bottom Line	Cloud	Innovation	Simulation
Vendor		Cobot	Integration	Smart
Value Chain		Collaboration	Internet of Service	Technology
		Cyber Physical System	Internet of Things	Traceability
		Data	Interoperability	Virtual
		Decentralization	Machine Learning	Wireless

4. Empirical Study

In the light of the discussions in the previous section, key words need to be identified in order to identify the content of new academic studies on SSCM in I4.0 and provide a roadmap. To identify the most important keywords and propose a road map about future research about SSCM in I4.0 literature. For this, we first selected the appropriate data set. Web of Science, Thompson and Reuters, and Scopus databases have coverage of high impact peer-reviewed journal articles. In these databases; between the years 2014–2019; printed in English; the first 100 articles with the most citations were selected. Each of the three concepts "supply chain", "sustainability" and "industry 4.0" should take the title as single, double or triple.

Later, we develop a Java code using a library provided by Apache Tika to count the number of occurrences of each predetermined word in the vocabulary within the peer-reviewed journal article. Lastly, the collected data that shows how many times each word was mentioned in each peer-reviewed journal article was analyzed using IBM SPSS Statistics 23 for Windows.

Based on desired criteria 100 articles were examined; three in both 2014 and 2015, seventeen in 2016, third-eight in 2018 and eighteen in the first eight months of 2019. The peer-reviewed journal articles varied in length from three page to over fifty-six pages (with an average of 15, 24 pages). In total, 1524 pages were analyzed to identify new keywords. In these articles, the average number of authors is three and the average number of keywords is five. In each article examined, 1616 of the words in Table 3 were used.

Table 4 shows the rank ordering of the twenty most frequent words based on how frequently their related issues are referred to in the peer-reviewed journal articles.

The descriptive results indicated that "industry 4.0" was the most commonly emphasized concept among hundred papers. Other commonly emphasized concepts were traceability, augmented reality,circular economy and artificial intelligence. Based on the results presented in Table 5, we can say that there were only slight differences among the years 2014–2016 in their emphasis of words concepts. *Industry 4.0* and *traceability* are ranked as top two words.

Looking at the studies conducted from 2017–2019, it is interesting to see that the first nine words are the same; even the order of the first five words is the same. After 2016, artificial intelligence issues especially began to gain importance.

Table 4: List of the Top 20 Most Used Words in the Academic Journals.

Rank	Words	#	Rank	Words	#
1	Industry 4.0	21854	11	Technology	4585
2	Traceability	21388	12	Data	3884
3	Augmented Reality	18138	13	Internet of Things	3633
4	Circular Economy	13432	14	Automation	3173
5	Artificial Intelligence	10841	15	Environment	2992
6	Infor. Com. Tech.	8312	16	Supply Chain Management	1588
7	Chain	7300	17	Smart	1483
8	Supply	6934	18	Digitalization	1333
9	Sustainability	6203	19	Industrial Internet of Things	1313
10	Supply Chain	5769	20	Big Data	1292

Table 5: List of Top 10 Most Used Words by Years in the Academic Journals.

Rank	2014		2015		2016	
	Words	#	Words	#	Words	#
1	Industry 4.0	440	Traceability	1225	Traceability	3078
2	Traceability	408	Industry 4.0	881	Industry 4.0	2784
3	ICT	122	Cloud	208	Augmented Reality	2218
4	Technology	88	ICT	190	Circular Economy	1482
5	Automation	60	Automation	148	AI	1181
6	Additive	31	Data	136	Supply	872
7	Augmented Reality	28	Additive	84	ICT	853
8	Smart	26	Big Data	75	Chain	787
9	Internet of Things	25	Innovation	73	Data	758
10	Data	22	Vendor	70	Supply Chain	654
Rank	2017		2018		2019	
	Words	#	Words	#	Words	#
1	Industry 4.0	3598	Industry 4.0	9821	Traceability	4405
2	Traceability	3459	Traceability	8813	Industry 4.0	4330
3	Augmented Reality	3246	Augmented Reality	8414	Augmented Reality	4197
4	Circular Economy	2567	Circular Economy	6457	Circular Economy	2926
5	AI	1873	AI	4991	AI	2778
6	ICT	1385	ICT	3609	Chain	2195
7	Chain	1187	Chain	3123	ICT	2153
8	Sustainability	1096	Supply	3097	Supply	1913
9	Supply	1042	Sustainability	2928	Sustainability	1802
10	Internet of Things	888	Supply Chain	2560	Supply Chain	1681

On the contrary, the least frequent 20 words are presented in Table 6. An interesting result is gained from that list that very popular technologies under I4.0 such as *Blockchain and 3-Dimension* printing appear in this list. Moreover, it has been noticed that I4.0 design dimensions are not so popular in the focus area of this chapter.

In a broader perspective, in Table 7 top five most used words are presented for SCM, sustainability and I4.0 separately. Under SCM, most common word that appeared is *chain,* under sustainability, surprisingly, circular economy is revealed as the most frequent word, which is twice as much than *sustainability.* Finally, as expected, *Industry 4.0* has the greatest proportion followed closely by *traceability.*

In the following section, results of the empirical study are discussed by pointing out the key findings and potential future study directions.

Table 6: List of Least Used Words in the Academic Journals.

Words	#	Words	#
Blockchain	399	Interoperability	175
Collaboration	376	Service orientation	162
Modular	354	Vendor	141
Integration	339	Simulation	135
Wireless	308	Supply Network	89
Factory of Future	292	Triple Bottom Line	82
Procurement	268	Demand Chain	81
3-Dimension	258	Decentral	68
Internet of Service	248	Human-Robot	48
Human-Machine	247	Cobot	7
Value Chain	211		

Table 7: List of Top 5 Most Used Words are presented for SCM, Sustainability and I4.0.

Rank	SCM	#	Sustainability	#	Industry 4.0	#
1	Chain	7300	Circular Economy	13432	Industry 4.0	21854
2	Supply	6934	Sustainability	6203	Traceability	21388
3	Supply Chain	5769	Environment	2992	Augmented Reality	18138
4	Supply Chain Man.	1588	Economic	921	Artificial Intelligence	10841
5	Supplier	730	Social	890	Infor. Com.Tech.	8312

5. Discussions

According to the results of the empirical study, as expected, I4.0 related topics are highly popular in the current literature. When it is specified for the words, it has been noticed that "Industry 4.0" is the most used word in last the 5 years, which is predictable since it is used to generalize the concepts related to the fourth industrial revolution. On the other hand, the second most important word "traceability", mostly refer to the main expectations from I4.0 technologies, especially in supply chains. However, surprisingly, "cloud" and "big data" appeared in the most important 10 words only in 2015, and one of the key concept "IoT" has only appeared in 2014 and 2017.

Furthermore, "circular economy" is the first word, which is categorized under sustainability, it appeared in the top 10 word lists in 2016, and has not changed its 4th rank in the analyzed time period. Based on the results, after 2016, "sustainability", "chain", and "supply" are also added to the most popular words in Table 5. Based on these results, it should be noticed that, although sustainability is an older concept, circular economy receives more attention nowadays and research directions are moved to concepts that are more circular when they are integrated to I4.0.

Surprisingly, according to Table 6, some of important concepts such as "blockchain", "3-Dimension", Human Machine" and I4.0 design principles, i.e., modularity, service orientation, interoperability, and decentralization are located at the bottom of the word list. Moreover, although all sustainability or triple bottom line dimensions namely, economic, environment, and social are listed in the most important concepts under sustainability in Table 6, "TBL" as an integrated concept does not appear and is in fact located in the 53th place.

According to these results, future research directions may be presented based on the current literature gaps. For instance, as mentioned before I4.0 design principles have not received the attention that they deserve so far. Moreover, in a more holistic view, sustainability dimensions may be covered under TBL approach, and integrated with I4.0 design principles. Based on the focus of this chapter, especially the circular supply chain concept can be studied more, keeping in mind the popularity of circular economy. With this view, future research directions may be shaped based on these suggested concepts as summarized in Figure 2.

In the following section, conclusion of the chapter is presented.

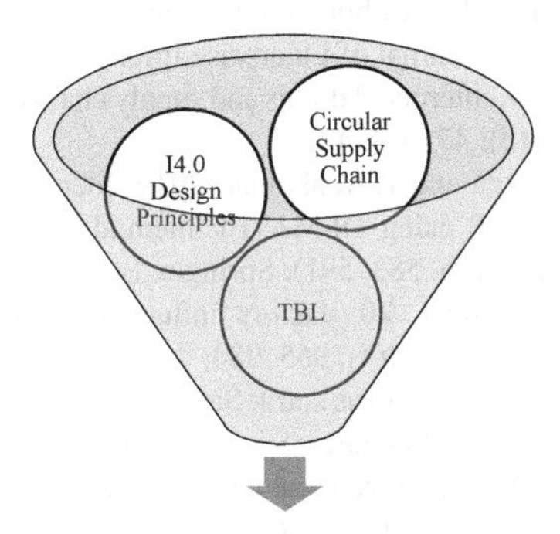

Fig 2: Future Research Directions of SSCM in I4.0 Literature.

6. Conclusion

Transformation to digitalization triggered the Fourth Industrial Revolution, and spread to industries in terms of altering manufacturing processes by high technologies and to the academic environment by highlighting the need of new business models for industrial and service systems. Some of the key technologies, components, and principles that change both the manufacturing processes and the organizational structures shape industry 4.0. With the technologies introduced by industry 4.0, the traditional supply chain structure is affected. Impacts of Industry 4.0 are not limited to the manufacturing level, but expands to all stages of the supply chain. Industry 4.0 can influence processes and business models for the supply chain dramatically. In the face of this change, companies need to build sustainable supply chain structures. Increase in environmental concerns reveals the importance of supply chain sustainability for organizations to improve their environmental performance while gaining economic and social benefits.

These issues are getting more and more attention from researchers. Especially in the last 3 years, there has been many academic publications about these issues. But, when the studies are examined, it is seen that researchers generally focus on the same subjects. However, as mentioned above, these issues have a very deep and wide area. From this point of view, the aim of this chapter is assessing the impacts of fourth industrial revolution on sustainable supply chain management by conducting a content analysis, based on a literature review, and proposing a framework for SSCM in I4.0. This chapter clearly gives the reader comprehensive insight about how to apply industrial 4.0 conceptual models to promote sustainable SCM. Also, this chapter identifies the most important keywords emerging from the SSCM in I4.0 literature. A road map has been produced based on a systematic literature review of the 100 most cited research articles conducted between 2014–2019.

Based on the results, future research directions may be shaped by merging I4.0 design principles, circular supply chain and triple bottom line dimensions. In addition, big data, factory of future, blockchain, 3-Dimensions and sensors, which are among the 4.0 technologies in the industry, are the areas that must be focused on. Moreover, industry 4.0 principles, such as virtualization, decentralization, real time, modularity, and service orientation are among the most important issues that must be considered by researchers.

Abbreviations:

SSC	:	Sustainable Supply Chain
SSCM	:	Sustainable Supply Chain Management

SCM : Supply Chain Management

I4.0 : Industry 4.0

References

Abdel-Basset, M., G. Manogaran and M. Mohamed. 2018. Internet of things (IoT) and its impact on supply chain: A framework for building smart, secure and efficient systems. Future Generation Computer Systems, 86: 614–628.

Ardito, L., A.M. Petruzzelli, U. Panniello and A.C. Garavelli. 2019. Towards Industry 4.0: mapping digital technologies for supply chain management-marketing integration. Business Process Management Journal, 25(2): 323–346.

Barata, J., P. Rupino Da Cunha and J. Stal. 2018. Mobile supply chain management in the industry 4.0 era: an annotated bibliography and guide for future research. Journal of Enterprise Information Management, 31(1): 173–192.

Ben-Daya, M., E. Hassini and Z. Bahroun. 2019. Internet of things and supply chain management: a literature review. International Journal of Production Research, 57(15-16): 4719–4742.

Bhagawati, M.T., E. Manavalan, K. Jayakrishna and P. Venkumar. 2019. Identifying key success factors of sustainability in supply chain management for industry 4.0 using DEMATEL method. In Proceedings of International Conference on Intelligent Manufacturing and Automation (pp. 583–591). Springer, Singapore.

Bienhaus, F. and A. Haddud. 2018. Procurement 4.0: factors influencing the digitization of procurement and supply chains. Business Process Management Journal, 24(4): 965–984.

Bonilla, S., H. Silva, M. Terra da Silva, R. Franco Gonçalves and J. Sacomano. 2018. Industry 4.0 and sustainability implications: A scenario-based analysis of the impacts and challenges. Sustainability, 10(10): 3740.

Braccini, A. and E. Margherita. 2019. Exploring organizational sustainability of industry 4.0 under the triple bottom line: the case of a manufacturing company. Sustainability, 11(1): 36.

Brinch, M. 2018. Understanding the value of big data in supply chain management and its business processes: Towards a conceptual framework. International Journal of Operations & Production Management, 38(7): 1589–1614.

Büyüközkan, G. and F. Göçer. 2018. Digital supply chain: literature review and a proposed framework for future research. Computers in Industry, 97: 157–177.

Chhetri, S.R., S. Faezi, N. Rashid and M.A. Al Faruque. 2018. Manufacturing supply chain and product lifecycle security in the era of industry 4.0. Journal of Hardware and Systems Security, 2(1): 51–68.

Dalenogare, L.S., G.B. Benitez, N.F. Ayala and A.G. Frank. 2018. The expected contribution of Industry 4.0 technologies for industrial performance. International Journal of Production Economics, 204: 383–394.

Dallasega, P., E. Rauch and C. Linder. 2018. Industry 4.0 as an enabler of proximity for construction supply chains: A systematic literature review. Computers in Industry, 99: 205–225.

Daú, G., A. Scavarda, L.F. Scavarda and V.J.T. Portugal. 2019. The healthcare sustainable supply chain 4.0: the circular economy transition conceptual framework with the corporate social responsibility mirror. Sustainability, 11(12): 3259.

De Sousa Jabbour, A.B.L., C.J.C. Jabbour, C. Foropon and M. Godinho Filho. 2018. When titans meet–Can industry 4.0 revolutionise the environmentally-sustainable manufacturing wave? The role of critical success factors. Technological Forecasting and Social Change, 132: 18–25.

Dolgui, A., D. Ivanov, D., S.P. Sethi and B. Sokolov. 2019. Scheduling in production, supply chain and Industry 4.0 systems by optimal control: fundamentals, state-of-the-art and applications. International Journal of Production Research, 57(2):411–432.

Dossou, P.E. 2018. Impact of Sustainability on the supply chain 4.0 performance. Procedia Manufacturing, 17: 452–459.

Fatorachian, H. and H. Kazemi. 2018. A critical investigation of Industry 4.0 in manufacturing: theoretical operationalisation framework. Production Planning & Control, 29(8): 633–644.

Frederico, G.F., J.A. Garza-Reyes, A. Anosike and V. Kumar. 2019. Supply chain 4.0: concepts, maturity and research agenda. Supply Chain Management: an International Journal.

Gamage, D.D. and T.D. Rupasinghe. 2017. A Simulation-based modelling approach for sustainable supply chains through smart collaboration. *In*: Proceedings of the International Postgraduate Research Conference 2017 (IPRC–2017), Faculty of Graduate Studies, University of Kelaniya, Sri Lanka.

Ganji, E.N., A. Coutroubis and S. Shah. 2018. DCM 4.0: integration of Industry 4.0 and demand chain in global manufacturing. In 2018 IEEE International Conference on Engineering, Technology and Innovation (ICE/ITMC) (pp. 1–7). IEEE.

Glas, A.H. and F.C. Kleemann. 2016. The impact of industry 4.0 on procurement and supply management: A conceptual and qualitative analysis. International Journal of Business and Management Invention, 5(6): 55–66.

Haddud, A., A. DeSouza, A. Khare and H. Lee. 2017. Examining potential benefits and challenges associated with the internet of things integration in supply chains. Journal of Manufacturing Technology Management, 28(8): 1055–1085.

Hahn, G.J. 2019. Industry 4.0: a supply chain innovation perspective. International Journal of Production Research, 1–17.

Iddris, F. 2018. Digital supply chain: survey of the literature. International Journal of Business Research and Management, 9(1).

Ivanov, D., A. Dolgui, B. Sokolov, F. Werner and M. Ivanova. 2016. A dynamic model and an algorithm for short-term supply chain scheduling in the smart factory industry 4.0. International Journal of Production Research, 54(2): 386–402.

Jayaram, A. 2016. Lean six sigma approach for global supply chain management using industry 4.0 and IIoT. In 2016 2nd International Conference on Contemporary Computing and Informatics (IC3I) (pp. 89–94). IEEE.

Korpela, K., J. Hallikas and T. Dahlberg. 2017. Digital supply chain transformation toward blockchain integration. In proceedings of the 50th Hawaii International Conference on System Sciences.

Krueger, R.A. 2014. Focus Groups: A Practical Guide for Applied Research. Sage Publications.

Lasi, H., P. Fettke, H.G. Kemper, T. Feld and M. Hoffmann. 2014. Industry 4.0. Business & Information Systems Engineering, 6(4): 239–242.

Lin, Y., P. Ieromonachou and W. Sun. 2016. Smart manufacturing and supply chain management. In 2016 International conference on Logistics, Informatics and Service Sciences (LISS) (pp. 1–5). IEEE.

Luthra, S. and S.K. Mangla. 2018. Evaluating challenges to Industry 4.0 initiatives for supply chain sustainability in emerging economies. Process Safety and Environmental Protection, 117: 68–179.

Ma, K., L. Wang and Y. Chen. 2018. A collaborative cloud service platform for realizing sustainable make-to-order apparel supply chain. Sustainability, 10(1): 11.

Majeed, A.A. and T.D. Rupasinghe. 2017. Internet of things (IoT) embedded future supply chains for industry 4.0: An assessment from an ERP-based fashion apparel and footwear industry. International Journal of Supply Chain Management, 6(1): 25–40.

Manavalan, E. and K. Jayakrishna. 2019. A review of Internet of Things (IoT) embedded sustainable supply chain for industry 4.0 requirements. Computers & Industrial Engineering, 127: 925–953.

Ojo, O.O., S. Shah, A. Coutroubis, M.T. Jiménez and Y.M. Ocana. 2018. Potential impact of industry 4.0 in sustainable food supply chain environment. In 2018 IEEE International Conference on Technology Management, Operations and Decisions (ICTMOD) (pp. 172–177). IEEE.

Oztemel, E. and S. Gursev. 2018. Literature review of Industry 4.0 and related technologies. Journal of Intelligent Manufacturing, 1–56.

Queiroz, M.M. and R. Telles. 2018. Big data analytics in supply chain and logistics: an empirical approach. The International Journal of Logistics Management, 29(2): 767–783.

Saberi, S., M. Kouhizadeh, J. Sarkis and L. Shen. 2019. Blockchain technology and its relationships to sustainable supply chain management. International Journal of Production Research, 57(7): 2117–2135.

Song, Z. and Y. Moo. 2017. Assessing sustainability benefits of cyber manufacturing systems. The International Journal of Advanced Manufacturing Technology, 90(5-8): 1365–1382.

Stock, T. and G. Seliger. 2016. Opportunities of sustainable manufacturing in industry 4.0. ProcediaCirp, 40: 536–541.

Stock, T., M. Obenaus, S. Kunz and H. Kohl. 2018. Industry 4.0 as enabler for a sustainable development: A qualitative assessment of its ecological and social potential. Process Safety and Environmental Protection, 118: 254–267.

Takhar, S. and K. Liyanage. 2018. The impact of Industry 4.0 on supply chains and sustainability. Presented at the 8th International Conference on Operations and Supply Chain Management (OSCM), Cranfield University, 10th September.

Tjahjono, B., C. Esplugues, E. Ares and G. Pelaez. 2017. What does industry 4.0 mean to supply chain? Procedia Manufacturing, 13: 1175–1182.

Tu, M. 2018. An exploratory study of Internet of Things (IoT) adoption intention in logistics and supply chain management: A mixed research approach. The International Journal of Logistics Management, 29(1): 131–151.

Tu, M., K.M. Lim and M.F. Yang. 2018a. IoT-based production logistics and supply chain system–Part 1: Modeling IoT-based manufacturing supply chain. Industrial Management & Data Systems, 118(1): 65–95.

Tu, M., K.M. Lim and M.F. Yang. 2018b. IoT-based production logistics and supply chain system–Part 2: IoT-based cyber-physical system: a framework and evaluation. Industrial Management & Data Systems, 118(1): 96–125.

Wang, G., A. Gunasekaran, E.W. Ngai and T. Papadopoulos. 2016. Big data analytics in logistics and supply chain management: Certain investigations for research and applications. International Journal of Production Economics, 176: 98–110.

Witkowski, K. 2017. Internet of things, big data, industry 4.0–innovative solutions in logistics and supply chains management. Procedia Engineering, 182: 763–769.

Wu, L., X. Yue, A. Jin and D.C. Yen. 2016. Smart supply chain management: a review and implications for future research. The International Journal of Logistics Management, 27(2): 395–417.

Xu, L.D., E.L. Xu and L. Li. 2018. Industry 4.0: state of the art and future trends. International Journal of Production Research, 56(8): 2941–2962.

Chapter 20

A New Collecting and Management Proposal Under Logistics 4.0 and Green Concept

Harun Resit Yazgan, Sena Kır, Furkan Yener* and *Serap Ercan Comert*

1. Introduction

The consumption has been increasing due to technological developments, industrialization, population growth and the increasing demand of this population. As a result, the amount of waste has also been increasing rapidly. This increase in the amount of waste causes environmental disasters and also reduces natural resources. Energy use, resource consumption and waste generation in the production activities of enterprises have negative effects on the environment (Beamon 1999; Kopicki 1993). As a result, the protection of the environment depends on developing environmental consciousness, and waste management holds an important place among the environmental protection policies in the world. The rapid increase of wastes, the inadequacy of disposal methods, and the presence of elements that would threaten the lives of the people have made the concept of recycling important (Beamon and Fernandes 2004). While recycling is important for human health and the environment due to the above reasons, it is also of great importance in terms of businesses. Decreasing sources of natural resources, rising raw material prices, manufacturers' desire to provide a competitive advantage, have led businesses to look for cheap sources. Raw materials obtained as a result of recycling are an inexpensive source.

In many countries, businesses are held liable for end of life products. Also, they are obliged to recycle or dispose of these wastes. In Turkey, special legislation for end of life tires (ELT) has been developed and the principles of recycling and disposal of ELT collection have been determined by these laws. In this context, producers must plan their activities taking into account the environmental factors during the life of a product and recycle or destroy the ELTs with minimal harm to the environment.

Wastes can be divided into different categories as below considering Waste Management Regulation in Turkey (Ministry of Environment and Urbanization 2015):

- Domestic wastes
- Construction waste
- Hazardous wastes
- Medical wastes
- Packaging wastes
- Waste batteries and accumulators
- Waste oils
- ELT
- Life-finished vehicles
- Electronic waste

Sakarya University, Department of Industrial Engineering, Sakarya/Turkey.
Emails: senas@sakarya.edu.tr; fyener@sakarya.edu.tr; serape@sakarya.edu.tr
* Corresponding author: yazgan@sakarya.edu.tr

In many countries, manufacturing enterprises are held liable for their wastes. Also, they are obliged to recycle or dispose of these wastes. One of the economically valuable wastes that is recycled is the ELT because of serious damages it causes to the environment in case its disposal is uncontrolled. In this context, usually, the manufacturer must plan their activities considering environmental factors during the life of a product, to recycle or destroy the ELT with minimal harm to the environment. Also,the ELT is currently the most efficacious waste material recycling in the world in solid waste recycling. According to the research, 84% of the ELT in the world and 95% in Europe are being recycled. The recycling and recovery of the ELT are provided by the Lifetime Completed Tire Control Regulation in Turkey.

In this study, a new ELT collecting and management system was proposed under logistics 4.0 and green concept for Turkey. The remainder of this chapter is organized as follows. In the second section, brief information about ELT management in the World and in Turkey is presented. In the third section, recent studies are presented about the end of life products (tires, vehicles). In the fourth and fifth sections, the ELT management problem is explained by dividing sub-problems and proposed solution methodologies are given step by step. In the sixth section, the ELT management in Turkey is presented. Finally, the findings of the study are presented in the conclusion section.

2. ELT Management

In this section, the tires and their properties, the management of ELT in the world and in Turkey, the existing system for ELT recycling and the proposed systems will be summarized.

2.1 Tire and Specifications

A tire consists of a combination of rubber, cord fabric, steel wire, and various chemical substances. The description made by JATMA (Japan Automobile Tire Manufacturers Association) about the shelf life and standby life of a tire is:

> "*Tires used or not used (including spare wheel) for passenger cars and light commercial vehicles must be replaced with new ones even if they are ten years old after the date of manufacture, even if they do not have any visible damage or deformity, or even if the depth of the tooth is sufficient for use.*"

ELT harms the environment and human health due to the chemicals contained in it. Therefore, the tires that are changed should not be unconsciously left to nature and must be recycled. An average of 95% of the ELT is made up of recyclable products.

2.2 ELT

These are the original or coated tires that have been removed from the vehicle by determining that they have completed their useful life and that they cannot be used again as a tire on the vehicle and therefore are discarded. In Turkey, an average of 30 million tires are consumed each year. There are two major environmental hazards in the places where the ELT is piled and thrown. The first one is the fires and the second is the bugs that find the opportunity to grow easily in these heaps. Diseases caused by insects breeding in the mass are especially seen after the rains (TMMOB-Union of Chambers of Turkish Engineers and Architects, Tire Industry and Petlas Sectoral Report Series 1994). In Ohio, 80% of the incidents that occur in children have been found to be near-the-pile heaps (TMMOB Union of Chambers of Turkish Engineers and Architects 1994).

Harmonious gases spread in the atmosphere in tons of places due to burning tires in areas where they are piled up. In a black cloud-like atmosphere, Metals such as carbon black, volatile organics, semi-volatile organics compounds, polycyclic hydrocarbons, oils, sulfur oxides, nitrogen oxides, nitrosamines, carbon oxides, volatile particles and As, Cd, Cr, Pb, Zn, Fe and, etc., are released. For these reasons, recycling of tires has become important.

2.3 Importance of ELT Management in the World

The current status of ELT in Europe is defined by European Tire and Rubber Manufacturers' Association (ETRMA). ETRMA demonstrates all the responsibilities of tire manufacturers. These are the development of sustainable economic and efficient recycling methods, the storage of the ELT, and the recovery in a way that does not disturb the ecological balance (ETRMA).

In Europe, the producer responsibility system is used. In this system, the cost of recycling a tire in Europe is also met by the contribution it receives from selling the same size tire. Countries implementing this system account for 64% of Europe's total waste tire production. This system is applied in Belgium, Bulgaria, Czech Republic, Estonia, Finland,

France, Greece, Hungary, Italy, Netherlands, Norway, Poland, Portugal, Romania, Slovenia, Spain, Sweden and Turkey (Karaağaç et al. 2017).

2.4 Current System of ELT Recycling in Turkey

Recycling of ELT provides the ATMs which is known as LASDER in Turkey. LASDER members are able to recycle the total tonnage of new tires sold as much as specified in the regulations. The points in the ELT collection process are primarily the areas indicated by municipalities and public institutions after the dealerships of LASDER members. LASDER members are about 9000 dealers. To determine how much ELT will go to the recycling plants, LASDER has been receiving demands (in tons) from the recycling plants. According to this, the contractors collect the ELT from the points of association and transfer it to the designated recycling plants. All decisions regarding the storage and transport of ELT are very important in terms of cost. This creates the motivation for our study.

3. Literature Review

In this section, a brief overview of relevant recent studies about end-of-life (EOL) product management and especially ELT and EOL vehicle management which are the sub-problem of waste management has been presented.

Dehghanian and Mansour (2009) proposed a multi-objective programming model for developing a sustainable recovery network of ELT. Their model maximizes total net profits of processing ELT, minimizes the total environmental impact of all activities and maximizes social benefits. They also used Eco-indicator and Analytic Hierarchical Processing methodologies respectively to quantify the environmental and social impact of ELT treatments. Sasikumar et al. (2010) designed a multi-echelon reverse logistics network for general product recovery and formulated a MINLP model to maximize the profit of the network regarding EOL truck tire management. They validated the proposed model on the real-life problem of India. Kannan et al. (2014) considered a regional ELT management problem similar to our study. They developed a structural 4 stage ELT management model for India. As a result of the analysis of the case study, with this interpretive structural model developed, the common drivers that affect the implementation of ELT management in an Indian context were identified and discussed. Unlike this study, Kannan et al. (2014) did not address the solution of any logistics problems encountered in ELT management. Costa-Salas et al. (2017) focused on the transportation problem of an ELT management system and they analyzed it in the reverse supply chain concept. They presented a leaner structural ELT management model than Kannan et al. (2014)'s and our study. Then, they handled a realistic case study in a Colombian city and proposed an algorithmic approach based on the combination of discrete-event simulation and optimization algorithms to solve a reverse supply chain network design problem. Fagundes et al. (2017) considered the ELT management problem in terms of the reverse logistics concept. Unlike Costa-Salas et al. (2017) they did not focused on the transportation problem. They presented a structural ELT collection model for Brazilian cities. Pedram et al. (2017) integrated both a forward and reverse supply chain to design a closed-loop supply chain network and formulated a MILP to minimize the waste and maximize the profit by providing the waste management. They also showed the applicability of the model in the tire industry. Another regional ELT management research was presented by Park et al. (2018) considering Columbia case. Firstly they focused on extended producer responsibility to understand the effects on the ELT management, as well as its operation in terms of the allocation of responsibilities among the key actors. This study shows that the Colombian extended producer responsibility governance model fails to incentivize other actors in the product chain to carry out their allocated tasks and responsibilities although it imposes full financial and operational responsibilities on tire producers and importers. Starting from this point the new policy recommendations were presented in this study.

In the literature other end-of-life (EOL) products management problems were presented that were similar to ELT problems. Schultmann et al. (2006) handled the EOL vehicle management problem in Germany and presented a closed-loop supply chain. They also studied the waste transportation problem and proposed a mathematical programming model and a Tabu Search Algorithm to solve the real-life vehicle routing problem (VRP). Cruz-Rivera and Ertel (2009) focused on establishing a closed-loop supply chain for the collection of EOL vehicles in Mexico. They modified the closed-loop supply chain presented by Schultmann et al. (2006). They also solved the uncapacitated facility location problem of the EOL vehicle collection process. Mora et al. (2014) considered the Italia EOL vehicle management problem and formulated a MILP model for EOL vehicle closed recovery network design. And then they identified the parameters most affecting the model outcomes by sensitivity analysis. One of the most similar research to our study was presented by Demirel et al. (2016) considering EOL vehicles management in Turkey. They designed a multi-stage network and proposed a MILP that optimized the costs associated with opening facilities, recovery processes, transportation of EOL vehicles and its components through the network and revenues. They focused on the logistic problems of EOL products management that was integrally similar to our study. Another EOL product management problem which is the allocation of EOL vehicles under uncertainty was handled by Simic (2016). A multi-stage interval-stochastic programming model was formulated for a solution and it was tested on semi-hypothetical problems to show that it was applicable to real-world problems. Ahmed et al. (2016)researched on the sustainability of EOL vehicle management systems and presented an integrated model to select the dimensions and criteria to evaluate the systems. They preferred decision making trial and evaluation laboratory and the fuzzy analytic hierarchy process method for the evaluation process. Zhou et al. (2019) focused on the EOL vehicle recycling management problem of China which has been the biggest country with vehicle production and sales since 2009. They investigated and identified the interpretive factors by interpretive the structural modeling approach. Zhou et al. (2019)'s study and Kannan et al. (2014)'s study are similar in terms of the way of handling the problem and the techniques used. Yang et al. (2019) studied on selecting the criteria for sustainable EOL vehicle management system and developed a group decision-making approach that utilized picture hesitant fuzzy entropy and similarity measurements to evaluate alternative EOL vehicle management systems with picture hesitant fuzzy information.

ELT management problem is one of the sub-problem of waste management problems. Concordantly analyzing the waste management literature is proper, too. Sahoo et al. (2005) focused on the cost of the waste collection operation because they thought that it obtains the highest revenue. So they reduced the problem to a routing problem and developed a route-management system to improve efficiency in operating the fleet which improves the bottom line. Aliahmadi et al. (2020)handled the logistic problems of waste management similar to our study. They presented a multi-component structural model and proposed a mathematical model with multiple depots and multiple intermediate facilities to minimize fixed and variable costs of waste collection. Because of the high complexity of the model, they also developed a genetic algorithm to solve a real-life case. Lastly, a case study on the vehicle routing of municipal solid waste was conducted in a district of Tehran, Iran.

4. The Problem

In this chapter, ELT collecting and transporting problems in Turkey are discussed. In other words, the problem of collecting ELTs from the tire dealers from 81 cities, transporting them to the collection centers which are established in determined cities and accumulating in these centers for some time and then transporting them to the authorized recycling facilities is a complex problem involving multiple sub-problems. Rather than handling the problem as a whole, it was preferred to solve the problem in three stages by separating it into three sub-problems. These sub-problems and the techniques used to solve them can be summarized as follows.

4.1 Sub-Problem #1

In the solution of this sub-problem, the following questions were answered:

1. The collection centers be established in which cities?
2. ELTs will be transported from which cities to the collection centers?
3. What will be the weekly ELT capacities of the collection centers?
4. ELTs accumulated in the collection centers will be transported to which recycling facilities?

A mixed-integer non-linear programming (MINLP) model was formulated to provide answers to all abovequestions in order to solve sub-problem #1.

Addressing the problem of recycling of ELTs is the product of an environmentalist perspective. In order to solve this problem, an environmentalist approach was taken and the cost of CO_2 emissions was minimized by the objective function of the proposed MINLP model.

Since it is not possible to optimize the problem in a finite time considering 81 cities, the clustering of cities was preferred before the optimization. 81 cities were divided into several clusters according to K-Means and K-medoids algorithms considering CO_2 emission costs and solutions were searched for these clusters. According to the CO_2 emission costs taken into consideration, the clustering of 81 cities by the K-Means algorithm with 8 clusters was found to be the most suitable method, and then these 8 clusters were optimized one by one. As a result, it was determined that the collection centers were established in which cities and their capacities. It was determined from which cities the collection centers will collect ELT. It was also determined that ELTs accumulated in the collection centers will be transported to which recycling facilities considering the demands of the recycling facilities.

4.2 Sub-Problem #2

In the solution of this sub-problem, it was determined by which routes the ELTs to be transported from cities to collection centers will be transported. This problem is a CVRP problem whose objective function is the CO_2 emission cost minimization. Toth and Vigo (2002)'s mathematical model was modified to solve the problem.

4.3 Sub-Problem #3

Up to this point, all components of the ELT collection system have been identified. In order to solve this sub-problem, it was determined which periods of operation of this system would be cost-efficient. The proposed ELT collection system was operated at different periods to minimize the transportation costs that would arise during the collection process. As a result of the simulation, the cost-efficient collection periods were determined. In this sub-problem, ELT supplies which were also considered in the solution of previous sub-problems were taken into consideration according to the Poisson process. Accordingly, transportation costs of scenarios in which collection is performed in 1, 3 and 6 month periods were compared.

5. Decision Support System for Tire Manufacturing Association Under Logistics 4.0 Concept

The Logistic 4.0 concept includes logistics operations in which data is collected, processed and automated in parallel with the Industry 4.0 concept. The Tire Manufacturers Association (TMA), which acts as an organizer in the proposed ELT collection and management system in this study, collects, processes and decides the data of the whole system. Firstly, ELT performs online tracking of vehicles loaded to collection centers. In this way, ELT can keep the flow and recording under control. In addition, the amount of ELT accumulated in the collection centers can be easily seen on the Turkey map. Collection centers are processed as the established sensor ELT amount online through the system. Officials who follow this map can easily monitor the flow of ELT from collection centers to recycling facilities. An example of the Turkey map indicating the fullness of the collection centers is given in Figure 1.

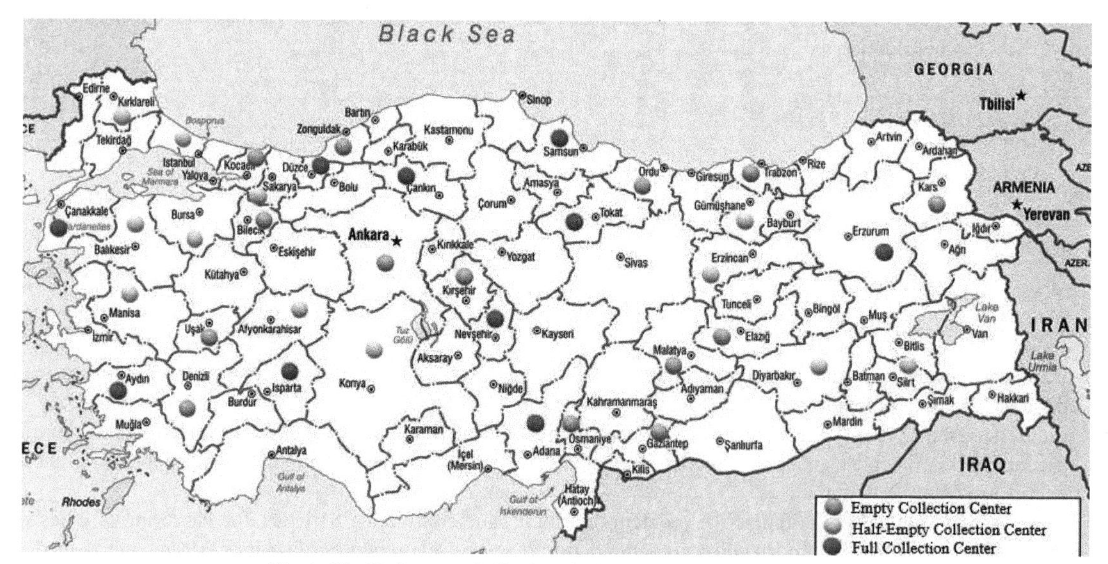

Fig. 1: The Turkey map indicating the fullness of the collection center.

5.1 Designed ELT Collecting System

The physical flow in the proposed system consists of the following steps:

1. Tire dealers receive ELTs from the ELT manufacturers (vehicle users).
2. Tire dealers collect and store ELTs for short time.

3. Authorized transportation companies collect the ELTs from the tire dealers periodically and bring them to the collection centers.
4. ELTs in collection centers are transported to the recycling plant by authorized transportation companies when reaching a certain amount.

Also, the data flow in the proposed system consists of the following phases:

1. The data of each unit of ELT, which is received from ELT manufacturers and delivered to a collection center, is entered into the database by tire dealers.
2. Intelligent sensors in the collection centers send signals to TMA and recycling plant when the collection center reaches the desired fullness.
3. Recycling plants periodically enter the actual requirement of ELT the database.
4. The entire data flow is monitored and controlled by the organizational unit which works under the umbrella of the TMA.

The physical and data flow of the proposed system is shown in Figure 2.

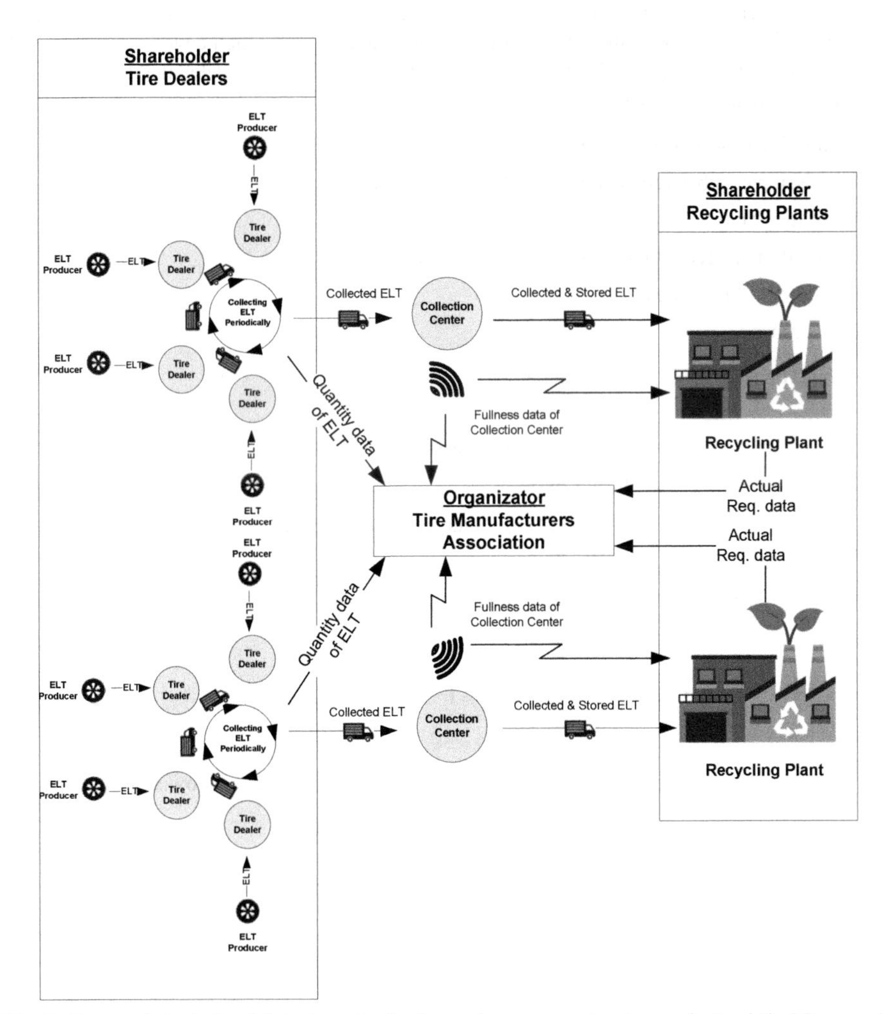

Fig. 2: Proposed physical and data flow of collecting and management system under Logistic 4.0 concept.

6. Proposed Solution Approach

In this section, details of the system designed for the solution of the problem and the solution methods used in each sub-problem are given step by step.

6.1 Methods Used to Solve Sub-Problem # 1

An MINLP model, which was proposed by Kır et al. (2019), was modified and adapted to the sub-problem # 1 for the solution. Since the problem was included in the NP-Hard complexity class, 81 cities discussed at this stage could not be optimized in finite time, thus clustering the cities were preferred. Two algorithms for clustering, K-Means and K-Medoids, are also described in this section. With these two algorithms, the most appropriate solution was obtained by creating 8 clusters in the K-Means algorithm.

6.1.1 Proposed MINLP

In the case study discussed in this chapter, ELTs of 81 cities were collected at determined collection centers and then transported to recycling facilities. It is not known where the collection centers will be established, what will be the capacity of these centers, ELTs of which cities will be transported to these centers and which collecting centers will send ELT to which recycling facilities. It is possible to identify these unknowns with the MINLP detailed below.

The parameters of the proposed MINLP are given below:

C_{ij} : CO_2 emission cost between i^{th} and j^{th} cities

CR_{ik} : CO_2 emission cost between i^{th} city and k^{th} recycling plant

O_j : Offer of the jth city

Dem_k : Demand of the kth recycling plant

N : Total number of the cities

M : A Big number

The variables of the proposed MINLP are given below:

$$\alpha_{ij} = \begin{cases} 1, \text{ if the } i^{th} \text{ toll center serves to } j^{th} \text{ city} \\ 0, \text{otherwise} \end{cases}$$

$$\beta_{ik} = \begin{cases} 1, \text{ if the } i^{th} \text{ toll center serves to } k^{th} \text{ recycling plant} \\ 0, \text{otherwise} \end{cases}$$

$$X_i = \begin{cases} 1, \text{ if a toll center is established in } i^{th} \text{ city} \\ 0, \text{otherwise} \end{cases}$$

$$Cap_i = \begin{cases} 0 \\ \sum_j \alpha_{ij} O_j, \text{ if } X_i = 1 \end{cases}$$

The objective function and constraints of the proposed MINLP were formulated as below:x

$$z_{min} = \sum_{i=1} \sum_{j=1} \alpha_{ij} C_{ij} + \sum_{i=1} \sum_{k=1} \beta_{ik} CR_{ik} \tag{1}$$

$$MX_i \geq Cap_i \qquad \text{for all I} \tag{2}$$

$$\sum_{j=1} \alpha_{ij} O_j = Cap_i \qquad \text{for all I} \tag{3}$$

$$\sum_{k=1} Dem_k \geq \sum_{i=1} Cap_i \tag{4}$$

$$MX_i \geq \sum_{j=1} \alpha_{ij} \quad \text{for all I} \tag{5}$$

$$MX_i \geq \sum_{k=1} \beta_{ik} \quad \text{for all I} \tag{6}$$

$$\sum_{k=1} \beta_{ik} \geq X_i \quad \text{for all I} \tag{7}$$

$$\sum_{i=1} X_i \leq N \tag{8}$$

$$\sum_{k=1} \beta_{ik} \leq 1 \quad \text{for all I} \tag{9}$$

$$\sum_{i=1} \alpha_{ij} = 1 \quad \text{for all j} \tag{10}$$

$$\sum_{i=1} \beta_{ik} Cap_i \leq Dem_k \quad \text{for all k} \tag{11}$$

$$X_i, \alpha_{ij}, \beta_{ik} \in \{0,1\}, Cap_i \geq 0$$

Equation (1) is the objective function of the model and minimizes the total CO_2 emission cost. Equation (2) and (3) provide to determine the capacity of the ith collection center if it is established in the ith city. Equation (4) provides that the total capacity of the collection centers to be established does not exceed the total capacity of the recycling plants. Equations (5), (6) and (7) determine the transportations that occur from cities to collection centers and from collection centers to recycling plants. Equation (8) limits the total number of collection centers by the number of cities. Equation (9) provides transportation from one collection center to only one recycling plant. Equation (10) provides transportation from one city to only one collection center. Equation (11) is non-linear and it limits the capacity of ith collection center by the demand (capacity) of related k^{th} recycling plant.

6.1.2 Clustering Analysis

Clustering is the process of grouping the information in the data set according to the similarity criterion. The name of the separated groups is the cluster. The data are separated from each other, provided that the elements within the cluster are similar in themselves and the groups are not alike. The main purpose of the clustering method is to minimize the similarities between the groups and to keep the similarities within the groups at the highest level (Berkhin 2002).

6.1.2.1 K-Means algorithm

The K-Means algorithm is one of the oldest clustering algorithms, and it was developed by MacQueen (1967). It is one of the most commonly used clustering techniques. Each data can belong to only one cluster and is, therefore, an incisive clustering algorithm. In this algorithm, the data set containing n data objects is divided into k sets. K-Means algorithm consists of 4 basic stages. Pseudo codes of the K-Means algorithm are given below:

Pseudo Codes of K-Means Algorithm

```
Begin:
        Determine the first set center
        Calculate the distance between the specified point and center.
        All objects are assigned to the set that is the closest to themselves
        while (if center points change)
         do
           {
New center points switch mean values of all objects in that set
           } end while
```

61.2.2 K-Medoids algorithm

The widely used K-Medoids algorithm was developed by Kaufman and Rousseeuw (1987). In this algorithm, the data set is divided into k sets. The main purpose of this clustering method is that the data in k clusters have high similarities and clusters are unique. The point closest to the center of the cluster is the representative object and is called the medoid. The representative object is the most central point of the cluster, which minimizes the average distance to the other elements of the cluster (Işık 2006). Pseudo codes of the K-Medoids algorithm are given below:

Pseudo Codes of K-Medoids Algorithm

```
Begin:
        Determine the number of k clusters
        Select k objects as initial medoids
        while (if there is a change)
        do
           {
           Assign the remaining objects to the nearest medoid x
           Calculate objective function
           Select randomly non-medoid y point
           }  end while
        If the displacement of x and y will minimize the objective function, exchange x, and y
```

6.2 Method Used to Solve Sub-Problem # 2

By solving sub-problem # 1, the locations of the collection centers and the cities from which they would collect ELT were found. However, it is not known which routes to follow while collecting ELTs from cities. By solving sub-problem # 2, the to follow when transporting from cities to collection centers were found out. The mathematical model developed by Toth and Vigo (2002) is used to solve the problem.
The parameters of the 0-1 ILP model are given below:

n : Number of customers

m : Fleet size

Q : Vehicle capacity

q(S) : Demand of customers in the subset S

e_{ij} : CO_2 emission between customer i and j

V : {0,…,n} and 0 is main depot

The variables of the 0-1 ILP model are given below:

$$x_{ij} = \begin{cases} 1 \text{ if a customer j immediately visits after a customer i using a vehicle} \\ 0 \text{ otherwise} \end{cases}$$

The objective function and constraints of the 0-1 ILP were formulated as below:

$$\text{Min } \sum_{i,j\in V} e_{ij}x_{ij} \tag{12}$$

$$\sum_{i\in V} x_{ij} = 1 \qquad \text{for all } j \in V/\{0\} \tag{13}$$

$$\sum_{j\in V} x_{ij} = 1 \qquad \text{for all } i \in V/\{0\} \tag{14}$$

$$\sum_{i\in V} x_{i0} \leq m \tag{15}$$

$$\sum_{j\in V} x_{0j} \leq m \tag{16}$$

$$\sum_{i,j\in S} x_{ij} \geq 2\gamma(S) \qquad \text{for all } S \subseteq V, S \neq \phi \tag{17}$$

$$\sum_{i,j\in S} x_{ij} \geq |S| - \gamma(S) \quad \text{for all } S \subseteq V, S \neq \phi \tag{18}$$

The objective function (12) minimizes total CO_2 emission. Constraints (13) and (14) guarantee that each customer is visited exactly by one vehicle. Constraints (15) and (16) impose a bound on a number of arrivals and departures for a depot. Naturally, this bound is a fleet size for customers and a depot. Constraint (17) is capacity constraints. $\gamma(S) = q(S)/Q$ and constraint (6) is called as fractional capacity inequalities. Constraint (18) eliminates all sub tours.

6.3 Method Used to Solve Sub-Problem # 3

In the proposed ELT collection system, the only question that wasn't answered until this stage is the period in which the system will be operated. For this purpose, the weekly average ELTs discussed in the first two sub-problems were defined according to random processes in this sub-problem and the ELT collection system was simulated at 1, 3 and 6 months periods. The assumptions considered in the simulation are as follows:

1. Inventory Holding Cost: Costs incurred during the physical storage of ELTs. The cost of having 1 ton of ELT in stock for 1 week in each city during the simulation was considered as an equal.
2. Transportation Cost: The ELT transportation operations are costed in accordance with the seasonal agreement also the transportation cost inside the city center is assumed as zero. The unit transportation cost was determined by taking the distances for each route into consideration.

In accordance with the assumptions described above, a cost-based simulation was performed. In the case of each route, ELT was collected in 1, 3 and 6 months periods, the total cost of ownership and transportation costs were compared.

7. Case Study: Proposed ELT Collecting and Management System Implementation in Turkey

The weekly ELTs that belong to 81 cities of Turkey were assumed as proportional to the population of the cities and the data in Table 1 was considered.

Depending on the distance and elevation of cities within the framework of the environmentalist approach and green concept, CO_2 emission cost calculation proposed by Nie and Li (2013) and detailed in Appendix 1 was used in this study. According to this, the carbon dioxide emission matrices of intercity and cities between recycling facilities are given in Appendix 2 and Appendix 3 respectively.

There are 26 recycling facilities that operate in tire recycling in Turkey. The weekly average ELT capacities of these facilities are given in Table 2.

In the previous section, the answer to the questions in which cities, how much capacity collection centers will be established, in which cities the ELTs will be collected and to which recycling facilities will be sent was searched. Before using the proposed MINLP, 81 cities were clustered into 8 different clusters by the K-Means algorithm. These 8 clusters are given in Table 3.

Each cluster given in Table 3 was solved one by one with the proposed MINLP. For example, it was found appropriate to establish a collection center with a capacity of 9 and 11 tons per week in Kars and Siirt cities, respectively, in 2 of the 13 cities given in cluster #1. Collected ELTs from 13 cities in cluster #1 would be transported to which one of the determined two collection centers and then they would be transported to which recycling facilities are shown in Table 4.

After the sub-problem # 1 of Cluster 1 was solved, sub-problem # 2 was solved and the routes where the ELTs would be collected by minimizing the CO_2 emission cost were determined. Since Cluster 1 has two different collection centers, there were two different routes. Route # 1 was determined as Kars-Igdir-Agri-Artvin-Ardahan-Kars for transportation to the collection center in Kars. Route # 2 was determined as Siirt-Bitlis-Muş-Van-Hakkari-Sirnak-Mardin-Batman-Siirt for transportation to the collection center in Siirt.

In Turkey the determined ELT collection centers and their capacities, the cities (81 cities) which these collection centers would collect ELTs from (with routes), and the recycling facilities to which these centers transport the ELTs are given in Table 5.

In the case of the proposed system being operated regularly, a simulation study has been carried out to determine which periods it would be appropriate to carry out the collection process with the lowest possible costs. For each route, 1, 3 and 6 months collection process was simulated one by one, and the most appropriate ELT collection period was determined by comparing the total costs of inventory holding and transportation costs.

Table 1: Weekly ELTs of 81 cities (Ton).

Cities	ELT	Cities	ELT	Cities	ELT	Cities	ELT
Adana	2	Edirne	1	Kutahya	1	Usak	1
Adiyaman	1	Elazig	3	Malatya	1	Van	2
Afyonkarahisar	1	Erzincan	2	Manisa	1	Yozgat	1
Agri	2	Erzurum	2	Kahramanmaras	2	Zonguldak	1
Amasya	2	Eskisehir	3	Mardin	1	Aksaray	1
Ankara	4	Gaziantep	2	Mugla	2	Bayburt	2
Antalya	1	Giresun	2	Mus	1	Karaman	1
Artvin	2	Gumushane	3	Nevsehir	1	Kirikkale	1
Aydin	1	Hakkari	2	Nigde	1	Batman	1
Balikesir	2	Hatay	1	Ordu	3	Sirnak	2
Bilecik	2	Isparta	1	Rize	1	Bartin	1
Bingol	3	Mersin	1	Sakarya	3	Ardahan	2
Bitlis	1	Istanbul	9	Samsun	3	Igdir	2
Bolu	2	Izmir	4	Siirt	1	Yalova	1
Burdur	1	Kars	1	Sinop	1	Karabuk	1
Bursa	3	Kastamonu	3	Sivas	1	Kilis	1
Canakkale	2	Kayseri	3	Tekirdag	1	Osmaniye	1
Cankiri	2	Kirklareli	1	Tokat	1	Duzce	1
Corum	1	Kirsehir	1	Trabzon	3	-	-
Denizli	1	Kocaeli	5	Tunceli	2	-	-
Diyarbakir	2	Konya	3	Sanliurfa	1	-	-

For example; cluster # 1 has two different routes. The total cost for route # 1 was found to be as 4950 USD. For this reason, it was deemed appropriate to carry out the collection for Route # 1 in 3-month periods. The total cost for route # 2 was found to be as 6484 USD. For this reason, it was deemed appropriate to carry out the collection for Route # 2 in 3-month periods. Costs obtained for each cluster for 1, 3 and 6 month periods are given in Table 6. The collection periods with the lowest cost were also indicated in the Table 6.

8. Conclusion

In this chapter, studied, firstly,the green concept and Logistic 4.0 for the collection and management of ELTs in Turkey was made a conceptual system design. In the proposed system, ELTs were thought to be delivered first by the end-users to tire dealers. ELTs collected at tire dealers were transported to collection centers to be established in certain cities. ELTs were collected in collection centers for a while and then transport to recycling facilities. In this process, data was collected and processed online and decision-making was done automatically. In the conceptual system, tire dealers and recycling plants were defined as shareholders, and The TMA was defined as an organizer. The TMA was also able to track and control all transportation data online, record, make automatic decisions and intervene when necessary.

These problems were divided into 3 different sub-problems according to their characteristics and were solved one by one. Sub-problem # 1 determines in which cities collection centers would be established, the capacity of these collection centers, ELTs would be transported from which cities to which collection centers, and ELTs would be transported from which collection centers to which recycling plants. Turkey's first solution to this problem for 81 cities was to divide them into 8 clusters, then each cluster dissolved CO_2 emissions costs were also taken into account and proposed one by one with MINLP. To solve this problem, firstly 81 cities of Turkey were clustered into 8 clusters, each cluster then was solved considering CO_2 costs by a modified MINLP. Sub problem # 2 attempts to determine the routes of the cities where ELTs would be collected from. To solve this problem, a CVRP model in the literature was modified to take into account CO_2 emissions. The sub-problem # 3 was to determine the most cost-efficient operation periods to ensure the sustainability of the system. This problem was solved using simulations under some cost assumptions.

Table 2: Recycling facilities and their capacities.

Recycling Plants (RP)		RP Cities	ELT Capacities (Week/Ton)
1	Barutcular	Konya#1	8
2	Tam Kauçuk	Aksaray	6
3	Naturel	Kocaeli # 1	8
4	Samsun Akın Rejenere	Samsun #1	3
5	Çetinkaya	Ankara	10
6	Katek Ağaoğlu	Usak	6
7	Hatko Dokuma	Osmaniye	10
8	Sami Çiftçi Lastik ve Kauçuk	Sakarya #1	7.5
9	Selçuk Kauçuk ve Plastik	Konya #2	12
10	Orbay	Izmir	4.5
11	Kahya Rejenere	Sakarya #2	6
12	Ün-Sal Danışmanlık	Adana	4.8
13	Çevre Hurda Kağıtçılık	Kocaeli #2	5.4
14	Ada Kauçuk San. Tic. Ltd. Şti.	Sakarya #3	3
15	Meta Kauçuk Endüstri	Kocaeli #3	8
16	Adk Kauçuk	Samsun #2	5
17	Ortadoğu Enerji	Malatya	6
18	Devasa	Kayseri	5
19	Tek	Bursa	4
20	Prokom	Erzincan #1	17
21	Kılınçkıran	Kahramanmaras	5
22	2 By	Kirikkale	30
23	Gazisan	Gaziantep	15
24	Prokom	Erzincan #2	5
25	Greenway	Canakkale	15
26	Akhisar Madencilik	Manisa	12

Table 3: Clustered cities.

Cluster #1	Cluster #2	Cluster #3	Cluster #4	Cluster #5	Cluster #6	Cluster # 7	Cluster # 8
Agri	Bingol	Amasya	Ankara	Adana	Bilecik	Afyonkarahisar	Aydin
Artvin	Diyarbakir	Giresun	Cankiri	Adiyaman	Bolu	Antalya	Balikesir
Bitlis	Elazig	Ordu	Corum	Gaziantep	Bursa	Burdur	Canakkale
Hakkari	Erzincan	Samsun	Kastamonu	Hatay	Eskisehir	Denizli	Edirne
Kars	Erzurum	Sinop	Kayseri	Mersin	Istanbul	Isparta	Izmir
Mardin	Gumushane	Sivas	Kirsehir	Malatya	Kocaeli	Kutahya	Kirklareli
Mus	Rize	Tokat	Konya	K.Maras	Sakarya	Usak	Manisa
Siirt	Trabzon		Nevsehir	Sanliurfa	Zonguldak		Mugla
Van	Tunceli		Nigde	Kilis	Bartin		Tekirdag
Batman	Bayburt		Yozgat	Osmaniye	Yalova		
Sirnak			Aksaray		Karabuk		
Ardahan			Karaman		Duzce		
Igdir			Kirikkale				

Table 4: Assignment of collection centers-cities-recycling facilities for cluster # 1.

Cluster #1	ELT	Collection Center	Recycling Plant
Agri	2		
Artvin	2		
Kars	1	Kars	Erzincan 1
Ardahan	2		
Igdir	2		
Bitlis	1		
Hakkari	2		
Mardin	1		
Mus	1	Siirt	Gaziantep
Siirt	1		
Van	2		
Batman	1		
Sirnak	2		

Table 5: Cities-collection centers-recycling facilities assignments.

Collection Centers	Cap.	Routes	RPs
Adana	3	Adana-Mersin-Adana	Adana
Afyonkarahisar	2	Afyonkarahisar-Kutahya-Afyonkarahisar	Usak
Ankara	4	Ankara	Ankara
Aydin	3	Aydin-Mugla-Aydin	Manisa
Balikesir	2	Balikesir	Canakkale
Bilecik	5	Bilecik-Eskisehir-Bilecik	Sakarya #2
Bursa	3	Bursa	Bursa
Canakkale	2	Canakkale	Canakkale
Cankiri	6	Cankiri-Corum-Kastamonu-Cankiri	Ankara
Denizli	3	Denizli-Antalya-Denizli	Manisa
Diyarbakir	5	Diyarbakir-Bingol-Diyarbakir	Kahramanmaras
Elzaig	5	Elazig-Tunceli-Elazig-	Malatya
Erzincan	2	Erzincan	Gaziantep
Erzurum	2	Erzurum	Gaziantep
Gaziantep	7	Gaziantep-Sanliurfa-Adiyaman-Kahramanmaras-Kilis-Gaziantep	Osmaniye
Gumushane	5	Gumushane-Bayburt-Gumushane	Erzincan #1
Isparta	2	Isparta-Burdur-Isparta	Usak
Istanbul	9	Istanbul	Kocaeli #3
Kars	9	Kars-Igdir-Agri-Artvin-Ardahan-Kars	Erzincan #1
Kirklareli	3	Kirklareli-Tekirdag-Edirne-Kirklareli	Canakkale
Kirsehir	3	Kirsehir-Kirikkale-Yozgat-Kirsehir	Konya #2
Kocaeli	6	Kocaeli-Yalova-Kocaeli	Kocaeli #1
Konya	4	Konya-Karaman-Konya	Konya #2
Malatya	1	Malatya	Malatya
Manisa	4	Manisa-İzmir-Manisa	Manisa
Nevsehir	6	Nevsehir-Kayseri-Nigde-Aksaray-Nevsehir	Konya #1
Ordu	5	Ordu-Sinop-Giresun-Ordu	Samsun #1
Sakarya	3	Sakarya	Sakarya #3
Samsun	4	Samsun	Samsun #2
Siirt	11	Siirt-Bitlis-Mus-Van-Hakkari-Sirnak-Mardin-Batman-Siirt	Gaziantep
Tokat	4	Tokat-Amasya-Sivas-Tokat	Kayseri
Trabzon	4	Trabzon-Rize-Trabzon	Erzincan #1
Usak	1	Usak	Usak
Zonguldak	3	Zonguldak-Karabuk-Bartin-Zonguldak	Sakarya #1
Osmaniye	2	Osmaniye-Hatay-Osmaniye	Osmaniye
Duzce	3	Duzce-Bolu-Duzce	Sakarya #1

Table 6: Lowest cost ELT collection periods for all routes.

Cluster #	Route #	Monthly			Quarterly			Once Every Six Months			Best Period
		IHC	TC	Total	IHC	TC	Total	IHC	TC	Total	
1	1	157	4,793	4,950	378	3,795	4,173	733	3,595	4,328	Quarterly
	2	181	7,132	7,313	450	6,035	6,485	850	5,760	6,610	Quarterly
2	1	75	850	925	183	784	967	349	719	1,068	Monthly
	2	33	0	33	81	0	81	145	0	145	Monthly
	3	68	794	862	165	672	837	295	611	906	Quarterly
	4	24	0	24	69	0	69	130	0	130	Monthly
	5	67	456	523	174	421	595	327	385	712	Monthly
	6	48	438	486	122	303	425	240	270	510	Quarterly
3	1	84	1,291	1,375	199	993	1,192	393	894	1,287	Quarterly
	2	44	0	44	106	0	106	190	0	190	Monthly
	3	108	2,044	2,152	274	1,887	2,161	543	1,730	2,273	Monthly
4	1	60	0	60	139	0	139	265	0	265	Monthly
	2	81	1,547	1,628	217	1,340	1,557	412	1,340	1,752	Quarterly
	3	123	1,277	1,400	302	1,094	1,396	558	1,094	1,652	Quarterly
	4	53	1,066	1,119	134	656	790	257	574	831	Quarterly
	5	58	695	753	148	481	629	291	481	772	Quarterly
5	1	39	1,116	1,155	94	687	781	183	601	784	Quarterly
	2	104	2,337	2,441	251	1,925	2,176	485	1,925	2,410	Quarterly
	3	27	696	723	71	348	419	126	290	416	Six Month
	4	13	0	13	36	0	36	66	0	66	Monthly
6	1	64	485	549	166	410	576	312	373	685	Monthly
	2	39	263	302	100	182	282	198	142	340	Quarterly
	3	40	0	40	103	0	103	193	0	193	Monthly
	4	122	0	122	307	0	307	559	0	559	Monthly
	5	73	409	482	178	350	528	329	321	650	Monthly
	6	43	0	43	102	0	102	201	0	201	Monthly
	7	44	806	850	98	372	470	204	310	514	Quarterly
7	1	40	1,285	1,325	95	890	985	183	692	875	Six Month
	2	25	298	323	66	115	181	117	115	232	Quarterly
	3	25	584	609	66	225	291	117	225	342	Quarterly
	4	13	0	13	33	0	33	61	0	61	Monthly
8	1	41	578	619	106	400	506	197	311	508	Quarterly
	2	27	0	27	76	0	76	143	0	143	Monthly
	3	28	0	28	73	0	73	137	0	137	Monthly
	4	46	943	989	101	435	536	197	435	632	Quarterly
	5	71	204	275	176	189	365	326	157	483	Monthly

In the light of the data taken into consideration in this chapter, it was determined that ELT collection centers would be established in 36 out of 81 cities based on their weekly capacities. It was determined that ELTs be transported to these collection centers according to which routes and the collection operations would be performed at the most cost-efficient periods. Finally, it was determined that ELTs were transported to which recycling plants from which collection centers. The operations in the whole system were defined step by step.

References

Ahmed, S., S. Ahmed, M.R.H. Shumon, E. Falatoonitoosi and M. A. Quader. 2016. A comparative decision-making model for sustainable end-of-life vehicle management alternative selection using AHP and extent analysis method on fuzzy AHP. Int. J. Sustain. Dev. World Ecol., 23: 83–97.

Aliahmadi, S.Z., F. Barzinpour and M.S. Pishvaee. 2020. A fuzzy optimization approach to the capacitated node-routing problem for municipal solid waste collection with multiple tours: A case study. Waste Manag. Res., 38: 279–290.

Beamon, M. 1999. Designing the green supply chain. Logist. Inf. Manag., 12: 333–342.

Beamon, B.M. and C. Fernandes. 2004. Supply-chain network configuration for product recovery. Prod. Plan. Control., 15: 270–281.

Berkhin, P. 2002. Survey of clustering data mining techniques, San Jose, California, USA: Accrue Software Inc.

Costa-Salas, Y., W. Sarache and M. Überwimmer. 2017. Fleet size optimization in the discarded tire collection process. Res. Transp. Bus. Manag., 24: 81–89.

Cruz-Rivera, R. and J. Ertel. 2009. Reverse logistics network design for the collection of End-of-Life Vehicles in Mexico. Eur. J. Oper. Res., 196: 930–939.

Dehghanian, F. and S. Mansour. 2009. Designing sustainable recovery network of end-of-life products using genetic algorithm. Resour. Conserv. Recycl., 53: 559–570.

Demirel, E., N. Demirel and H. Gökçen. 2016. A mixed integer linear programming model to optimize reverse logistics activities of end-of-life vehicles in Turkey. J. Clean. Prod., 112: 2101–2113.

ETRMA End of Life Tyres. A valuable resource with growing potential. http://www.etrma.org/tyres/ELTs.

Fagundes, L.D., E.S. Amorim and R. da Silva Lima. 2017. Action research in reverse logistics for end-of-life tire recycling. Syst. Pract. Action Res., 30: 553–568.

Işık, M. 2006. Data mining applications using partitional clustering methods. M.S. Thesis, Marmara University, Istanbul, Turkey.

Kannan, D., A. Diabat and K.M. Shankar. 2014. Analyzing the drivers of end-of-life tire management using interpretive structural modeling (ISM). Int. J. Adv. Manuf. Technol., 72: 1603–1614.

Karaağaç, B., M. Ercan Kalkan and V. Deniz. 2017. End of life tyre management: Turkey case. J. Mater. Cycles Waste Manag., 19: 577–584.

Kaufman, L. and P. Rousseeuw. 1987. Clustering by means of medoids. *In:* No. 87 in Reports of the Faculty of Mathematics and Informatics.

Kir, S., S.E. Comert, F. Yener, H.R. Yazgan and G. Candan. 2019. Hazardous waste recycling: end of life tires case. ACTA Phys. Pol. A., 135: 681–683.

Kopicki, R., M.J. Berg and L. Legg. 1993. Reuse and recycling—reverse logistics opportunities, United States: N. p.

LASDER. Tyre Industrialists Association. http://www.lasder.org.tr/turkiyede-otl/.

MacQueen, J. 1967. Some methods for classification and analysis of multivariate observations. Proc. Fifth Berkeley Symp. Math. Stat. Probab., 1: 281–297.

Ministry of Environment and Urbanization. 2015. Waste Management Regulations. https://www.resmigazete.gov.tr/eskiler/2015/04/20150402-2.htm.

Mora, C., A. Cascini, M. Gamberi, A. Regattieri and M. Bortolini. 2014. A planning model for the optimisation of the end-of-life vehicles recovery network. Int. J. Logist. Syst. Manag., 18: 449–472.

Nie, Y. and Q. Li. 2013. An eco-routing model considering microscopic vehicle operating conditions. Transp. Res. Part B Methodol., 55: 154–170.

Sahoo, S., S. Kim, B.I. Kim, B. Kraas and A. Popov. 2005. Routing optimization for Waste Management. Interfaces (Providence), 35: 24–36.

Sasikumar, P., G. Kannan and A.N. Haq. 2010. A multi-echelon reverse logistics network design for product recovery-a case of truck tire remanufacturing. Int. J. Adv. Manuf. Technol., 49: 1223–1234.

Schultmann, F., M. Zumkeller and O. Rentz. 2006. Modeling reverse logistic tasks within closed-loop supply chains: An example from the automotive industry. European Journal of Operational Research, 171: 1033–1050.

Simic, V. 2016. A multi-stage interval-stochastic programming model for planning end-of-life vehicles allocation. J. Clean. Prod., 115: 366–381.

TMMOB Union of Chambers of Turkish Engineers and Architects. 1994. Tire Industry and Petlas Sectoral Report Series. http://www.atikyonetimi.cevreorman.gov.tr.

Toth, P. and D. Vigo. 2002. The vehicle routing problem. *In:* Discrete Mathematics and Applications, Philadelphia.

Park, J., N. Díaz-Posada and S. Mejía-Dugand. 2018. Challenges in implementing the extended producer responsibility in an emerging economy: The end-of-life tire management in Colombia. J. Clean. Prod., 189: 754–762.

Pedram, A., N. Bin Yusoff, O.E. Udoncy, A.B. Mahat, P. Pedram and A. Babalola. 2017. Integrated forward and reverse supply chain: A tire case study. Waste Manag., 60: 460–470.

Yang, Y., J. Hu, Y. Liu and X. Chen. 2019. Alternative selection of end-of-life vehicle management in China: A group decision-making approach based on picture hesitant fuzzy measurements. J. Clean. Prod., 206: 631–645.

Zhou, F., M.K. Lim, Y. He, Y. Lin and S. Chen. 2019. End-of-life vehicle (ELV) recycling management: Improving performance using an ISM approach. J. Clean. Prod., 228: 231–243.

Appendix

Appendix 1—Fuel and CO_2 Emission Model

The comprehensive modal emission model (CMEM) is the emission model used in the calculation of the fuel consumption and consequently, the CO_2 emission of the vehicles, in which the emission amount increases or decreases depending on vehicle's physical and operational properties (weight, drag coefficient, air conditioning, etc.). When fuel consumption rates are known, It allows estimating CO_2 emissions based on carbon balance and the relationship between CO/HC emissions and fuel consumption.

CMEM converts the vehicle's total engine power (P) to fuel ratio (f). The conversion between these two parameters is shown as Equation (1).

$$f = \phi P / \lambda \tag{1}$$

The total engine power (P) consists of three components. These are as follows; P_t is the tractive power used to move the vehicle, P_w provides engine power against friction force and P_a provides the power of vehicle accessories such as air conditioning.

$$P = \Sigma_{i=0}^{3} \alpha_i . v^i + \beta . a . v \tag{2}$$

The methods of calculating the factors that affect P in Equation (2) are as follows;

$$a_0 = \frac{P_a}{\eta} \tag{3}$$

η is a constant parameter representing motor efficiency in Equation (3).

$$\alpha_1 = Z.g.\frac{G + c_1}{\eta.\varepsilon} + c_4.K_0.V.\theta.(\underline{r} + c_3.v^2); c_3 = \frac{\bar{r} - \underline{r}}{v^2}; c_4 = (1 + 0.0001.(N_1 - 33)^2) \tag{4}$$

$$\alpha_2 = Z.g.\frac{c_1}{c_2.\eta.\varepsilon} - 2c_3.c_4.K_0.V.v.\theta; \quad \theta\frac{1}{\pi.d.(1-j)} \tag{5}$$

$$\alpha_3 = \frac{\rho.c_d.A}{2.\eta.\varepsilon} + c_3.c_4.K_0.V.\theta; \quad \beta\frac{z.(1+e_0)}{\eta.\varepsilon} \tag{6}$$

The coefficients α_1, α_2 and α_3 used in the determination of total motor power are calculated according to Equations (4), (5) and (6). The value of e_0 given in Equation (6) is about 0.1.

If the vehicle is in the idling state (v = 0 and a = 0), the value of P_t is 0. Equation (7) is used to calculate the P_w parameter.

$$P_w = K_1.C.N_1 \tag{7}$$

In this case, the amount of power required in the idling state is of is calculated as in Equation (8).

$$P_1 = K_1.V.N_1 + (P_a / \eta) \tag{8}$$

The values obtained after the calculation of the total engine power are converted to the fuel rate and the CO_2 emission rate shown as e_{CO2} is estimated based on the carbon balance.

$$e_{CO2} = Ar(CO_2).\left(\frac{f - e_{HC}}{Ar(C) + \mu} - \frac{e_{CO}}{Ar(CO)}\right) \tag{9}$$

In Equation (9), the atomic weights of element C are expressed by Ar (CO_2), Ar (CO), Ar (C), CO_2 and CO compounds. The emission rates of HC and CO compounds are shown as e_{HC} and e_{CO}. μ is the hydrogen-to-carbon rate of fuel. The e_{HC} and e_{CO} values are obtained using the relationships in Equations (10) and (11).

$$e_{HC} = c_7.f + c_8 \tag{10}$$

$$e_{CO} = (c_5.(1 + \phi^{-1}) + c_6).f \tag{11}$$

Equation (12), which expresses a new relationship for the CO_2 emission rate, is obtained by regulating Equation (9). γ_0 and γ_1 in the Equation are shown in Equation (13).

$$e_{CO2} = \gamma_1.f + \gamma_0 \tag{12}$$

$$\gamma_0 = -\left(\frac{Ar(CO_2).c_8}{Ar(C)+\mu}\right);\ \gamma_1 = Ar(CO_2).\left(\frac{1-c_7}{Ar(C)+\mu} - \frac{c_5(1-\phi^{-1}).c_6}{Ar(CO)}\right) \tag{13}$$

Equations (1) and (9) are the fuel and CO_2 emission values in the unit of grams per second. It is often more appropriate to measure these emissions by distance. Equations (14) and (15) are used to calculate equations in the unit of grams per meter.

$$F(v,a) = \frac{f}{v} = \frac{\phi}{\lambda}(\Sigma_{i=0}^{3}\alpha_i v_i - 1 + \beta.a) \tag{14}$$

$$ECO_2(v,a) = \frac{e_{CO2}}{v} = \gamma_1.F(a,v) + \frac{\gamma_0}{v} \tag{15}$$

Vehicles that distributing and collecting release CO_2 on a set of nodes such as G (N,A) until they complete their tours. In this statement, N represents the customers for distribution and collection, and A is the distance between the two customers. Each of the links between the two nodes is a length l_b and the speed of the vehicle in this link is v_b. Accordingly, the vehicle's travel time and fuel consumption are as given in Equation (16).

$$t_b = \frac{l_b}{v_b};\quad F_b = F(V_b,0).l_b \tag{16}$$

CO_2 emission on link b is calculated as in Equation (17)[32].

$$E_b = F_b.y_1 \tag{17}$$

Appendix 2—The Carbon Dioxide Emission Matrices among Intercities

Cities	1	2	3	. . . .	79	80	81
1	0	752572	1291728	. . . .	552635	195444	1651166
2	752572	0	2044301	. . . .	471761	557128	2203801
3	1291728	2044301	0	. . . .	1844364	1487173	842431
.	.	.	.	. . . .	.	.	.
.	.	.	.	. . . .	.	.	.
.	.	.	.	. . . .	.	.	.
.	.	.	.	. . . .	.	.	.
.	.	.	.	. . . .	.	.	.
.	.	.	.	. . . .	.	.	.
.	.	.	.	. . . .	.	.	.
79	552635	471761	1844364	. . . .	0	357191	2143146
80	195444	557128	1487173	. . . .	357191	0	1846610
81	1651166	2203801	842431	. . . .	2143146	1846610	0

Appendix 3—The Carbon Dioxide Emission Matrices among Cities and Recycling Facilities

Recycling Facilities/Cities	1	2	3	. . . .	68	71	80
1	0	1105270	1891540	. . . .	599811	1071573	195444
2	752572	1666891	2502584	. . . .	1302961	1507391	557128
3	1291728	575100	622276	. . . .	819966	761558	1487173
.	.	.	.	. . . .	.	.	.
.	.	.	.	. . . .	.	.	.
.	.	.	.	. . . .	.	.	.
.	.	.	.	. . . .	.	.	.
.	.	.	.	. . . .	.	.	.
.	.	.	.	. . . .	.	.	.
.	.	.	.	. . . .	.	.	.
79	552635	1606236	2439682	. . . .	1152446	1446736	357191
80	195444	1300714	2086984	. . . .	795255	1267017	0
81	1651166	530170	514445	. . . .	1051354	705396	1846610

SECTION 12

Management of Digital Transformation in SCM

CHAPTER 21

The Roles of Human 4.0 in the Industry 4.0 Phenomenon

Nurcan Deniz

1. Introduction

Increasing numbers of products, shorter product life cycles, one lot sizes, fulfilling customer expectations, and increasing pressure for innovative and global supply networks are the major challenges of manufacturing companies. Digital economy and society research program was launched by German Federal Government to cope with the current and future business challenges as a vision in the Hannover Fair (2011) by three German engineers—a physicist Henning Kagermann, an artificial intelligence professor Wolfgang Wahlster and another physicist Wolf-Dieter Lukas (Kirazli and Hormann 2015; Deniz 2017). World Economic Forum's (2016) 'Mastering the Fourth Industrial Revolution' motto shows the quick spread of Industry 4.0 (Pfeiffer 2017).

Computer Integrated Manufacturing (CIM) and Lean Production that emerged in the third revolution are accepted as the messengers of the fourth industrial revolution (Kirazli and Hormann 2015). Wahlster (2014) states that there is a transformation from the era of mathematization to the era of informatization in medical, media, energy, law, automotive, computational linguistics, biotechnology, and neuroinformatics sciences. Schuh et al. (2014) stress that the fourth revolution's source is society contrary to the other revolutions that took place in the industry (Magruk 2016). In the fourth industrial revolution, manufacturers begin to integrate automation, robotics, and other data-driven technologies into their workflows (Wuest and Romero 2017). In this context, smart chains are developed based on the communication between production, products, components, plants, and human (Magruk 2016).

To create a common understanding is seen one of the greatest challenge in the Industry 4.0. Because, as a young term it is used for the wrong purpose and in the wrong context generally. Internet of Things (IoT) and Internet of Services (IoS), embedded systems, cyber-physical systems, smart factory and big data are the terms that Industry 4.0 is confused about

Eskisehir Osmangazi University, Faculty of Economics and Administrative Sciences, Department of Business Administration, Eskisehir, TURKEY, Email: nurcanatikdeniz@gmail.com, ndeniz@ogu.edu.tr

(Kirazli and Hormann 2015). The 'smart' term is used to demonstrate intelligence and knowledge with reference to the applications of Industry 4.0 (Lu 2017). Furthermore, practical implementation is an uncertain issue (Kirazli and Hormann 2015). Magruk (2016) claims that there is a high level of uncertainty from economic, technological, social, and legal aspects.

As a subclass of digital transformation (Shamim et al. 2016) Industry 4.0 is defined as "*the systematic development of an intelligent, real-time capable, horizontal, and vertical networking of human, objects and systems*" (Kirazli and Hormann 2015). According to this definition, Industry 4.0 combines digital, physical, and biological systems (Strozzi et al. 2017). Industry 4.0 is based on four key components as cyber-physical systems (CPS), IoT, IoS, and smart factory. Aforementioned technologies are based on perpetual communication via internet. Human and human, human and machine, machine and machine interaction, and exchange of information become possible in this context (Roblek et al. 2016). The major features of the Industry 4.0 are described as digitization, automation and adaptation, optimization, customization of production, human-machine interaction, value added services and businesses, and automatic data exchange and communication (Lu 2017).

In another classification human, systems and management are the basic components in Industry 4.0. Intelligence, knowledge, creativity, decision making, judgment, innovation, brain ware, intuition, and human capital represents the human component. Systems component is characterized by data, information, structure, optimization, organization, resource allocation, communication, information and communication technology (ICT). Finally, coordination, knowledge management, teamwork, goal setting, strategy, tradeoffs, leadership, self-management, and motivation signify the management component (Mládková 2018). Especially, the interaction between physical and the cyber elements in CPS is extremely important. Smart manufacturing, smart electric grid, autonomous cars, robotic surgery, intelligent buildings, and implanted medical devices are some of the examples to show the life changing potential of CPS (Monostori et al. 2016).

Industry 4.0 is not only related with internet technologies and advanced algorithms, but also it deals with value adding and knowledge management (Lu 2017). Unfortunately, most of the recent studies performed are based on the technological perspective of Industry 4.0 (Shamim et al. 2016). By the same token, Lu (2017) indicates that the human side is misanalysed; in the way the development of industry is defined as complexity and agility integration. Richert et al. (2016a) arouse interest on preparing and training people for Industry 4.0 (Deniz 2017). It is believed that the education is the most important step in this transformation (Lu 2017). As a fresh and highly complex issue (Magruk 2016), Industry 4.0 is still a vision and the concept has to be extended according to the future needs (Kirazli and Hormann 2015).

In spite of the full automation concept, human is still one of the most important dimension in this debate. Richert et al. (2016) define human as a swimmer along the wave of Industry 4.0 but human related studies are still limited (Kadir et al. 2019). As a consequence, this chapter aims to arouse interest of the researchers related with Industry 4.0 for the human element in a holistic way. Human's position and importance in terms of a citizen, customer, and worker will be studied and there will be a discussion about the future. Smart people, Customer 4.0, Patient 4.0, and Operator 4.0 are the different roles of the Human 4.0 in Industry 4.0. According to this fact, human element needs to be integrated by the researchers in their projects in the early stages of the implementation. It is important not to replicate the same ignorance of human element as in the third industrial revolution. To realize this, an interdisciplinary understanding for Industry 4.0 is needed (Gorecky et al. 2014).

In the second part of the chapter, there is a short literature review about the studies especially focusing on the human element in digital transformation. In the third part, different contexts in which human play a role are analyzed. Smart people as a citizen of smart cities, Customer 4.0 in smart supply chains and Marketing 4.0, Operator 4.0 in Factory 4.0 and Patient 4.0 in Healthcare 4.0 contexts are the four dimensions discussed in the third part. The conclusion and the future studies are discussed in the last part.

2. Literature Review

After the launch of Industry 4.0, researchers started to study this concept. Though there is a large interest on its technological side, there are fewer studies related to the human element. Some of them are listed in a chronological order in this section to summarize the related literature to the readers.

Jara et al. (2012) presented the key elements from the book Marketing 4.0, such as several technologies from the IoT to enable the interaction of the user with the products and internet. Solanas et al. (2014) introduced the new concept of smart health. This concept complements mobile health within smart cities. Gorecky et al. (2014) sought the solutions for technological assistance of workers. An augmented reality system that supports human workers in a rapidly changing production environment is presented by Paelke (2014).

Richert et al. (2016b) made an experimental study about factors that influence hybrid team (robots and human) development. The article of Shamim et al. (2016) differs from the recent articles as being one of the initial attempts about management practices in Industry 4.0. These practices are playing an important role in the development of effective learning, dynamic capabilities, and innovation climate. Magruk (2016) analyzed economic, social, technological, and legal dimensions of uncertainty in Industry 4.0. Hecklaua et al. (2016) described a strategic approach for employee qualification. This approach deals with shifting employee's capacities to workspaces with more complex processes. The conceptual article of Roblek et al. (2016) aimed to synthesize theory and practices of Industry 4.0. Moreover, they investigate the changes related with the development of the IoT. The industrial engineering curriculum was analyzed according to the Industry 4.0 in the South African context by Sackey and Bester (2016). Schumacher et al. (2016) proposed an empirically grounded novel model to assess the Industry 4.0 maturity of industrial enterprises. This model consists of 62 items under 9 dimensions: technology, products, operations, customers, strategy, leadership, culture, governance, and people. The role of the human in Industry 4.0 is considered using a best practice approach in Nelles et al.'s (2016) study. Grigoriadis et al. (2016) analyzed the application of health 4.0 to multiple sclerosis patients.

Longo et al. (2017) proposed SOPHOS-MS in which the human is in the center of the framework. Benesova and Tupa (2017) collected qualifications and skills for both IT and production job profiles. Prasetyo and Arman (2017) studied on designing a group management system implementation in smart society. Wang and Zhang (2017) provided solutions to society-centered Intelligent Transportation Systems (ITS) called Transportation 5.0. Theummler and Bai (2017) showed the usability of different Health 4.0 applications such as massive IoT (mIoT) and 5G for asthma treatment.

The paper of Ardito et al. (2018) presented innovative efforts undertaken to manage the interface between marketing and supply chain management processes in the context of digital transformation. Galletta et al. (2018) focused on customer loyalty programs and proposed cloud-based software to store and analyze big data related to purchases and products' ranks. It is possible to provide a list of recommended products for customers by this way. Smaradottır et al. (2018) developed a decision support tool based on International Normalized Ratio (INR) called 'Warfarin Guide' value to help cardiovascular patients. Patients declared their desire to use this mobile phone application after the study. Dubgorn et al. (2018) implemented a telemedicine project in the Healthcare 4.0 context. Pang et al. (2018) provided a brief history of key enabling technologies of Industry 4.0, and its revolution in healthcare. Kumari et al. (2018) proposed a patient-driven healthcare architecture as a combination of real-time data collection, processing, and transmission. Additionally, they gave insights to the end users for the applicability of gateways and fog devices. The strategic position of such gateways was exploited by Rahmani et al. (2018). Razaghi and Finger (2018) incorporated governance literature with socio-technical systems and systems theory. Mládková (2018) discussed the relationship between humans and technology in the aviation industry. Ten Bulte (2018) conducted a document analysis consists of 38 policy documents from 10 different European Industry 4.0 initiatives to investigate the impact of Industry 4.0 on Human Resources (HR) practices. Furthermore, interviews were conducted with experts to examine the impact of Industry 4.0 on HR practices. Taylor et al. (2018) expanded the ideas behind Industry 4.0 with the aim of enhancing human work simultaneously manufacturing innovation and safety.

Kadir et al. (2019) presented the findings of a systematic literature review, consisting of both qualitative and quantitative data, about the integration between human factors and Industry 4.0. Ramingwong et al. (2019) focused on investigating the human resource and the human resource management related to Industry 4.0. They explored Thai industry's organization and management potential according to the indicators selected from the World Economic Forum's Global Competitiveness Report. Ardanza et al. (2019) presented a hardware and a software architecture to build flexible advanced Human Machine Interfaces (HMI) that provide adaptable and useful information to the machine operators.

3. The Roles of Human 4.0 in the Industry 4.0

Human 4.0 have different roles in the context of Industry 4.0. First of all, Human 4.0 are smart citizens living in a smart city and smart society. Secondly, they are Customer 4.0 for Marketing 4.0. In the smart factory context a smart worker or Operator 4.0 gain new skills to survive. Lastly, patients become smarter in Healthcare 4.0. All of these roles are studied under subsections.

3.1 Smart People as Citizens of Smart Cities

Smart city is one of the applications of the smart environment. Smart homes, smart transportation, smart grids, and smart healthcare are the other applications. Improved quality of life, sustainability, resilience, governance, intelligent management of city facilities and natural resources are some of the main smart environment characteristics (Hashem 2016).

Bakıcı et al. (2012) define the smart city as "*a high-tech intensive and advanced city that connects people, information, and city elements using new technologies in order to create a sustainable, greener city, competitive and innovative commerce, and an increased life quality.*" It is stated as an "*instrumented, interconnected, and intelligent city*" in an IBM corporate document (Albino et al. 2015). On the other side, Roblek et al. (2016) define the smart city as a combination of smart economy, smart environment, smart mobility, smart living, smart people, and smart governance factors. By the way, there is not a consensus on the definition of 'smart city'. Intelligent, digital, virtual, and ubiquitous are some of the alternative adjectives used for this fuzzy concept. But in all of these alternatives 'humans' are ignored (Albino et al. 2015).

According to the association of the components of a smart city in Lombardi et al.'s (2012) study; smart economy is related with economy, smart governance is related with e-democracy, smart mobility is related with logistics and infrastructure, smart people is related with education, smart environment is related with efficiency and sustainability, and smart living is related with security and quality. Although the quality of life is seen as a dimension of smart city, Shapiro (2006) indicates that the quality of life is an overall objective of smart cities (Albino et al. 2015).

Albino et al. (2015) show that the smart city concept looks not only like diffusion of ICT, but also of community and people needs. When the term smart city is applied to 'hard domains', ICT plays a decisive role in the functions buildings, water management, logistics and mobility, natural resources, energy grids, and waste management. Subsequently education, policy innovations, culture, government, and social inclusion are the 'soft domains' where the applications of ICT are not usually decisive (Albino et al. 2015). Shapiro (2006) and Holland (2008) indicate that rather than blindly assuming that ICT can automatically create a smart city, smarter cities start from the human capital side (Albino et al. 2015). In a similar vein Caragliu et al. (2011) present quality and availability of social infrastructure and knowledge communication for urban performance along with the physical infrastructure (Hashem 2016).

Thuzar (2011) explains that a smart city is, "*where all residents, including the poor, can live well.*" This statement incorporates the ultimate goal. To emphasize on creativity as a key driver of smart city is another detail in this chapter. According to this finding, knowledge, education, and learning are introduced as the central roles in a smart city. Intellectual and social capitals are the social infrastructures presented in this context (Albino et al. 2015). Dirks et al. (2010) identify that being smart, clever, creative, skillful, interactive, competitive, and connected become key ingredients of knowledge based urban development (Albino et al. 2015).

The aims of the smart city are ordered by Lu (2017) as to make cities sustainable, to improve life quality, to improve citizen's safety, and to provide energy efficiency. Smart people—as the citizens of smart city—are influenced by the applications of the IoT. Smart infrastructure is one of the applications of IoT that reduces costs, enhances safety and reduces manpower requirements. For example, smart people can gain control over door locks with remote devices from any internet connected source, control the supply of food in the refrigerator, and adjust a thermostat with smartphone applications (Roblek et al. 2016). As a consequence of the transformation from traditional city to smart city, the citizen's role is shifted from users to key stakeholders (Lu 2017). According to Nam and Pardo (2011) and Albino et al. (2015), flexibility, affinity to lifelong learning, social and ethnic plurality, open-mindedness, cosmopolitanism, creativity, and participation in public life are some of the important characteristics of the smart people. Availability of data as well as cheap and strong processing power, occurrence of new data processing technologies and processing logics, emergence of new channels of expression and communications of opinions are the technological aids in the digital transformation (Razaghi and Finger 2018).

Another important dimension and bottleneck of the smart city is governance. To deal with increasingly multifaceted urban challenges and to make a collaboration among a large number of actors, governance is presented as a main hindrance for utilizing the benefits of technology. Citizens can reach their voices with new communication channels and share their personal opinions instantly (Razaghi and Finger 2018). Citizen-centric and citizen-driven structure is the spirit of e-governance (Albino et al. 2015).

Transportation is a worthy example to show the importance of governance. With the effect of the change from traditional to smart cities, transportation becomes a smart service with advanced planning, effectiveness, and efficiency (Lu 2017). Sharing economy principles have started to be applied in transportation area (Razaghi and Finger 2018). According to the 'mobility' doctrine, transportation services need to be integrated. Thanks to the technological benefits of Industry 4.0, it is possible to integrate different modes of transportation—bus, taxi, private cars, regional rail network, and metro- contrary to the traditional silos. Major bottlenecks of this integration are ticketing and scheduling. Also, real time data and control tools give chance to make a dynamic traffic control based on traffic situation. As an example, if traffic congestion leads to distortion in the bus sequence, it is possible to send additional fleets to the site. Additionally, passengers have chance to change their route according to the introduction about delay (Razaghi and Finger 2018).

Intelligent Transportation Systems (ITS) consist of dual perspectives of both society and technology. Automatic traffic management systems, traveler information systems, transportation information services, electronic payment systems, fleet management and location systems, and cooperative vehicular systems are some of the services ITS include. Intelligent parking systems and traffic flow prediction are the other activities in this system (Wang and Zhang 2017). Reducing the number of accidents, providing alternative routes, reducing the environmental impact and increasing safety are some of the other benefits of smart transportation (Hashem 2016).

Hong Kong and Seoul are the pioneer cities in which passengers use all modes of public transportation with a single transportation card and are kept informed with apps or public screens about different modes of transportation. Helsinki is another example in which 'mobility as a service' is discussed (Razaghi and Finger 2018). Wang and Zhang (2017) state that 'Parallel Transportation Management System' in Qingdao Shandong Province of China has received 'IEEE, ITS Outstanding Application Award' as a best practice in Transportation 5.0. Transportation 5.0 is defined as a paradigm shift from technology-centered ITS to society-centered ITS.

To determine a city's smartness is a complicated issue. The capacity of clever people to generate clever solutions to urban problems is determined as a performance indicator (Albino et al. 2015). On the other side, The University of Vienna developed an assessment metric to rank 70 European medium-sized cities. Foreign language skills, percentage of population with secondary-level education, individual level of computer skills, participation in life-long learning, patent applications per inhabitant are some of the indicators of smart people (Albino et al. 2015). Intelligent Community Forum's assessment is based on broad band connectivity, innovation, digital inclusion, a knowledgeable workforce, marketing, and advocacy. Zygiaris' (2013) measurement system differs from the others as it uses six layers to identify the smartness: green city, interconnection, open integration, instrumentation, innovation, and application. The exclusion of the smart mobility dimension is criticized by Lombardi et al. (2012). The Global Power City Index developed by Japanese Institute for Urban Strategies, Natural Resources Defense Council of Unites States' Smarter Cities Ranking and IBM's Smart City are the other initiatives to assess smartness of cities (Albino et al. 2015).

Stockholm started to use smart applications to address environmental and traffic issues. Helsinki opened public data such as transportation, economics, employment, conditions, and well-being to all stakeholders—government, citizens, academicians, businesses, and research institutions. The eighth smarter European city Copenhagen, aims to become the world's first carbon-neutral capital by 2025 (Hashem 2016). Masdar City was built according to the eco-city paradigm. Seattle in United States, Quebec City in Canada, Friedrichshafen in Germany, Songdo in Korea, and Beijing in China are the other leading smart cities (Albino et al. 2015). Approximately 143 ongoing or completed self-designated smart city projects were recorded at the beginning of 2013.

Nonetheless, there are some research challenges in the smart city context. These can be related to business challenges, planning, sustainability, cost of acquiring smart city, source of customer and markets, technological challenges, cloud computing integration, privacy, big data analytics, data integration, Geographic Information System (GIS) based visualization, and computational intelligence algorithms for smart city (Hashem 2016). On the other side, numerous systemic feedback loops, difficulty to collect relevant data from each subsystem, difficulty of imposing certain courses of action on citizens, adaptive behaviors in the system, difficulty to collect relevant data from the overall system, and emergent behavior of the system are the other challenges of smart cities (Razaghi and Finger 2018).

Beyond smart city, the term 'smart community' needs to be defined. In 2016, the Japanese Cabinet proposed an initiative called 'Society 5.0' in 5th Science and Technology Basic Plan. The creation of a 'Super Smart Society' was the vision of this plan. After hunter/gatherer, pastoral/agrarian, industrial, and information stages, the super smart society is placed as the fifth developmental stage (Shiroishi et al. 2018). Society 5.0's ultimate goal is not to improve technology alone, but also to improve the quality of human life. In this context, smart society's aim is to enhance the quality of human life with the potential of digital technology, digital devices, and networks (Prasetyo and Arman 2017). The promotion of job growth and economic development are the other important goals of smart communities (Shiroishi et al. 2018).

Utilization of remote sensing, e-learning system, smart agriculture, smart food, empowerment of women, early warning alert system, smart grid system, and smart cities are some of the sustainable development goals in Society 5.0 (Shiroishi et al. 2018). ITS, energy value chains, new manufacturing systems, regional inclusive care systems, infrastructure maintenance and updates, society resiliency against natural disasters, new business and services, hospitality systems, global environment information platforms, integrated material development systems, smart food chain systems, and smart manufacturing systems are the service platforms to create a Super Smart Society (Shiroishi et al. 2018).

3.2 Customer 4.0 in the Smart Supply Chains and Marketing 4.0

Human also play a key role in supply chain management (SCM) and marketing. Formerly, SCM and marketing were accepted as independent functions. This misunderstanding caused efficiency problems both in production and distribution. Thanks to the Porter's (1985) value chain framework, a close relationship between SCM and marketing functions is noticed. At this point, digitization of firm processes is accepted as the most effective solution to strengthen this relationship. Cutting-edge digital technologies such as cloud computing, and big data analytics give the firms opportunity to improve their capacity to acquire, analyze, and distribute market/operational knowledge. Improving service levels, lowering procurement costs, forecasting the demand and replenishment quantity, and reducing inventories are SCM related functions. Marketing functions are improved customer relationship management, predictive analytics, customer profiling, and targeted marketing (Ardito et al. 2018).

Smart customers gain more power on the smart supply chains with the aid of digital transformation. Schrauf and Berttram (2016) define smart supply chains as "*the chain that becomes a completely integrated ecosystem which is fully transparent to all the players involved—from the suppliers of raw materials, components, and parts, to the transporters of those supplies and finished goods, and finally to the customers demanding fulfillment*" (Ardito et al. 2018). Customer co-creation/open innovation under the supply/demand match value driver is one of the levers of human factor. The automation's important role in closed loop systems to manage the data-driven process quality control and management systems is also important (Sackey and Bester 2016).

Supply chains/logistics application of IoT give a chance to the smart customer to trace products with real time information (Roblek et al. 2016). The digital transformation affects the relationship between customers, producers, and suppliers. Instead of pushing systems in which producing decisions are given by manufacturers and retailers, customers are more involved in quality related decisions and the customization of products in pull systems (Lu 2017). Acquisition, exchange, and elaboration of market/operational knowledge in a well-timed mode are the key factors that became successful both in SCM and marketing (Ardito et al. 2018).

From the product perspective, it is possible to produce in a more flexible, cost-efficient, and time efficient way in the context of Industry 4.0 (Lu 2017). Based on demand oriented production, smart products can be defined as "*highly differentiated customized products, well-coordinated combination of products and services, the value added services with the actual product or service, and efficient supply chain*" (Shamim et al. 2016). It is indicated that smart products need to be created with high technologies in physical and digital processes (Lu 2017). Smart products are subcomponents of the cyber-physical systems and are not considered without the machine to machine interaction—an enabler of the IoT (Roblek et al. 2016).

Customer 4.0 experience innovations about smart products through the digitization processes and full automation, and the use of information technologies (IT) and electronics in Industry 4.0. First of all, Customer 4.0's awareness about quality and reliability will increase. 3D printing technology make possible the development of online sale services such as car services, ordering food directly sent from the store to the refrigerator and medical examinations from home (Roblek et al. 2016). Zhang et al. (2014) point out that user's participation in production transform from partially to fully (Magruk 2016). Companies gain the chance to produce new value added products that can rapidly reach the market with the Industry 4.0 (Galletta et al. 2018).

Dominici et al. (2016) provide that Customer 4.0's behavior change with the Human 4.0 and CPS, IoT, and IoS integration (Roblek et al. 2016). According to the evolution of marketing, it is easily seen that user requirements are changed. The only requirement was the 'needs' in Marketing 1.0. In this product-centric era, the marketing was basically focused on selling products. Marketers mainly focused on the art to persuade the customers and to increase sale. Thanks to the information and communications age, 'wants' came to the order in Marketing 2.0. By this way, consumers started to be well informed and compare several values offerings of similar products. 'Anxieties', 'desires','creativity', and 'values' are added to the list in Marketing 3.0 which is defined as the human-centric era. Lastly, 'participate' and 'validate' are the new requirements that exist in Marketing 4.0. Marketing moves around the customers not around the product as in the past, in accordance to the smart customers' desire to be a part of the smart products (Jara et al. 2012).

In addition to interaction of customer and product, thanks to the social networks,the interactions between customers are increased (Jara et al. 2012). Smart customers get a chance to share their unique experiences about the products. The capacity to consult the experiences from other customers in this customer-centric era is not the only benefit. The capacity to confirm, check and validate the features and promises from the product is another benefit of this interaction. Value is created in a collaborative way, where the customer is able to check and confirm. Brand identity, brand image, brand interaction brand integrity, and reputation are the important marketing terms affected from the digital transformation in marketing. Internet is the key driver in the marketing evolution but it is not the only technological development. Near Field Communications (NFC), Radio Frequency Identification (RFID), and Low Power Area Networks (6LoWPAN) are some of the main technologies utilized in IoT for products (Jara et al. 2012).

3.3 Operator 4.0 in the Factory 4.0

Dujin et al. (2014) define CPSs and marketplace, big data, smart robots and machines, energy efficiency and decentralization, new quality of connectivity, skills and virtual industrialization as the key determinants of the new industrial landscape (Magruk 2016). Human factor is seen as the magnitude of the smart manufacturing in which physical and the cyber contexts act synergistically. This importance is stated as being in the center of a virtuous closed-loop chain with a valuable feedback system. This system makes the whole manufacturing system grow and evolve over time. Cyber context is related with cyber twin by means of Cloud Computing and IoT (Longo et al. 2017). However, human integration in smart factories is scarcely studied. There is need to draw attention to human factors as a key element to address new and volatile behaviors in Intelligent Manufacturing Systems (IMS) (Longo et al. 2017).

Labor is accepted as one of the Industry 4.0 value drivers. Human-robot collaboration, digital performance management, remote monitoring control, and automation of knowledge work are the levers of labor (Sackey and Bester 2016). Human operator's experience is required to be highly flexible and they need to exhibit adaptive capabilities in a very dynamic and complex working environment (Longo et al. 2017). Weyer et al. (2015), specify the Industry 4.0 through the smart machine, the smart product, and the augmented operator (Magruk 2016). Super-strong, super-safe, super-informed, and constantly connected are the features of Operator 4.0—tech-augmented human workers. Operator 4.0's strength can be enhanced by wearing robotic exoskeletons. As a consequence,it can be possible to control the large robot's physical power and become super-strong. Super strong Operator 4.0 can handle enormously heavy objects easily without losing natural human's flexibility and still have energy to spend time with friends or to play with the kids at the end of a day. In addition, Operator 4.0 will suffer severe injuries from accidents or overwork.

An 'augmented operator' is super informed with augmented reality. Smart glasses are the tools by which Operator 4.0—no matter where he or she is looking—can receive step-by-step and individualized instructions that are displayed (Wuest and Romero 2017). It can be possible for operators to accomplish the unfamiliar tasks like assembly of new products with augmented reality system and spatially registered information (Paelke 2014). Wearable sensors track pulse rate, chemical exposure, body temperature or other factors. By this way, the risks of injury can be estimated to make the Operator 4.0 healthier and super safe (Wuest and Romero 2017). Today, thanks to the increased digitization and networking, people access any information in daily life and react accordingly. Smartphones and fitness trackers, vehicle navigation, and driving assistance systems are some of the products Operator 4.0 use (Nelles et al. 2016).

Teams that consist of human and machines in Industry 4.0 are called 'hybrid teams'. Robots started to support the human physically in third industrial revolution. It is expected that the meaning of teamwork will be redefined by smart robots and hybrid human-robot teams. In the fourth revolution it is expected that the robots are able to identify and adapt to any individual strengths/weaknesses. There are differences between regular human teams and hybrid teams. Defining certain conditions where cooperation is necessary to sustain the continuous coordination is important in hybrid teams whereas the heterogeneity of members, different backgrounds and ways of thinking and acting increases the productivity of the teams in former. Cooperation behavior, reciprocal influencing and joint reasoning are the main research areas in human-robot interactions. Sproull (1996) indicates that social and communication perspective of this interaction should be analyzed (Richert et al. 2016). Robots and Human 4.0 work more closely together than ever before in hybrid teams. Human 4.0 use their unique abilities to innovate, collaborate, and adapt to new situations. They need to handle challenging tasks with knowledge-based reasoning. A "collaborative operator" may be linked to "collaborative robots" (Wuest and Romero 2017). Supported by big data and cloud computing in the industrial value chains, connected human and robots networks interact and work together (Lu 2017). The conditions for the successful implementation of hybrid teams, transferring the knowledge from the development of human teams into the design of hybrid teams and shaping of human-computer-interaction are the issues emphasized on to design Work 4.0 (Richert et al. 2016). Modern human machine interactions can change the decision making process, involving the operators who access real time data and become empowered with appropriate tools (Ardanza et al. 2019). There is an important point to stress that, human can solve failures and problems of technology using the appropriate knowledge. On the other side, even the best technology cannot eliminate human mistakes (Mládková 2018).

Not only Industry 4.0 creates many new opportunities for companies but also there are emerging challenges in the Industry 4.0 context. These challenges can be classified under social, economic, environmental, technical, legal, and political challenges. Markets have become increasingly unpredictable and heterogeneous with the higher level of flexibility and customization, in the context of economic challenges. Increasing need for innovation, ongoing globalization, demand for higher service orientation, growing need for cooperative and collaborative work are the other economic challenges. Demographic change is one of the most influencing social challenges because less young people enter the labor market to replace retired ones. Complex processes improved jobs qualifications and employees need to be qualified to realize more strategic, coordinating, and creative tasks with higher responsibilities. Increasing virtual work, changing social values, and growing complexity of processes are the other social challenges. Exponential growth of technology/data usage and growing collaborative work on platforms are classified in technical challenges. Environmental challenges covers resource scarcity and climate change. Standardization, data security, and personnel privacy are the political and legal challenges (Hecklaua et al. 2016).

To cope with these challenges, Operator 4.0 needs to develop technical, methodological, social and personal competencies. Technical skills, state-of-the-art knowledge, media skills, process understanding, coding skills, and understanding information technology security are the technical competencies. On the other hand entrepreneurial thinking, creativity, decision making, problem solving, conflict resolution, research skills, analytical skills, and efficiency orientation are classified in methodological competencies. Social competencies are related with intercultural, language, communication,leadership, and networking skills. Additionally, ability to work in a team, to be compromising and cooperative, to transfer knowledge and skills are classified in social competencies. Lastly, flexibility, motivation to learn, ambiguity tolerance, sustainable mind set, and ability to work under pressure are the personal competencies (Hecklaua et al. 2016). There are very critical factors for Industry 4.0 including aging society, lack of skilled workforce, mass customization, resource efficient and clean urban production, increasing product variability, shorter product life cycle, dynamic value chain, volatile markets, and cost reduction pressure (Shamim et al. 2016).

There is a change in the fields of education and employee development (Roblek et al. 2016). The broader literature points to an increasing rate in adoption of ICT to support engineering education, to make teaching and learning more collaborative and virtual. Teaching/learning demonstration format combinations are virtual laboratory, a flexible production system laboratory, and Industry 4.0 learning factory (Sackey and Bester 2016).

The other issue related with Operator 4.0 is the technological change's negative impact on employment. Hungerland et al. (2015) state that new production technologies and processes started to destroy jobs with redundancy effect and cause 'technological unemployment' (Roblek et al. 2016). This term is not a recent phenomenon and it was exaggerated in the past revolutions (Ten Bulte 2018). In the fourth revolution, manual repetitive tasks and many jobs of a cognitive nature are expected to be eliminated (Sackey and Bester 2016). On the other hand, job profiles change and new jobs emerge. Sackey and Bester (2016) define this situation as double-edged.

Big data, advanced analytics, human-machine interfaces, and digital to physical transfer are important in all sectors of industry. Operator 4.0's capabilities are enhanced to use digital devices and systems (Sackey and Bester 2016). Tasks and demands for the human 4.0 in the factory are expected to be changed. Human 4.0 is defined as the most flexible entity in cyber-physical production systems. There will be a large variety of jobs ranging from specification and monitoring to verification of production strategies for Human 4.0. Human 4.0 need to adopt the strategic decision-maker and flexible problem-solver role (Gorecky et al. 2014). Language skills, autonomy, flexibility, communicativeness, and creativity are the common skills that need to be developed (Benesova and Tupa 2017). There are some jobs emerged both in IT and production in Industry 4.0 context. Informatics specialist, Programmable Logic Controller (PLC) programmer, robot programmer, data analyst, cyber security, and software engineer are some of the IT related jobs; electronics technician, automation technician, production technician, and manufacturing engineering are some of the production related jobs (Benesova and Tupa 2017). Robot coordinator is another emerged job in the human robot cooperation context (Sackey and Bester 2016).

Industry 4.0 needs both the social and technical skills in an interdisciplinary thinking. Production thinking is transformed to design thinking in this context (Magruk 2016). Kadir et al. (2019) highlight not only the physical and cognitive side of human factor but also organizational issues. It is stated that hybrid production systems are bridging the gap between human and machines, human-centered designs benefit workers, new human–machine interactions affect work organization and design, the combination of new technology and work organization determine future skills' development and work organization is expanding across departments.

One of the skills that Human 4.0 need to gain is 'digital thinking' to manage the process in a new way: to read the data, analyze them, and determine their nature independently. Human 4.0 still have to use their brains and need more autonomy (Roblek et al. 2016). The important point is, smart people need to maintain humanity while enhancing job efficiency with high-tech know-know. Small and Vorgan (2008) point out the other side as values, expectations, aspirations, and personal experiences are different between younger and older generations. Millenials (AKA Generation Y-1981–2000 born) and Generation Xers (1965–1980) are the subgroups of digital natives and Babyboomers (1946–1964) and Seniors (before 1946) are the subgroups of digital immigrants. They indicate that the time in 2008 as a brain evolution and they thought that to move forward and thrive, digital immigrants and digital natives need to share one another's knowledge and experience. It is stated that tech-savvy and emotionally intelligent minds can also complement one another's abilities within a generation (Small and Vorgan 2008).

It seems that human resource management activities will differ according to the Industry 4.0. More technical tools like Artificial Intelligence (AI) are started to use in recruitment and selection. Big Data is used to assess people's performance and rewards are given in a more flexible and individualized way (Ten Bulte 2018). Finally, Taylor et al. (2018) foresee that the Operator 4.0 transform to a Maker 1.0—a person who works alongside the automated production system but with a different role. Maker 1.0 is essentially creative, rather than assisting or monitoring non-discretionary work flow steps or processes especially in agile manufacturing.

3.4 Patient 4.0 in the Healthcare 4.0

Healthcare is another application of IoT that will help smart people. The Health 4.0 concept is a medicine specific interpretation of the Industry 4.0 concept. It's aims are to decrease expenditures, to increase efficiency, and to serve customized services for patients (Dubgorn 2018). This is a continuous but disruptive transformation process of the entire healthcare value chain ranging from hospital care, non-hospital care, medicine and medical equipment production, healthy living environment, healthcare logistics to financial and social systems (Pang et al. 2018). There is a shift in the system design in healthcare paradigm from open, small, and single loop to closed, large, and multiple loop systems (Pang ve ark. 2018).

Sensors integrated in the house or smartphone applications can monitor the patients and send information to doctors. T-shirts that measure calories burned, heart rate, movement sensing, and so on are innovations in textile industry that smart people wear (Roblek et al. 2016). An enormous amount of data has been generated in the past decade. According to the proper analytics of big healthcare data, it is possible to predict epidemics, cures, and diseases (Hashem 2016). AI approximates human cognition capability in the analysis of complex health or medical. Thanks to AI-based tools, health or medical oriented applications that can be employed on wearable and networked smart devices are able to sense, comprehend, learn, and act. As a consequence, they can perform administrative and clinical functions (Pang et al. 2018).

In an evolutionary perspective, some basic and passive modern medical tools such as piston syringe, portable clinical thermometer, and flexible tube stethoscope were invented and applied in clinics in the Healthcare 1.0. In Healthcare 2.0, more complex medical equipment such as sphygmomanometer, X-ray imaging, and electrocardiograph were invented and applied. Thanks to the advancement of microelectronics, automation, computer science, and biomedical engineering; more complex medical systems, such as implantable pacemaker, brightness mode ultrasonography, magnetic resonance imaging, X-ray computed tomography, artificial heart, and Positron Emission Tomography were introduced in Industry 3.0 (Pang et al. 2018). In the period of Healthcare 3.0 (2006–2015) the system was hospital centric and to use of electronic health records (EHR) was popular. This means patients of long-lasting sickness suffered a lot from the multiple hospital visits for their routine checkups. Due to the recent technological advancements such as cloud computing and fog in Healthcare 4.0, the entire healthcare segments have achieved significant progress. The ultimate vision of 8-P Healthcare is presented as patient-centered, predictive, preventive, participatory, personalized, pre-emptive, precision, and pervasive healthcare (Kumari et al. 2018).

Of course, these developments cannot be achieved without difficulty. Challenges for Health 4.0 are time and resource constraints, priority to aging persons, lack of monitoring, location based services, limited healthcare workforce, expensive healthcare services, and high-tech services with high touch (Kumari et al. 2018). Automated medical production, healthcare Big Data, human–robot-symbiosis, healthcare robotics, smart and unobtrusive sensing are the emerging research topics to cope with these challenges (Pang et al. 2018).

Pang et al. (2018) believe that 'Care-giving Home' is the new generation concept of Healthcare 4.0. Smart devices are integrated in the home environment in the 'Care-giving Homes' by heterogeneous but interoperable communication networks. By this way comprehensively detailed data about human health and condition of infrastructure can be collected through smart sensors.

There are several services (developer-centric) and applications (user-centric) in the context of Healthcare 4.0. Ambient assisted living, m-Health, adverse drug reaction, medical implants, and patients with special need are the examples for services. Applications are categorized as individual and in clustered ways. Body temperature and blood pressure monitoring, ECG monitoring, speech monitoring, oxygen saturation monitoring and glucose level sensing are individual applications. On the other hand, diagnosis app, medical calculator, clinical communications app, and medical education app are clustered applications (Kumari et al. 2018).

IoT enables Electronic Health (e-Health), Mobile Health (m-Health) and Ambient Assisted Living (AAL) that allow remote monitoring and tracking of patients living alone at home or treated in hospital. But to design just standalone wearable devices is not sufficient; to create a complete ecosystem becomes vital (Rahmani et al. 2018). Healthcare industry can dramatically increase its efficiency, deliver treatment, decentralize its services, and manage chronic diseases outside hospitals or healthcare centers with the evolution of 5G ecosystem opportunities. Individualized medicine and smart pharmaceuticals, wearable devices are some of the new opportunities for Patient 4.0. There is an important detail about sensors. Body sensor data are need to be supplemented with environmental information—date, time, humidity, temperature—to identify unusual patterns and make more precise suggestions about the situation (Firouzi et al. 2018).

Body area sensor network, internet connected smart gateways (fog layer) or a local access network, cloud, and big data support are three main components of healthcare IoT systems (Health-IoT). Smart e-Health Gateways act as the bridging point with receiving data from different sub networks, perform protocol conversion, and provide other higher level services. Data aggregation, filtering, mining, local notification, encryption, and dimensionality reduction are some of these high level services (Firouzi et al. 2018). Blockchain technology is a promising technology which gives an opportunity to cope with mistakes and missing data (Wong et al. 2018). Telemedicine is an application area of the modern medicine and one of the main trends in digitalization of health care service put into practice within the Health 4.0 concept. IoT technology, M2M connectivity, predictive analytical systems, computer-based education, mobile networks 5G, cyber-physical systems

(CPS), mathematical modeling, statistical analysis and simulation modeling are the methods, approaches and technologies which are supposed to be used in order to implement the telemedicine system services (Dugborn et al. 2018).

In spite of significant advances in m-health, it is still in its early stages and is evolving in parallel with smart cities (Solanas et al. 2014). Transforming the traditional hospitals to smart hospitals is different from that of factories. It is more risky because it deals with human life (Thuemmler and Bai 2017). Graduate programs, such as e-Health, biomedical engineering, and smart systems engineering, are needed to support the Healthcare 4.0 vision (Pang et al. 2018).

As a more general term 'Smart health (s-health)' is proposed as the natural complement of mobile health (m-health) in the smart city context as a novel and richer ubiquitous concept. It is defined as "*the provision of health services by using the context-aware network and sensing infrastructure of smart cities*." To promote health to a higher position within society in a distributed, secure, private, sustainable, and efficient way by reusing the principles of smart cities and m-health is the ultimate goal of s-health. The source of m-health and s-health are different. The data come from a completely independent new source (i.e., the smart city sensing infrastructure) in s-health, where it comes from the patients in m-health. Also s-health is city-centric, not only user-centric (personalized) as in m-Health. S-health has an impact on society, governments, and research. It improves healthcare services, contributes to the creation of a healthier society and significantly helps to reduce the healthcare costs. Patients require fewer treatments thanks to the early detection and prevention mechanisms (Solanas et al. 2014). In smart health, technologies and infrastructure of smart cities can be leveraged and mixed with the concepts of telemedicine and m-health. An allergic patient has a chance to get information from an interactive information pole to check the pollution level and according to this information patient can avoid areas that could be dangerous with the support of s-health. In the next step, in m-Health augmented with s-Health context, when a cyclist wearing a bracelet with accelerometers and vital constants monitoring capabilities has an accident, the body sensor network detects the fall and sends an alert to the city infrastructure. After the alert and the analysis of traffic condition, an ambulance is dispatched through the best possible route. The traffic lights of the city can be dynamically adjusted in order to reduce the time needed by the ambulance to reach the cyclist in addition to this scenario (Solanas et al. 2014).

4. Conclusion

As it is indicated in various papers, Industry 4.0 is still a vision and the concept has to be extended for the future needs. To create a common understanding is one of the greatest challenges facing Industry 4.0. There is a general uncertainty about how it is to be implemented in practice (Kirazli and Hormann 2015). To accomplish Industry 4.0, the technology roadmap is still not clear and the gap between current manufacturing systems and Industry 4.0 requirements shows that there is still a long way to (Longo et al. 2017). The Connected Enterprise Maturity Model (2014), IMPULS—Industrie 4.0 Readiness (2015), Strategy for Industry 4.0 (2016), and Industry 4.0/Digital Operations Self-Assessment (2016) are some of the Industry 4.0 readiness and maturity models proposed by different institutions (Schumacher et al. 2016).

2008 was the milestone year in which more than 50 percent of all people started to live in urban areas. United Nations Population Fund predicts that in 2050, 70 percent of people will be living in urban areas and will consume between 60 to 80 percent of energy worldwide. Utilization of the potential benefits of technologies and innovations, will enhance the three main pillars that define the urban performance: efficiency, resilience, and sustainability (Razaghi and Finger 2018). Climate change and scarcity of resources are mega-trends that will affect all Industry 4.0 players. These mega-trends leverage energy decentralization for plants, triggering the need for the use of carbon-neutral technologies in manufacturing (Magruk 2016). 'Smart city' concept is important to cope with these challenges (Albino et al. 2015).

'Factory 4.0' or 'Smart Factory' implies the application of the vision of Industry 4.0 in the factory and places at the hearth of industry 4.0 (Kirazli and Hormann 2015; Magruk 2016) but it is a limited term. Work 4.0 (Richert 2016) is a more suitable term because it includes both the service and manufacturing sector. Worker 4.0 who works both in the service and manufacturing sector need to be studied along with Operator 4.0. As Kotler et al. (2016) indicate in their 'Marketing 4.0' book, more humankind institutions gain the game. This perspective shows the importance of the unique properties of humans. Humans are naturally smart and flexible diversely from machines. So to make factories more powerful and efficient, it is important putting the operators in the digital loop. Due to the mass customization, it is important to produce high precision tasks and the low volume products. Hence it is expected that numerous complex, high precision processes are still managed manually and with human-robot collaboration (Peruzzini et al. 2018).

As a summary, the human dimension of the Industry 4.0 phenomenon is discussed in this chapter in a conceptual way. Smart people, Customer 4.0, Operator 4.0 and Patient 4.0 are the different roles of Human 4.0 in Industry 4.0. It is foreseen that human sustain their key position as they always do. According to this fact, the researchers should put the human element in their projects in the early stages of implementation. It is important not to replicate the same ignorance of human element as in the third industrial revolution (Gorecky et al. 2014). Ten Bulte (2018) asks managers to take a pro-active stance and try to anticipate the changes that are necessary for HR Practices as according to the findings, there is a high level of uncertainty around the topic of Industry 4.0.

In conclusion, it should be stated that system scientists, engineers, urbanists, mathematicians, political scientists, economists, and professionals from many other disciplines need to work in an interdisciplinary approach (Razaghi and Finger 2018). 'Education' is the most important step of this transformation (Lu 2017).

Acknowledgments

This chapter is the upgraded and extended version of the proceeding called "Human 4.0 in the Industry 4.0" presented in "International Symposium on Industry 4.0 and Applications (ISIA 2017), Karabuk, Turkey".

References

Albino, V., U. Berardi and R. M. Dangelico. 2015. Smart cities: Definitions, dimensions, performance, and initiatives. Journal of Urban Technology, 22(1): 3–21.

Ardanza, A., A. Moreno, A. Segura, M. de la Cruz and D. Aguinaga. 2019. Sustainable and flexible industrial human machine interfaces to support adaptable applications in the Industry 4.0 paradigm. International Journal of Production Research, 57(12): 4045–4059.

Ardito, L., A.M. Petruzzelli, U. Panniello and A.C. Garavelli. 2018. Towards industry 4.0: Mapping digital technologies for supply chain management-marketing integration. Business Process Management Journal, 25(2): 323–346.

Benesova, A. and J. Tupa. 2017. Requirements for education and qualification of people in industry 4.0. Procedia Manufacturing, 11: 2195–2202.

Deniz, N. 2017. Human 4.0 in the Industry 4.0. In the Proceedings of the International Symposium on Industry 4.0 and Applications (ISIA 2017), Karabuk, Turkey.

Dubgorn, A., O. Kalinina, A. Lyovina and O. Rotar. 2018. Foundation architecture of telemedicine system services based on Health 4.0 concept, SHS Web of Conferences, 44(32): 1–10.

Firouzi, F., A.M. Rahmani, K. Mankodiya, M. Badaroglu, G.V. Merrett, P. Wongg and B. Farahani. 2018. Internet-of-Things and big data for smarter healthcare: From device to architecture, applications and analytics. Future Generation Computer Systems, 78: 583–586.

Galletta, A., L. Carnevale, A. Celesti, M. Fazio and M. Villari. 2018. A cloud-based system for improving retention marketing loyalty programs in Industry 4.0: A Study on big data storage implications. IEEEAccess, 6: 5485–5492.

Gorecky, D., M. Schmitt, M. Loskyll and D. Zühlke. 2014. Human-machine-interaction in the Industry 4.0 era. In the Proceedings of the 12th IEEE International Conference onIndustrial Informatics (INDIN).

Grigoriadis, N., C. Bakirtzis, C. Politis, K. Danas and C. Thuemmler. 2016. Health 4.0: The case of multiple sclerosis. In the Proceedings of the IEEE 18th International Conference on e-Health Networking, Applications and Services (Healthcom).

Hashem, I.A.T., V. Chang, N.B. Anuar, K. Adewole, I. Yaqoob, A. Gani, E. Ahmed and H. Chiroma. 2016. The role of big data in smart city centre. International Journal of Information Management, 36: 748–758.

Hecklaua, F., M. Galeitzkea, S. Flachsa and H. Kohl. 2016. Holistic approach for human resource management in Industry 4.0. Procedia CIRP, 54: 1–6.

Jara, A.J., M.C. Parra and A.F. Skarmeta. 2012. Marketing 4.0: A new value added to the marketing through the Internet of Things. In the Proceedings of the 6th International Conference on Innovative Mobile and Internet Services in Ubiquitous Computing.

Kadir, B.A., O. Broberg and C.S. da Conceicao. 2019. Current research and future perspectives on human factors and ergonomics in Industry 4.0. Computers & Industrial Engineering, 137: 1–12.

Kirazli, A. and R. Hormann. 2015. A conceptual approach for identifying Industrie 4.0 application scenarios, 862–871. In the Proceedings of the Industrial and Systems Engineering Research Conference.

Kotler, P., H. Kartajaya and I. Setiawan. 2016. Marketing 4.0: Moving from Traditional to Digital, Wiley. USA.

Kumari, A., S. Tanwar, T. Sudhanshu and N. Kumar. 2018. Fog computing for Healthcare 4.0 environment: Opportunities and challenges. Computers and Electrical Engineering, 72: 1–13.

Longo, F., L. Nicoletti and P. Antonio. 2017. Smart operators in industry 4.0: A human-centered approach to enhance operators' capabilities and competencies within the new smart factory context. Computers & Industrial Engineering, 113: 144–159.

Lu, Y. 2017. Industry 4.0: A survey on technologies, applications and open research issues. Journal of Industrial Information Integration, 6: 1–10.

Magruk, A. 2016. Uncertainty in the sphere of the Industry 4.0—Potential areas to research. Business, Management and Education, 14(2): 275–291.

Mládková, L. 2018. Industry 4.0: Human-technology interaction experience learned from the aviation industry, In the Proccedings of the 19th European Conference on Knowledge Management (ECKM 2018).

Monostori, L., B. Kadar, T. Bauernhans, S. Kondoh, S. Kumara, G. Reinhart, O. Sauer, G. Schuh, W. Sihn and K. Ueda. 2016. Cyber-physical systems in manufacturing. CIRP Annals - Manufacturing Technology, 65: 621–641.

Nelles, J., S. Kuz, A. Mertens and S.M. Schlick. 2016. Human-centered design of assistance systems for production planning and control: The role of the human in Industry 4.0. In the Proceedings of the International Technology (ISIT) IEEE Conference.

Paelke, V. 2014. Augmented reality in the smart factory: Supporting workers in an industry 4.0. Environment, Emerging Technology and Factory Automation (ETFA), IEEE. 1–4.

Pang, Z., G. Yang, R. Khedri and Y.T. Zhang. 2018. Introduction to the Special Section: Convergence of automation technology, biomedical engineering, and health informatics toward the Healthcare 4.0. IEEE Reviews in Biomedical Engineering, 11: 249–259.

Peruzzini, M., F. Grandi and M. Pellicciari. 2020. Exploring the potential of operator 4.0 interface and monitoring. Computers& Industrial Engineering, 139: 1–19.

Pfeiffer, S. 2017. The vision of 'Industrie 4.0' in the making—A case of future told, tamed, and traded. Nanoethics, 11: 107–121.

Prasetyo, Y.A. and A.A. Arm. 2017. Group management system design for supporting Society 5.0 in smart society platform. In the Proceedings of the International Conference on Information Technology Systems and Innovation (ICITSI).

Rahmani, A.M., T.N. Gia, B. Negash, A. Anzanpour, I. Azimi, M. Jiang and P. Liljeber. 2018. Exploiting smart e-Health gateways at the edge of healthcare Internet-of-Things: A fog computing approach. Future Generation Computer Systems, 78: 641–658.

Ramingwong, S., W. Manopiniwes and V. Jangkrajarn. 2019. Human factors of Thailand toward Industry 4.0. Management Research and Practice, 11(1).

Razaghi, M. and M. Finger. 2018. Smart governance for smart cities. In the Proceedings of the IEEE, 106(4): 680–689.

Richert, A., M. Shehadeh, L. Plumanns, K. Groß, K. Schuster and S. Jeschke. 2016a. Educating engineers for industry 4.0: Virtual worlds and human-robot-teams: Empirical studies towards a new educational age, 142–149. In the Proceedings of the Global Engineering Education Conference (EDUCON).

Richert, A., M. Shehadeh, S. Müller, S. Schröder and S. Jeschke. 2016b. Robotic workmates: Hybrid human-robot-teams in the Industry 4.0, 127–131. In the Proceedings of the 11th International Conference on e-Learning, Kuala Lumpur, Malaysia.

Roblek, V., M. Meško and A. Krapež. 2016. A complex view of Industry 4.0, SAGE Open. April–June 2016: 1–11.

Sackey, S.M. and A. Bester. 2016. Industrial engineering curriculum in Industry 4.0 in a South African context. South African Journal of Industrial Engineering, 27(4): 101–114.

Schumacher, A., S. Erol and W. Sihna. 2016. A maturity model for assessing Industry 4.0 readiness and maturity of manufacturing enterprises. Procedia CIRP, 52: 161–166.

Shamim, S., S. Cang, H. Yu and Y. Li. 2016. Management approaches for Industry 4.0: A human resource management perspective, 5309–5316. In the Proceedings of the IEEE Congress on Evolutionary Computation (CEC).

Shiroishi, Y., K. Uchiyama and N. Suzuki. 2018. Society 5.0: For human security and well-being. Computer, 51(7): 91–95.

Small, G. and G. Vorgan. 2008. iBrain Surviving the Technological Alteration of the Modern Mind, Collins Living, New York.

Smaradottır, B., S. Martinez, E. Borycki, G. Loudon, A. Kushniruk, J. Jortveit and R. Fensli. 2018. User evaluation of a smartphone application for anticoagulation therapy. Studies in Health Technology Informatics, 247: 466–470.

Solanas, A., C. Patsakis, M. Conti, I.V. Viachos, V.R. Ramos, F.F. Falcone, O. Postolache, P.A. Perez Martinez, R.P. Di Pietro and A.M. Martinez Balleste. 2014. Smart health: A context-aware health paradigm within smart cities. IEEE Communications Magazine, 1(8): 74–1.

Strozzi, F., C. Colicchiai, A. Creazza and C. Noe. 2017. Literature review on the 'Smart Factory" concept using bibliometric tools. International Journal of Production Research, 55(22): 6572–6591.

Taylor, M.P., P. Boxall, J.J. John, X.X. Chen, A. Liew and A. Adeniji. 2020. Operator 4.0 or Maker 1.0? Exploring the implications of Industrie 4.0 for innovation, safety and quality of work in small economies and enterprises. Computers & Industrial Engineering,139: 1–5.

Ten Bulte, A. 2018. What is Industry 4.0 and what are its implications on HRM Practices? The Proceedings of the 11th IBA Bachelor Thesis Conference, Enschede, The Netherlands.

Thuemmler, C. and C. Bai. 2017. Health 4.0: Application of Industry 4.0 design principles in future asthma management. pp. 23–37. *In*: Thuemmler, C. and C. Bai [eds.]. Health 4.0: How Virtualization and Big Data are Revolutionizing Healthcare. Springer, USA.

Wang, F.Y. and J.J. Zhang. 2017. Transportation 5.0 in CPSS: Towards ACP-based society-centered intelligent transportation. 762–767. In the Proceedings of the 20th IEEE International Conference on Intelligent Transportation Systems (ITSC).

Wong, M.C., K.C. Yee and C. Nohr. 2018. Socio-Technical considerations for the use of blockchain technology in healthcare. Studies in Health Technology Informatics, 247: 636–640.

Wuest, T. and D. Romero. 2017. Value Walk: Introducing 'Operator 4.0', a tech-augmented human worker" Newstex Global Business Blogs, Chatham: Apr 23, 1–3 [Online]. Available: http://search.proquest.com/docview/1890676896?pq-origsite=summon.

Chapter 22

Lean Manufacturing and Industry 4.0

A Framework to Integrate the Two Paradigms

Batuhan Eren Engin,[1] *Ehsan Khajeh*[2,*] and *Turan Paksoy*[1]

1. Introduction

Lean manufacturing is a customer-value focused method which is widely regarded as a great tool to cope with any type of waste generated during the manufacturing, and originated as part of the Toyota Production Systems. The method encourages the identification and elimination of any practices that do not directly add value to the product or system (Rosin 2019). Ohno (1988) defined seven potential sources of waste as overproduction, waiting, transportation, inappropriate processing, unnecessary inventory, unnecessary motion and defects in manufacturing. The most often used practices commonly associated with lean production that aims to reduce these mud as are: Just-in-time philosophy, pull flow, cellular manufacturing, lead/set-up/total time reduction, continuous improvement programs such as Kaizen, 5S, Kanban, 6 Sigma, mistake proofing Poka Yoke, quick changeover, Total Quality Management/Quality Management System, maintenance optimization such as Total Productive Management, waste management, elimination and reduction, JIT delivery, lot sizing, order consolidation, courier and transport modes optimization, inventory control and reduction (Rosin 2019).

Industry 4.0 is the fourth wave of technological advancement that enables the usage of advanced application of information and communication systems in manufacturing. These technologies are Internet, additive manufacturing, advanced robotic, augmented and virtual reality, Internet of Things (IoT), big data and analytics, cloud computing, machine learning and artificial intelligence, simulation and horizontal and vertical system integration (including Information Technology (IT) and Operational Technology (OT) integration. Through the application of these technologies, manufacturing environment becomes smarter and more efficient than ever before). The ability to connect devices, sensors, machines and software enables the companies to collect big data in real-time that gives them an opportunity to improve processes or predict failures before they occur, meanwhile machines can automatically optimize themselves, diagnose problems or configure more efficiently (Sullivan et al. 2002).

There is a consensus that Industry 4.0 is equipped with high-end solutions which possess several tools to help lean manufacturing (Sanders et al. 2016), and the company size should not be seen as an impediment for the concurrent deployment of both Industry 4.0 and lean manufacturing (Tortorella and Fettermann 2018). Both paradigms target operational excellence via different type of tools, however, attention and the literature seem to be increasing the knowledge about how to implement both paradigms holistically to achieve synergetic benefits, rather than separately. Buer et al. (2018) carried out the first systematic literature review on the relationship between Industry 4.0 and lean manufacturing for establishing a future research agenda. Their finding also supported the fact that there is no implementation framework for an Industry 4.0 and lean manufacturing integration in the literature. They stated that it is important to gain a more in-depth understanding of how these two domains interact before an implementation framework can be proposed. Since 2018 when their article was published, the number of studies on this issue nearly doubled in the literature. Besides, another literature survey by Brito et al. (2019) based on reviewing the literature on the relationship between lean production systems and Industry 4.0 in terms of occupational ergonomic conditions, as well as on workers' well-being. This growing number of studies

[1] Department of Industrial Engineering, Konya Technical University, Konya, Turkey, Email: erengn@gmail.com; tpaksoy@yahoo.com
[2] Department of Management, Kingston Business School, Kingston University, London, United Kingdom.
* Corresponding author: e.khajeh@kingston.ac.uk

encouraged us to carry out a systematic literature review on studies addressing the relationship between Industry 4.0 and lean manufacturing that has been published up to 2020. The method that has been used to find and assess the articles in the literature is systematic literature review. Its aim is to help the researchers to evaluate the existing literatures and improve the existing body of knowledge more deeply (Tranfield et al. 2003). To increase the quality of analysis and validate knowledge, just peer-reviewed articles and book chapters were included in the research. The process of Preferred Reporting Items for Systematic Reviews and Meta-Analyses has been used to collect materials (Moher et al. 2009). Considering this, the main research questions addressed in this review are:

- ✓ What are the most frequent issues in research papers addressing the relationship between Industry 4.0 and lean manufacturing?
- ✓ What could be the future directions for researchers and practitioners willing to integrate some of the components of Industry 4.0 into lean manufacturing?

This paper is organized in the following manner. After discussing the lean and green paradigms the research methodology and our approach on the selection of the papers are expressed. A descriptive analysis of selected articles is announced here. In the next section detailed analysis of studies is presented through classification of studies and the announcement of possible interactions between industry 4.0 components and lean manufacturing. The last section provides a conclusion and future directions.

2. Research Methodology

The methodology that was used to conduct the literature review for the purpose of this research to answer the research questions is a systematic literature review. In this methodology the researchers looks for the articles in the databases and analyse and codes them to find their gaps and trends. This method increases the reliability of the research by decreasing bias. It also makes the process more transparent (Tranfield et al. 2003; Denyer and Tranfield 2009). A systematic search of Scopus database was undertaken to identify relevant studies and reviews.

2.1 The Systematic Literature Review Protocol

The different stages that were followed while conducting the systematic literature review, is summarised in Table 1 and described in more detail below.

- First stage: In this step, the Title, Abstract and Keyword field of the Scopus database for the combination of keywords of "lean" OR "industry* 4.0" OR "logistic* 4.0" AND "lean manufacturing" OR "lean production" AND "smart" OR "internet of things" OR "big data" have been searched. The 779 relevant articles were selected from peer-reviewed articles of scholarly journals (Academic Journal Guide (AJG 2018) and Financial Times Top 50 Journals) and book chapter in English language published up to March 2020.
- Second stage: Papers were scanned manually to check that they were related to our scope and those articles that did not match with the objective of the research was eliminated. It helps to secure the reliability of findings and resulted with the selection of 103 articles.
- Third stage: It is about selection of papers base on quality and relevance criteria (Denyer and Tranfield 2009). Researchers read the abstract of each article and choose each one that discussed the lean manufacturing and Industry 4.0 while excluding those with no relationship between lean manufacturing and industry 4.0. According to Tranfield et al. (2003) more than one researcher should be involved in selection of articles for short listing as this decision is relatively subjective. After reading abstracts and excluding the non-applicable papers, 68 articles were short listed for the next stage.
- Fourth stage: In this stage, researchers reviewed articles that have been selected in stage three and identified any relevant cited paper is by Denyer and Tranfield (2009) confirmed the process that the result of the findings should be in-depth and complete thus answering the research question. After excluding and including the relevant high-quality papers, 56 articles remained in the database for the analyses.
- Fifth stage: In the last stage of the systematic literature review, a database in Excel spreadsheet was made to analyze articles and find the links and relation of each paper in order to provide insight into our research questions.

Table 1: Five efficient stages for conducting the systematic literature review.

First Stage	Second Stage	Third Stage	Fourth Stage	Fifth Stage
Keywords Search 779 articles →	Duplicates eliminated 103 articles →	Short-Listing of articles 68 articles →	56 articles →	Full paper analysis
Scopus database used			Abstract Analysis	Excel database
Search in Title, Abstract and Keywords Scopus Period: up to March 2020			Excluding and Including cited article/s in the main ones	Classification of articles

2.1.1 Analysis/Coding articles

To analyse and code the articles to find the links between them a Microsoft excel has been made, which makes it possible to scan each article for content analysis. For the purpose of reliability and to avoid any type of error each researcher coded articles separately. To finalise the coded paper and decide on the right selection, some articles were exchanged between researchers for the agreement on the codes. Table 2 presents the descriptive analysis of the coded papers.

Detailed Analysis of Studies

The focus of this section is to analyze the identified articles from the last stage of systematic literature review. In the first part an overview on the conceptual framework will be discussed and after that the articles based on empirical studies will be analyzed.

2.1.2 Conceptual frameworks

These studies analyze some theoretical and practical factors to develop a framework that will ease the process of adopting both paradigms simultaneously. A concise summary of related articles is declared below.

Wagner et al. (2017) presented a framework to start design and develop Industry 4.0 integrated applications in which includes a matrix representing their impacts on the elements of lean production systems. Lean practices were taken as 5S, Kaizen, Just-in-Time, Jidoka, Heijunka, Standardisation, Takt Time, Pull flow, man-machine separation, waste reduction and people and teamwork, while Industry 4.0 technologies were taken as sensors and actuators, cloud computing, big data, analytics, vertical integration, horizontal integration, virtual reality and augmented reality. Their estimated impact on lean practices were rated by eight lean production experts. Sanders et al. (2017) investigated the co-existence of lean manufacturing tools and Industry 4.0 technologies, and how lean manufacturing metrics are impacted by Industry 4.0 technologies through interdependence matrix. The estimated values in interdependence matrix were allocated based on authors' perception. The authors reported that the used technologies in Industry 4.0 benefits the TPM, Kanban, production smoothing, automation and waste elimination.

Leyh et al. (2017) aimed to analyze the existing architectural/reference models of Industry 4.0 which they characterized in terms of Lean management/production. They stated that the lean production principles were not often addressed in Industry 4.0 models, yet the most frequent integration of lean production and Industry 4.0 was found to be the vertical integration model. In this respect, Sony (2018) proposed research propositions for future research investigating the effect of integration models of Industry 4.0 and lean management, which are vertical, horizontal and end-to-end integration models.

Meanwhile some researchers investigated the relationships between these two issues in manufacturing systems through several cases. Satoglu et al. (2018) emphasized the relationship between Industry 4.0 and lean manufacturing by presenting several cases that combined lean production and Industry 4.0 components. They included how Industry 4.0 components can help reducing seven wastes defined in terms of Lean manufacturing. In this way, Mayr et al. (2018) analyzed how Industry 4.0 technologies can support existing lean practices using an electric drives production use case and provided a matrix depicting which Industry 4.0 technologies could assist specific lean methods, namely JIT, heijunka, Kanban, VSM, TPM or single minute exchange of die (SMED).

Powell et al. (2018) studied an automotive company in Italy in order to highlight several ways and abilities of Industry 4.0 technologies to support lean production constructs. According to the result from the case study, data analytics support the reduction of cost through elimination of waste and overproduction and levelling of production (known as Heijunka in

Table 2: Descriptive Analysis.

Authors	Qualitative	Quantitative	Conceptual Framework	Survey/ Questionnaire/ Interview	Case Study	Keywords
Wagner et al. (2017)	•		•		•	Cyber physical production system; connected industry; Industry 4.0; cyber physical system; Lean Production; technology management
Sanders et al. (2017)	•		•		•	Industry 4.0; Lean Management; Production Management
Leyh et al. (2017)	•					Industries; Lean production; Companies; Databases
Sony (2018)	•					Industry 4.0; lean management; model; lean automation; cyber-physical systems
Satoglu et al (2018)	•		•			Internet of Things; Operations and Technology Management; Value Chains guide to industry 4.0; Framework for industry 4.0; Industry 4.0 roadmap
Mayr et al. (2018a)	•		•		•	Industry 4.0; lean management; production management; cyber physical systems; internet of things
Powell et al. (2018)	•		•		•	Lean production; Digital lean manufacturing; Cyber-Physical Production Systems; Industry 4.0
Bakator et al. (2018)	•		•			Youth entrepreneurship; Industry 4.0; Lean startups; Serbia
Tortorella et al. (2019)	•			•		Emerging economies; Lean production; Industry 4.0; Operational performance improvement
Kolla et al. (2019)	•		•			Industry 4.0; lean; maturity model; self-assessment tools; digital transformation
Brusa (2018)	•		•			Industry 4.0; Model Based Systems Engineering; Lean Manufacturing; Smart Manufacturing; Product lifecycle development; System Design
Bandara et al. (2018)	•		•			Industry 4.0; Lean Management; Operational Performance Improvement; Banking Sector
Ilangakoon et al. (2018)	•		•			Healthcare; Lean Management; Industry 4.0; Operational Performance; Pre-medical diagnosis of diseases
Duarte and Cruz-Machado (2017)	•		•			Industry 4.0; Lean paradigm; Green paradigm; Supply chain
Duarte and Cruz-Machado (2018)	•		•			Industry 4.0; Lean paradigm; Green paradigm; Supply chain
Duarte et al. (2019)	•		•			Industry 4.0; Business model canvas; Lean management; Green management
Teixeira et al. (2018)	•		•			Industry 4.0; Lean Thinking; Information management; Information systems
Dogan and Gurcan (2018)	•		•			Lean six sigma; Industry 4.0; Big data; Data mining; Quality improvement; Process mining
Goienetxea et al. (2018)	•		•			Lean; Simulation; Optimization; Industry 4.0; Simulation-based optimization; Decision-making;
Lai et al. (2019)	•		•			Industry 4.0; Cyber Physical Systems (CPS); 7 wastes; lean manufacturing
Slim et al. (2018)	•		•			Industry 4.0; Lean; Design process; Production systems; Contradiction
Sharma and Gandhi, (2018)	•					Industry 4.0; Lean Automation; Computer Integrated Machining

Table 2 contd. ...

...Table 2 contd.

Authors	Qualitative	Quantitative	Conceptual Framework	Survey/ Questionnaire/ Interview	Case Study	Keywords
Rosin et al. (2019)	•		•	•		Industry 4.0; lean management; capability levels
Beifert et al. (2017)		•			•	Industry 4.0, lean manufacturing, shipbuilding
Tortorella and Fettermann (2018)		•		•		Industry 4.0; lean manufacturing; manufacturing management; lean production; emerging economies; empirical research
Kamble et al. (2019)		•		•		Industry 4.0; lean manufacturing; sustainability; organizational performance; manufacturing companies
Rossini et al. (2019)		•		•		Industry 4.0; Lean production European manufacturers; Survey; Lean 4.0
Ghobakhloo and Fathi (2019)		•			•	Information technology; Digitization; Lean manufacturing; Manufacturing performance; Industry 4.0
Varela et al. (2019)		•		•		Lean Manufacturing; Industry 4.0; Sustainability; economic; environmental; and social; structure equations modeling

the context of Lean management), and automated coordinate measuring machines and subsequent digitalization of quality control documentation support the control of abnormality. Besides those, e-learning platform support the full utilization of workers capabilities, workers' safety and developing workers' knowledge and skills.

Bakator et al. (2018) provided an entrepreneurship model that guides the young firms towards the application of Industry 4.0 technologies along with lean manufacturing. Tortorella et al. (2019) aimed to investigate the moderating effect of Industry 4.0 technologies on the impact of LP practices on operation performance indicators through surveying 147 Brazilian manufacturers that had implemented LP practices as well as Industry 4.0 technologies. Their findings supported that the adoption of Industry 4.0 technologies supporting product and service development improved the operational performance of flow practices, however, the result indicated that process-related technologies negatively moderate the effect of low setup practices on operational performance. Kolla et al. (2019) derived the essential components of lean and Industry 4.0 and mapped them with the specific characteristics of small and medium scale manufacturing enterprises, which helps them to reach their goals using lean and Industry 4.0 technologies, separately.

Brusa (2018) link the enabling technologies that come with Industry 4.0 with Lean manufacturing components or goals, such as process management, visual management, continuous improvement, improving quality, elimination of muda, customer, stakeholder and operators, flexible production line, etc. On other hand, Bandara (2018) provided a conceptual framework describing the relationship between lean and Industry 4.0 concepts in the banking sector in order to improve operational performance in the aspects of cost reduction, quality, productivity, profitability, etc. Lean tools are utilized to streamline the processes end-to-end, eliminating the unnecessary practices that will result in shorter time and cost to serve. In other aspect, Ilangakoon et al. (2018) presented a conceptual framework for the healthcare sector to enhance their operational performance, i.e., patient throughput, reduced waiting times and efficient allocation of resources, through premedical diagnosis of diseases that integrates Industry 4.0 technologies, such as big data analytics, Internet of Things and cloud computing and lean techniques, such as virtual stream mapping.

Sustainability and green supply chain concepts are another important aspect that have been linked by industry 4.0 and lean management. For instance, Duarte and Cruz-Machado (2017 and 2018) aimed to establish the link between lean and the green supply chain and Industry 4.0 by developing a conceptual model. Their model included several characteristics of the lean and green supply chain, namely: manufacturing, "logistics and supply, product and process design, product, customer, supplier, employee, information sharing and energy", which were linked to Industry 4.0 concepts (Ustundag and Cevikcan 2018). Duarte et al. (2019) presented a conceptual relationship model between the concepts of "Business Model, Lean and Green Management, and Industry 4.0". The Business model canvas is composed of nine elements interacting with each other which represent the business, i.e., "value proposition, customer segments; customer relationships, customer channels, revenue streams, key activities, key resources, key partners and cost structure" (Ustundag and Cevikcan 2018). This model and its elements are linked with lean and green paradigms and the concepts of Industry 4.0 by the authors.

Information management is another key issue that has a strong relation with lean management and industry 4.0 concepts. Teixeira et al. (2018) discussed lean thinking and Industry 4.0 with regard to information management process and presented the lean information management framework in an attempt to eliminate the waste associated with the information management process. Dogan and Gurcan (2018) developed a lean six sigma method utilizing data collection techniques and analysis methods from Industry 4.0 in each of its step which makes the lean six sigma method easier, faster and more reliable.

There are several coherent and well-written articles that have dealt with industry 4.0 and lean management subjects. For instance, Goienetxea Uriarte et al. (2018) discussed a conceptual framework that integrates simulation, optimization and lean practices for enhanced decision-making process and supported organizational learning to make the lean goals more efficient and achieve better system performance indicators. The authors also wrote a handbook to describe in detail the framework. Lai et al. (2019) analyzed the literature on the impact of Industry 4.0 technologies on Lean manufacturing effectiveness in terms of seven wastes in lean philosophy. They determined the eight most cited papers on the subject and associated the papers with seven wastes via the expert thoughts.

Slim et al. (2018) analyzed the literature to link lean tools to smart concepts from Industry 4.0 for concurrent implementation of both. They also provided a table presenting the convergences and contradictions between lean and Industry 4.0 based on three dimensions: technical, management and people system. Also, Sharma and Gandhi (2018) analyzed the literature to find the relationship between lean automation and Industry 4.0, and they suggested that the Industry 4.0 is not making lean obsolete. Finally, Rosin et al. (2019) highlighted the type of techniques that makes a relation between Industry 4.0 and methods in lean management. The focus of their study was on the effect of the technology's capability level for the development of lean approaches base on Industry 4.0. Their findings show that development of lean principles based on the tools and technologies of Industry 4.0 need to be tracked.

2.1.3 Empirical studies

Besides conceptual studies, there are several researches that use empirical evidence to investigate the interactions between industry 4.0 concept and lean management.

Beifert et al. (2017) investigated if several Industry 4.0 approaches could be efficiently applied in the lean ship-building sector and discussed the shortcomings and problems of implementing them using body of empirical data collected from shipbuilding companies and suppliers. They also presented the possible implementation models of lean production in the sector through Industry 4.0. Tortorella and Fettermann (2018) investigated the interaction between lean manufacturing tools and Industry 4.0 technologies through acquired data from a survey carried out with 110 companies of different sizes and sectors at different stages of their LP implementation. They reached to a conclusion that LP practices are positively associated with Industry 4.0 technologies, which means improved performance merits, i.e., productivity, delivery service level, inventory level, workplace safety and quality (scrap and rework), achieved through the concurrent implementation of both. Kamble et al. (2019) investigated the indirect effects of Industry 4.0 components on sustainable organizational performance indicators with lean manufacturing practices already implemented. The hypotheses were tested on data collected from 205 managers, working in 115 manufacturing firms. Their findings suggested that lean manufacturing practices carried a mediating role in the effect of Industry 4.0 technologies, leading to enhancement of sustainable organizational performance indicators, i.e., economic, social and environmental indicators.

Rossini et al. (2019) carried out an empirical study to investigate the impact of interrelation between the adoption of Industry 4.0 technologies and the implementation of lean production practices on operation performance indicators of European manufacturers; i.e., using a survey with 108 different manufacturers that have been using lean manufacturing practices along with Industry 4.0 technologies. Operational performance indicators were selected as productivity, delivery service level, inventory level, workplace safety and quality (scrap and rework). Their findings emphasized that companies that aim to achieve higher levels of Industry 4.0 must have previously implemented a certain level of LP practices. This is due to the fact that LP practices allow companies to operate under well-designed and robust processes to which the addition of Industry 4.0 technologies can make a bigger impact.

Ghobakhloo and Fathi (2019) tracked and analyze the five years of a case company which underwent a digital transformation by integrating IT-based technology trends of Industry 4.0 with the firm's core capabilities and competencies. Their case study provided information on how recent IT tools can lead to the development of the lean-digitized manufacturing system. During the company's transition, JIT production system was supported by electronic Kanban, which enhanced the flow of work by enabling work-in-process limits, tracking lead times and analysis of workflow. To control quality, the company implemented a real-time statistical process control software, and using the computers located in the shop floor, it provided real-time information about the process by creating a number of charts, such as X-bar and S, X-bar and R, Median (R)-R, Z-bar and S, Pareto charts and frequency histograms, which also automatically evaluate control charts based on the real-time data and provide quality managers with the real-time clear stop/go instruction. The key finding in their case study is that the use of IT in manufacturing have significantly lowered the defect rates since 2013, and the use of preventive maintenance has resulted in improved average mean time between failure, maintenance breakdown severity and mean time to repair. Another significant improvement was seen in the time of receiving raw materials and spare-parts/dies from international suppliers due to the use of virtual communication, several IT tools, cloud ERP solution and electronic-banking. Varela et al. (2019) tested six hypotheses questioning the effects of lean manufacturing and industry 4.0 on sustainability (economic, environmental and social metrics) using a survey with 252 valid responses. Their result indicated a positive correlation between Industry 4.0 and sustainability but not for lean management, contrary to the popular belief.

3. Industry 4.0 Components and Possible Interaction with Lean Manufacturing

There are many performance evaluation factors that have been used by researches, e.g., cost, flexibility, productivity, quality, reduced inventory and reliability. The survey of researches that collected through the systematic literature review showed more than 31 percent of these studies used flexibility as performance evaluation factor (Figure 1).

In the following sub sections, some of the new tools that have emerged out of the conjunctive use of Industry 4.0 and lean thinking are covered.

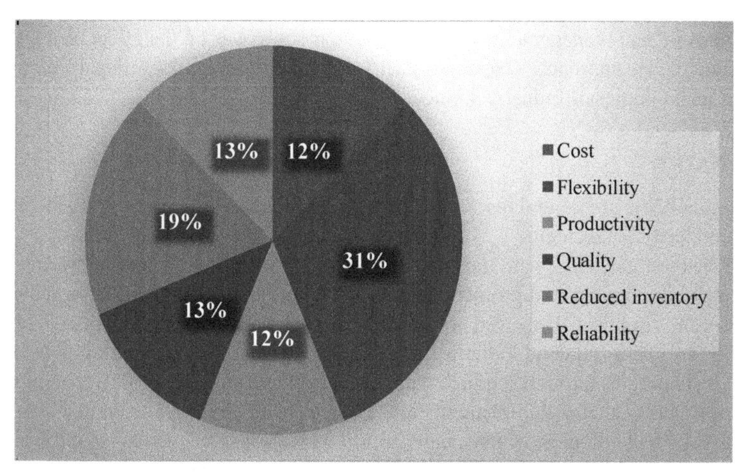

Fig. 1: Industry 4.0 and Lean manufacturing Performance evaluation.

3.1 *Just-In-Time 4.0*

Lean manufacturing has two main pillars as described by Ohno (1988), Just-in-time (JIT) delivery and automation. JIT delivery denotes a manufacturing system in which materials or components are delivered immediately before they are required in order to minimize storage costs. Cyber-Physical Systems (CPS) are one of the key components of Industry 4.0 that enables the physical processes to be controlled by microcontrollers with communication interfaces and connected field devices with feedback loops where physical processes affect timing and computations, and vice versa. It enables highly customized products in mass production as it adds modularity and changeability to the production line. In an automotive industry that has integrated lean manufacturing, some CPS are called "cyber-physical Just-In-Time delivery" which is an IT-system designed to support a lean manufacturing material flow (Wagner et al. 2017).

Automation describes the machine design in which the machines do specific tasks that humans would find difficult or repetitive, if any abnormality is detected by machines, they will stop so that the operator can investigate and fix the situation that caused the condition. One advantage of using automation in production line is that it significantly decreases the defect rate (Sullivan et al. 2002), which is one of the waste defined by Ohno (1988). The implementation of CPS in production line supports the automation thanks to the intelligent machines.

Meanwhile, Hofmann and Rüsch (2017) reported that as far as JIT systems are concerned, the increasing use of Auto-ID and the virtual ERP system using clod or distributed ledger technology may reduce bullwhip effects and pave the way for highly transparent and integrated supply chains as well as improvements in production planning. The interviewed experts see the majority of implications on the operative level of logistics management. Also Şenkayas and Gürsoy (2018) investigated the digitalization project requiring the installation of manufacturing execution and monitoring system in a rim manufacturing company which has lean production lines and they realized that the overall equipment efficiency increased from 75 percentage to 91 percentage.

3.2 Value Stream Mapping 4.0

Value Steam Mapping (VSM) is a lean manufacturing tool to demonstrate the flow of materials and information through the system within the value creation chain to identify waste in the system. The main objective is to shorten the lead time and facilitating a flow through production. In this context, components of Industry 4.0 provides a variety of technologies to enable the development of a digital and automated manufacturing environment as well as the digitization of the value chain. For instance,real time data can be transmitted through cencors and via Auto-ID to instantly localize the object and access the information about the state of the product (Mayr et al. 2018).

Two aspect of visualization and data analytics use to check and monitor the quality of data and process of softwares regularly, support the identification of causes and waste such as high downtime, cycle time, failure rate, supplier and customer fluctuation, etc, that point to problems in the value stream, which is the main benefit of VSM 4.0, i.e., the transparency through a real-time display of value streams. Meudt et al. (2017) offered a VSM 4.0 framework that allows companies to seize the opportunities offered by digitalisation and Industry 4.0 to develop their lean production approach to the next level using VSM 4.0, in which they provided new notations and symbols to visually represent the collected data. Bosch (the Software Innovation company) based in Germany also shared an article about VSM 4.0, and a research study at Siemens Healthcare. Haschemi and Roessler (2017) focused on enhancing Value Stream Mapping to analyse and improve material and control flows by means of Lean and Digital Manufacturing levers.

3.3 Kanban 4.0

Kanban is an efficient pull manufacturing method that retains a continuous material flow by maintaining a predefined stock level, which uses consumption-based replenishment philosophy instead of forecasting based replenishment philosophy, to minimize unnecessary inventory level. The traditional Kanban method uses a bin system in which the materials are stocked, and when any bin becomes empty, a demand signal is triggered. New information and communication systems that are brought by Industry 4.0, such as Radio Frequency Identification (RFID), allowed Kanban systems to undergo digital transformation, which is now called Electronic Kanban systems. Compared to the traditional Kanban, all movements of the Kanban cards are actually digitized, which are recognized by barcode readers. When a bin becomes empty, Kanban signals are sent automatically. E-Kanban is reported to help reducing the production lead times, financial costs, effective and efficient work processes and waste (Jarupathirun et al. 2009). The advantages of using Electronic-Kanban are that lost cards will be solved, the demand need is delivered right on time, the card handling is eliminated and it improves the supply chain transparency (Kouri et al. 2008).

In this regards, Kolberg et al. (2017) derived the requirements of the Kanban method and determined the requirements at the interface of workstations to support the Kanban method. After the technology-independent overall architecture was designed, CPS have been taken as the technology from Industry 4.0 to realize the interface for digitalized usage. Meanwhile, Hofmann and Rüsch (2017) reported that the Kanban 4.0 improves the demand assessment due to less human interaction required, while shortening the cycle times in production due to the data being transferred closer to real time. This helps the lean cause as lean management seeks to increase the customer satisfaction and reduce cycle time.

3.4 Total Productive Maintenance 4.0

Total Productive Maintenance (TPM) is a system for optimizing maintenance in an attempt to reach the state of perfect efficiency devoid of the defects, short stoppages, sub-optimal production rates, downtime or accidents, through minimizing all kind of losses or inefficiencies. There are eight main pillars of TPM, i.e., autonomous maintenance, planned maintenance, quality management, focused improvement, new equipment management, education and training, safety, health and environment, and last but not least, administration. Turanoglu Bekar et al. (2018) sought answers to the question "Which key technologies of Industry 4.0 have the highest statistically significant impacts on TPM" by performing conjoint analysis. They found their models were statistically significant to forecast the effect of industry 4.0 on total production maintenance, especially simulation, internet of things and additive manufacturing. With Industry 4.0, predictive maintenance by means of machine learning can change the timing of planned maintenance to avoid downtime. Industry 4.0 also brings to light the availability of information retrieval through organized data collection and artificial intelligence algorithms, correlations between root causes and defects and downtime can be revealed. Regarding the training of employee, Digital Twin visualization provides an opportunity for the employee to get more familiar with the product, processes and manufacturing, from components and machines to production lines. Sensors and early detection systems can help decrease the risk of injury and health issues through measuring air quality, radiation, temperature and other environmental conditions.

4. Conclusion and Future Directions

With the aim of providing an efficient study in lean manufacturing, industry 4.0 and the interplay between the two paradigms, 56 published researches till March 2020 have been collected systematically. For the purpose of analysis, book chapters and peer-reviewed articles from the most reliable database was considered. The search period was till March 2020 and during this period 779 publications were narrowed down to 56 final selected researches. Two main types of conceptual frameworks and empirical studies were considered in order to pursue a proper analyses. Then all 56 researches were studied more deeply by using case study, survey, conceptual framework and quantitative/qualitative descriptive levels. Moreover, some of the new tools that emerged to provide proper interplay in the industry 4.0 with lean thinking were demonstrated such as: Just-In-Time 4.0, Value Stream Mapping 4.0, Kanban 4.0 and Total Productive Maintenance 4.0.

Among the technologies such as IoT, simulation, big data and robots that are used in manufacturing to develop the lean principles, Internet of thing (IoT) is the most well-known. Also, during the monitoring phase, IoT is the most used technology. Regarding to the company size, findings show that 20% of companies that used any type of industry 4.0 in their manufacturing were large companies and 18% of them were SMEs. Other 62% of papers did not categories the type of manufacturing that they used industry 4.0.

The proposed research guides researchers for future research efforts focusing on how lean manufacturing can benefit from the industry 4.0. More empirical studies are needed to look at the advantages of these relation between industry 4.0 and lean manufacturing in more details. Additional research is needed to study the consequence of industry 4.0 on conventional lean manufacturing.

References

Bakator, M., D. Đorđević, D. Ćoćkalo, M. Nikolić and M. Vorkapić. 2018. Lean startups with industry 4.0 teschnologies: Overcoming the challenges of youth entrepreneurship in Serbia. Journal of Engineering Management and Competitiveness (JEMC), 8(2): 89–101.

Bandara, O., V. Tharaka and A. Wickramarachchi. 2018. Industry 4.0 and Lean based Operational Performance Improvement Approach: A Conceptual Framework for the Banking Sector.

Beifert, A., L. Gerlitz and G. Prause. 2017. Industry 4.0–for sustainable development of lean manufacturing companies in the shipbuilding sector. Paper presented at the International Conference on Reliability and Statistics in Transportation and Communication.

Brito, M.F., A.L. Ramos, P. Carneiro and M.A. Gonçalves. 2019. Ergonomic analysis in lean manufacturing and industry 4.0—a systematic review. In Lean Engineering for Global Development. Springer, 95–127.

Brusa, E. 2018. Synopsis of the MBSE, Lean and Smart Manufacturing in the Product and Process Design for an Assessment of the Strategy" Industry 4.0". Paper presented at the CIISE.

Buer, S.V., J.O. Strandhagen and F.T. Chan. 2018. The link between Industry 4.0 and lean manufacturing: mapping current research and establishing a research agenda. International Journal of Production Research, 56(8): 2924–2940.

Denyer, D. and D. Tranfield. 2009. Producing a systematic review.

Dogan, O. and O.F. Gurcan. 2018. Data perspective of Lean Six Sigma in industry 4.0 Era: a guide to improve quality. Paper presented at the Proceedings of the international conference on industrial engineering and operations management Paris.

Duarte, S. and V. Cruz-Machado. 2017. An investigation of lean and green supply chain in the Industry 4.0. Paper presented at the 2017 International Symposium on Industrial Engineering and Operations Management (IEOM).

Duarte, S. and V. Cruz-Machado. 2018. Exploring Linkages Between Lean and Green Supply Chain and the Industry 4.0, Cham: 1242–1252.

Duarte, S., M. do Rosário Cabrita and V. Cruz-Machado. 2019. Business Model, Lean and Green Management and Industry 4.0: A Conceptual Relationship. Paper presented at the International Conference on Management Science and Engineering Management.

Ghobakhloo, M. and M. Fathi. 2019. Corporate survival in Industry 4.0 era: the enabling role of lean-digitized manufacturing. Journal of Manufacturing Technology Management.

Goienetxea Uriarte, A., A.H. Ng and M. Urenda Moris. 2018. Supporting the lean journey with simulation and optimization in the context of Industry 4.0. Paper presented at the Procedia Manufacturing.

Haschemi, M. and M.P. Roessler. 2017. Smart value stream mapping: an integral approach towards a smart factory. Paper presented at the Proceedings of the 3rd International Congress on Technology–Engineering & Science: 273–279.

Hofmann, E. and M. Rüsch. 2017. Industry 4.0 and the current status as well as future prospects on logistics. Computers in Industry, 89: 23–34.

Ilangakoon, T., S. Weerabahu and R. Wickramarachchi. 2018. Combining Industry 4.0 with Lean Healthcare to Optimize Operational Performance of Sri Lankan Healthcare Industry. Paper presented at the 2018 International Conference on Production and Operations Management Society (POMS).

Jarupathirun, S., A. Ciganek, T. Chotiwankaewmanee and C. Kerdpitak. 2009. Supply chain efficiencies through e-kanban: a case study. International Journal of the Computer, the Internet and Management, 17(1): 55–1.

Kamble, S., A. Gunasekaran and N.C. Dhone. 2019. Industry 4.0 and lean manufacturing practices for sustainable organisational performance in Indian manufacturing companies. International Journal of Production Research, 1–19.

Kolberg, D., J. Knobloch and D. Zühlke. 2017. Towards a lean automation interface for workstations. International Journal of Production Research, 55(10): 2845–2856.

Kolla, S., M. Minufekr and P. Plapper. 2019. Deriving essential components of lean and industry 4.0 assessment model for manufacturing SMEs. Procedia CIRP, 81: 753–758.

Kouri, I.A., T.J. Salmimaa and I.H. Vilpola. 2008. The Principles and Planning Process of An Electronic Kanban System. Dordrecht.

Lai, N.Y.G., K.H. Wong, D. Halim, J. Lu and H.S. Kang. 2019. Industry 4.0 Enhanced Lean Manufacturing. Paper presented at the 2019 8th International Conference on Industrial Technology and Management (ICITM).

Leyh, C., S. Martin and T. Schäffer. 2017. Industry 4.0 and Lean Production—A matching relationship? Ananalysis of selected Industry 4.0 models. Paper presented at the 2017 Federated Conference on Computer Science and Information Systems (FedCSIS).

Mayr, A., M. Weigelt, A. Kühl, S. Grimm, A. Erll, M. Potzel and J. Franke. 2018. Lean 4.0-A conceptual conjunction of lean management and Industry 4.0. Procedia CIRP, 72: 622–628.

Meudt, T., J. Metternich and E. Abele. 2017. Value stream mapping 4.0: Holistic examination of value stream and information logistics in production. CIRP Annals, 66(1): 413–416.

Moher, D., A. Liberati, J. Tetzlaff, D.G. Altman and P. Group. 2009. Preferred reporting items for systematic reviews and meta-analyses: the PRISMA statement. PLoS Med., 6(7).

Ohno, T. 1988. Toyota production system: beyond large-scale production.crc Press.

Powell, D., D. Romero, P. Gaiardelli, C. Cimini and S. Cavalieri. 2018. Towards Digital Lean Cyber-Physical Production Systems: Industry 4.0 Technologies as Enablers of Leaner Production. Cham.

Rosin, F., P. Forget, S. Lamouri and R. Pellerin. 2019. Impacts of industry 4.0 technologies on Lean principles. International Journal of Production Research.

Rossini, M., F. Costa, G.L. Tortorella and A. Portioli-Staudacher. 2019. The interrelation between Industry 4.0 and lean production: an empirical study on European manufacturers. The International Journal of Advanced Manufacturing Technology, 102(9): 3963–3976.

Sanders, A., C. Elangeswaran and J.P. Wulfsberg. 2016. Industry 4.0 implies lean manufacturing: Research activities in industry 4.0 function as enablers for lean manufacturing. Journal of Industrial Engineering and Management (JIEM), 9(3): 811–833.

Sanders, A., R.K. Subramanian, T. Redlich and J. Wulfsberg. 2017. Industry 4.0 and Lean Management—Synergy or Contradiction? In IFIP International Conference on Advances in Production Management Systems, 341–349. Springer, Cham.

Satoglu, S., A. Ustundag, E. Cevikcan and M.B. Durmusoglu. 2018. Lean Production Systems for Industry 4.0. In Industry 4.0: Managing The Digital Transformation: pp. 43–59. Cham: Springer International Publishing.

Şenkayas, H. and Ö. Gürsoy. 2018. Industry 4.0 Applications And Digitilization of Lean Production Lines. The Annals of the University of Oradea:124.

Sharma, S. and P.J. Gandhi. 2018. Journey readiness of industry 4.0 from revolutionary idea to evolutionary implementation: a lean management perspective. International Journal of Information and Communication Sciences, 3(3): 96.

Slim, R., H. Rémy and C. Amadou. 2018. Convergence and contradiction between lean and industry 4.0 for inventive design of smart production systems. Paper presented at the International TRIZ Future Conference.

Sony, M. 2018. Industry 4.0 and lean management: a proposed integration model and research propositions, 6(1): 416–432.

Sullivan, W.G., T.N. McDonald and E.M. Van Aken. 2002. Equipment replacement decisions and lean manufacturing. Robotics and Computer-Integrated Manufacturing, 18(3): 255–265.

Teixeira, L., C. Ferreira and B.S. Santos. 2018. An information management framework to industry 4.0: ALean Thinking Approach. Paper presented at the International Conference on Human Systems Engineering and Design: Future Trends and Applications.

Tortorella, G.L. and D. Fettermann. 2018. Implementation of Industry 4.0 and lean production in Brazilian manufacturing companies. International Journal of Production Research, 56(8): 2975–2987.

Tortorella, G.L., R. Giglio and D. van Dun. 2019. Industry 4.0 adoption as a moderator of the impact of lean production practices on operational performance improvement. International journal of operations & production management.

Tranfield, D., D. Denyer and P. Smart. 2003. Towards a methodology for developing evidence-informed management knowledge by means of systematic review. British Journal of Management, 14(3): 207–222.
Turanoglu Bekar, E., A. Skoogh, N. Cetin and O. Siray. 2018. Prediction of Industry 4.0's Impact on Total Productive Maintenance Using a Real Manufacturing Case. Proceedings of the International Symposium for Production Research, 136–149.
Ustundag, A. and E. Cevikcan. 2018. Industry 4.0: Managing The Digital Transformation. Cham: Springer.
Varela, L., A. Araújo, P. Ávila, H. Castro and G. Putnik. 2019. Evaluation of the relation between lean manufacturing, industry 4.0, and sustainability. Sustainability, 11(5): 1439.
Wagner, T., C. Herrmann and S. Thiede. 2017. Industry 4.0 impacts on lean production systems. Procedia CIRP. 63: 125–131.

Index

U

V

W

9780367634285